PAPUA NEW GUINEA

MATHEMATICS

OXFORD

Oxford University Press is a department of the University of Oxford. It furthers the University's objective of excellence in research, scholarship, and education by publishing worldwide. Oxford is a registered trademark of Oxford University Press in the UK and in certain other countries.

Published in Australia by
Oxford University Press
253 Normanby Road, South Melbourne, Victoria 3205, Australia

First published 2010
Reprinted 2011, 2012, 2013, 2014, 2018, 2020, 2022

ISBN 978 0 19 556545 4

Edited by Brenda Hamilton
Illustrated by Uramina and Nelson
Series design by Domenika Fairy
Typeset by Palmer Higgs
Printed in China by Golden Cup Printing Co. Ltd

Contents

Unit

1 Managing your money

Unit summary

This unit focuses on the mathematics needed for the financial transactions commonly used in our day-to-day living. It revisits the topic of percentages and their application to shopping in situations such as selecting the 'best buys', calculating discounts or working out profit and loss.

We will look at different options for saving money, such as bank savings accounts or fixed term accounts, and how interest is calculated on our savings.

We will also look at options for borrowing money and examine how interest is calculated when we buy goods on credit or take out a loan to pay off a house.

Lastly, we will look at the different ways of earning money and being paid for our work, as well as how to calculate our taxes and plan budgets.

The unit will be assessed by tests of your mathematical skills in using money in the following ways:

- calculations of percentages
- application of percentages to problems
- earning and saving money
- operations involving large amounts of money
- basic ideas of budgeting
- basic ideas of borrowing money
- calculating discounts, interest and repayments.

The unit will also be assessed by a group project in which you will be investigating the practical application of those skills to a personal or business situation.

Syllabus references

10.1.1 Apply percentages to a range of financial transactions and be aware of whether or not the result is reasonable

10.1.2 Determine the costs and benefits of simple credit and investment or saving schemes

10.1.3 Communicate mathematical processes and results both orally and in writing

10.1.4 Undertake investigations individually and cooperatively in which mathematics can be applied to solve problems

Managing your money: Topics

Week	Content	Page
1 and 2	**Spending money** • Using percentages • Sales, discounts and 'best buys' • Mark-up, profit and loss • Budgets • Depreciation and appreciation • Working with large sums of money • Revision and assessment	4
3 and 4	**Saving money** • Different ways of saving • Simple interest • Savings accounts • Term deposits • Compound interest • Inflation, appreciation and depreciation • Comparing investment options • Revision and assessment	30
5	**Borrowing money** • Ways of borrowing money • Loans and simple interest • Hire purchase • Credit cards • Reducing balance loans	48
6	**Earning money** • How people earn • Calculating pay using rates and scales • Tax and other deductions • Revision and assessment (Borrowing and earning money)	59
7, 8, 9 and 10	**Option A: Three monthly budgets** • Planning a monthly personal, family or business budget • Recording and estimating income and expenses • Examining cash flow **Option B: Obtaining and using a mini loan** • Identifying sources for a small loan (K100 to K500) • Comparing loans from at least two sources • Completing a loan application • Calculating the interest and repayments	71 76

SPENDING MONEY

About percentages

Percentages are used in many areas of our lives. Some of the most common applications are related to money.

We need a good understanding of percentages to help us manage our money, for instance, when we are shopping for goods and services or working out a family budget.

With practice, we can make a reasonable estimate of simple percentages without pen and paper.

Example

An estimate of the percentage of fuel remaining in the tank, as shown on the fuel gauge, is:

A 10% **B** 20%
C 50% **D** 70%
E 80%

Answer

When the fuel tank is full it is at 100%. Half full is 50% and quarter full is 25%. The level is just below quarter full, therefore an estimate is 20% so the answer is **B**.

Giving information in the form of a percentage is often more useful than using a fraction or decimal. For instance, a reduced fat milk advertises that a 250 mL glass provides 18% of your daily protein needs and 39% of your daily calcium needs, and contains only 1% fat. Compare this to the information in the nutrition table below.

protein	9 g
calcium	315 mg
fat	2.5 g

Discussion

1 Explain why information stated as a percentage can be easier to understand than raw figures.

2 Give some examples you know of where information is given as a percentage rather than as a fraction or decimal.

3 Can you describe situations where percentages might be misunderstood or not be useful?

4 Thinking back to the mathematics used last year, identify which of these statements are true:

A Some percentages can be written as fractions.
B All fractions can be written as a percentage.
C Percentages can only take values between 0 and 100.
D All percentages have a decimal and fraction equivalent.
E Some fractions can be written as a percentage.

EXERCISE

1 Choose the best estimate of the percentage of liquid in each of these tanks.

a A 50%
B 25%
C 19%
D 40%
E 56%

b A 30%
B 38%
C 62%
D 55%
E 49%

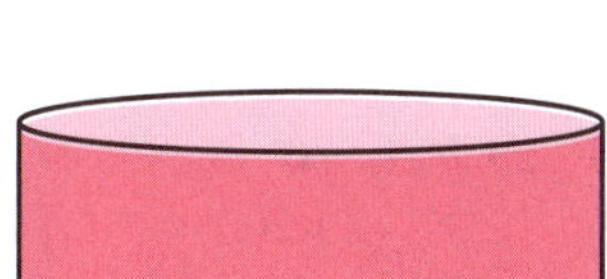

c A 75%
B 69%
C 82%
D 78%
E 85%

d A 99%
B 95%
C 88%
D 90%
E 85%

e A 15%
B 20%
C 40%
D 35%
E 28%

2 If 10% of the money in Caleb's savings account is K30, how much is:

a 5% b 15% c 30% d 2% e 100%

3 Change these percentages into:

i a decimal ii a simplified fraction

a 70% b 40% c 38% d 85%
e 125% f 66% g 132% h 37.5%
i 6.25% j $5\frac{2}{3}$%

Remember

To change a fraction or decimal into a percentage, multiply by 100.

4 Change the following into percentages.

a $\frac{4}{5}$ b $\frac{3}{8}$ c $1\frac{1}{4}$ d $\frac{7}{12}$
e $2\frac{7}{10}$ f $3\frac{1}{2}$ g $\frac{7}{8}$ h $\frac{19}{20}$
i $1\frac{5}{8}$ j $\frac{41}{50}$

5 Change the following into percentages. Give your answer correct to two decimal places.

a $\frac{5}{6}$ b $\frac{3}{11}$ c $\frac{2}{3}$ d $\frac{1}{9}$
e $\frac{5}{19}$ f $\frac{3}{7}$ g $4\frac{1}{3}$ h $2\frac{1}{6}$
i $\frac{2}{13}$ j $1\frac{1}{7}$

Lesson 2

Finding a percentage of a given quantity

Working out a percentage of any quantity is the same as finding a fraction of that quantity. We need to calculate percentages in working out many practical things, such as the size of a 10% deposit on buying a car or how much we save when we buy goods with 15% off the price.

If we run a business, we need to work out percentages such as the Goods and Services Tax (GST) to pay on our sales, and the amount of tax to deduct from wages.

Example

Find:

a 25% of K30

b 12.5% of K200

c 1.5% of K2.8 million

Answer

a $\frac{25}{100} \times 30 = 7.5$

So 25% of K30 = K7.50

b $\frac{12.5}{100} \times 200 = \frac{125}{1000} \times 200$

$= 25$

So 12.5% of K200 = K25

c $\frac{1.5}{100} \times 2.8 \times 1\,000\,000 = 42\,000$

So 1.5% of K2.8 million = K42 000

EXERCISE

1 Find (to the nearest toea):

a 20% of K3000
b 5% of K82.40
c 8% of K55
d 70% of K230
e 15% of K299
f 125% of K450
g 12% of K8300
h 22% of K17 000 000
i 6% of K12 500
j 3.5% of K10 000
k 11.3% of K4000
l 73.8% of K35

2 Millie earns K40.50 per week as a hairdresser. She saves 5% of this and spends 75% of the remainder on rent.

a How much does she save?
b How much rent does she pay?

3 The Malolo family won a competition with K10 000 prize money. They plan a holiday to an amusement park. The family spends 68% of the money on air fares and sets aside 55% of the remaining money for accommodation and meals. How much do they have left to spend at the amusement park?

4 A warehouse holds 50 computers valued at K2399 each. A cyclone causes damage to 85% of the computers so that they are now worthless.

a What is the value of the goods before the cyclone?
b What is the cost of the damage caused by the cyclone?

5 An importer is charged 12.5% duty on consignments brought into the country. What duty would be payable on these consignments?

a K4800
b K24 320
c K160 400

6 A large cannery has an annual wages bill of K542 780.

a If the government charges a training levy of 2% of the wages bill, how much does the cannery have to pay for the levy?

b The cannery has to contribute 7% of its wages bill to a superannuation fund. How much does it pay to the fund?

7 Full-time employees are required to contribute 5% of their gross pay to a superannuation fund. How much will these employees contribute out of their fortnightly pay (to the nearest kina)?

a Leo K135.45 b Damian K493.80 c Arnold K365.80

d Kiteni K648.90 e Nialel K1208.30

8 An electrical goods store has sales for July totalling K14 560.

a How much GST will have to be paid if the rate of GST is 10%?

b If wages amount to 45% of the sales earnings (after GST has been paid), what is the store's wages bill for July?

9 Paul receives 13.5% holiday pay for his 4 weeks of annual leave, worked out on his weekly earnings. If he earns K480 per week, how much does he receive as holiday pay?

10 A person buys a valuable painting at auction for K4.6 million. He has to pay a 'buyer's premium' to the auction house of 15% on the first K50 000 and 10% on the remainder. How much does he pay the auction house in total for the painting?

Smart shopping—sales, discounts and 'best buys'

We can save money on shopping for food and other items if we know how to work out 'best buys' or calculate the savings from sales and discounts. These are important skills to help stretch our money further.

Example

a A store offers a 5% discount to its staff. What would a staff member pay for goods that normally sell for K349?

b Which is the better buy, and by how much? A TV normally priced at K499 at Store A on sale at 15% off, or one normally costing K529 at Store B advertised at $\frac{1}{6}$ off?

Answers

a Discount = $\frac{5}{100} \times 349$

= 17.45 (or 10% = K34.90 so 5% = half of K34.90 = K17.45)

Discounted price = 349 – 17.45

= K331.55

b Store A: discount = $\frac{15}{100} \times 499 = 74.85$

TV now costs 499 – 74.85 = K424.15

Store B: $\frac{1}{6} \times 529 = 88.17$ (rounded to two decimal places)

TV now costs 529 – 88.17 = K440.83

Store A has the better buy as the TV is K16.68 cheaper than at Store B.

Discussion

1 Give some reasons why businesses have sales.
2 Do you think people spend more money when there is a sale?
3 Are sales always good for the customer?
4 Why do you think some businesses give their employees a discount? What examples do you know where this happens?
5 What are some of the advantages and disadvantages for a business of running a sale?
6 What are some of the advantages and disadvantages for a business of offering a discount to certain customers? Give some examples to support your argument.

EXERCISE

1 Good Value store is having a '15% off' sale. What is the sale price of these items?

- **a** camera, normally K549
- **b** mobile phone, normally K199
- **c** hi fi system, normally K799
- **d** DVD player, normally K599
- **e** TV normally K499

2 Anton gets a discount of 8% when he buys goods from the building supplies store where he works. What does he pay for each of these purchases? Give your answer to the nearest 5 toea.

- **a** nails K14.50
- **b** screws K19.30
- **c** rope K24.43
- **d** chain K29.65
- **e** spade K56
- **f** bush knife K36.79
- **g** sand K9.73
- **h** timber K57.40
- **i** barrow K239

3 In Question 2 above, how much does Anton save in total as a result of getting the discount?

4 Which is the better buy, tins of fish at Store A for K1.65 or at Store B marked at K2 on sale for 25% off?

5 Three different stores had the same brand of mobile phone for sale. The price at Store A was K379, the price at Store B was K429 with a special offer of K45 off, while the price at Store C was K410 with a 10% discount.

   a Which is the best buy?

   b How much is saved by buying the phone at the cheapest store rather than the most expensive?

6 Nema wants to make copies of a one-page leaflet to advertise her shop. She can photocopy the leaflet for 8t per page.

   a How many pages can she photocopy for K4?

   b If she wants 500 copies, how much will it cost?

   c The printer charges a fixed cost of K25 and 4t per page to print the leaflets. Is printing a more economical choice for the 500 copies than photocopying?

   d Which is cheaper method if Nema wants 750 copies, and by how much?

7 Rene pays K43.40 for clothes in a sale. If the original price of the clothes was K54.25:

   a how much has she saved?

   b what percentage discount did she receive on the original price?

8 Manus is allowed a staff discount of 5% at the co-op where he works. The co-op is having a '10% off' sale. Manus buys goods originally worth K96.

   a How much does Manus save with the staff discount?

   b If he is also given the sale discount of '10% off' after the staff discount, how much does he pay for the goods?

   c How much has he saved altogether?

9 a Calculate 15% of K96.

   b Why is this a different amount to the total Manus saves in Question 8 above?

Lesson 4

Percentage increase and decrease

There are two methods to calculate the answer when a quantity is increased or decreased by a percentage. These can be used in many situations, such as finding the cost of items on sale at a 15% discount or working out the price to charge customers in a store to give a 40% profit.

Percentage decrease

Example

A TV set is on sale for 15% off the marked price of K450. What is its sale price?

Answer

Method 1

This method finds the percentage saved and subtracts it from the original quantity.

Amount saved = 15% of 450

$= \frac{15}{100} \times 450 = 67.5$

So sale price = K450 – 67.50

= K382.50

Method 2

This method works out what percentage of the original quantity value will be remaining and applies that to the original quantity.

After a reduction of 15%, the value of the television will be (100 – 15)% = 85%

85% is called the multiplying factor.

So sale price $= \frac{85}{100} \times 450 = 382.50$

= K382.50

A rule for the method shown in the above example is:

For any quantity, Q, decreased by n%,

the multiplying factor is $1 - \frac{n}{100}$

and the decreased quantity is $Q(1 - \frac{n}{100})$.

We sometimes need to increase a quantity by a percentage, for instance, when prices go up, a population increases, or production improves. We can use a similar approach to the above methods to find the increased quantity.

Percentage increase

Example

The cost of air freight increases by 12%. What would it now cost to send a box that originally cost K75?

Answer

Method 1

Increase = 12% of 75

$= \frac{12}{100} \times 75 = 9$

So increased amount = 75 + 9 = 84

New cost = K84

Method 2

Multiplying factor = $(1 + \frac{12}{100}) = 1.12$

New cost = K75 × 1.12

= K84

For any quantity, Q, increased by n%, the multiplying factor is $1 + \frac{n}{100}$

and the increased quantity is $Q(1 + \frac{n}{100})$.

Percentage change

We can also find the percentage change when we are given a quantity and its new value after an increase or decrease.

Example

What is the percentage change when:

a K50 increases to K55

b K20 decreases to K17.50?

Answers

a the increase is K5 on the original value of K50, so

% change $= \frac{5}{50} \times 100$

= 10%

b the decrease is K2.50 on the original value of K20, so

% change $= \frac{2.50}{20} \times 100$

= 12.5%

EXERCISE

1 Give the multiplying factor for:

a an increase of:

i 10% ii 15% iii 20% iv 50% v 2.5%

b a decrease of:

i 12% ii 6% iii 25% iv 30% v 8.5%

2 For the following multipliers,

i state whether this represents an increase or decrease, and

ii give the percentage increase or decrease:

a 1.25 **b** 0.95 **c** 1.08 **d** 2 **e** 0.875

3 What is the percentage change to the following. Answer to the nearest whole percent.

a K80 increases to K100
b K900 decreases to K720
c K450 decreases to K399
d K4000 increases to K4895
e K2.45 increases to K2.69

4 Using either method of calculation, find the sale price of items with the given savings:

a set of glasses K79, 10% off
b toaster K149, 5% off
c steam iron K149, 12% off
d mobile phone K199, 5% off
e electric kettle K149, 8% off
f drill K239, 12% off
g computer game K49, 30% off
h coffee table K199, 15% off
i mower K2300, 12% off
j bed K399, 25% off

5 What is the saving on the following items

i in kina?
ii as a percentage of the original price (to the nearest whole number)?

a bookcase was K559, now K499
b TV was K699, now K649
c CD player was K199, now K189
d jeans were K39, now K33
e washing machine was K1289, now K1199

6 Due to a fall in the value of the kina, a shipping company has increased its prices for export goods by 12%. If the company charged K3.50 per kilogram before the increase, find:

i the original cost of the consignment
ii the cost after the increase, to the nearest kina

a 30 kg
b 125 kg
c 7650 kg
d 4 tonnes
e 520 tonnes

Remember
1 tonne = 1000 kg

Challenge

7* Josephine and Kialou have K3800 in savings and wedding gifts to buy goods for their house.

a Do they have enough money to buy all of the items in Question 4 above

i before the sale?
ii when they are on sale?

b If they purchased all of the items at the sale prices, what percentage savings would they make overall?

Mark-up, profit and loss

A business needs to make enough money from its sales or services to enable it to pay all of the running costs of the business. Otherwise, it will not be able to continue trading. This is true even for the smallest business. When the costs of a business are higher than its earnings, the business is running at a loss. Unless it can borrow money or increase its earnings, it will go broke and have to stop operating.

A business where the costs are exactly equal to the income is said to be 'breaking even'.

Discussion

1 Give examples of the running costs of businesses such as a trading store, a courier service or a boarding house.

2 What sorts of costs would a person have if they are working for themselves, for instance, making bilums, growing food to sell or raising chickens?

Mark-up

Retail businesses buy supplies (e.g. from the manufacturer) and add a mark-up to determine the selling price we customers will pay. The mark-up represents the profit the retailer will make from the sale.

Percentage profit or loss

When a company sells goods, the profit is usually calculated as percentage.
We can find the profit as a percentage of the cost price or as a percentage of the selling price.

Percentage profit (based on cost price) $= \frac{\text{profit}}{\text{cost price}} \times 100$

Percentage profit (based on selling price) $= \frac{\text{profit}}{\text{selling price}} \times 100$

Example

a Mr Lau buys electrical goods for K4660 and puts a 45% mark-up on the goods. What will the selling price of the goods be?

b A company imports a sawmill costing K8520. It sells the sawmill for K11 320. Calculate the profit as a percentage of:

i the cost price

ii the selling price.

Answers

a A 45% mark-up is equal to a multiplying factor of 1.45.

So selling price = 1.45 × 4660
= 6757

Selling price is K6757.

b Profit = 11 320 – 8 520 = 2800

i Profit % $= \frac{2800}{8520} \times 100 = 32.9\%$

ii Profit % $= \frac{2800}{11\,320} \times 100 = 24.7\%$

EXERCISE

1 A store buys CDs for K12 each and wants to make a profit of 60% on the cost price. Find:

a the mark-up　　b the selling price

of the CD.

2 A furniture store imports goods and marks up the prices by 30%. What will be the:

i mark-up　　ii selling price

of goods imported at the following price?

a coffee table K150　　b bunk beds K600　　c chest of drawers K850

d desk K290　　e shelves K575　　f dressing table K240

3 Mairie makes dumplings to sell at a market stall. The ingredients to make a batch of 50 dumplings cost her K2.55.

a If she sells the dumplings for 6 toea each, how much profit will she make when she has sold all of them?

b What is her percentage profit?

c If she only sells:

i 45　　ii 42　　iii 40

of the dumplings, how much profit or loss does she make?

d Mairie is considering increasing the price of the dumplings to 7t. What percentage profit will she then make if she sells all her dumplings?

e What is the minimum number of dumplings Mairie needs to sell at 7t to cover her costs?

4 Ilikis buys an old bicycle for K30. He spends K10.50 on new tyres and brakes then sells the bicycle for K57.50.

a What profit does he make on the sale?

b What is the profit as a percentage of the cost (to the nearest %)?

5 Ulato is having a closing down sale and offers 25% off all goods.

a What will I pay for a vase originally marked at K14?

b If the vase cost Ulato K12, what profit or loss is she making on the sale?

c Express the profit or loss as a percentage of the cost price.

6 A supermarket bought 50 kg of taro at K2.50 per kg. It sold 40 kg at K3.15 but the remaining 10 kg was damaged and sold at a reduced price of K1.40. Find the percentage profit or loss on the sale of the whole 50 kg.

7 A store advertises a box of soft drink cans on special.

a By how much has the box of cans been reduced?

b Express the saving as a percentage of the original price (to the nearest %).

c Calculate a 20% reduction on K9.95.

d Is the advertisement correct? Why or why not?

18 cans of soda

Was K9.95

Now K8.45

20% saving

8 A manufacturer makes a drill costing K48 to produce. The manufacturer sells to a wholesaler at a profit of 40%. The wholesaler sells to a retailer at a profit of 25%, and the retailer sells it to customers for a profit of 25%. Find the selling price of the drill.

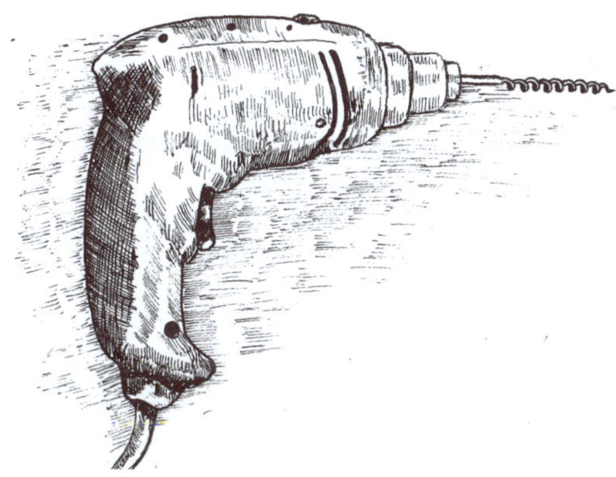

9 A sports store buys runners from the manufacturer at K112 and marks up the cost by 15%. At sale time, the selling price is marked down by 15%.

a What is the mark-up?

b What is the selling price of the runners?

c What is the discounted price of the runners?

d What profit or loss does the sports store make on the sale?

10 An importer of machinery buys a machine from overseas costing K37 000. He must pay duty of K5000 plus K500 for every K10 000 or part thereof of the machine's cost price.

a How much duty does the importer pay on the machine?

b What does the machine cost to import, in total?

c If the importer wants to make a profit of 30%, what price tag must he put on the machine?

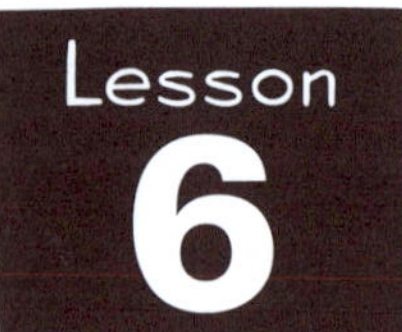

Working backwards

If we are given a certain percentage of an amount, we can find 100% of the amount (or any other percentage). One useful way of doing this is to find the value of 1%, or one unit, and use this to find other percentages. This is called the unitary method.

The unitary method

Example

Vera sells used clothing at the market. She tells her friend Lillian that she marks up the cost of clothes by 60%. If a pair of jeans has a selling price of K40, what did they cost Vera originally?

Answer

If Vera has marked up the cost by 60%, the selling price must represent 160%. We need to find the original cost price, that is, 100%.

Now 160% = K40

So $1\% = \frac{40}{160}$ (this is the value of a 'unit')

$100\% = \frac{40}{160} \times 100$ (i.e. 100 units)

$= 25$

So the jeans cost Vera K25 originally.

Check: Mark-up = 60% of K25 = 15

selling price = 25 + 15 = K40✓

A general rule:

To find 100% given x% of any quantity N:

$$100\% = \frac{N}{x}$$

We can also work out problems such as the one above using the multiplication factor, as the next example shows.

The multiplication factor

Example

Safea buys a rice cooker costing K73.48. The price includes 10% GST. What is the cost of the rice cooker before GST, that is, the retailer's price?

Answer

Let the retailer's price be x.

The selling price is equal to 110% of the retailer's price. This is a multiplication factor of 1.10. Therefore:

$x \times 1.1 = 73.48$

$x = 73.48 \div 1.1$

$x = 66.8$

So the retailer's price is K66.80.

When we are working with goods whose price includes GST, it is helpful to remember:

Price including GST = 1.10 × retailer's price

Retailer's price = price including $\frac{\text{GST}}{1.10}$

GST paid = retailer's price ÷ 11

Using two different methods

Example

Yams are sold at K4.35 a bag to make a profit of 30%. Find the cost price of the bag.

Answers

Selling price = 130% of cost

$\therefore$ 4.35 = 1.30 × cost price

$\frac{4.35}{130}$ = cost price

cost price = K3.35

OR

130% of cost price = K4.35

$\therefore$ 1% of cost price = $\frac{4.35}{130}$

100% of cost price = $\frac{4.35}{130} \times 100$

cost price = K3.35

More than one percentage operating

Example

A retailer adds 35% to the wholesale price of goods but offers a 10% trade discount. A tradesman pays K261 for the goods.

a What was the retail price?

b How much discount was the tradesman given?

c What was the wholesale price of the goods?

d What percentage profit did the retailer make on the goods sold to the tradesman?

Answers

a 10% discount means a multiplication factor of 0.9 (i.e. $1 - \frac{10}{100}$)

retail price × multiplying factor = discounted price

retail price × 0.9 = 261

retail price = $\frac{261}{0.9}$

= 290

b Discount = 290 – 261 = K29

c The retail price of K290 includes the 35% mark-up (increase) so the multiplying factor is 1.35.

wholesale price × 1.35 = retail price

wholesale price × 1.35 = 290

wholesale price = $\frac{290}{1.35}$

= 214.81

So wholesale price was K214.81.

d The goods were bought for K214.81 and sold to the tradesman for K261.

Multiplying factor = $\frac{261}{214.81}$

= 1.215 or 121.5%

So the percentage profit is 21.5%.

(The final multiplying factor can be found by multiplying the two previous multiplying factors, i.e. 1.35 × 0.9.)

EXERCISE

1 If 24% of the Kiowa family's budget is spent on food and the food costs for a fortnight for the family total K72, what is the total family budget?

2 Membership of a business club costs K385 per annum, including GST. What is the cost excluding GST?

3 A business imports a printing machine that costs K37 800. If this price includes 8% duty, what is its original cost?

4 A CD was discounted by 10% and sold for K27.50.
  a What was the marked price?
  b If the profit on the sale was 14% of the cost, what price did the storekeeper pay for the CD?

5 Homoka gets a 6% discount at the store where he works. If he buys goods that cost him K79.90, what was their original selling price?

6 The price of a house in Lae has increased by 12% in the last year, and it is sold for K280 000. What was its value one year ago?

7 A builder's supply store has a 15% off sale. If a mower sells for K675 in the sale, find:
  a its marked price
  b the price the storekeeper paid for the mower if the profit on the sale was 20%.

8 The scales on a weighbridge need adjusting as they record a weight 2% more than the correct weight. A docket from the weighbridge shows the weight of gravel in a truck as 8.6 tonnes.
  a What is the correct weight of the gravel?
  b If the gravel is sold to a customer for K27 per tonne, and the weighing error is not taken into account, how much has the customer been overcharged? Answer to the nearest toea.

9 The manager at Discount City buys a tool set from the manufacturer for K24. He wants to make a profit of 20%, but he always offers customers a 20% discount. What price should he mark on the tool set so that after it is sold with 20% discount, he still makes 20% on the sale?

10 A lathe is sold to a storekeeper for K462 after the manufacturer has included a 40% profit margin plus 10% GST. What is the manufacturer's cost of producing the lathe?

Budgets

Lesson 7

A budget is a plan to help us manage our money. Working out a budget involves listing our income (the money we earn) and our expenses (the money we need to pay for different living costs). That way we can plan for the bills we need to pay, and cut back on unnecessary expenses at times when our essential living costs are highest.

A budget can be worked out for any time period—one week, a fortnight, a month or the whole year. The Government prepares a budget for the whole year so that it can allocate money to services such as health and education.

It is important for a business to plan a budget so that it has sufficient cash flow to pay wages, buy stock, cover costs such as power bills and rent, and many other things.

Discussion

1 Give examples of essential living costs and non-essential costs that a family might include in its budget.
2 Take a survey to find out if students in your class, or their families, plan their spending using a budget.
3 Do you think even wealthy people need a budget?
4 What do you know about the PNG National Budget?
5 What expenses can you suggest these different businesses would have:
   a someone who sells their home-grown fruit and vegetables at a market?
   b someone who delivers parcels and documents on their motor scooter?
   c someone who runs a store in a village?
   d someone who runs a factory canning fish?

EXERCISE

1 Gima is a forklift operator. His budget for the week is:

Income (K)		Expenses (K)	
Wages	50	Board	25
		Food	12.50
		PMV	4.50
		Savings	…
		Other (clothes, entertainment etc.)	3.50

a How much does Gima save each week?
b Which of his expenses is closest to 10% of his wages?
c What percentage of his wages does Gima spend on:
   i board? ii food? iii other items?

- **d** Gima is saving for a bike costing K60. How many weeks will he need to save?
- **e** If Gima's board increased by 5%:
  - **i** how much board would he be paying then?
  - **ii** would he still be able to cover his expenses?

  Explain your reasons.

2 Isabel is a receptionist in a big company. Draw up a table for her budget given the following fortnightly figures: rent K182, HP payments K80, salary K372, food K54, clothes K12, power and other utilities K8, telephone plan K15, other costs K15.

- **a** How much is Isabel able to save each fortnight?
- **b** Which expense is closest to 15% of her salary?
- **c** What fraction of her salary is spent on clothes? Write this as a percentage, correct to two decimal places.
- **d** If Isabel's salary increases by 8%, how much more will she be paid?

3 A store uses 332 units of electricity per quarter on average. It is charged K0.55 per unit plus 10% GST.

- **a** What is its average electricity bill per quarter?
- **b** How much should the store owner allow in the monthly budget for electricity?
- **c** If the cost of electricity increases by 5%, what quarterly bill can the store owner expect?

4 A shipping company has telephone costs which average K26.50 each workday, Monday to Friday.

- **a** How much should the company allow in its annual budget for telephone costs, assuming the company trades every week of the year?
- **b** An economy drive reduces telephone use by 6%. How much will the annual telephone costs of the company fall?

5 In her monthly budget, Mariana allows K14 for her mobile phone costs.

- **a** What average weekly amount can she spend on phone calls?
- **b** If the phone charges for the first three weeks of July were K3.65, K2.89, K4.39, how much can Mariana afford to spend on phone calls in the fourth week to stay within her budget?

6 Golonso earns K286 per fortnight as a driver. He estimates that he needs to allow the following amounts in his budget: rent 30%, food 40%, bills 20%, school fees 2%, savings 5% and other costs 3%.

- **a** Draw up a monthly budget showing Golonso's income and the amount each month, to the nearest toea, he should allow for each expense.
- **b** For two weeks in August, Golonso's bills were K79.80, K7.34, K11.56, K17.30 and K3.41. How much more or less were his bills than budgeted for?

7 Reginald works for the government and earns an annual salary of K55 890 after tax. He also receives an annual spouse allowance of K612 and K972 for each of his three children.

a What is Reginald's total pay for the year?

b What fortnightly pay is this equivalent to? Give your answer to the nearest kina.

c Copy and complete Reginald's fortnightly budget table.

d Which of the family's expenses would you consider non-essential? In what categories could savings possibly be made? Explain your answer.

Income per fortnight (K)		Expenses per fortnight	%	K
Salary and allowances	......	House payment		1028.25
		Food	35	
		Electricity		45.70
		Sewage & garbage	0.5	
		Entertainment	2.5	
		Clothes & household goods		91.40
		School fees	2	
		Newspapers & magazines		22.85
		Savings		
		Telephone	1.5	
		Medical expenses	3	

Depreciation and appreciation

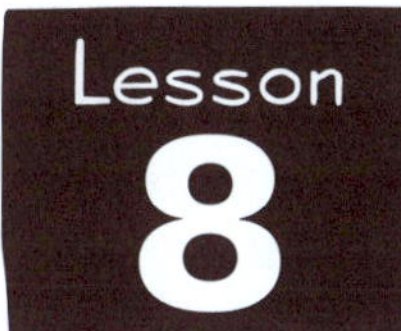

The value of most of the goods we own—like clothes, bicycles or CD players—decreases as time goes on because the goods become worn. Eventually they will wear out altogether or stop working. This loss of value is called **depreciation**.

Understanding depreciation is important in a business as the machinery and other assets of the business will need to be replaced when they are worn out, or the business will not be able to continue. When the business prepares its financial reports each year, it will show the depreciated value of its assets. The business is able to claim the depreciation as a tax deduction.

The opposite to depreciation is **appreciation**, where the value of something becomes greater as time goes by. Works of art, old coins and property may increase in value year by year.

Discussion

1 Why do you think antiques, old coins, stamps, artworks and other things can increase in value (appreciate) with age?

2 What items in your area do you think are already in this category or will appreciate as time goes by?

When the value of an item depreciates by a constant percentage for each year of its useful life, its value is calculated by the same method as a percentage decrease. Similarly, when the value of an item appreciates by a constant percentage for each year of its useful life, its value is calculated by the same method as a percentage increase.

Depreciation

Example

A lathe costing K9000 depreciates at a rate of 15% every year. What is its value at the end of three years, to the nearest whole number?

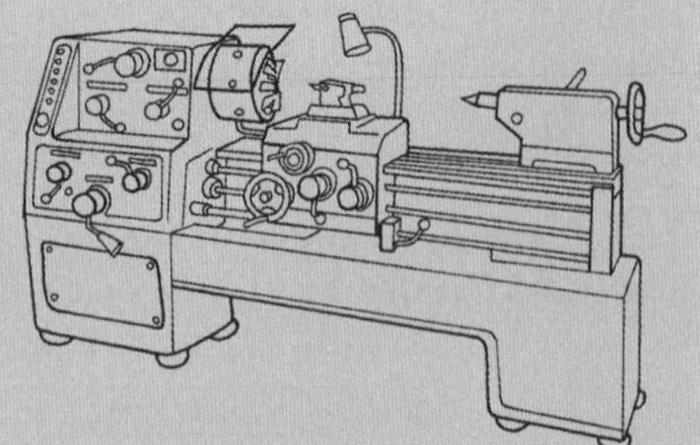

Answer

Method 1

$$9000 \times \frac{15}{100} = 1350$$

At the end of 1 year, the lathe is worth $9000 - 1350 = 7650$

$$7650 \times \frac{15}{100} = 1147.50$$

At the end of 2 years, the lathe is worth $7650 - 1147.50 = 6502.50$

$$6502.50 \times \frac{15}{100} = 975.38$$

At the end of 3 years, the lathe is worth $6502.50 - 975.38 = 5527.12$

So the value after 3 years is K5527.

Method 2

A 15% decrease means that $(100 - 15) = 85\%$ remains.

This equals a multiplication factor of 0.85.

At the end of 3 years the value of the lathe $= 9000 \times 0.85^3$

$$= 9000 \times 0.614 = 5527.125$$

So the value of the lathe is K5527.

Appreciation

Example

Merien's family owns a collection of old coins that has been valued at K2000. If the value of the coins appreciates at the rate of 5% every year, what is the value of the collection at the end of 3 years?

Answer

Using the multiplication factor method

A 5% increase corresponds to a multiplication factor of $(100 + 5)\%$, that is, 1.05.

The value of the collection at the end of 1 year is

$$2000 \times 1.05 = 2100$$

The value of the collection at the end of 2 years is

$$2100 \times 1.05 = 2205$$

The value of the collection at the end of 3 years is

$$2205 \times 1.05 = 2315.25$$

(or calculating in one step, $2000 \times 1.05^3 = 2315.25$)

So at the end of 3 years the value of the coins is K2315.25.

Tables of value

We can use a calculator or a table of values such as those below to find the value of items that appreciate or depreciate over a longer period. This means we don't have to calculate step-by-step. Table 1 shows the value of one kina when it depreciates at the rate of 5% or 10% per year. Table 2 shows the value of one kina when it appreciates at the rate of 5% or 10% per year.

Table 1 Value of K1 when depreciating by a constant percentage

Number of years	Depreciation rate	
	5%	10%
1	0.95	0.9
2	0.9025	0.81
3	0.8574	0.729
4	0.8145	0.6561
5	0.7738	0.5905
6	0.7351	0.5314
7	0.6983	0.4783
8	0.6634	0.4305
9	0.6302	0.3874
10	0.5987	0.3487

Table 2 Value of K1 when appreciating by a constant percentage

Number of years	Appreciation rate	
	5%	10%
1	1.05	1.1
2	1.1025	1.21
3	1.1576	1.331
4	1.2155	1.4641
5	1.2763	1.6105
6	1.3401	1.7716
7	1.4071	1.9487
8	1.4775	2.1436
9	1.5513	2.3579
10	1.6289	2.5937

Example

Use the tables to find what an item valued at K400 is worth after 7 years if it:

a appreciates 5% per year

b depreciates 10% per year

Answers

a Value $= 400 \times 1.05^7$

$= 400 \times 1.4071$

$= 562.84$

So its value would be K562.84.

b Value $= 400 \times 0.9^7$

$= 400 \times 0.4783$

$= 191.32$

So its value would be K191.32.

Flat rate depreciation

Flat rate depreciation means that an item decreases in value by the same fixed amount each year. For instance, a machine bought for K30 000 might be considered to have a useful life of 6 years and then be scrapped. Its depreciation would be spread over the 6 years so that it would lose K5000 a year, or 20% of its value.

Example

A printer bought for K2000 depreciates at a flat rate of 12% each year. Find its value after 3 years.

Answer

The printer loses $\frac{12}{100} \times 2000 = 240$ for each of the 3 years, i.e. K720 in total.

Its value is $2000 - 720 = $ K1280.

EXERCISE

Use the tables on page 23 where appropriate.

1 Hirim buys a new computer to use in his business for K2500. If the rate of depreciation is 20%, find its value to the nearest kina at the end of:
 a 1 year b 2 years c 3 years d 4 years

2 The present value of the cash register used in Mrs Toua's store is K600. If it depreciates at the rate of 10%, what is its value at the end of:
 a 3 years b 7 years c 10 years?

3 Mr Rudani owns a hotel in Lae which is currently valued at K3 500 000. If the property appreciates in value at the rate of 8%, what is the value of the hotel at the end of 3 years?

4 Edris has a collection of rare stamps with a value of K800. If the collection appreciates at the rate of 5% each year, what is its value at the end of:
 a 4 years b 9 years?

5 Ryan buys a set of golf clubs for K2600 and plans to trade them in after 6 years. If they depreciate by 10% of their value each year, how much will Ryan receive when he trades them in?

6 If a car is bought for K22 000 and depreciates at a flat rate of 10% per year:
 a what is the depreciation for one year?
 b what will the car be worth after 4 years at this rate?
 c what is the difference after 4 years between the car's value calculated as a flat rate compared with its value calculated as a constant percentage?

7 Matthew buys a lathe for K5000. It depreciates at the rate of 5% each year. When will the value of the lathe first drop below K4000? (Use trial and error to work out the answer.)

8 A mining company buys a machine valued at K275 000. It estimates the machine has a useful life of 15 years before it will be scrapped.
 a What constant amount would it lose in value each year?
 b What flat rate of depreciation does this represent?

Working with large sums of money

Most of our everyday calculations with money involve small sums. However, very large sums of money are involved in the national budget as well as in the operations of companies such as an import–export business, an airline or shipping firm, or a mining company. It helps to use estimates when working with percentages of such big amounts.

Example 1

Trade between PNG and China increased from K382 million in 2002 to K823 million in 2008. What percentage increase does this represent? Answer correct to one decimal place.

Answer

The increase in millions of kina is 823 – 382 = 441.

So the increase is 441 000 000, and the original amount is 382 000 000.

(An estimate of the percentage increase is more than 100% because 441 is greater than 382.)

$$\text{Percentage increase} = \frac{441\,000\,000}{382\,000\,000} \times 100$$
$$= 115.4\%$$

Example 2

In 2007, the profit of an airline company was K4.56 million. The profit rose by 15% in 2008 then dropped by 12% in 2009. What was the profit in 2009?

Answer

A rise (increase) of 15% is equal to a multiplication factor of 1.15.

A drop (decrease) of 12% is equal to a multiplication factor of 0.88.

(As an estimate: the decrease is not as great as the increase so the answer should be a little higher than K4.56 million.)

So final profit (K million) = 4.56 × 1.15 × 0.88 = 4.614 72

Profit is K4 614 720.

EXERCISE

1 The Big Ships Trading Company reported the following figures in its financial statements for 2006 and 2007.

Item	2006	2007
Sales	336 million	407 million
Depreciation of assets	31.8 million	35.6 million
Interest on borrowings	1.47 million	1.17 million
Capital expenditure	70.1 million	77.7 million

- **a** Write the 2007 sales figure in scientific notation.
- **b** What is the difference between the 'depreciation of assets' figures in 2006 and 2007?
- **c** By how much has the interest on borrowings decreased?
- **d** What is the percentage increase on capital expenditure?
- **e** Which item has increased in value by approximately 10%?
- **f** Which item has the greatest percentage increase in value?

2 The net profit of Big Ships Trading Company for 2007 was K74 157 000. If the net profit for the previous year was K47 479 000:

a how much has the net profit increased?

b what is the percentage increase in profit? Answer to the nearest whole number.

3 In 2008 the PNG government spent K355 000 on new school buildings in one province. If the buildings depreciate by 8% each year, what are they worth at the end of 3 years?

4 A mobile phone company sets a sales target of K100 000 per month in Buka.

a If its sales for August 2008 totalled K84 500, what percentage of its target did it achieve?

b Sales in December 2008 totalled K103 450. By what percentage did sales increase from the August figure?

c If the sales for September, October and November 2008 were K89 040, K92 500 and K98 250 respectively, what is the average sales figure for the five-month period, to the nearest hundred kina?

Remember
1 billion
= 1000 million
or 1 000 000 000

5 A copper and gold mine project is being developed in the highlands.

a The project has an estimated cost of K7 billion which is expected to be recovered in 3.5 years through the earnings of the project. How much of the cost would be recovered per year?

b Two of the three companies developing the mine contribute 75% and 15% of the costs of development.

i What percentage does the third company contribute?

ii If the profits for year 4 are K5.65 billion, find the share of profits for each of the companies.

Remember
A pound is a unit of weight in the Imperial system. Pounds are still used in the United States and some other countries instead of the metric weight units of grams, kilograms and tonnes.

c It is estimated that 7.5 million tonnes of copper can be extracted from the mine. If 1 tonne = 2204.6 pounds, calculate the millions of pounds that can be mined.

d Copper sells for K5.60 per pound. What is the value of the copper from the mine? Answer to the nearest million kina.

Revision and assessment

Lesson **10**

1 Which is closest to 8% of K35 600?

A K280 B K2800 C K28 000
D K4400 E K440

2 Which is closest to 155% of K490?

A 7000 B 300 C 70
D 3000 E 700

3 The calculation to find a 12% increase on K234 000 is:

A K234 000 × 0.12 B K234 000 × 0.88
C K234 000 + 1.12 D K234 000 × 1.12
E K234 000 + 12

4 Choose the correct answer. The selling price in kina of a pair of jeans marked at K48 after a discount of 8% is:

A 48 × 0.08 B 48 + 48 × 0.08
C 48 × 0.92 D 48 + 48 × 0.92
E 48 – 0.08

5 A coffee grower receives 75% of the wholesale price of coffee. If the wholesale price is K5.60 per kilo and 850 000 kilos of coffee are sold, the amount the coffee grower receives is:

A 5.60 × 850 000
B 5.60 × 850 000 × 75
C $5.60 \times 850\,000 \times \frac{75}{100}$
D $5.60 \times 850\,000 \times \frac{0.75}{100}$
E 5.60 × 75

6 Find what percentage:

a 20 is out of 50 b 15 is out of 60
c 25 is out of 80 d 11 is out of 88
e 20 is out of 110

7 State the percentage increase or decrease equal to the following multiplication factors.

a 0.95 b 1.1 c 0.83 d 1.22 e 3

8 Increase these prices by the given %.

a K349 by 15% b K680 000 by 6%
c K5.75 million by 25%

9 Decrease these costs by the given %.
  a K620 by 35%
  b K9235 by 11%
  c K7.45 billion by 80%

10 Find the percentage increase or decrease, to the nearest whole number.
  a K80 increases to K90
  b K9000 decreases to K5850
  c 76 toea increases to 93 toea
  d 95t increases to K1.20
  e K234 500 decreases to K215 500

11 A tradesman is given a 12% discount on paint.
  a How much will he save on paint costing K58?
  b What will he pay for tins of paint costing K95?

12 Which is the best buy, and by how much? A CD player costing K1629 in Store A marked down K25, or K1729 in Store B marked down by 12%?

13 Rosa sets up a garden business. Her costs are: plants and seeds K35, fertiliser K12 and tools K13.
  a What is the total cost of setting up the business?
  b What percentage of her costs is:
    i plants and seeds?
    ii fertiliser?
  c Later Rosa buys a wheelbarrow for K140. What is her total outlay for the business then?
  d What percentage of the new total has she spent on fertiliser?

14 Employers contribute 8.2% of the company's payroll to the national superannuation fund. If a company's payroll totals K15 245, what is its total contribution to the fund?

15 A factory puts a mark-up of 35% on the bread it sells to storekeepers. If it costs the factory 69t to bake a loaf of bread, what will it cost the storekeeper, to the nearest toea?

16 Ken runs an office supply shop and sells a brand of printer for K1595. If the printer cost him K1145:

a what is the mark-up on the printer?

b what percentage profit does Ken make on the printer? Give your answer correct to one decimal place.

17 A 25 m roll of chicken wire was bought wholesale for K41 and sold for K2.50 a metre.

a Find the profit when all the wire is sold.

b Express this as a percentage of:

i the cost   ii the selling price

18 A jeweller's shop has a '25% off' sale.

a What is the multiplication factor in the sale?

b If the marked price of an item is K69.95, what is its sale price?

c Jacinta buys a watch priced at K82.50 in the sale. What was its original retail price?

19 The duty on imported whisky is 35% of its value. Find:

a The duty to be paid on a consignment of whisky valued at K26 000.

b How much duty was paid if the storekeeper paid K26 000 for a consignment on delivery?

20 A gift store is closing down and sells a mirror for K246.50. If the store makes a loss of 15% on the sale, what was the cost price of the mirror?

21 Ivan's monthly expenses are shown in the table below.

a How much does he need to earn a month to cover his expenses?

b To one decimal place, find what percentage of his expenses is spent on:

i food   ii payments on his motor scooter

c If the cost of his board increases by 8%, how much more does he need to earn to cover his expenses?

Board	K85
Food	K103
Payments on motor scooter	K28.70
Clothes & household goods	K9.60
Other	K5.60

22 A motor scooter costing K650 new depreciates 20% in the first year, 15% in the second year and 10% in the third year. What is its value at the end of three years?

23 Ken buys a block of land for K75 000. If it appreciates in value by 10% each year, what is the value of the land after 5 years?

24 PNG debt was K8365 million in 2002 and K7341 in 2006.

a By how much as the level of debt decreased?

b What is the percentage decrease?

SAVING MONEY

Lesson 1

Different ways of saving money

There are a number of different ways we can save up our money to buy expensive items such as a car or a house. One way is to deposit money into a bank account.

Discussion

1 Suggest some ways people in your community could be encouraged to save money.
2 What are some of the reasons people find it difficult to save?
3 What are some of the things you would like to save money for?
4 How do you think you could achieve this?

Class investigation

1 Investigate the different financial institutions in your area, such as banks or credit unions, by completing the following:
   - collect information on the types of account you are able to open (savings account, cheque account, fixed-term deposit and so on).
   - find out any requirements for opening the account—do you need to be over a certain age or have a minimum deposit?
   - what is the interest rate, and how often is interest paid?
2 Collect any brochures or other printed information from these sources and display them in the classroom.
3 Produce wall posters showing the different saving options listing any special conditions and the advantages and disadvantages of each option.

Simple interest calculations

Lesson 2

If we put money into an account at a bank or building society, we are paid interest on the money we have deposited.

The amount of interest paid depends on:

- the amount borrowed or invested (called the **principal**)
- the **term** (length of time) of the loan or investment
- the interest **rate**.

Situations where simple interest is paid include savings accounts, fixed-term deposits and government bonds.

Simple interest is calculated as a fixed percentage of the amount invested. The interest rate is given as a percentage per annum, for example, 4% p.a.

Example

What is the interest paid on a deposit of K1000 for 4 years if simple interest of 8% p.a. is paid?

Answer

The amount of interest paid in 1 year is $1000 \times \frac{12}{100} = 80$

This amount is paid for each of the 4 years,

so the interest $= 4 \times 80$

$= 320$

Total interest paid is K320.

Simple interest formula

The steps in the example above can be summarised as a formula:

$$I = \frac{Prt}{100}$$

where: I is the simple interest earned

P is the principal invested

r is the rate of interest p.a.

t is the term in years.

Example

a Calculate the interest earned if I deposit K2460 for 4 years into an account paying 6.5% simple interest.

b How much would I have in my account at the end of 4 years?

Answer

a In this case, $P = 2460$, $r = 6.5$ and $t = 4$

Substituting into the formula $I = \frac{Prt}{100}$

$= \frac{2460 \times 6.5 \times 4}{100}$

$= 639.60$

So interest would be K639.60.

b The amount in my savings account would be $2460 + 639.60 =$ K3099.60.

Discussion

1 What financial institutions are there in your area or the nearest larger town where you can invest money and earn interest?
2 What is the rate of interest being offered in these places?
3 Survey the class to find out who has money in a savings or other account.

EXERCISE

1 Calculate the simple interest on the following investments:
   a K1200 invested for 4 years at 5% p.a.
   b K25 000 invested for 8 years at 5.75% p.a.
   c K2800 invested for 3.5 years at 6.7% p.a.
   d K800 invested for 30 months at 8.5% p.a.
   e K15 000 invested for 205 days at 6% p.a.

2 Meroe deposits K8500 in an investment account that pays 6.25% simple interest p.a.
   a How much interest will she earn over 2 years?
   b How much interest will she earn if she takes the money out after 300 days?

3 Yawie invests K3800 for 3 years and is paid simple interest of 8%. How much interest will he earn?

4 Tessie has K6500 from selling her car and deposits it in a credit union account for 5 years. The account pays interest of 12% p.a. calculated quarterly.
   a How much interest is paid each quarter?
   b What is the total interest paid over 5 years?
   c How much will Tessie have altogether in her account at the end of 5 years?

5 George buys savings bonds worth K84 000. He holds them for a period of $3\frac{1}{2}$ years. The bonds pay a flat rate of interest of 8.2% p.a, payable monthly.
   a What interest will George be paid each month?
   b Calculate the total amount of interest he will be paid.

6 Veru has K550 in an account paying simple interest of 3.5% p.a.
   a How much interest will she earn in 3 years?
   b How much more interest will she earn if the interest rate is increased to 3.75% after the first year?

More simple interest calculations

We can also calculate the rate of interest or the number of years if we know how much interest has been earned on a certain amount of money. To do this, we can substitute the known values into the simple interest formula and solve the resulting equation, as shown in this example.

Example

Mata deposits K1000 into a savings account paying 4.5% interest p.a. When she withdraws her money she is paid K225 interest. How long was her money invested?

Answer

Here we have $P = 1000$, $r = 4.5\%$, $I = 225$. We want to find the value of t.

Substituting into the simple interest formula:

$$I = \frac{Prt}{100}$$

$$225 = \frac{1000 \times 4.5 \times t}{100}$$

$$t = \frac{225 \times 100}{1000 \times 4.5}$$

$$= 5$$

So Mata's money was invested for 5 years.

Alternatively, we can first transpose the simple interest formula to make P, r or t the subject of the formula:

Since $I = \frac{Prt}{100}$

To make P the subject, multiply both sides by 100 and divide by rt, so $P = \frac{100I}{rt}$

To make r the subject, multiply both sides by 100 and divide by Pt, so $r = \frac{100I}{Pt}$

To make t the subject, multiply both sides by 100 and divide by Pr, so $t = \frac{100I}{Pr}$

EXERCISE

1 If I earn K350 interest in 1 year on an investment of K7000, what is the rate of interest being paid per annum?

2 Interest is being paid at the rate of 2.5% p.a. If I earn K1675 on an investment of K13 400, for how long has my money been invested?

3 An investment held for 3 years earned K108 interest. If the interest rate was 4% p.a., how much had been invested?

4 At the end of three years, Ethan received K487.50 interest on his investment, which was earning 6.5% simple interest. How much did Ethan have invested?

5 At the end of 16 months Darusilla earned K225 on her investment of K3750. What rate of interest was being paid?

6 Thomas put K7500 in an investment and withdrew his money 9 months later. If he received K8000, what was the rate of interest being paid per annum, correct to one decimal place?

7 How many years will it take for an investment of K500 to double if it is earning interest at the rate of 6% per annum? Give your answer correct to one decimal place.

8 Gei invests K10 000 and earns K400 after 292 days. What rate of interest per annum is the investment earning?

9 Francis invests K3000 for 3 years and earns simple interest of 12% p.a. If the interest is paid quarterly, how much interest does Francis receive each payment?

10 An investment of K10 000 earns simple interest of K2762.50. If the interest rate is 6.5% p.a., for how long was the investment held?

Savings accounts

As we found from the class investigations in Lesson 1, one way to manage our money is to open a savings account with a bank, credit union or similar financial institution. One advantage of using a savings account is that our money is safe, but we can make deposits and withdrawals from the account whenever we need.

We might be charged a fee by the bank for keeping the account. We will also be paid interest on the money in our account.

Most banks pay interest on the '**minimum monthly balance**'—the lowest figure in our account for that month. The interest rate is usually quoted as a percentage per annum, so we need to divide this figure by 12 to calculate the monthly rate of interest.

Some special cash management accounts calculate the interest daily although they generally pay the interest monthly.

Example

Walo's bank book showed the following entries for May 2009.

Date	Debit	Credit	Balance
01 May			K380.00
09 May	K45.00		
15 May		K73.50	
28 May	K23.65		
31 May		K73.50	

If the bank pays simple interest of 3.5% p.a. on the minimum monthly balance, how much interest would Walo be paid for the month?

Answer

First we need to work out the balance after each transaction by completing the entries.

Date	Debit	Credit	Balance
01 May			K380.00
09 May	K45.00		K335.00
15 May		K73.50	K408.50
28 May	K23.65		K384.85
31 May		K73.50	K458.35

The minimum balance for May was K335 on 9 May.

$$\text{Interest} = \frac{335 \times 3.5}{100 \times 12}$$
$$= 0.977$$

So interest of 98 toea would be paid into the account on 1 June.

EXERCISE

1 Find the balance in this savings account at the end of July.

Date	Debit	Credit	Balance
01 July			K129.45
07 July	K21.55		
16 July		K43.19	
18 July	K11.67		
23 July		K67.90	
28 July		K51.23	
31 July	K9.72		

2 What monthly interest rate is equivalent to:

a 6% p.a. **b** 4.5% p.a. **c** 3%?

3 Find the interest payable for a savings account with a minimum monthly balance of K7350 if interest of 3.5% p.a. is paid.

4 What annual interest rate is being paid if my minimum monthly balance is K5450 and I earn K10.90 interest for the month?

5 The balance in Tabitha's savings account on 1 January is K3000. A deposit of K1200 on 5 May is the only transaction on the account for the year. If interest of 4.5% p.a. is paid on the minimum monthly balance, how much interest will Tabitha be paid for the year?

6 Interest on Mata's savings account is paid on 1 July at a rate of 6% p.a. calculated on the minimum monthly balance. Mata opened the account on 4 February with a deposit of K4800. What will be the balance of the account on 1 July after interest has been added?

7 Golonzo's bank book shows the following transactions during one financial year (from 1 July to the following 30 June).

Date	Debit	Credit	Balance
1 July			K3296.00
4 Sept		K120.00	
18 Dec	K784.00		
22 Mar	K235.70		
19 Apr		K400.00	
17 Jun		K1415.30	

a Complete the balance for all entries.

b If interest of 3.5% per annum is paid on the minimum monthly balance, calculate the total interest Golonzo will earn for the year.

8 On 3 September, Tapas deposits K20 000 in a cash management account that pays 4.5% p.a. interest. The interest is calculated daily and paid into the account at the end of the month.

a How many days in September will the money earn interest?

b How much interest is earned in September?

c If Tapas leaves the interest in the account, what is the balance in the account on 1 October?

d How much interest is earned in October?

Term deposits

Most banks operate a form of investment called a **fixed-term deposit**. Fixed-term deposits are savings accounts that offer higher rates of interest than a standard savings account. In return, the investor has to commit the funds to the account for a set period. In this investment, the bank guarantees a certain rate of interest for a set period of time. Usually, the longer the customer is prepared to save (i.e. the longer the term), the higher the rate of interest paid to them. The trade-off for this higher rate is that funds saved in a term deposit account cannot be withdrawn until the term has ended.

A range of terms is usually available from as little as 1 week (7-day term) or 1 month (30-day term) up to 5 years. Sometimes a minimum deposit is required, such as K1000. The interest may be paid monthly, quarterly, annually or at maturity (at the end of the time).

Example

A bank offers interest of 3.4% p.a. on a 2-year fixed-term deposit. If I invest K5000, what will be the balance at the end of the term?

Answer

$$I = \frac{Prt}{100}$$
$$= \frac{5000 \times 3.4 \times 2}{100}$$
$$= 340$$

So I will have 5000 + 340 = K5340 in the account at the end of 2 years.

EXERCISE

1 A bank offers the following interest rates for fixed-term deposits.

Term	Interest rate (% p.a)
30 day	3.5
90 day	4.2
180 day	4
1 year	3.50
2 year	3.40

a Calculate the interest earned on the following deposits if the interest is paid at the end of the term in each case.

i K2000 for a 2-year term
ii K5000 for a 30-day term
iii K10 000 for 180 days
iv K250 000 for 1 year

b The K5000 investment in ii above is reinvested, together with the interest earned, and put into a 1-year term deposit. What will be the total value of the investment at the end of 1 year?

2 One bank has a special offer running on a 36-month fixed-term deposit. Interest is paid yearly at the rates shown below. Assume the interest is not reinvested.

Investment (K)	1000–9999	10 000–24 999	25 000–49 999	50 000+
Rate of interest p.a.	3.50	4.70	6.00	6.00

Find the interest paid over 36 months if I invested:

a K15 000 b K62 000 c K9000 d K24 999 e K32 500

3 A bank offers the following interest rates for fixed-term deposits. A minimum investment of K5000 is required.

Term	Interest rate (% p.a)
90 day	3.6
180 day	3.75
1 year	3.5
2 year	3.4

a You will see that the rate of interest being offered increases and is highest for terms of 90 or 180 days, but then it falls. Give a reason why the bank might offer a lower rate of interest for the 1- or 2-year terms.

b Benson invests K6000 for 180 days. How much interest will he earn at the end of the term? Answer to the nearest kina.

c Benson wonders whether it would be better to invest his K6000 for 90 days then, at the end of that term, reinvest his K6000 as well as the interest earned for another 90-day term.

  i Calculate the interest he would earn on the first 90-day investment.

  ii How much would he be reinvesting into the second 90-day term?

  iii How much interest would be paid at the end of the second 90-day term?

  iv Which is the better option and by how much: investing in two 90-day terms as described above, or investing in the 180-day term as in **b** above?

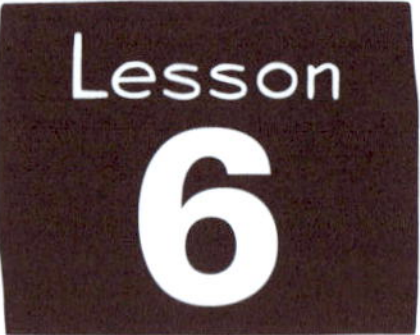

Compound interest

Compound interest is different to simple interest. With compound interest, the interest is added to the principal each time it is paid so that the investment increases (compounds). This means that each interest payment will always be greater than the one before.

With simple interest, an investment will earn exactly the same amount in year 1, year 2, year 3 and so on. With compound interest, an investment will earn more in year 2 than in year 1, and more in year 3 than in year 2.

A comparison of simple and compound interest

Let's say we have K1000 to invest and we deposit it in an account that pays 5% interest each year. The calculation of simple interest versus compound interest for 4 years is shown in the table below.

Time	Simple interest		Compound interest	
	Interest	Balance (K)	Interest	Balance (K)
Start of Yr 1		1000		1000
End of Yr 1	0.05 × 1000 = 50	1050	0.05 × 1000 = 50	1050
End of Yr 2	0.05 × 1000 = 50	1100	0.05 × 1050 = 52.50	1102.50
End of Yr 3	0.05 × 1000 = 50	1150	0.05 × 1102.50 = 55.12	1157.62
End of Yr 4	0.05 × 1000 = 50	1200	0.05 × 1157.62 = 57.88	1215.50

At the end of Year 1, the interest earned is the same whether it is simple interest or compound interest. However after that, the amount of compound interest increases each time it is paid (in this case, each year) while the amount of simple interest is fixed at K50.

Calculating compound interest step by step

The best way to calculate compound interest is to set up a table, as shown in the following example. This way, you can keep track.

Example

a Find the value of my investment of K1000 at the end of 4 years if it is earning compound interest, paid annually, at a rate of 8% p.a.? Answer to the nearest kina.

b Find the total interest earned.

Answer

Time	Interest (K)	Value of investment (K)
Start of Yr 1		1000
End of Yr 1	0.08 × 1000 = 80	1000 + 80 = 1080
End of Yr 2	0.08 × 1080 = 86.40	1080 + 86.40 = 1166.40
End of Yr 3	0 08 × 1166.40 = 93.31	1166.40 + 93.31 = 1259.71
End of Yr 4	0.08 × 1259.71 = 100.78	1259.71 + 100.78 = 1360.49

a The value of my investment is K1360.

b Total interest = 1360 – 1000 = K360

EXERCISE

1 If the interest rate is 8% per year, what is the rate:

a per half year? **b** monthly? **c** quarterly?

d per week? **e** per day?

Give your answer as a decimal correct to three decimal places.

2 Set up a table similar to the above example to find the value of the following investments:

a K125 000 invested for 2 years at 6.5% p.a. compound interest paid annually

b K2000 invested for 5 years at 7% p.a. compound interest paid annually

c K8000 invested for 1 year at 6% p.a. compound interest paid six-monthly (remember to adjust the rate of interest)

d K12 000 invested for 2 years at 5.5% p.a. compound interest paid six-monthly

e K10 000 invested for 3 years at 8% p.a. compound interest paid quarterly

f K200 000 invested for 1 year at 8.5% p.a. compounded six-monthly

3 **a** How much interest would be earned on an investment of K5000 for 3 years earning 7% p.a. compound interest paid annually?

b How much less would be earned if K5000 was invested for 3 years in an account paying 7% p.a. simple interest?

4 Tessie and Jack want to buy a refrigerator costing K1500. They can either save K100 a month and earn interest of 6.5% p.a. compounded at the end of every six months, or buy the refrigerator on terms at 22% p.a. simple interest over 24 months.

a Calculate the monthly payments if they buy the refrigerator on terms.

b Set out your calculations as shown below to find the value of their savings at the end of one year if they start saving on 1 January.

Amount of savings at the end of six months =

Interest on savings =

Balance in account =

Amount of additional savings at the end of year 1 =

Total in account =

Interest on savings =

c In what month will Tessie and Jack have enough in their account to buy the refrigerator for cash?

d What are the advantages of the two options for buying the refrigerator?

Compound interest formula

We can obtain a formula for compound interest by considering the value of an investment, P, earning interest at the rate of r% p.a. compound interest for n years. At the end of one year, the interest on the investment is equal to $\frac{r}{100} \times P$. When this is added to the original investment, P, the amount at the end of the first year is given by:

$$\begin{aligned} A &= P + \frac{r}{100} \times P \\ &= P\left(1 + \frac{r}{100}\right) && \text{as } P \text{ is a common factor of both terms} \\ &= PR && \text{where } R = 1 + \frac{r}{100} \end{aligned}$$

At the end of the second year, the interest is equal to $\frac{r}{100} \times PR$ and the amount at the end of the second year is given by

$$\begin{aligned} A &= PR + \frac{r}{100} \times PR \\ &= PR\left(1 + \frac{r}{100}\right) && \text{as } PR \text{ is a common factor of both terms} \\ &= PR^2 \end{aligned}$$

Similarly, the amount at the end of the third year is equal to PR^3, the amount at the end of the fourth year is PR^4 and so on.

Compound interest formula

$A = PR^n$ where A is the amount accumulated (the value of the investment)
P is the principal (the amount invested)
$R = 1 + \frac{r}{100}$ where r is the interest rate
n is the number of years

You should recognise that the value $R = 1 + \frac{r}{100}$ is the multiplying factor for a percentage increase of r%.

If interest is paid at different times, such as quarterly, monthly etc, we need to adjust the values of R and n accordingly.

Example

Find the value of an investment of K1000 after 4 years if compound interest of 10% p.a. is paid:

a annually

b every six months.

Answer

a Interest of 10% is paid annually

so $R = 1 + \frac{10}{100} = 1.1$

Number of years $n = 4$

$A = PR^n$
$= 1000 \times 1.1^4$
$= 1000 \times 1.4641$
$= 1464.1$

So the investment has a value of K1464.10.

b 10% p.a. is equivalent to 5% every six months

so $R = 1.05$

In four years there will be $4 \times 2 = 8$ interest payments

so $n = 8$

$A = PR^n$
$= 1000 \times 1.05^8$
$= 1000 \times 1.47746$
$= 1477.46$

So the investment has a value of K1477.46.

EXERCISE

Note: If you do not have a calculator, use the table of values on page 42 to answer the following questions. Give all answers to the nearest kina.

1 Find the value of these investments earning compound interest:
   - **a** K5000 after 3 years earning 6% p.a.
   - **b** K10 000 after 6 years earning 5.5% p.a.
   - **c** K7500 after 4 years earning 8% p.a.
   - **d** K10 000 after 12 years earning 4.5% p.a.
   - **e** K1500 after 10 years earning 2.5% p.a.

2 How much interest will be earned after 3 years on an investment of K5000 paying 6% p.a. compound interest?

3 How much interest would be earned after 5 years on deposit of K20 000 in an account paying 7.5% p.a. compound interest?

4 Moses invests K1000 for 2 years in an account earning compound interest of 12% p.a. paid quarterly.

a What percentage interest is paid each quarter?

b How many interest payments will be received in 2 years?

c What is the value of the investment at the end of 2 years?

d How much interest has been earned in this time?

5 Find the value of these investments earning compound interest.

a K2000 after 3 years earning 10% p.a. paid quarterly

b K40 000 after 6 years earning 7% p.a. paid half-yearly

c K8000 after 6 months earning 12% p.a. paid monthly

6 a How much interest will be paid on an investment of K300 000 for 12 months earning 15% p.a. paid every two months?

b How much greater is this than the interest which would have been earned if it were paid every six months?

7* Find the value of an investment of K10 000 after 12 months if it earns interest of 5% p.a. compounded:

a annually? b quarterly? c six-monthly?

8* How much interest will be paid on an investment of K8500 at a rate of 8.4% compounded monthly?

9* How much should I invest now into an account paying 5% p.a. compound interest if I want the investment to have a value of K10 000 in 2 years' time?

10* An investment of K450 has grown to K1065 in 10 years. What rate of compound interest p.a. is this equal to?

Compounding interest

Interest rate	Number of years										
	2	3	4	5	6	7	8	9	10	11	12
1.005	1.01003	1.01508	1.02015	1.02525	1.03038	1.03553	1.04071	1.04591	1.05114	1.05640	1.06168
1.01	1.0201	1.03030	1.04060	1.05101	1.06152	1.07214	1.08286	1.09369	1.10462	1.11567	1.12683
1.015	1.03023	1.04568	1.06136	1.07728	1.09344	1.10984	1.12649	1.14339	1.16054	1.17795	1.19562
1.02	1.0404	1.06121	1.08243	1.10408	1.12616	1.14869	1.17166	1.19509	1.21899	1.24337	1.26824
1.025	1.05063	1.07689	1.10381	1.13141	1.15969	1.18869	1.21840	1.24886	1.28008	1.31209	1.34489
1.03	1.0609	1.09273	1.12551	1.15927	1.19405	1.22987	1.26677	1.30477	1.34392	1.38423	1.42576
1.035	1.07123	1.10872	1.14752	1.18769	1.22926	1.27228	1.31681	1.36290	1.41060	1.45997	1.51107
1.04	1.0816	1.12486	1.16986	1.21665	1.26531	1.31593	1.36857	1.42331	1.48024	1.53945	1.60103
1.045	1.09203	1.14117	1.19252	1.24618	1.30226	1.36086	1.42210	1.48610	1.55297	1.62285	1.69588
1.05	1.1025	1.15763	1.21551	1.27628	1.34010	1.40710	1.47746	1.55133	1.62889	1.71034	1.79586
1.055	1.11303	1.17424	1.23882	1.30696	1.37884	1.45468	1.53469	1.61909	1.70814	1.80209	1.90121
1.06	1.1236	1.19102	1.26248	1.33823	1.41852	1.50363	1.59385	1.68948	1.79085	1.89830	2.01220
1.065	1.13423	1.20795	1.28647	1.37009	1.45914	1.55399	1.65500	1.76258	1.87714	1.99151	2.12910
1.07	1.1449	1.22504	1.31080	1.40255	1.50073	1.60578	1.71819	1.83846	1.96715	2.10485	2.25219
1.075	1.15563	1.24230	1.33547	1.43563	1.54330	1.65905	1.78348	1.91724	2.06103	2.21561	2.38178
1.08	1.1664	1.25971	1.36049	1.46933	1.58687	1.71382	1.85093	1.99900	2.15893	2.33164	2.51817
1.085	1.17723	1.27729	1.38586	1.50366	1.63147	1.77014	1.92060	2.08386	2.26098	2.45317	2.66169
1.09	1.1881	1.29503	1.41158	1.53862	1.67710	1.82804	1.99256	2.17189	2.36736	2.58043	2.81266
1.095	1.19903	1.31293	1.43767	1.57424	1.72379	1.88755	2.06687	2.26322	2.47823	2.71366	2.97146

Inflation, appreciation and depreciation

Lesson 8

As well as enabling us to work out the interest earned on an investment, the compound interest formula can be applied to three other calculations involving money: inflation, appreciation and depreciation.

Inflation

The cost of goods and services increases over time, so that, for instance, the particular model of a machine that now costs K6000 might be expected to cost K7000 if we purchased it in two years' time. This natural increase in costs or values is called inflation.

Inflation is usually stated as a percentage per year, and the inflated value of an item can be worked out using the formula for compound interest, $A = PR^n$.

In this case, A represents the inflated amount (value)

P = original value

$R = 1 + \frac{r}{100}$ where r is the interest rate

n = the number of years (or periods of time).

Example

The inflation rate in a particular country is 2% p.a. If a house is currently valued at K500 000, what will its value be at the end of 5 years, assuming the rate of inflation does not change?

Answer

$A = PR^n$ where $A = 500\,000$, $R = 1.02$ and $n = 5$

$= 500\,000 \times 1.02^5$

$= 552\,040$

So the inflated price of the house is K552 040.

Appreciation and depreciation

If we look again at the examples in Lesson 8 in the 'Spending Money' unit, we will see that the compound interest formula can be adapted to carry out these calculations.

To calculate the value of an item that **appreciates** over time, we can use $A = PR^n$:

where A = the new value

P = original value

$R = 1 + \frac{r}{100}$ where r is the rate of appreciation

n = the number of years.

When we apply this to calculate the value of an item that is **depreciating**, we use $R = 1 - \frac{r}{100}$ where r is the rate of depreciation per annum) as the value is decreasing by a percentage instead of increasing.

We can rewrite the compound interest formula as $V = V_0R^n$

where V = the value after n years

V_0 = the original value

$R = 1 - \frac{r}{100}$.

Appreciation

Example

A stamp collection is valued at K5000. If the value of the collection appreciates at the rate of 3% every year, what is the value of the collection at the end of 8 years?

Answer

Using $A = PR^n$ where $A = 5000$, $R = 1 + 0.03$, $n = 8$

$A = 5000 \times 1.03^8$

$= 6333.85$

So the value is K6333.85.

Depreciation

Example

Mr Li buys a new delivery truck for K35 000. How much value has the truck lost at the end of 4 years if it depreciates at the rate of 20% p.a.?

Answer

Using $V = V_0R^n$ where $V_0 = 35\,000$, $R = 1 - 0.20 = 0.80$, $n = 4$

$= 35\,000 \times 0.80^4$

$= 14\,336$

So the truck is valued at K14 336. It has lost 35 000 – 14 336 = K20 664.

EXERCISE

Note: If you do not have a calculator, use the table of values on page 42 to answer the following questions where possible. Give all answers to the nearest kina.

1 I buy a computer for K3000 that depreciates in value at the rate of 25% p.a. The calculation to find how much its value has decreased at the end of 5 years would be:

A 3000×0.25^5 **B** $3000 - 3000 \times 0.25$ **C** 3000×0.75^5
D $3000 - 3000 \times 1.25^5$ **E** $3000 - 3000 \times 0.75^5$

2 A collector of rare coins has a collection valued at K450. If the collection appreciates at the rate of 5% p.a., the value of the collection at the end of 6 years is given by:

A $450 + 450 \times 1.05^6$ **B** 450×1.05^6 **C** 450×0.95^6
D $450 - 450 \times 0.95^6$ **E** 450×0.05^6

3 Cyril buys a motorbike for K4500. If its value depreciates at the rate of 10% p.a.:

a what is the bike's value at the end of 4 years?
b how much value has the bike lost?

4 A painting currently valued at K15 000 appreciates at the rate of 6.5% p.a. What will its value be after 8 years?

5 A loaf of bread costs K3.20. Assume that the price of bread increases at a yearly inflation rate of 5%. Find the expected cost of the bread, to the nearest t, after:

a 1 year **b** 2 years **c** 4 years **d** 10 years

6 An item costs K300 today. How much will its cost have increased after 10 years if inflation averages 4.5% p.a?

7 Puka and Pelik bought a house for K150 000. They sold it 6 years later for K175 000.

   a Find the inflated value of the house, to the nearest kina, if the rate of inflation was 3% p.a. over the 6 years.

   b By how much has the house value increased due to inflation?

   c How much profit did Puka and Pelik make on the sale?

   d How much real profit or loss does this represent when compared with the inflated value?

8 A loaf of bread costs K2.30 today. If the inflation rate has been constant at 6% p.a., the calculation to find its cost 8 years ago would be:

   A $2.30 \times (1.06)^8$ **B** $2.30 \times (1.6)^8$ **C** $\frac{2.30}{1.6^8}$ **D** $\frac{2.30}{1.06^8}$

9 A CD costs K15 today. Assuming an inflation rate of 5% p.a., what would it have cost 4 years ago?

10 After how many years will K1000 have doubled in value, assuming an inflation rate of 9% p.a?

Comparing investments

It is clear that an account paying compound interest will pay us a greater amount than one paying simple interest (at the same rate of interest and over the same time). This is because we earn 'interest on our interest'. However, we may need to look more closely to compare our options if the interest rates or time periods differ between the accounts we are considering.

Example

Milton has K15 000 to invest. He can deposit the money into an account paying 6% p.a. compounding monthly or an account paying simple interest of 7.5% p.a. Which is the better option?

Answer

Compound interest option

6% p.a. is equivalent to 0.5% per month.

$$A = PR^n$$
$$= 15\,000 \times 1.005^{12}$$
$$= 15\,925.20$$

Amount in account is K15 925.20 so interest is K925.20.

Simple interest option

$$I = \frac{Prt}{100}$$
$$= \frac{15\,000 \times 7.5 \times 1}{100}$$
$$= 1125$$

Interest on the account = K1125

Milton is better off investing in the account paying simple interest (by K199.80).

EXERCISE

Set out all of your working clearly, as in the previous example.

1 Mata has invested K7500 for 2 years at 6% p.a. compound interest paid quarterly. How much more interest will she have earned than if she had put the money in an account paying 5% p.a. simple interest? Answer to the nearest K10.

2 What is the difference between the compound interest on an investment of K6000 for 1 year at 12% p.a. paid quarterly, and the compound interest on the same investment at the same rate but paid monthly?

3 Ignatius and Karen want to invest their savings of K3000 for three years. They can choose an account paying 7.5% p.a. simple interest or one paying 6% p.a. compound interest paid quarterly.
   a Find the interest earned in each case.
   b Decide which is the better financial option, and by how much (to the nearest kina).

4 In Question 3 above, if the interest rate in the account paying simple interest drops to 6% in year 2, how much simple interest will be earned in the three years? Which account now is the better option, assuming all other factors stay the same?

5 Kema has invested K7500 for 3 years at 8% p.a. interest compounded quarterly.
   a What is the value of his investment at the end of 3 years?
   b How much interest has he earned?
   c What simple interest rate would produce the same amount of interest over this time, to the nearest %?

Lesson 10

Revision and assessment

1 Choose the correct answer. The simple interest on an investment of K430 for 9 years at 3% p.a. is:

A $430 \times 9 \times 0.03$ B $430 \times 9 \times 1.03$ C $430 + 9 + 3$

D $\frac{430 \times 9 \times 0.03}{100}$ E $0 \times 10 \times 1$

2 Choose the correct answer. The value after 3 years of an investment of K8000 at 8% p.a. interest compounded quarterly is:

A $8000(1.08)^3$ B $8000(1.08)^{12}$ C $8000(1.2)^3$

D $8000(1.2)^{12}$ E $8000(1.02)^{12}$

3 Interest on a savings account is calculated at 0.25% per month based on the minimum monthly balance. Peter's bank book shows the following transactions.

Date	Debit	Credit	Balance
01 Feb			3057
25 Feb		63	
03 Mar	140		
15 Mar		80	
03 Apr	65		2995

The interest in kina for the month of March would be calculated by evaluating:

A 3060×0.0025 B 2980×0.0025 C 2980×0.025

D 2995×0.025 E 2995×0.0025

4 Find the simple interest on K500 at 10% for 3 years.

5 Janet invests K6000 for 8 years in an account paying simple interest of 4.5% p.a. How much interest does she earn?

6 Gari has earned simple interest of K640 in 5 years in an account that paid 5% p.a. simple interest. How much had he invested in the account?

7 Find the compound interest on K5000 for 2 years at 5% p.a.

8 If you put K500 in an account paying 2% p.a. compounded quarterly, how much would be in the account after 2 years?

9 a Complete the balance column in this table for Safea's savings account with the credit union.

Date	Debit	Credit	Balance
01 May			312.67
08 May		103.07	
23 May		67.90	
26 May	59.23		
31 May	29.52		

b The credit union pays 3% p.a. interest on the minimum monthly balance. How much interest will be paid into the account at the end of May?

10 A bank offers fixed-term deposit accounts with the interest rates shown in the table below.

Term	Interest rate (% p.a.)
30 day	2.5
90 day	3.0
1 year	3.50
2 year	3.25

Calculate the interest earned on the following deposits. Assume the interest is paid at the end of the term.

a K12 000 for a 1-year term

b K900 for a 30-day term

c K400 000 for a 2-year term

11 A new car costs K45 000 today. What will it cost in 7 years' time if the rate of inflation is 2.5% p.a.?

12 Waisina has a collection of historical photographs valued at K480. If the collection appreciates at the rate of 4.5% p.a., use the table on page 42 to find the value of her collection at the end of 7 years. Answer to the nearest kina.

13 What is the difference between the compound interest on an investment of K10 000 for 1 year at 8% p.a. paid quarterly and the compound interest on the same investment at the same rate but paid six-monthly?

14 Which account earns the higher interest on an investment of K500 for 2 years, and by how much: a fixed-term deposit account paying 3.5% p.a. compound interest, or a savings account paying 4.5% p.a. simple interest?

BORROWING MONEY

Ways of borrowing money

Learning to manage our money is a very important life skill: it is much easier for people to get into debt than to get out of it. We need to be careful not to borrow money without considering the real cost of any loan, and we should be aware of other alternatives. Similarly, if we buy goods on hire purchase or using a credit card we need to make sure we can afford the repayments.

Discussion

1 What are some of the reasons people take out a loan?
2 For each of those reasons, identify an alternative approach.
3 Explain what you know about buying on terms or with a credit card.

Class investigation

1 Investigate the loans available from the different financial institutions in your area, such as banks, cooperatives or credit unions by completing the following:
- find out any requirements for taking out a loan, e.g. do you need to be over a certain age or have a certain income?
- collect information on the interest rate charged, e.g. what percentage, how is it calculated?
- what happens if you can't make a loan repayment?

2 Collect any brochures or other printed information from these sources and display them in the classroom.
3 Produce wall posters showing the different loan options listing any special conditions and the advantages and disadvantages of each option.

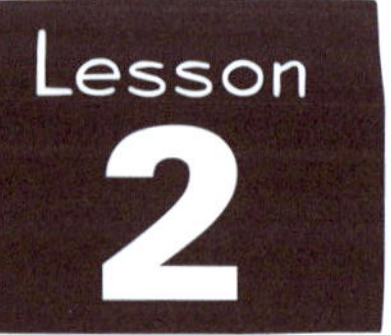

Loans and simple interest

When we borrow money, the lender is entitled to charge a fee for using its money. Banks, building societies, credit unions and other financial institutions all charge interest on the amount borrowed.

Just as we found when working out the interest earned from an investment, the amount of interest paid depends on three things:

- the amount borrowed or invested (called the **principal**)
- the **term** (length of time) of the loan or investment
- the interest **rate**.

Again, as we found with investments, there are two types of interest: simple interest and compound interest. Here we will investigate loans that operate with simple interest.

Simple interest is a flat rate of interest. This means that the same amount of interest is charged every year. It is calculated using the formula:

$$I = \frac{P \times r \times t}{100}$$

where I = interest, P = amount of the loan, t = term of the loan (years) and r is the interest rate percent per year (r% p.a.).

Example

Calculate:

a the simple interest paid on a loan of K2460 for 4 years if the interest is 6.5% p.a.

b the total amount repaid over 4 years.

Answer

$$I = \frac{P \times r \times t}{100}$$

where $P = 2460$, $r = 6.5$ and $t = 4$

$$I = \frac{2460 \times 6.5 \times 4}{100}$$
$$= 639.60$$

a The interest charged is K639.60.

b The amount repaid would be 2460 + 639.60 = K3099.60.

EXERCISE

1 Calculate the simple interest on the following loans:
   - a K800 for 4 years at 8.3% p.a.
   - b K12 000 for 18 months at 6.5% p.a.
   - c K56 000 for $3\frac{1}{4}$ years at 7.75% p.a.

2 Find the total amount owing on a loan of K18 000 over 3 years 7 months if it is charged 8.5% simple interest p.a.

3 Paul and Eva borrow K2500 from the credit union to start a motorbike courier service. If they have the loan for 185 days and interest is charged at 9.25% p.a., how much will they pay back to the credit union?

4 Hoi has borrowed K3800 at 7.25% p.a. for 150 days. How much interest will he be charged, to the nearest kina?

5 Luther borrows K20 000 over 8 years to set up a printing business. The bank charges simple interest of 8.35% p.a.
   - a What interest does he pay over 8 years?
   - b What is the total Luther must repay for the loan?
   - c If he makes a regular monthly payment to cover this over 8 years, how much is each monthly payment? Give your answer correct to the nearest 5t.

6 A company takes out bridging finance of K220 000 for 20 days to purchase an office building. If the interest on the loan is charged at 8.35%, how much interest is paid on the loan?

Did you know?

Bridging finance is a short-term loan, usually only a few months, which is either paid back or replaced with a longer term loan. Bridging finance is often arranged to complete the purchase of a property before the borrower receives payment from the sale of another property.

7 Martin needs to borrow K8000 to buy a used car.

a Which of the loans below charges lesser interest?

b How much does Martin save by taking the cheaper loan?

People's Bank	Nationwide Bank
K8000	K8000
pay over $3\frac{3}{4}$ years	pay over $3\frac{1}{2}$ years
14.25% p.a. simple interest	14.5% p.a. simple interest

8 Janet was charged K252 simple interest on a loan of K4500 taken out for 5 months. What annual rate of interest is she being charged?

9 For how long have I had the loan if I pay K290 interest on a loan of K10 000 charging interest at a flat rate of 5.8% p.a.?

Hire purchase

You will remember from Grade 9 work that hire purchase is a way of buying goods by instalments. It is also called 'buying on terms'. Many stores offer hire purchase or time payment plans. We may be required to pay a deposit and then make regular repayments to cover the balance of the cost. Interest is also charged on the balance over the time we are paying off the goods.

Discussion

1 What are the advantages and disadvantages of buying goods by hire purchase?

2 How does this differ from simply hiring goods?

3 What happens if payments are not made?

4 How does this option compare with buying by: cash, lay-bye, credit card, or a loan?

5 Review this discussion after you have completed the calculations in the following exercise.

Example

Salome buys a washing machine advertised for K499 by hire purchase. She pays a deposit of K49 and weekly instalments of K6.50 for 2 years. How much interest has she been charged over that time? What percentage interest is this?

Answer

The balance owed on the washing machine is K499 – K49 = K450.

Salome pays a total of 6.50 × 52 × 2
(instalment × weeks per year × no. of years) = K676

The interest over 2 years is K676 – K450 = K226

So the interest per year is K113.

The rate of interest is $\frac{113}{450} \times 100 = 25.111$

So she is being charged 25% p.a.

EXERCISE

1 Flora buys a new stove on hire purchase. The stove costs K2199 but can be bought over 12 months with a deposit of K199 and payments of K99 every fortnight.

  a What is the balance owing on the stove after the deposit is paid?
  b How many fortnightly payments will be made over 12 months?
  c What total amount will Flora pay for the stove?
  d How much interest has been charged?
  e What percentage rate of interest is being charged?

2 A sound system costing K1549 can be bought with a deposit of K49 and monthly payments of K81.25 for 2 years.

  a What is the balance owing on the sound system?
  b How many monthly payments will be made?
  c What total amount will be paid for the sound system?
  d What interest has been charged over 2 years?
  e Calculate the rate of interest being charged.

3 Solomon earns K15 per month in a part-time job. He wants to buy a CD player costing K200. He has saved K40 and plans to use the money he earns to pay off the balance in equal monthly instalments.

  a What is the balance owing on the CD player?
  b If the interest being charged on the balance is 12% p.a., will he have paid off the CD player at the end of 12 months? Show your working in full.

4 Ravi buys a computer costing K5400 on hire purchase from his uncle's store. He pays a deposit of K1400 and is charged a flat rate of 8% interest. He has a choice of making monthly payments for 12 months or 18 months. Calculate the amount he will be paying each month if he makes:

  a 12 payments   b 18 payments.

5 Gari is setting up his business and wants to buy equipment costing K10 000 on hire purchase with no deposit. The store offers him the choice of making fortnightly payments of:

    i K953 for 6 months,   ii K495 for 12 months or
    iii K398 for 18 months.

  a How many fortnights are there in:
    i one year?   ii 6 months?
  b Calculate the total repayments for each option.
  c Work out the rate of interest per annum being charged in each case.
  d Why do you think the interest rates are not the same?
  e What advice would you give to Gari in selecting which repayment plan to use? What other information would he need to consider?

6 A hire purchase agreement on goods costing K300 require the purchaser to pay a deposit of K60 and make eight monthly payments of K32.85. What is the flat rate of interest being charged p.a.?

7 Manni wants to buy an electric guitar advertised for K485. After paying a 10% deposit he will be charged interest of 20% p.a. If Manni chooses to pay off the guitar over 5 years, what will be the monthly repayments?

8 A business buys a car costing K45 000 under a hire purchase agreement. The terms of the agreement are a deposit of 20% and monthly repayments of K1180 for 5 years. Find:
  a the deposit
  b the total paid for the car
  c the annual interest rate, correct to one decimal place.

Credit cards

Buying goods using a credit card is similar to taking a loan from a bank. We are charged interest on the amount owing on our credit card, just as we would be charged interest on a loan. This is because the credit card company or bank that has issued the card has paid the store for our purchase immediately but some time will pass before we pay the money back. Using a credit card is an alternative to buying goods with cash or through hire purchase.

Credit cards from different companies have different conditions of operation. For instance, some charge an application fee and/or an annual fee. There are differences in whether they allow any 'interest-free' days. Most importantly, there is a large range in the interest rate charged by different companies. Because of these differences, it could be the case that a card with a lower interest rate could end up charging you more interest than a card with a higher rate. Consumers should find out all of the conditions before signing up for a credit card.

Discussion

1 What are the advantages and disadvantages of the three methods of purchasing—cash, hire purchase and credit card?
2 What companies do you know of that issue credit cards? Find out the rate of interest and how it is charged, for example, are there any interest-free days?

No interest-free days

In this case, simple interest on the amount of each transaction is charged for the total number of days for which that amount is owed.

Example

What interest will be charged on Tamara's credit card statement that shows the following information.

Statement period 27 March to 26 April

Date of transaction	Transaction details	Amount	Balance
12 April	DVD store	28	28

Interest rate on purchases is 17.5% per year (0.04795% daily).

Find the interest charged if Tamara pays the balance in full on 30 April.

Answer

Simple interest is calculated using the formula:

$$I = \frac{P \times r \times t}{100}$$

Now P is the amount owing (K28), $r = 0.04795\%$ and t = number of days before the amount is paid, i.e. 19.

$$\text{So } I = \frac{28 \times 0.04795 \times 19}{100}$$

$$= 0.255094$$

Interest of 25t would be charged.

Remember

To calculate the number of days (inclusive) between 12 April and 30 April, subtract 12 from 30 and add 1.

Many transactions

Where a number of transactions have occurred, interest is charged on each transaction for the number of days that amount is owing, as shown in the following example.

Example

Genevieve buys a number of items using her credit card during December. She pays the amount in full on the due date of 18 January.

Find the interest she will be charged if the credit card has no interest-free days and charges an annual rate of 18.5% p.a. (0.05068% daily).

Statement period 04 December to 04 January

Date of transaction	Transaction details	Amount	Balance
3 December	Jeans Only	45	45
11 December	Hardware	115	160
18 December	Novelty & gift shop	58	218
22 December	Phone cards	32	250

Payment due by 18 January.

Answer

We need to identify the different values of P and t for each transaction and substitute those values into the simple interest formula.

She owes K45 for 8 days $I = \frac{45 \times 0.05068 \times 8}{100} = 0.182448$

She owes K160 for 7 days $I = \frac{160 \times 0.05068 \times 7}{100} = 0.567616$

She owes K218 for 4 days $I = \frac{218 \times 0.05068 \times 4}{100} = 0.441930$

She owes K250 for 28 days $I = \frac{250 \times 0.05068 \times 28}{100} = 3.5476$

Total interest = 4.7395, which is K4.73.

(Note that bank calculations usually drop the third and subsequent decimal places instead of rounding.)

Minimum payments

A credit card statement will usually identify a minimum payment amount, unless the card has been set up to be paid automatically by a direct debit from a linked bank account. The amount is only a fraction of your outstanding balance, and it can be as low as 2.5% or K25 for every K1000 owed.

It is important to be aware that making only the minimum payment will not pay off the balance in a reasonable time. For instance, it has been calculated that it would take you 11 years to pay off a balance of K1000 supposing you only pay the minimum amount calculated on the figures above.

EXERCISE

1 Find the amount of interest paid on the following credit card balances:
   - a K246 paid after 10 days, interest rate 18.5% p.a.
   - b K945 paid after 24 days, interest rate 14.5% p.a.
   - c K2605 paid after 28 days, interest rate 21% p.a.

2 Kashmir has to pay a minimum of K10 or 5% of the balance owing, whichever is the greater.
   - a What will Kashmir pay in each of the following months?
     - i July 2008, balance outstanding K68.11
     - ii August 2008, balance outstanding K142.36
     - iii February 2009, balance outstanding K352.99
   - b For what balance would the minimum payment figure and 5% of the balance owing be the same?

3 Find the total interest payable at the due date for the following credit card accounts if there are no interest-free days.

Statement period 12 March to 11 April
Payment due by 28 April.

Date of transaction	Transaction details	Amount	Balance
16 March	Hong's Homewares	62	62
21 March	Allphone	95.56	157.56
28 March	Car hire	118.04	275.60
06 April	Pharmacy	49	324.60

Interest rate on purchases is 17.5% per year (0.04795% daily).

4 a Complete the missing values at A, B and C.

b Find the total interest payable on the account. There are no interest free days and the account is paid in full on 31 July.

Statement period 19 June to 18 July
Payment due by 31 July.

Date of transaction	Transaction details	Amount	Balance
22 June	Top Fashions	78.90	78.90
26 June	Quality Jewellers	113.64	A
08 July	Camera Shop	208.99	B
12 July	Karla's Travel Agency	732.45	C

Interest rate on purchases is 21% per year (0.05753% daily).

Compound interest and reducing balance loans

It is more usual to be charged compound interest on a loan than to be charged simple interest, which we practised calculating in Lesson 2.

Calculating compound interest on a loan is similar to calculating compound interest on an investment.

The formula for working out compound interest is $A = PR^n$ where A = amount owed after n periods and $R = 1 + \frac{r}{100}$ where r is the interest rate.

Example

Simon borrows K15 000 to buy a second-hand van for his business. He has to repay the money over 5 years and is charged 9% p.a. compound interest on the whole amount of the loan. Use the formula $A = PR^n$ to calculate:

a the amount Simon will have to repay

b his monthly instalments over the 5 years.

Answer

a $A = PR^n \quad r = \frac{9}{100} = 0.09$ and $R = 1 + 0.09 = 1.09$

$= 15\,000 \times 1.09^5$

$= 15\,000 \times 1.53862$

$= 23\,079.30$

So Simon will have to repay K23 079.30.

b There will be $5 \times 12 = 60$ instalments in 5 years.

Each instalment will be K23 079.30 ÷ 60 = K384.655 = K384.65

(Note that the final payment will need to be adjusted because of the rounding off.)

Reducing balance interest

A more realistic situation exists than the calculation of Simon's loan and interest payments shown in the example above. When a loan is paid back over a period of time, the amount owed—the balance—decreases each time a payment is made. The bank usually calculates the interest on the balance after the payment has been deducted, rather than the original loan amount. This means that the interest charged every time period (each month, fortnight or quarter) is less than for the period before, as the next example shows.

Example

I borrow K4000 at a reducing balance interest rate of 15% p.a. and make monthly repayments of K800. Calculate the interest payments for the first four months of the loan.

Answer

15% per year is the same as 15 ÷ 12 = 1.25% per month.

End of month	Balance of loan (K)	Interest (1.25% per month)	Repayment	Closing balance (amount owing)
1	4000	0.0125 × 4000 = 50	800	4050 – 800 = 3250
2	3250	0.0125 × 3250 = 40.62	800	2490.62
3	2490.62	0.0125 × 2490.62 = 31.13	800	1721.75
4	1721.65	0.0125 × 1721.75 = 21.52	800	943.27

If the interest had been charged at a fixed rate, I would have paid K50 every month. As the table shows, with a reducing balance loan only the first month's interest is K50 and the months that follow are charged less than that.

Mortgages

House mortgage repayments are calculated using the reducing balance method. Reducing balance repayment is much cheaper than flat rate interest repayment for the same interest rate.

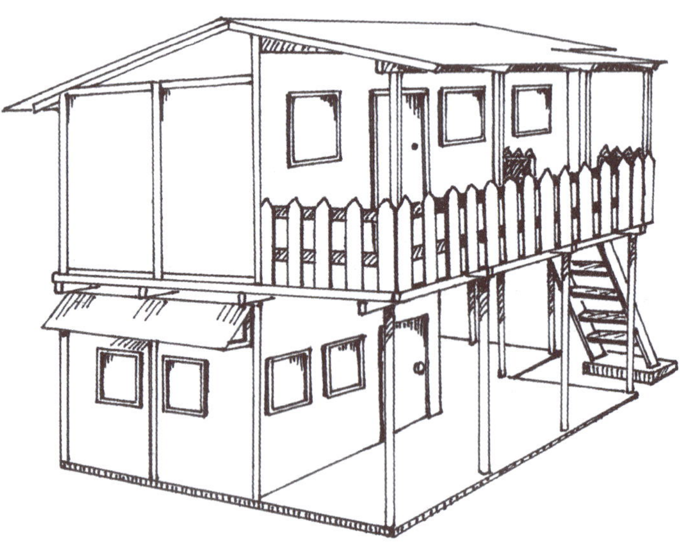

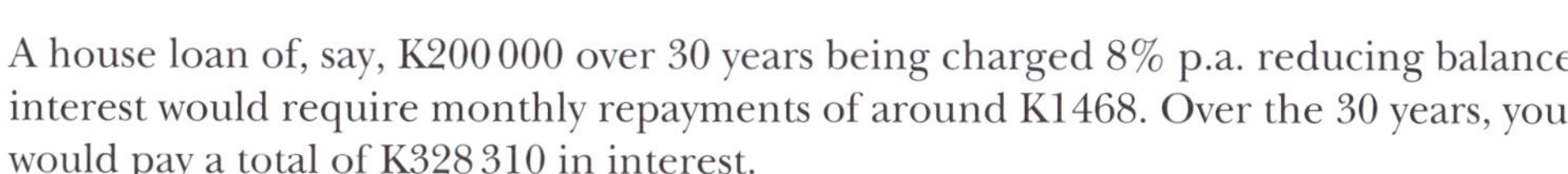

A house loan of, say, K200 000 over 30 years being charged 8% p.a. reducing balance interest would require monthly repayments of around K1468. Over the 30 years, you would pay a total of K328 310 in interest.

Making payments more frequently significantly reduces the loan term. For instance, on the same mortgage of K200 000 the loan term would be reduced by 7 years and 3 months if fortnightly payments of K734 were made.

Did you know?

'Mortgage' comes from an old word meaning 'dead pledge'. When you take out a mortgage on, say, a house, you transfer an interest in the property to the lender as security. The mortgage ends ('dies') when the loan has been repaid. If the borrower is not able to continue repaying the loan, the lender is entitled to sell the property to recover their loan.

Discussion

1 Imagine someone is repaying a loan of K100. Explain the difference between flat rate interest (simple interest), interest compounded monthly on the full loan, and interest compounded monthly on the reducing balance.

2 Which would be the:

a most expensive **b** least expensive option?

EXERCISE

1 Calculate:

i the amount owing **ii** the total interest to be paid

on the following loans, assuming the interest is calculated on the whole loan.

a K1000 for 3 years at 9% p.a. compounded annually

b K20 000 for 30 months at 8% p.a. compounded every six months

c K10 000 for six months at 12% compounded monthly

d K2 million for 4 years at 8.5% compounded annually

2 Therese has borrowed the sum of K5000. She is paying 9% p.a. compound interest and is making regular payments of K1200 at the end of the year, after the interest has been added. Copy and complete the following table showing the reducing balance on her loan.

End of year	Balance of loan (K)	Interest	Repayment	Amount owing at end of year
1	5000.00	$0.09 \times 5000 = 450$	1200.00	4250
2				
3				
4				

Find:

a the amount owing after 4 years

b the amount of interest paid in 4 years.

3 Copy and complete the table showing step-by-step calculations for the first year of a loan of K7500 with 9.6% p.a. with repayments of K200 quarterly.

End of quarter	Balance of loan (K)	Interest	Repayment	Amount owing at end of quarter
1	7500.00	$\ldots \times 7500 = 180$	200.00	7480
2				
3				
4				

4 A loan of K15 000 is to be repaid with quarterly payments of K1000. Interest is charged at 8.4% pa. on the reducing balance.

 a What is the quarterly rate of interest?

 b Show step-by-step calculations to find the amount still owing at the end of the first year.

 c Find the interest charged in the first year.

5 A loan of K10 000 is being charged interest at the rate of 7% p.a. Regular repayments of K2000 are made at the end of each year, after the interest has been added.

 a Set up a table to show the calculations needed to find the amount still owing after 5 years.

 b How long does it take before the loan is completely repaid?

 c What is the final payment?

 d What is the total interest paid on the loan?

6 A house mortgage of K200 000 is being repaid with interest of 7.5% p.a. and payments of K2400 monthly.

 a What percentage interest is being charged monthly?

 b What is the amount owing on the mortgage at the end of 3 months?

EARNING MONEY

How people earn

People can earn money in many different ways. A plumber might set up their own business and charge for their work. A shopkeeper might earn their living by selling goods at a profit. However, most people work for an employer and are paid depending on the type of work they do and the type of contract or agreement that exists about their pay.

Two ways an employee might receive an income are:

a from a fixed salary paid in regular instalments (such as every week or fortnight), or

b from wages based on the number of hours worked.

Discussion

1 Apart from fixed salary or wages, how else can an employee be paid?

2 Identify different jobs people carry out in your area and the way they are paid.

3 The government sets an award rate of pay for some jobs. Find some examples of jobs with an award rate.

4 What is the basic wage? How is this set?

EXERCISE

1 The workers in a store are paid K7.90 per hour. How much will these workers earn for the week if they have worked the following hours?

a Atholie 5 hours
b Tomas 12 hours
c Reginald 28 hours
d Myra 21.5 hours
e David 17.25 hours
f Elizah 22.5 hours

2 Veitu's rate of pay is K8.55 per hour for normal working hours. She is paid double this rate ('double time') when she works the night shift. What will she be paid in a fortnight when she has worked 18 hours normal time and 4 hours at double time?

3 A forklift operator is paid K6.35 per hour on weekdays and time-and-a-half on weekends. If he works 30 hours during the week and 12 hours on the weekend, what will he be paid in total for the 7 days?

4 Motu delivers timber to the mill and earns K14.55 per tonne of timber delivered. How much will he be paid for delivering 90 tonnes?

5 Frank is paid a retainer of K150 per week and 15% commission on sales. If his sales for four weeks are K650, K724, K248 and K502, what will be his total pay for the month?

6 Seth is paid an annual salary of K46 000. How much does he earn (to the nearest kina):

a per month

b per week?

7 A shipping agent is paid by commission on the value of contracts, earning 2% on the first K150 000, 1.5% on the next K100 000 and 1% on anything over that. What will he be paid on a contract valued at:

a K330 000

b K4.5 million?

8 Mekere works for the government and earns K580 per week. He is also paid an annual allowance of K972 per child for each of his three children, and an allowance of K450 for his wife.

a What total payment does he receive for the year?

b What is his monthly pay, to the nearest kina?

9 A saleswoman is paid an annual retainer of K17 200 (regardless of how many sales are made) and a commission of 6.5% on the value of sales she makes.

a Find her annual salary if the sales for the year totalled K365 000.

b Last year her annual salary was K42 322.50. What was the total value of sales made?

c If she wants to earn K50 000 for the year, what value of sales will she need to achieve?

10 Nerilee earned K78.60 for a 30-hour week. Her husband earned K158.90 for a fortnight when he worked 56 hours.

a What was Nerilee's hourly rate of pay?

b Is Nerilee paid more per hour than her husband? Show your reasoning.

11 Anna's annual salary of K35 000 increased to K42 000. Find:

a the percentage increase

b Anna's new weekly pay, to the nearest toea.

12 Sam is paid 50t for each kilo of coffee beans he picks. If he earns K126.75 in one week, how many kilos of coffee beans did he pick?

More about earnings

Lesson 2

Some pay scales are based on age and experience, so that an older or more experienced worker will earn more for their work than someone new to the job. Also, a promotion at work can lead to a pay rise, as can taking a job that pays a higher wage with another company.

On the other hand, we might work fewer hours because we need time to care for children or to study, so our earnings will drop. Or it might be necessary to take a pay cut when business is going through a difficult period.

All of these situations are illustrated in the following exercise.

Discussion

1. What are some of the part-time jobs available in your area that can be performed by people your age?
2. What rates do they pay?
3. What jobs are the members of your class interested in getting when you leave school?
4. What do you know about the rate of pay for those jobs and opportunities to move up to higher levels?

EXERCISE

1. Maria is paid 75t for each shirt she sews. If she makes 35 shirts in one week, what will she be paid?
2. Robert is paid K7.80 per hour. If he works 32 hours in a fortnight, what will his fortnightly pay be?
3. Jacob has just been given a pay rise of 6% p.a. and his fortnightly pay is now K1696. Calculate:
   - a his pay before the increase
   - b the value of the pay rise.
4. Oscar has a part-time job as a security guard earning K215 per fortnight. He moves to a new company paying K260 per fortnight.
   - a What percentage increase is the new wage equal to? Give your answer correct to one decimal place.
   - b What annual salary does he earn:
     - i in the old job
     - ii in his new job?
5. Toksi is an administrative officer level 2, earning K72 106 a year. When he advances to level 3 he will be paid K74 526 per year.
   - a How much will his pay have gone up?
   - b What percentage pay increase does this represent? Answer to the nearest whole per cent.
6. Nema is paid an annual salary of K45 000. The company allows her to work part time while she is studying, so Nema agrees to a salary reduction of 40%. Find:
   - a her annual salary after the reduction
   - b her monthly pay then.

7 Apprentice plumbers working in a shipyard are paid according to the table below.

Level	Wages per 38-hour week (K)
1st year	251
2nd year	357
3rd year	461
4th year	539

a Sebu is a 3rd year apprentice. What is his weekly pay?

b Nigel is a 1st year apprentice. How much less per week does he earn than Sebu?

c Vero is a 4th year apprentice. What will his fortnightly pay be?

d Vero also receives an allowance of 83 toea per hour when he is working inside the ship's hull. If he works 12 hours inside the hull in one week, what wages will he take home that week?

e How much more does a 4th year apprentice earn than a 1st year apprentice:

i per fortnight? **ii** per year?

Give your answer to the nearest kina.

8 Assuming a 40-hour week, convert the following annual salaries to:

i monthly **ii** weekly **iii** hourly amounts,

to the nearest toea.

a K18 500 **b** K25 600 **c** K55 400 **d** K136 900

Help box

Redundant means no longer needed. If a worker has been employed for some time, the employer is required to make a payment when putting the worker off.

9 Kenai works as a salesman for office equipment. He has a choice of being paid:

A 20% commission on sales only, or

B a retainer of K45 000 plus 1% commission on sales, or

C a fixed monthly salary of K5000.

Which is the best option, **A**, **B** or **C**, if he sells equipment to the value of K250 000 over the year?

10 Edris earns weekly wages averaging K158. He is made redundant and his employer has to give him a redundancy payout of 34.7 weeks of salary. How much will Edris receive in his final weekly pay packet if it includes his redundancy pay?

Tax and other deductions

Lessons 3 4

When we enter the workforce, we are expected to pay income tax on the money we earn in our job. This is collected by the Government and is used to pay for everything it provides for society and running the country, that is, things like building roads and bridges, printing currency, running hospitals and schools, training soldiers and policemen.

Personal income tax

The amount of income tax we pay depends on how much we earn. It is calculated as a percentage of our earnings over a certain level or threshold. A different rate applies to low income, middle income and high income earners. As a result, those people who earn very little money pay no income tax, those who earn an average amount pay some tax, those who earn more pay a slightly higher rate of tax, and those who earn a great deal pay the highest rate of income tax.

The amount of personal tax we are required to pay is worked out on the basis of our 'assessable income'. This includes the income we receive from employment (our wages or salary), any interest we earn on investments or bank accounts, and any business income or money that we receive as rent. At the same time, some deductions are allowed for expenses incurred in earning our income, especially if we are self-employed. For instance, a builder who works for themself would be able to deduct the cost of tools used for building or special protective clothing

Tax rates

The Government determines the tax rate each year and this will change from time to time. The tax rates for 2009 are shown in Table 1 below. These rates apply to residents, that is, people whose main home is in Papua New Guinea or who live in PNG for more than half the year.

Table 1 Personal income tax 2009, residents

1	2	3	4
Income exceeding	**But not exceeding**	**Tax on column 1**	**% tax on balance**
0	**7000**	**0**	**0**
7000	18000	0	22
18000	33000	2420	30
33000	70000	6920	35
70000	250000	19870	40
250000		91870	42

Non-residents, that is people who are working in Papua New Guinea but whose main residence is overseas, have a different tax scale that does not allow any tax-free income. Their tax scale is shown in Table 2.

Table 2 Personal income tax 2009, non-residents

1	2	3	4
Income exceeding	but not exceeding	Tax on column 1	% tax on balance
0	18 000	0	22
18 000	33 000	3 960	30
33 000	70 000	8 460	35
70 000	250 000	21 410	40
250 000		93 410	42

Example

Calculate the tax that would be paid on an income of K20 000 by:

a a resident

b a non-resident.

Answer

a $\text{Tax} = 2420 + 30\% \text{ of } (20\,000 - 18\,000)$
$= 2420 + 0.30 \times 2000$
$= 3020$

So the resident would pay tax of K3020.

b $\text{Tax} = 3960 + 30\% \text{ of } (20\,000 - 18\,000)$
$= 3960 + 0.30 \times 2000$
$= 4560$

So the non-resident would pay K4560.

Remember

'Gross pay' refers to the total of an employee's regular remuneration, including overtime pay, commissions and bonuses, before any deductions are made.

'Net pay' refers to the remaining amount of an employee's gross pay after deductions such as taxes and superannuation contributions are made.

Allowances and deductions

Depending on their job conditions, some employees are entitled to a car allowance, a housing allowance or some other allowance on top of their hourly earnings. This is added to their gross pay before tax.

At the same time, compulsory superannuation is deducted from the employee's gross pay. Other deductions might be made for things such as rent for company housing, or private use of a company car or office phones.

Superannuation

All employees contribute a percentage of their gross income to a national superannuation fund. In addition, employers contribute a certain percentage of the employee's basic pay each fortnight. These percentages of gross income are set by the Government.

Example

Bernard is paid at the rate of K8.13 per hour. He worked 63.58 hours in one fortnight. A deduction is made for his compulsory 6% payment to the national superannuation fund. He is paid an allowance of K55 for housing.

a What is his gross pay for the fortnight before any adjustment for superannuation and housing?

b To the nearest kina, how much does he contribute to superannuation?

c What is Bernard's taxable income after the adjustments for his superannuation payment and housing allowance?

d Bernard's tax for the fortnight is K37.76. He also has to pay K10.20 for private use of the company car and K2.80 for private use of the office telephone. What will be his net pay?

e His employer has to pay 8.4% of Bernard's gross income (before adjustments) into the national super fund. How much does the employer pay, to the nearest kina?

Answers

a $8.13 \times 63.58 = 516.90$

Before adjustment, Bernard earns K516.90.

b Superannuation $= \frac{6}{100} \times 516.90 = 31$

Bernard pays K31 into the superannuation fund.

c Bernard's taxable income = gross pay less superannuation plus housing allowance

$= 516.90 - 31 + 55$

$= 540.90$

d Net pay $= 540.90 - 37.76 - 10.20 - 2.80$

$= 490.14$

Bernard's net pay is K490.14.

e $\frac{8.4}{100} \times 516.90 = 43.419$

His employer has to pay K43 into the super fund.

Other tax deductions

A deduction is allowed for dependants, that is, a person who is supported by the taxpayer and does not earn over a certain amount each year. Dependants include a husband or wife, children under 16 years, full-time students aged 16 to 24, an invalid relative or a parent of the taxpayer or their spouse who lives with them. There are limits on the number of dependants who can be claimed for, and this is set when the tax rates are determined.

In addition, deductions are allowable for things like the cost of tools used in earning an income, school fees paid to non-government schools, and donations to sporting bodies and charities.

The deduction for dependants is accounted for in the salary and wages tax tables used to calculate an employee's pay each fortnight. A section of a typical table is shown in Table 3. Different amounts are deducted for non-residents, residents who have not completed a declaration*, and residents who have completed a declaration and who may be able to claim for one or more dependants.

* 'Declaration' means a form lodged by an employee in accordance with the Income Tax Regulations.

Table 3 Income tax (salary or wages tax) rates

Column 1		Column 2	Column 3	Column 4			
Salary per fortnight		Non-resident taxpayers	Where no declaration is lodged	Where a declaration is lodged Number of dependants			
Exceeding	Not exceeding			0	1	2	3 or more
309	311	68.42	130.62	7.50	5.77	4.61	3.46
311	313	68.86	131.46	7.94	6.21	5.05	3.90
313	315	69.30	132.30	8.38	6.65	5.49	4.34
⋮	⋮	⋮	⋮	⋮	⋮	⋮	⋮
619	621	136.62	260.82	75.70	64.34	56.77	49.20
621	623	137.06	261.66	76.14	64.72	57.10	49.49
623	625	137.50	262.50	76.58	65.09	57.43	49.78

Example

Use the tax table to find:

- **i** the amount of tax that would be deducted from each of the following employee's pay packet
- **ii** their net pay.

a Rosewitha earns a salary of K311.80 per fortnight. She has signed a tax declaration and has no dependants.

b Simeon earns a salary of K310.90. He has not lodged a declaration.

c Materua earns a salary of K621.44. He has five dependants.

Answers

a

311	313	68.86	131.46	7.94	6.21	5.05	3.90

relevant figure

- **i** Rosewitha's tax is K7.94.
- **ii** Her net pay is 311.80 – 7.94 = K303.86.

b

309	311	68.42	130.62	7.50	5.77	4.61	3.46

relevant figure

- **i** Simeon's tax is K130.62.
- **ii** His net pay is 310.90 – 130.62 = K180.28

c

621	623	137.06	261.66	76.14	64.72	57.10	49.49

relevant figure

- **i** Materua's tax is K49.49.
- **ii** His net pay is K571.95.

Discussion

1. What is superannuation? Why is it important to the individual and the country as a whole?
2. Find out what compulsory superannuation payment rates currently apply to both employees and employers.
3. Obtain copies of the current salary or wages tax rates. Find out how these have changed in the last few years.
4. What do you think about the allowances the Government makes for income earners to claim for their dependants? Would you increase or decrease these allowances if you were in government?

Company income tax

It is not only individuals who pay tax. Companies also pay tax to the Government, which is calculated on the company's income from its activities. For instance, a mining company will pay tax on its income from selling the minerals it has mined, and a tourist company will pay tax on its income from running tours. The resident company tax rate is 30% for most companies.

The taxable income of a company is calculated by working out the assessable income after certain deductions have been made, such as for the depreciation of assets. The calculation of company income tax is beyond the scope of this unit. However, if you plan to set up a business you need to be aware that you will have tax obligations.

EXERCISE

Use Tables 1, 2 and 3 on pages 63–66 as appropriate to work out your answers.

1 Based on the 2009 tax tables, how much can I earn in wages per:
   a month **b** fortnight
   before I have to pay any tax? Answer to the nearest kina.

2 Calculate the tax payable by:
   i a resident **ii** a non-resident
   earning the following annual salaries after all allowable deductions have been made. Give your answer to the nearest kina.
   a K14 667 **b** K70 239 **c** K6549 **d** K234 100
   e K31 560 **f** K389 200 **g** K22 670 **h** K103 560

3 After deductions and allowances, Mapu's taxable income is K16 500 per year.
   a How much tax will he pay?
   b What will his net salary be for the year?
   c What is the equivalent monthly net pay, to the nearest toea?

4 In one fortnight Arthur works 76 hours plus 3 hours overtime. His job pays K5 an hour plus time-and-a-half for overtime.
   a Calculate Arthur's gross pay for the fortnight.
   b How much will Arthur have to pay into the national superannuation fund, based on 6% of his ordinary earnings?

5 Minnie is a cleaner in an office. She is paid at the rate of K2.50 per hour. If she works 72.7 hours in one fortnight, find:
   a her gross pay
   b the amount of her 6% compulsory payment to the national superannuation fund
   c her taxable income (allowing a deduction for the superannuation payment)
   d the equivalent annual taxable income
   e the tax payable on that income.

6 Minnie's employer is required to make an 8.4% contribution to Minnie's superannuation fund. How much does she contribute this fortnight?

7 Employees incurring more than K200 in expenses in earning their income in any financial year can claim a tax rebate calculated at 25% of the extent to which the expenses claimed exceed K200. If Timothy had expenses totalling K465, what tax rebate would he receive?

8 Violet is paid at the rate of K5.63 per hour. She worked 80.88 hours in one fortnight. A deduction is made for her compulsory 6% payment to the national superannuation fund. She is paid an allowance of K55 for housing.

  a What is her gross pay for the fortnight before any adjustment for superannuation and housing?
  b To the nearest kina, how much does she contribute to superannuation?
  c What is Violet's taxable income after the adjustments for her superannuation payment and housing allowance?
  d Violet's tax for the fortnight is K38.54. She also has to repay a loan of K39 as well as pay K10 for private use of the company car, K10 for use of the photocopier and K4.80 for private use of the office telephone. What will her net pay be?
  e Her employer has to pay 8.4% of Violet's gross income (before adjustments) into the national super fund. How much does the employer pay, to the nearest kina?

9 Richard earns a salary of K661 per fortnight. He has lodged a declaration and has two dependants.

  a If his superannuation payment is 6% of his gross salary, how much is his superannuation contribution?
  b What is his taxable income per fortnight after super has been paid?
  c How much tax does Richard pay?
  d What is his net pay after tax has been deducted?

10 Two employees in a factory have lodged a tax declaration. Both have two dependants. Compare the tax paid by employee A, with a taxable income of K310, and the tax paid by employee B, who has a taxable income that is twice as much.

  a How much tax is paid by each employee?
  b What percentage of their taxable incomes does each pay in tax?
  c Is the amount of tax in the same proportion as the employees' salaries? Discuss your result.

11 Lena earns a gross fortnightly salary of K663.46. She has lodged a tax declaration and has two dependants. What pay will she take home after deductions for the compulsory 6% superannuation contribution, tax and K11.78 to her employer for the use of the office phone?

12 Rose is paid K4 per hour. In one fortnight, she works 80 hours at this rate. She earns an additional K12.50 in overtime pay. She has no dependants.

  a What is her gross salary?
  b What is the value of her 6% super contribution?
  c What is her taxable income?
  d How much tax does Rose pay if she has lodged a declaration?
  e If she receives a car allowance of K125, what is her net pay after super and tax?

Remember

Superannuation payments are deducted from gross wages to give an employee's taxable income.

Revision and assessment (Borrowing and earning money)

Lesson 5

1 Ethan and Milla borrow K8500 and pay simple interest of 16.5% p.a. They repay the loan plus interest in monthly instalments over 4 years. To the nearest kina, the monthly instalment is:

A 206 B 177 C 349 D 294

2 A computer was purchased for K2500 and depreciated over five years at a flat rate, by which time its value was K500. By how much did the computer depreciate each year?

A K200 B K300 C K400 D K500

3 Rosetta earns a fortnightly salary of K310. Her compulsory superannuation payment is 6%. The expression representing her taxable income is:

A 0.94×310 B $310 - 0.94 \times 310$

C 310 D $310 + 0.94 \times 310$

4 Neema borrows K5000 for 4 years. Over that time she repays K7500. What rate of simple interest is she being charged on the loan?

5 Using the table on page 42 or otherwise, calculate the interest on these loans:

a K5000 for 11 years at 8.5% p.a. compound interest

b K15 000 for 1 year at 6% p.a. compounded monthly

c K6500 for 2 years at 24% p.a. compounded quarterly.

6 A business owner borrows K60 000 to set up his shop. He is charged 6.24% p.a. compounded monthly and makes a payment of K650 per month on the loan. The loan account shows the interest and balance owing at the end of the first month.

Month	Amount (beginning of month)	Interest	Repayment	Balance (end of month)
1	60 000	$60\,000 \times 0.0052$	650	59 662
2	59 662	A	650	B
3				

a Explain why 0.0052 is used to calculate the monthly interest.

b Find the missing amounts at A and B.

c Complete the entries for month 3.

7 Phillip buys building tools costing K2450 through hire purchase. After paying a deposit of K450, he makes 18 monthly payments of K125.

a What is the balance owing on the tools after the deposit?

b What is the total cost of the tools?

c How much interest does Phillip pay?

d What is the rate of interest being charged per annum?

8 Moses wants to buy furniture with a total cost of K4200. Two different stores offer a hire purchase agreement.

Store A: No deposit and payments of K330 for 18 months.

Store B: A deposit of K200 and payments of K130 per fortnight for 18 months.

a How much interest will Moses pay if he buys the goods from Store A?

b How many payments will Moses make if he buys the goods from Store B?

c Find the total cost of the furniture at:

i Store A   ii Store B.

d Which store offers the better option and what rate of interest is being charged?

9 Find the amount of interest on a credit card balance of K529.86 paid after 15 days if the interest rate is 17.5% p.a.

10 What amount will be owing on a loan of K7000 after 5 months if it is being charged 18% p.a. interest compounded monthly and no repayments are made?

11 A loan of K10 000 is to be repaid with payments of K800 every six months. Interest is charged at 12.4% pa. on the reducing balance. Show step-by-step calculations to find the amount still owing at the end of 2 years.

12 Michaeline earns K1.25 per hour in the garden. What will she be paid for 47 hours' work?

13 Olivia earns K242.30 per week while Glen earns K1075 per month. Who earns more per week, and by how much?

14 Simeon earns an annual salary of K32 560 before tax. If his deductions are K5230 for tax and K346 for other items, what will be his monthly pay?

15 The fortnight's payroll list for a store shows the following information.

Name	Rate	Hours	Overtime (time and a half)	Extras
Apelis	4.50	70	4	2.50
Joab	8.75	63	0	0
Justina	2.25	76	2	0
Patti	3.00	52	0	7.24
Bob	5.50	73	1	0
Victor	2.50	22	0	3.28

a Calculate the pay for each worker.

b What is the total payroll bill for the store?

16 Neema pays K15.42 tax on her fortnightly pay of K345.60, while Raymond pays K22.90 on his fortnightly pay of K380.09. Who is paying the higher percentage of tax, and by how much?

17 Fredrica is a resident paid an annual salary of K47 218. Calculate the tax she will pay.

18 Michael is a non-resident working in Papua New Guinea. His annual assessable salary is K78 250.

a Calculate the tax he will pay.

b Calculate his monthly pay after tax.

OPTION A: THREE MONTHLY BUDGETS

This option is a research project conducted as part of a small group of three to six students. Your teacher will instruct you about setting up the group. Some roles that individual students could take are:

- coordinating the project
- conducting the research
- collecting and organising data
- presenting data as posters or charts
- preparing a written report
- planning a role-play
- making an oral presentation to the class.

Your teacher will also explain how the project is to be carried out, and how the group is to present what you have found from your research.

You will present monthly budgets for at least three different situations, for example, a personal or family budget, a budget for the school canteen or local organisation, and a budget for a small business.

Assessment

Your project will be assessed on the extent to which you can:

- demonstrate appropriate investigation skills
- choose and apply relevant mathematical techniques
- make an effective communication of the project results.

The task is looking for evidence that you can use percentages and other mathematics in a real-life situation dealing with money. It aims to help you understand the importance of budgeting.

1 Planning a monthly personal or family budget

In order to plan a personal or family budget, you will need to determine both the expected income and the expected expenditure.

Income

Discuss how you would identify the 'expected income'. What assumptions are being made, if any?

Decide how to account for variations in income, for example as a result of working longer or shorter hours in a fortnight.

Expenditure

You will also need to identify the 'expected expenditure'. Add to the list below some of the expenses a typical personal or family budget might include. Then decide how to group items together into general categories that apply to most individuals or families.

- rent
- food and drink
- household items (soap, fuel etc)
- loan or HP payments
- school fees
- transport costs
- vehicle running costs
- medicines
- clothing and shoes
- repairs and maintenance
- electricity bills
- garbage
- doctor or clinic visit
- newspapers and magazines
- school books
- insurance
- lottery tickets
- telephone card
- sports club
- DVD rental
- birthday gifts
- garden tools
- savings
- cigarettes
- donations to fundraising
- insurance
- tools and equipment

Work out how you are going to allocate the whole year's expenses to the monthly budget.

You will find that these expenses occur over different time periods. For instance, families will spend money every week on items such as food and PMV tickets. The bill for electricity may only come every quarter (three months), rent might be paid each month, and school fees might be paid twice a year.

Practice family monthly budget

Mr and Mrs Baki and their three children live in Lai. They keep a record of their expenses for a period of time and draw up a budget as follows.

Wages for Mr Baki	K441.56 net per fortnight
Wages for Mrs Baki	K345.17 net per fortnight

Expenses

Rent	K962 per month
Food	K86 per week
Medicine	K14 per month
Household goods	K17 per month
Hire purchase payment	K16 per fortnight
School fees	K600 per year
Clothing	K22 per month
PMV	K7.45 per week
Electricity	K6.50 per fortnight
Waste collection	K5.50 per fortnight
Raffle tickets	K2 per week
Papers and magazines	K4.50 per fortnight
Gifts	K5.20 per month
Savings	K5 per week
Telephone	K27.50 per month
Film night	K5 per week

a What is the family income for the fortnight?

b Separate the expenses into 'essential' and 'non-essential' items.

c Is the Baki family income enough to cover all of the expenses?

2 Planning a monthly business budget

The basic process of planning a business budget involves listing the business's fixed and variable costs on a monthly basis. However, we need to distinguish between a business that is already up and running, and a business that has not yet been set up.

Start-up costs

A person starting a business has to pay a number of setting-up costs. Here is a sample of start up expenses:

- business registration fees
- business licensing and permits
- starting inventory (stock)
- rent deposits or down payments on property
- purchase or down payments on equipment
- utility set-up fees
- insurance
- printing of stationery.

Discussion

What other start-up expenses can you identify?

Operating costs

Once the business has been established, the business operator needs to plan a budget for the operating costs. Operating expenses are the costs of keeping your business running—the costs the business will have to pay each month. A business's list of operating expenses may include:

- salaries (the owner's and staff salaries)
- rent or mortage payments
- telecommunications
- utlities
- raw materials
- storage
- distribution
- promotion (advertising)
- loan payments
- office supplies
- maintenance.

Discussion

What other possible operating expenses can you identify for different businesses?

Thinking about business budgets

Discuss the following examples of small businesses that might be set up in your area.
List the start-up and operating costs you would expect to find in these businesses.

Example 1

Moimo is 21 years old. He wants to set up a tour business to take tourists on a trip around an island in his motorboat. He is paying off the boat.

Looking at the list of possible business expenses, identify which ones you think Moimo should plan for.

Are there any other expenses he might need to cover?

If Moimo has only a little money in his budget, which do you think are the most important costs for him to cover?

Example 2

Elizabeth is 25 years old. She plans to open a hairdressing shop and run it from a room in her parents' house. She would like to have an apprentice work with her so that she can have plenty of customers.

Looking at the list of possible business expenses, identify which ones you think Elizabeth should plan for.

Are there any other expenses she might need to cover?

Example 3

Puka is 18 years old. He wants to start a small business doing gardening in his area. He does not have any tools but can ride his bicycle or walk to gardens in the area.

Looking at the list of possible business expenses, identify which ones you think Puka should plan for.

Are there any other expenses he might need to cover?

Can you suggest how Puka could avoid the cost of buying tools?

Using the budget

While planning a budget can occur at any time, for many businesses planning a budget is an annual task, where the past year's budget is reviewed and budget projections are made for the next three or even five years.

Discussion

Explain how the past year's budget can be reviewed. What would you look at?

How do you manage unexpected expenses?

How can families or businesses be helped to stick to their budget?

Presenting your information

When you have selected the three examples of monthly budgets and collected all the relevant information, you are ready to prepare your report.

You will need to show your budgets neatly set out in columns, showing income and expenditure. It would be useful to separate the expenditure into two sections: 'essential' and 'non-essential'. This way it would be possible to discuss and demonstrate how the budget could be varied for different circumstances, such as meeting unexpected bills or increasing savings for a special project.

Consider the different ways you could present your findings in an informative way—for instance, identifying the percentage of income spent in each category and showing the different areas of expenditure as a pie graph.

Role-play

Prepare a role-play showing parents or older siblings explaining to younger members of the family how the family budget has been planned, and why there is a limit on what can be spent on things like toys and entertainment.

OPTION B: OBTAINING AND USING A MINI LOAN (K100–K500)

This option is a research project conducted as part of a small group of three to six students. Your teacher will instruct you about setting up the group. Some roles that individual students could take are:

- coordinating the project
- conducting the research
- collecting and organising data
- presenting data as posters or charts
- preparing a written report
- planning a role-play
- making an oral presentation to the class.

Your teacher will also explain how the project is to be carried out and how the group is to present what you have found from your research.

Assessment

Your project will be assessed on the extent to which you can:

- demonstrate appropriate investigation skills
- choose and apply relevant mathematical techniques
- make an effective communication of the project results.

The task is looking for evidence that you can use percentages and other mathematics in a real-life situation dealing with money. It aims to help you understand the real cost of credit, and to consider alternatives such as better financial planning and saving strategies.

Investigation

You will investigate the process for obtaining a small loan of between K100 and K500 from at least two different sources. Your project should include:

- advertising pamphlets, such as flyers, from banks or stores
- calculations showing the cost of credit or a loan compared to the costs of savings
- documentation required
- a written account of the steps involved in obtaining a mini-loan
- a financial plan showing the intended use of the funds from the loan
- completed loan application forms
- oral presentation, which can include a role-play.

Types of loan

Depending on what you plan to use the money for, there are a number of different options. For instance, if you want to use the money to buy goods, you could do this by entering a hire purchase agreement or using a credit card. If you want to obtain a personal loan, you can do so from different financial institutions.

What to consider

Some of the things you need to investigate—and to consider in your calculations—are:

- the term of the loan or agreement
- the rate of interest
- whether the interest is simple, compound or reducing balance
- whether the interest rate is fixed for the term of the loan
- how frequently the interest is calculated

- how frequently you are required to make repayments
- whether the repayments are a set amount or open to negotiation.

Most personal loans require a repayment each month with some allowing you to choose to make weekly or fortnightly repayments. Usually the repayment amount will be calculated by your loan provider, and it will depend on the interest rate, fees and term of your loan.

In addition, there may be other costs or considerations to investigate, such as:

- Do you need to provide some sort of security?
- Is there a cost to establish the loan?
- Do you have to pay any fees during the loan?
- Are you able to pay out the loan earlier than planned? Is there any penalty for doing so in the form of a termination fee or exit fee?

Security for loans

Personal loans can be split into two groups: secured and unsecured. Secured means that the loan provider takes ultimate control over an asset until the loan is paid out. Usually the item subject to control by the loan provider is the asset purchased with the loan. This is quite often the case with car or equipment loans. The benefit of this type of loan is that the interest rate can be significantly cheaper because, in the event of the borrower being unable to make repayments, the lender can recover their costs from the secured asset.

Loan application and documentation

When you apply for a loan, you can expect that the lender will ask you questions about yourself, where you live, your employment history and so on. You will need to provide details of your income and expenses as well as any assets and liabilities.

These facts help the lender judge whether you are able to repay the loan and whether you are a good 'credit risk'.

You can expect you will have to provide proof of your details by supplying various documents to prove what you have said. Documents typically include:

- personal identification items (a car licence)
- your employment pay slips or records
- records of bank accounts and other loan accounts.

The lender may also contact your employer to confirm some details.

For your project:

1 Find out the procedure for applying for a loan from the sources you have selected.
2 Obtain copies of any application forms to include with your project report and show how these will be completed.
3 Prepare a financial plan to show the lender how you intend to use the loan and how you will be able to meet the loan repayments.

Calculating the cost of credit

In earlier lessons in this unit we looked at how to calculate the interest on loans. We calculated simple interest over different time periods and for different interest rates, using the formula:

$$I = \frac{Prt}{100}$$

where I = interest, P = amount borrowed (principal), $\frac{r}{100}$ = % interest rate and t = time.

We also calculated compound interest over different time periods and for different interest rates, using the formula:

$$A = PR^n$$

where A = amount owed after n periods, P = amount borrowed and $R = 1 + \frac{r}{100}$, where r is the interest rate (for that time period).

We also looked at 'reducing balance' calculations that take into account the fact we have made payments to reduce the loan, and calculated these step-by-step.

A formula can be used to work out the amount owing on a loan that is being paid off at regular intervals. This is:

$$A = PR^n - \frac{Q(R^n - 1)}{R - 1}$$

Here, A is the amount owed after n periods, P is amount borrowed, Q is the regular repayment amount, $R = 1 + \frac{r}{100}$ where r is the interest rate (for that time period), and n is the number of time periods.

Example

If I borrow K5000, how much will I owe after 4 years if I am being charged 20% p.a. compound interest quarterly, and making regular quarterly repayments of K200? Use the list showing compound interest on K100 on page 42 and give your answer to the nearest kina.

Answer

$A = PR^n - \frac{Q(R^n - 1)}{R - 1}$

where $P = 5000$, $Q = 200$, $r = 5\%$ $(20\% \div 4)$, $R = 1.05$ and $n = 16$

$A = 5000 \times 1.05^{16} - \frac{200(1.05^{16} - 1)}{1.05 - 1}$

$= 5000 \times 2.18 - \frac{200 \times 1.18}{0.05}$

$= 10900 - 7200$

$= 3700$

So K3700 is still owing on the loan.

We can also use this formula to find out how long it will take us to pay off a loan (i.e. reduce the amount owing to zero).

In the example above, we would have:

$$A = PR^n - \frac{Q(R^n - 1)}{R - 1}$$

where $A = 0$, $P = 5000$, $Q = 200$, $r = 5\%$ $(20\% \div 4)$, $R = 1.05$ and n = unknown.

$$0 = 5000 \times 1.05^n - \frac{200(1.05^n - 1)}{1.05 - 1}$$

The solution of this equation is beyond the scope of this course. However, a 'guess and check' method can be used to give an approximate value.

For your project compare the cost of taking out your loan using all three formulas:

- simple interest
- compound interest and
- reducing balance interest.

Presenting your project

When you have selected at least two sources of your loan and collected all the relevant information, you are ready to prepare your report.

For each loan, you will need to show a financial analysis of the loan costs and the repayment schedule. You should also include completed application forms for the loan.

In order to illustrate the cost of obtaining a loan, include a comparison with saving for the purchase. Assume you are able to save the equivalent of the loan repayment and invest this in an interest-bearing account.

Discussion

How long would it take to save up the amount of the loan? What are the advantages of saving versus obtaining credit?

Role-play

Prepare a role-play showing a couple visiting the local bank branch to arrange for a loan of K500 to buy household goods.

Compound interest

Compound interest on K1			
Number of years	Interest rate %		
	5%	10%	20%
1	1.05	1.1	1.2
2	1.1	1.21	1.44
3	1.16	1.33	1.73
4	1.22	1.46	2.07
5	1.28	1.61	2.49
6	1.34	1.77	2.99
7	1.41	1.95	3.58
8	1.48	2.14	4.3
9	1.55	2.36	5.16
10	1.63	2.59	6.19
11	1.71	2.85	7.43
12	1.8	3.14	8.92
13	1.89	3.45	10.7
14	1.98	3.8	12.84
15	2.08	4.18	15.41
16	2.18	4.59	18.49
17	2.29	5.05	22.19
18	2.41	5.56	26.62
19	2.53	6.12	31.95
20	2.65	6.73	38.34
21	2.79	7.4	46.01
22	2.93	8.14	55.21
23	3.07	8.95	66.25
24	3.23	9.85	79.5
25	3.39	10.83	95.4
26	3.56	11.92	114.48
27	3.73	13.11	137.37
28	3.92	14.42	164.84
29	4.12	15.86	197.81
30	4.32	17.45	237.38

Unit

2 Functions and graphs

Unit summary

This unit has two parts:

- **i** the core, which all students do as part of their course
- **ii** the option, from which students do either Option A Algebra and graphs, or Option B Quadratic equations.

The core part of Unit 2 focuses on indices, algebra and graphs, and it will take five weeks. It will be assessed with tests that require the students to demonstrate understanding of mathematical concepts, and correctly choose, and apply, mathematical techniques.

Topics covered in the core part of the unit include:

- directed numbers and their application
- indices including negative, fractional and zero indices, and scientific notation
- basic algebra including expanding and factorising, solving simple equations, and rearranging and substituting in formulae
- graphs, including plotting points on the Cartesian axes, graphing linear functions and applying linear relations to modelling problems.

The chosen option will take the remaining five weeks. It will be assessed with an individual directed investigation that is presented in a written report.

Syllabus references

10.2.1 Interpret and develop linear and quadratic equations from information provided in a given context

10.2.2 Plot and sketch graphs of linear and quadratic equations

10.2.3 Communicate mathematical processes and results in writing

10.2.4 Undertake investigations individually in which mathematics can be applied to solve problems

Functions and graphs: Topics

REVIEW OF DIRECTED NUMBERS AND INDICES

Revision of directed numbers

- Positive numbers and negative numbers together with zero make up the set of **directed numbers**.
- Negative numbers occur often in practical situations. For example, the temperature in a refrigerator is given as $^{-}4^{\circ}$C.
- Directed numbers have both size and direction and can be shown on a number line.
- The number line below shows the directed numbers from $^{-}5$ to $^{+}5$. The arrows on the ends of this number line indicate that the number line extends to infinity in both directions.

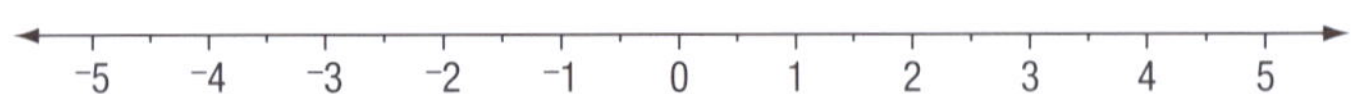

- Zero lies at the centre of the number line. Every number on the **right** of 0 has a 'partner' on the **left**. The pairs of numbers are called opposites.
- Positive numbers can be written with a positive sign (+) before the number, or with no sign at all (when we assume the number is positive).
- The position of numbers on a number line makes it easy to compare their size and arrange them in order. The number further to the right is a larger number, for example $^{+}3$ is bigger than $^{-}1$ because it is further to the right. In symbols, we can write $^{+}3 > {}^{-}1$. In a similar manner we can write $^{-}2 > {}^{-}7$.
- Ascending means going up in size. Descending means going down in size.
- **Addition**
  Adding a positive number is the same as 'adding', for example $^{+}(^{+}3) = {}^{+}3 = \mathbf{3}$.
  Adding a negative number is the same as 'subtracting', for example $^{+}(^{-}4) = \mathbf{{}^{-}4}$.
- **Subtraction**
  Subtracting a positive number is the same as 'subtracting', for example $^{-}(^{+}5) = \mathbf{{}^{-}5}$.
  Subtracting a negative number is the same as 'adding', for example $^{-}(^{-}2) = {}^{+}2 = \mathbf{2}$.
- **Multiplication**
  Multiplying a positive number by a positive number gives a positive number, for example $^{+}2 \times {}^{+}4 = {}^{+}8 = \mathbf{8}$.
  Multiplying a positive number by a negative number gives a negative number, for example $^{+}4 \times {}^{-}3 = \mathbf{{}^{-}12}$ or $^{-}5 \times {}^{+}3 = \mathbf{{}^{-}15}$.
  Multiplying a negative number by a negative number gives a positive number, for example $^{-}5 \times {}^{-}2 = {}^{+}10 = \mathbf{10}$.

In summary, when multiplying two numbers:
If the signs are different then the product is negative.
If the signs are the same then the product is positive.

- **Division**
  Dividing a positive number by a **positive** number gives a positive number, for example $\frac{^+10}{^+2} = {}^+5 = \mathbf{5}$.
  Dividing a positive number by a **negative** number gives a negative number, for example $\frac{^+15}{^-3} = \mathbf{^-5}$.
  Dividing a negative number by a **positive** number gives a negative number, for example $\frac{^-16}{^+4} = \mathbf{^-4}$.
  Dividing a negative number by a **negative** number gives a positive number, for example $\frac{^-21}{^-7} = {}^+3 = \mathbf{3}$.

In summary, when dividing one number by another:
If the signs are different then the answer is negative.
If the signs are the same then the answer is positive.

Example

1 Write the numbers 2.8, $^-3$, $^-2.4$, 1.7 and $^-3.2$ in ascending order.

2 Write each of the following as a positive or negative number:

a $^-0.75 \times 4$ **b** $^-(^-4)$ **c** $\frac{24}{^-8}$.

3 Replace the pairs of signs with a single sign and evaluate the following:
$^-3 + {}^-5 - {}^-4 - {}^+6 + 10$.

Answers

1 Ascending order means from smallest to largest: $^-3.2$, $^-3$, $^-2.4$, 1.7, 2.8

2 **a** The signs are different so the answer will be negative: $^-0.75 \times 4 = {}^-3$.
b Subtracting a negative creates a positive: $^-(^-4) = 4$.
c The signs are different so the answer will be negative: $\frac{24}{^-8} = {}^-3$.

3 $^-3 + {}^-5 - {}^-4 - {}^+6 + 10 = {}^-3 - 5 + 4 - 6 + 10 = 0$.

EXERCISE

1 Arrange these numbers in ascending order:

a 5, $^-1$, 0, $^-7.2$ **b** $^-99$, $^-98$, $^-101$, $^-50$ **c** 0.56, $^-0.26$, $^-0.05$, $^-0.25$, 0.5

2 Add a > or < sign to make each of the following true.

a $^-0.67$ …… 0.6 **b** 0 ….. $^-0.82$ **c** $^-0.983$ …… $^-0.97$

3 Write the next three numbers for these number patterns.

a $^-4$, $^-1$, 2, … **b** 98, $^-1$, $^-100$, … **c** 64, $^-32$, 16, $^-8$, …

4 Replace the pairs of signs with a single sign and then evaluate the following.

a $7 - {}^-5$ **b** $^-8 - {}^+3$ **c** $^-8 \times {}^-2$

d $\frac{^-15}{^+3}$ **e** $^-6 + {}^-5$ **f** $^-99 - {}^-100$

g $0.2 \times {}^-0.5$ **h** $\frac{0.33}{^-0.3}$ **i** $\frac{^-54}{^-9}$

Remember
Evaluate means 'find the value of'.

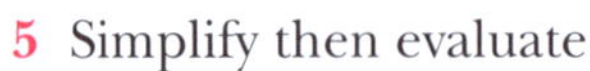

5 Simplify then evaluate.

a $3 - {}^{-}2 + 4$ **b** $10 + {}^{-}6 - {}^{+}7$ **c** $0.5 \times {}^{-}0.3 \times 1.2$

d $\frac{{}^{-}32}{0.4 \times {}^{-}5}$ **e** ${}^{-}8 - {}^{-}7 - {}^{-}5$ **f** ${}^{-}6 \times {}^{-}5 \times {}^{-}3$

g ${}^{-}23 + {}^{-}21 - {}^{-}19$ **h** $\frac{64}{{}^{-}8} \div {}^{-}2$ **i** $13 \times {}^{-}2 \times {}^{-}0.4$

6 Copy and complete this multiplication table.

×	${}^{+}4$	${}^{-}11$	80
${}^{-}6$			
${}^{+}0.4$			
${}^{-}1.2$			

7 Fill in the blank spaces in this multiplication table.

×	${}^{+}7$		${}^{-}0.5$
	${}^{-}63$		
			${}^{-}0.03$
${}^{-}1.2$		${}^{+}6$	

8 **a** Divide 24 by:

i ${}^{-}3$ **ii** ${}^{+}6$ **iii** ${}^{-}0.12$ **iv** ${}^{+}80$

b Divide ${}^{-}144$ by:

i ${}^{-}3$ **ii** ${}^{+}6$ **iii** ${}^{-}0.12$ **iv** ${}^{+}80$

9 Complete the following division table in which the number in the first row is divided by the number in the first column.

÷	${}^{+}72$	${}^{-}48$	${}^{+}0.36$
${}^{-}6$			
${}^{+}40$			
${}^{-}1.2$			

10 Complete the blank spaces in this division table in which the number in the first row is divided by the number in the first column.

÷	${}^{+}10$		${}^{-}0.24$
	${}^{-}5$		
			${}^{-}0.03$
${}^{-}1.2$		${}^{+}6$	

Lesson 2

Mixed operations and practical problems using directed numbers

When there is more than one operation to be done in a calculation, the order of operations is important.

Operations are carried out from **left** to **right** in this order:

- **brackets** first
- **multiplication and division** from left to right
- **addition and subtraction** from left to right.

Example 1

Use the correct order of operations to evaluate the following:

a $^-5 \times 4 + ^-12 \div 3$

b $^-2(7 - 12) \times (23 - 5) - 6$

c $\frac{3(^-2 + 6)}{4(^-3 - 6)} + 5(^-7 - ^-6)$

Answers

a $^-5 \times 4 + ^-12 \div 3 = ^-20 + ^-4$ multiplications and divisions are done before additions

$= ^-20 - 4$

$= ^-24$

b $^-2(7 - 12) \times (23 - 5) - 6 = ^-2 \times ^-5 \times 18 - 6$ brackets first

$= 10 \times 18 - 6$ multiplication before addition

$= 180 - 6$ multiplication before addition

$= 174$

c $\frac{3(^-2 + 6)}{4(^-3 - 6)} + 5(^-7 - ^-6) = \frac{3 \times 4}{4 \times ^-9} + 5 \times ^-1$ brackets first

$= \frac{12}{^-36} - 5$ multiplication before subtraction

$= -\frac{1}{3} - 5$ simplifying the fraction

$= ^-5\frac{1}{3}$ subtraction

Practical problems using directed numbers

When solving practical problems involving directed numbers, we need to identify which of the numbers in the problem should be written as positive numbers and which should be written as negative numbers.

Example 2

Joseph's parents deposit K130 in his account each month and Joseph withdraws K30 each week for expenses. After 6 weeks how much will be in the account?

Answer

Identifying deposits as positive numbers and withdrawals as negative numbers the amount in the account after 6 weeks will be:

$130 + ^-30 + ^-30 + ^-30 + ^-30 + 130 + ^-30 + ^-30 = 130 - 30 - 30 - 30 - 30 + 130 - 30 - 30$

$= 80$

There will be K80 in the account after 6 weeks.

EXERCISE

1 Evaluate each of the following showing your working.

a $15 \div ^-3 + 4$

b $7 - 15 \div 3$

c $11 + 20 \div ^-5$

d $3(2 - 5) + 9$

e $(3 \times ^-2) \times (4 - 7)$

f $24 \div ^-3 - 15 \div 5$

g $6(2 \times 4 - 3) \div 5$

h $^-5(7 - ^-3 - 15) + 2$

i $2(4 - 5)(6 \times 2 - 8)$

j $\frac{^-2 \times 13 + 2 \times 15}{7 - 4 + 1}$

k $\frac{3(6 - 4)}{5} + \frac{^-2(^-3 - ^-8)}{50}$

l $\frac{^-2(6 + ^-3)}{2} + \frac{24}{^-6} - ^-6$

2 Rufus is trying to lose weight. His initial weight is 86 kilograms and his progress for the first 5 weeks is 1.2 kg loss, 1.4 kg loss, 0.8 kg gain, 2.1 kg loss, 0.3 kg loss. Write an expression for his weight after 5 weeks and evaluate this expression to give his weight after five weeks.

3 Moreina deposits K5 each week in a savings account. After 3 weeks he withdraws K8 and after 4 weeks he withdraws half the balance.

a Write an expression for the amount in the account after 3 weeks and evaluate.

b Write an expression for the amount in the account after 4 weeks and evaluate.

4 Maxine has a bank account with an overdraft facility. This means she can temporarily withdraw more money than is in the account. For May the transactions in Maxine's account are as shown. Complete the balance column and calculate the balance in the account at the end of May.

Date	Deposit	Withdrawal	Balance (K)
01/05/09			504.82
08/05/09	112.00		
11/05/09		385.50	
19/05/09		400.00	
22/05/09	112.00		
25/05/09	200.00		
28/05/09		150.00	
31/05/09	100.00		

5 A building has 17 floors above ground level and 3 floors below ground level.

a How many floors, including ground level, are there in the building?

b If ground level is 'level 0' what is the level of:

i the 12th floor above ground level?

ii the lowest floor below ground level?

c Write an expression for the number of floors travelled on the following lift journey: The lift starts at level 15, goes down 7 floors, up 2 floors, down 12 floors.

d What floor will the lift finish on for the journey in part **c**?

6 A bus company is monitoring the number of passengers that travel on a particular route to town. For a particular trip to the town these figures are collected.

a How many people arrive at the town (stop 11)?

b At which stop are there first at least 30 people on the bus?

c Are there ever 40 or more people on the bus?

d How many fares does the bus driver collect?

Stop	Get on	Get off
1	10	0
2	9	0
3	14	0
4	9	2
5	7	8
6	7	5
7	8	2
8	4	1
9	0	0
10	0	0
11	0	0

Index form and the index laws

You will remember that we can write some numbers as a **power** of another number, for example $8 = 2 \times 2 \times 2 = 2^3$.

We can write $8 = 2^3$ where '2' is called the **base** and '3' is called the **index** (**power** or **exponent**).

Writing numbers as a product of their prime factors

To write a number as a product of its prime factors, we divide the number by its smallest prime factor as many times as we can then move on to the next prime number that is a factor of the number and so on.

Example 1

Write 168 as a product of its prime factors.

Answer

2 is the smallest factor of 168 that is a prime number.
Starting with 168 we divide by 2 as many times as we can, then divide by 3, 5, 7, 11 … (all prime numbers) if they are a factor.

$168 \div 2$
$84 \div 2$
$42 \div 2$
$21 \div 3$
$7 \div 7$ (*Note*: 5 is not a factor of 168)
1

So $168 = 2 \times 2 \times 2 \times 3 \times 7 = 2^3 \times 3 \times 7$.

Remember

Prime numbers have only themselves and 1 as factors.
In ascending order the prime numbers are 2, 3, 5, 7, 11, 13, 17, 19, 23, …
Note: 1 is not a prime number.

In algebra we often write products of pronumerals in index form, for example:

$p \times p \times p \times p \times p = p^5$: in this case p is the base and '5' is the index

$w \times w \times w \times x \times x \times y = w^3 \times x^2 \times y = w^2x^2y$

Week 1

Remember

The index laws apply to multiplication and division. There are no laws that apply to addition and subtraction.

The index laws

The index laws are used to simplify expressions in algebra to make the process of solving equations easier.

You will recall that the index laws apply to numbers (or pronumerals) written with the **same base**. The first three index laws are given below.

Law 1: $a^m \times a^n = a^{m+n}$ When **multiplying** numbers, **add** the indices.

Law 2: $\frac{a^m}{a^n} = a^{m-n}$ When **dividing** numbers, **subtract** the indices.

Law 3: $(a^m)^n = a^{m \times n}$ When **raising to a power**, **multiply** the indices.

Example 2

Simplify the following using the index laws.

a $2^3 \times 2^4$

b $a^5 \times b^3 \times a^5 \times b$

c $\frac{x^6}{x^2}$

d $(y^3)^4$

e $\frac{(m^2)^4 \times n^3}{m^5 \times n^2}$

f $2p^3 \times 8p^2 \times 9q$

g $(2x^2y)^3$

Remember

$b = b^1$.

Answers

a Using the first law: $2^3 \times 2^4 = 2^{3+4} = 2^7$

b Using the first law: $a^5 \times b^3 \times a^5 \times b = a^{5+5} \times b^{3+1} = a^{10}b^4$

c Using the second law: $= \frac{x^6}{x^2} = x^{6-2} = x^4$

d Using the third law: $(y^3)^4 = y^{3 \times 4} = y^{12}$

e Using the third and second laws: $\frac{(m^2)^4 \times n^3}{m^5 \times n^2} = \frac{m^8 \times n^3}{m^5n^2}$

$= m^{8-5} \times n^{3-2}$

$= m^3n^1$

$= m^3n$

f $2p^3 \times 8p^2 \times 9q = 2 \times 8 \times 9 \times p^{3+2} \times q$

$= 144p^5q$

Note: This answer can also be given as $2^43^2p^5q$.

g $(2x^2y)^3 = (2 \times x^2 \times y)^3$

$= 2^3 \times (x^2)^3 \times y^3$

$= 8 \times x^6 \times y^3$

$= 8x^6y^3$

EXERCISE

1 Write in index form:

a $b \times b \times b \times b$ **b** $2 \times 2 \times x \times x \times y$

c $ab \times ab \times ab$ **d** $m \times n \times n \times 5 \times n \times m \times n$

2 State the base and the index for each of the following.

a 2^8 **b** $(ab)^5$

3 Write these numbers as a product of their prime factors.

a 24 **b** 150 **c** 102 **d** 429

4 Copy and complete this table of powers of 2, 3 and 10.

n	2^n	3^n	10^n
1	$2^1 = 2$		
2		$3^2 = 9$	
3			
4			$10^4 = 10\,000$
5			
6			

5 Simplify the following using the index laws, then use the table from Question **4** to evaluate.

a $\frac{2^5}{2^2}$ **b** $\frac{2^3 \times 3^8 \times 2^4}{2^5 \times 3^6}$ **c** $\frac{(2 \times 3)^4 \times 10^{10}}{10^8 \times 2^2 \times 3}$ **d** $\frac{2 \times 3^{10}}{3^7} - \frac{3 \times 2^{10}}{2^7}$

6 Simplify the following.

a $\frac{20a^5}{4a^2}$ **b** $\frac{x^2}{15} \times 21x^2 \times 10x$ **c** $a^2b^2 \times 4ab^7 \times {}^-8a^2$

d $\frac{x^8 \times (x^3)^4}{(x^7)^2}$ **e** $\frac{(m^5)^2 n^6 p^2}{(m^2 n^3 p)^3}$ **f** $\frac{(3k)^4}{(2klm)^2} \times \frac{12k^3lm^2}{18(km)^2}$

g $\frac{2p^3 \times (p^5)^3}{4(p^4)^3 \times p^5}$ **h** $\frac{3^4 \times (3p^2q)^5 \times pq}{p^3q^2 \times 3^7(p^3q^2)^2}$ **i** $\frac{x^6yz^5 \times x^3(y^5z^2)^4 \times xz}{(xz)^2 \times (x^2y^5z)^2 \times y^4}$

7 Simplify the following giving your answers as powers of 2 and 3.

a 18×8 **b** $6^2 \times 12^6$ **c** $\frac{36 \times 8^2 \times 3}{4^2 \times 9}$

Working with the index laws

Lesson **4**

From the first three index laws we can deduce some other important laws for indices.

Law 4

We know that, for example, $\frac{a^3}{a^3} = \frac{a \times a \times a}{a \times a \times a} = 1$.

However if we use the second index law to simplify $\frac{a^3}{a^3}$, we get $a^{3-3} = a^0$.

Hence we can say $a^0 = 1$.

This is the fourth index law:

$a^0 = 1$ Any number raised to the power zero is equal to 1.

Law 5

To simplify $\frac{1}{a^n}$ we need to replace the 1 in the numerator by a^0 and use the second law:

$\frac{1}{a^n} = \frac{a^0}{a^n} = a^{0-n} = a^{-n}$.

Fifth index law:

$\frac{1}{a^n} = a^{-n}$: a^{-n} is the reciprocal of a^n.

Law 6

You will recall that $\sqrt{a} \times \sqrt{a} = a = a^1$.

Both the $\sqrt{a}$ terms in $\sqrt{a} \times \sqrt{a} = a^1$ are the same and if $\sqrt{a}$ is to be written in index form then we must find an index such that $a^? \times a^? = a^1$.

The sum of the indices must be 1 so $\sqrt{a} = a^{\frac{1}{2}}$.

Similarly, we know that $\sqrt[3]{a} \times \sqrt[3]{a} \times \sqrt[3]{a} = a^1$ so $\sqrt[3]{a} = a^{\frac{1}{3}}$.

Sixth index law:

$\sqrt[n]{a} = a^{\frac{1}{n}}$ The nth root of a number, a, written in index form is $a^{\frac{1}{n}}$.

Note: An extension to Law 6, $\sqrt[n]{a^m}$ written in index form is $a^{\frac{m}{n}}$.

Examples

1 Simplify:

a $m^{-1} \times m^0 \times m^{\frac{1}{2}}$

b $\dfrac{(\sqrt{y})^4 \times y^5}{y^{-2}}$

c $(2x)^3 \times (x^{\frac{2}{3}})^{-6} \times \dfrac{x^0}{x^2}$

2 Simplify using the index laws then evaluate:

a $2 \times (2^{-\frac{1}{2}})^6 \times 2^2$

b $8 \times 3^{-2} \times (\frac{3}{2})^4$

Answers

1 a Using the first law:

$m^{-1} \times m^0 \times m^{\frac{1}{2}} = m^{-1+0+\frac{1}{2}} = m^{-\frac{1}{2}}$

This could also be written as $\sqrt{\frac{1}{m}}$ or $\frac{1}{\sqrt{m}}$.

b $\dfrac{(\sqrt{y})^4 \times y^5}{y^{-2}} = \dfrac{(y^{\frac{1}{2}})^4 \times y^5}{y^{-2}}$

$= \dfrac{y^{\frac{1}{2} \times 4} \times y^5}{y^{-2}}$

$= \dfrac{y^2 \times y^5}{y^{-2}}$

$= y^{2+5-(-2)}$

$= y^9$

c $(2x)^3 \times (x^{\frac{2}{3}})^{-6} \times \dfrac{x^0}{x^2} = 2^3 \times x^3 \times (x^{\frac{2}{3}})^{\times -6} \times \dfrac{1}{x^2}$

$= 8 \times x^3 \times x^{-4} \times \dfrac{1}{x^2}$

$= 8 \times x^{3+(-4)-2}$

$= 8 \times x^{-3}$

(This can also be written as $\frac{8}{x^3}$.)

2 a $2 \times (2^{-\frac{1}{2}})^6 \times 2^2 = 2^1 \times 2^{-\frac{1}{2} \times 6} \times 2^2$

$= 2^1 \times 2^{-3} \times 2^2$

$= 2^0$

$= 1$

b $8 \times 3^{-2} \times (\frac{3}{2})^4 = 2^3 \times 3^{-2} \times (3 \times 2^{-1})^4$

$= 2^3 \times 3^{-2} \times 3^4 \times [(2^{-1})]^4$

$= 2^3 \times 3^{-2+4} \times 2^{-1 \times 4}$

$= 2^{3-4} \times 3^2$

$= 2^{-1} \times 9$

$= \frac{1}{2} \times 9$

$= 4.5$

EXERCISE

1 Copy and complete.

a $\dfrac{1}{x^7} = x^{\square}$

b $y^{-4} = \dfrac{1}{\square}$

c $\dfrac{6}{\square} = 6z^{-3}$

d $\dfrac{5^2}{5^3} = 5^{\square}$

e $\dfrac{w^3}{w^5} = \square$

f $\dfrac{1}{x^3 y^4} = x^{\square}y^{\square}$

2 Simplify, then evaluate leaving your answers as fractions.

a 7^{-1}

b 10^{-2}

c $(5^{-2})^0$

d $(2^{-2})^{-3}$

e $4^{-3} \times 4$

f $\dfrac{1}{3 \times 2^{-2}}$

g 4×8^{-2}

h $\dfrac{3^{-4}}{6^{-2}}$

i $(5^3)^{-2} \times 5^4$

j $(\frac{2}{3})^{-1}$

k $7^0 + 5^{-1}$

l $\dfrac{3^5 \times 3^{-3}}{(3^3)^2}$

3 Express each of the following as a power of 2.

a 2 b 32 c $\frac{1}{32}$ d $\frac{1}{256}$

e 0.25 f $\sqrt{2}$ g $2\sqrt{2}$

4 Express each of the following as a power of 10.

a 1000 b 100 c 10 d 1

e $\frac{1}{10\,000}$ f 0.001 g $\frac{1}{100}$ h 0.1

5 Copy and complete.

a $25^{\frac{3}{2}} = (5^{\square})^{\frac{3}{2}} = 5^{\square}$ b $8^{\frac{2}{3}} = (\square^{3})^{\frac{2}{3}} =$

c $64^{\frac{3}{2}} = (2^{\square})^{\frac{3}{2}} = 2^{\square}$ d $27^{\frac{4}{3}} = (\square^{3})^{\frac{4}{3}} =$

6 Evaluate each of the following.

a $4^{\frac{1}{2}}$ b $27^{\frac{1}{3}}$ c $9^{-\frac{1}{2}}$ d $\sqrt[5]{32}$ e $(\frac{1}{8})^{\frac{1}{3}}$ f $\sqrt[4]{81}$

Scientific notation

Lesson 5

Scientists often have to work with extremely small or extremely large numbers. **Scientific notation** is a convenient way to write and do calculations with these numbers. Scientific notation is also called **standard form**. For example:

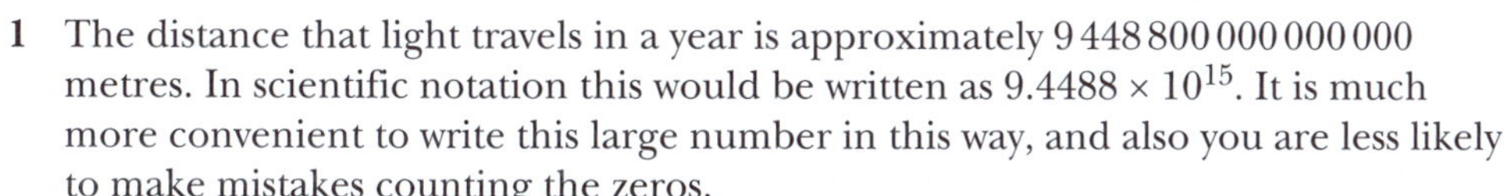

1 The distance that light travels in a year is approximately 9 448 800 000 000 000 metres. In scientific notation this would be written as 9.4488×10^{15}. It is much more convenient to write this large number in this way, and also you are less likely to make mistakes counting the zeros.

To write 9 448 800 000 000 000 in scientific notation we count the number of places that we need to move the decimal point to get to a number from 1 up to 10.

← 15 places →

$9\,448\,800\,000\,000\,000 = 9.4488 \times 10^{15}$

↑ Final position of the decimal point.

2 The radius of a single atom is approximately 0.000 000 000 053 04 metres. In scientific notation this is written as 5.304×10^{-11}.

← 11 places →

$0.000\,000\,000\,053\,04 = 5.304 \times 10^{-11}$

↑ Final position of the decimal point.

A number written in scientific notation (standard form) is made up of *(a number from 1 up to 10)* × *(a power of 10)*.
The power of 10 will be positive for very large numbers and negative for very small numbers.

Working with numbers written in scientific notion

We can use the index laws to simplify the multiplication and division of numbers written in scientific notation.

Examples

1 Write the following list of numbers in ascending order:

$$0.00513,\ 5.29 \times 10^{-2},\ 0.005,\ \frac{1}{500},\ 28.35 \times 10^{-3}$$

2 Find the product of the numbers 3.23×10^{11} and 4.75×10^{5}, writing your answer in standard form.

3 Divide the number 6.24×10^{-5} by the number 1.2×10^{-8}, giving your answer in standard form and in decimal form.

Answers

1 If all the numbers are written in standard form then it will be easier to order them, smallest to largest.

$0.00513 = 5.13 \times 10^{-3}$ in standard form.

5.29×10^{-2} is already in standard form.

$0.005 = 5 \times 10^{-3}$ in standard form.

$\frac{1}{500} = 0.002 = 2 \times 10^{-2}$ in standard form.

$28.35 \times 10^{-3} = 2.835 \times 10^{-1} \times 10^{-3} = 2.835 \times 10^{-4}$ in standard form.

When comparing numbers in standard form, the smallest number will be the one that has the smallest power of 10 so 2.835×10^{-4} will be the smallest number.

There are three numbers that have the same power of ten (10^{-3}) so we will order these by the other part of their standard form number. In ascending order these will be: 2×10^{-3}, 5×10^{-3}, 5.13×10^{-3}.

The largest of the numbers will be 5.29×10^{-2}.

In ascending order the numbers are 2.835×10^{-4}, 2×10^{-3}, 5×10^{-3}, 5.13×10^{-3}, 5.29×10^{-2}.

2 $3.23 \times 10^{11} \times 4.75 \times 10^{5} = (3.23 \times 4.75) \times 10^{11} \times 10^{5}$

$= 15.3425 \times 10^{11+5}$ multiplying the numbers and using the first index law

$= 15.3425 \times 10^{16}$ The number 15.3425 is not between 1 and 10 so we need to write this in standard form.

$= 1.53425 \times 10^{1} \times 10^{16}$

$= 1.53425 \times 10^{17}$

3 $\frac{6.24 \times 10^{-5}}{1.2 \times 10^{-8}} = \frac{6.24}{1.2} \times \frac{10^{-5}}{10^{-8}}$

$= 5.2 \times 10^{-5-(-8)}$

$= 5.2 \times 10^{3}$ in standard form

$= 5200$ in decimal form

EXERCISE

1 Which of the following numbers are not written in standard form?
32.65, 5.782, 7.36×10, 56.23×10^2, 10×10^{-2}, 0.2348×10^{-1}

2 Write the following numbers as decimals.
- a 2.785×10^{-1}
- b 8.001×10^3
- c 1.754×10^{12}
- d 9.883×10^{-7}

3 Write the following numbers in standard form.
- a $0.009\,54$
- b $156\,700\,000$
- c $0.000\,003\,401$
- d 0.55
- e 6700

4 Write the following numbers in ascending order:
7.856×10^4, 5600, 2.468×10^{-1}, 0.00653, 4.8609

5 Write the following numbers in ascending order:
$\frac{1}{2000}$, $0.009\,64$, 1.254×10^{-9}, $1\,278\,000$, 6.780×10^6, $0.000\,499\,8$, $0.000\,000\,011\,11$

6 Evaluate each of the following giving your answer in standard form.
- a $0.000\,000\,12 \times 0.000\,06$
- b $(5.6 \times 10^{-5}) \div (8 \times 10^{-8})$
- c $860\,000\,000 \times 240\,000$
- d $720\,000\,000 \times 0.000\,000\,001\,2$
- e $0.0048 \div 16\,000\,000$

7 Evaluate each of the following giving your answer in standard form.
- a $\frac{0.000\,99 \times 0.03}{0.000\,001\,1}$
- b $\frac{1.6 \times 10^{-2} \times 5 \times 10^4}{4 \times 10^{-3}}$
- c $\frac{7\,200\,000 \times 560\,000}{42\,000\,000 \times 9600}$
- d $\frac{0.000\,006\,5}{1.3 \times 10^{-5}} \div 50\,000$

8 The diameter of a spherical molecule is $0.000\,000\,008\,33$ m.
- a Find the radius, in millimetres, of the molecule, giving your answer in standard form.
- b Find the volume of the molecule, in cubic millimetres, giving your answer in standard form. Use the formula $V = \frac{4}{3}\pi r^3$ and $\pi = \frac{22}{7}$.

9 The diameter of the Earth is approximately 12 756 km. Find the surface area of the Earth, in square kilometres, using the formula $A = 4\pi r^2$ and $\pi = \frac{22}{7}$. Give your answer in standard form.

ALGEBRA

Review of simplifying and expanding algebraic expressions

Simplifying algebraic expressions

- A **term** consists of the product of numbers, pronumerals or both. For example, a **term** could be $4x^2y$ or $\frac{3abc}{2}$ or m^2n or 6. The first three terms are **algebraic terms** and the term 6 is called a **constant term** because it does not have any pronumeral parts.
  In the term $4x^2y$ the number 4 is called the **co-efficient** of the term; it is the number that multiplies the pronumerals in that term.
- An algebraic **expression** contains one or more terms that are separated by addition or subtraction signs. For example, an **expression** could be $5x^2y^2 - 2xy + 3$. This is an expression with three terms; two algebraic terms and one constant term.
- **Like terms** contain the same pronumeral parts. For example: x^2y and $\frac{3x^2y}{2}$ are like terms as they both have the product of x^2 and y in their term. The term $3xy$ would not be a like term of x^2y and $\frac{3x^2y}{2}$.
- An expression is **simplified** by collecting (adding or subtracting) like terms.
  For example, the expression $ab + 3a^2b - 2b + 4ab - a^2b$ can be simplified to $5ab + 2a^2b - 2b$.
  like terms like terms
- An expression is **expanded** when the brackets are removed from the expression. For example, $2x + 8$ is the expanded version of $2(x + 4)$. This was found by multiplying each term in the bracket by the number outside the bracket: $2(x + 4) = 2 \times x + 2 \times 4 = 2x + 8$.

Examples

1 **a** For the expression $15a^2bc - 12ab + 3ab^2c + ab - 45$, write down:
- **i** the number of terms
- **ii** the constant term
- **iii** the like terms
- **iv** the co-efficient of the second term.

b Simplify the expression.

2 For the expression $4(a - 4) + 3(2a + 1)$:
- **a** expand the terms of the expression
- **b** simplify the expression.

3 Expand and simplify the expression $5a(a - 5) - a(2a - 3)$.

Answers

1 **a** **i** Terms are separated by + or – signs; there are five terms in this expression.

ii The constant term has no pronumeral parts so the constant terms is $^{-}45$.

iii The like terms have the same pronumeral parts; $^{-}12ab$ and ab are like terms.

iv The co-efficient of a term is the number in front of the pronumeral parts so the co-efficient of $^{-}12ab$ is $^{-}12$.

b To simplify an expression we collect like terms; $^{-}12ab$ and ab are the only like terms $15a^2bc - 12ab + 3ab^2c + ab - 45 = 15a^2bc + (^{-}12ab + ab) + 3ab^2c - 45 = 15a^2bc - 11ab + 3ab^2c - 45$

2 **a** $4(a - 4) + 3(2a + 1) = 4 \times a + 4 \times {}^{-}4 + 3 \times 2a + 3 \times 1 = 4a - 16 + 6a + 3$

b $4a - 16 + 6a + 3 = 10a - 13$

3 $5a(a - 5) - a(2a - 3) = 5a \times a + 5a \times {}^{-}5 + {}^{-}a \times 2a + {}^{-}a \times -3$

$= 5a^2 - 25a - 2a^2 + 3a$

$= (5a^2 - 2a^2) + (^{-}25a + 3a)$ collecting like terms

$= 3a^2 - 22a$

EXERCISE

1 For each of the following expressions, state:

i the number of terms

ii the constant term

iii the coefficient of the x term.

a $3x - 5$ **b** $5x^2 - 24x + 6$ **c** $3xy + 15 - 2x^2y - 15x$

2 For each of the following expressions, state:

i the number of terms

ii the constant term(s)

iii the like terms.

a $18 - 2bc + 3b^4 + 5bc$ **b** $d^2 + 2d + e^2$

c $\frac{5f^2}{7} + 5g^2 - \frac{5f}{7} + 5f^2g^2 - \frac{5}{7}$ **d** $2hj^2 + 3 - 4h^2j + j^2h - 3j^2 + 4h^2$

3 Simplify each of the following expressions by collecting like terms.

a $13d - d$ **b** $13d - 13$

c $cd + 15 - 6dc - 7$ **d** $7b^2cd + 6bc^2d - 9 - b^2cd + 6$

e $d + 3 - 4d + 9$ **f** $5d + 4d^2 - cd + 6d^2 - 7d + cd$

g $^{-}4d + 3 - 3 + 5d$ **h** $15 - 2bcd + 4b^2cd - 5bcd + 3$

i $5d^2 + d + 5 - 2d^2 - 5$ **j** $15bcd - 5bc^2d + bcd$

k $d^2 + d - 4d^2 - 2d$ **l** $11c + 14cd - 5c - 2 + 6dc + 5 - 6c$

m $^{-}17cd + 2cd - 3cd$ **n** $7b^2c^3d^2 + 5b^2c^2d^2 - b^2c^3d^2 - 3b^2c^2d^2$

4 Expand:

a $3(a + 5)$ **b** $7(m - 2)$ **c** $7(x + y)$ **d** $12(4 - c)$

e $5(2a + 1)$ **f** $10(5 + 4m)$ **g** $3(15 - 2y)$ **h** $9(3a - 2b)$

5 Expand:

a $a(a+2)$ b $m(m-9)$ c $x^2(x+6)$ d $c(4-c)$
e $3a(2a+5)$ f $m(5m+4n)$ g $7x(x-2y)$ h $9a(3-2a)$

6 Expand:

a $^-3(a+6)$ b $^-(a+b)$ c $^-c(4-c)$
d $^-2(3a-4b)$ e $^-2y(15-2y)$

7 Expand and simplify.

a $3(a+5)+2(a+4)$ b $7(m-2)+3(m+2)$
c $7(x+y)-4(x+y)$ d $12(4-c)-3(2c+7)$
e $5(2a+1)-4(3a-1)$ f $m(5-4m)-(6+5m)$
g $3y(15-2y)+4(3y+1)$ h $^-6(6+7c)-5(1-4c)$

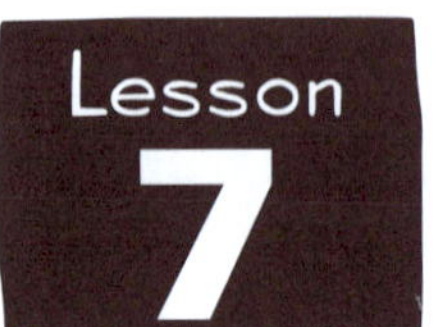

Lesson 7

Factorisation of algebraic expressions

Factorisation is the reverse process to expanding. For example, $2(x+4)$ is the factorised version of $2x+8$.

To factorise an expression the common factors of all terms need to be considered. For example, the terms of the expression $4ab+2a^2$ have many factors in common, shown below.

Factors of $4ab$ are 1, 4, 2, a, b, $4a$, $4b$, $2a$, $2b$, ab, $4ab$, $2ab$.

Factors of $2a^2$ are 1, 2, a, a^2, $2a$, $2a^2$.

The **highest common factor** is **2*a***.

The **factorised** version of $4ab+2a^2$ is $2a(2b+a)$.

To **factorise** an expression:
1 The highest common factor of all the terms being considered must be identified and placed outside the bracket.
2 The remaining factors of each term are collected to form an expression inside the bracket.

Examples

1 Factorise fully the following expressions.

a $5x-10y$ b $8m^2+12m$ c $3p^3q-2pq^2+6pq$

2 Factorise fully the following expressions using a negative highest common factor.

a $^-5x-10$ b $^-8x^2+4xy$ c $^-6pq^3+24p^2q^2-15p^3q^2$

Answers

1 a The HCF of the terms $5x$ and ^-10y is 5.
The first term is $5\times x$ and the second term is $5\times{}^-2y$ so $5x-10y=5(x-2y)$.

b The HCF of $8m^2$ and $12m$ is $4m$.
$8m^2=4m\times 2m$ and $12m=4m\times 3$ so $8m^2+12m=4m(2m+3)$.

c The HCF of all terms is pq.
$3p^3q=pq\times 3p$, $^-2pq^2=pq\times{}^-2q$, $6pq=pq\times 6$
So $3p^3q-2pq^2+6pq=pq(3p^2-2q+6)$

2 **a** The HCF of ^-5x and $^-10$ is $^-5$.
$^-5x = {}^-5 \times x$ and $^-10 = {}^-5 \times {}^+2$ so $^-5x - 10 = {}^-5(x + 2)$.
b The HCF of $^-8x^2$ and $4xy$ is ^-4x.
$^-8x^2 = {}^-4x \times 2x$ and $4xy = {}^-4x \times {}^-y$ so $^-8x^2 + 4xy = {}^-4x(2x - y)$.
c The HCF of all terms is $^-3pq^2$.
$^-6pq^3 = {}^-3pq^2 \times 2q$, $24p^2q^2 = {}^-3pq^2 \times {}^-8p$ and $^-15p^3q^2 = {}^-3pq^2 \times 5p^2$
So $^-6pq^3 + 24p^2q^2 - 15p^3q^2 = {}^-3pq^2(2q - 8p + 5p^2)$.

EXERCISE

1 Copy and complete with the missing factor.

a $3 \times \ldots = 9a$ **b** $2m \times \ldots = 4m^2$ **c** $4c \times \ldots = 4bc^2$
d $^-2p \times \ldots = {}^-8pq$ **e** $5m \times \ldots = {}^-20m^2n$ **f** $^-p \times \ldots = p^2$
g $ab \times \ldots = 7a^3b^2$ **h** $(x - 2) \times \ldots = 5(x - 2)$ **i** $2ab \times \ldots = 4ab^2$

2 Find the highest common factors.

a 12 and 18 **b** $2a$ and $4a$ **c** $35b$ and $14ab$
d $8a$ and $4a^2$ **e** x^2 and $5x$ **f** ^-6y and $^-3$
g $2a^2b$ and $5ab^2$ **h** $15m^3n$ and $12mn$ **i** ^-10pq and $4p^2$

3 Copy and complete the following factorisations.

a $6a + 3 = 3(\ldots)$ **b** $a^2 + 5a = a(\ldots)$
c $4m^2 - 2m = 2m(\ldots)$ **d** $3pq - 6q = 3q(\ldots)$
e $^-2x - 2y = \ldots(x + y)$ **f** $ab - 2a^2b = \ldots(1 - 2a)$
g $18uv - 15u^2v^2 = \ldots(6 - 5uv)$ **h** $3xy - 2y = y(\ldots)$

4 Factorise fully.

a $2a + 6$ **b** $9x - 6$ **c** $12 - 3m$
d $24d + 18$ **e** $45 - 63m$ **f** $20x - 25$
g $2a^2 + a$ **h** $m^2 - 3m$ **i** $xy - yz$
j $xy - 3x$ **k** $3pq - pr$ **l** $2xy - x + zx$

5 Factorise fully.

a $2ab + 4a^2$ **b** $9x^2 - 3x$ **c** $12mn^2 - 3n^2$
d $8p - 2pq$ **e** $7ab + 28a$ **f** $6x^3 - 2x$
g $m - m^2$ **h** $ab^2 - ab$ **i** $2a - ab + ac$
j $a^3 + a^2 + a$ **k** $15xy - 9x^2y^2$ **l** $p^3 - p^2$

6 Factorise fully using a negative HCF.

a $^-3a - 3b$ **b** $^-2m^2 - 2m$ **c** $^-5a + 10$
d $^-10m - 4n$ **e** $^-6b^2 + 2b$ **f** $^-p - pq$
g $^-xy + xz - 3x$ **h** $^-4x^2 + 16$ **i** $^-15c^2 + 12c$

7 Factorise completely.

a $2a^2 + ab$ **b** $^-5a - 15ab$ **c** $2x - 8x^2 - 4x^3$
d $m^2n - 2mn + 5mn^2$ **e** $^-3a - 6ab - 9a^2$ **f** $63m^2 - 45m^3$
g $pq - p^2 + p$ **h** $8a^3 + 4a^2 + 2a$ **i** $x^2 - x$
j $^-2x^3 - 5x^2$ **k** $20a^2b^3 + 12ab^2 - 8b^2$ **l** $^-6y^2 - 3xy - 3$

Solving equations

An **equation** is a mathematical sentence with an equals sign that indicates that two expressions have the same value. For example, $2x - 3 = 7$ is an equation with one unknown quantity x.

To **solve** an equation we need to isolate the unknown quantity on one side of the equals sign. We can do this by applying the same operation to both sides of the equation so that the equation remains 'balanced'. The operations that we apply are the inverse operations to those applied in the equation.

To solve the equation $2x - 3 = 7$, the first step would be to add 3 to both sides of the equation. Adding 3 is the inverse operation to the $- 3$ in the equation: $2x - 3 + 3 = 7 + 3$.

Collect like terms and the equation is changed to $2x = 10$.

To isolate x we need to divide both sides of the equation by 2.

Dividing by 2 is the inverse operation to the multiplying by 2 in the equation: $\frac{2x}{2} = \frac{10}{2}$.

Simplifying this gives $x = 5$ and we have solved the equation.

The answer can be checked by substituting $x = 5$ into the left side of the equation $2x - 3 = 7$.

$2x - 3 = 2 \times 5 - 3 = 10 - 3 = 7$ which is the same value as the right side of the equation; so our answer is correct.

Remember

The inverse operations are carried out in the reverse order to the operations used to form the equation.

Remember

Operation	Inverse operation
+	−
−	+
×	÷
÷	×

Examples

1 Solve for p: $6p + 3 = 4$.

2 Solve the equation $5(3 - y) = 20$.

3 Solve $\frac{3(b+7)}{5} = 9$.

4 Solve for x: $3(x - 2) - 2(x - 5) + 4x = 26$.

Answers

1 $6p + 3 = 4$ subtract 3 from both sides of the equation

$6p + 3 - 3 = 4 - 3$

$6p = 1$ divide both sides of the equation by 6

$\frac{6p}{6} = \frac{1}{6}$

$p = \frac{1}{6}$

2 $5(3 - y) = 20$ expand the bracket

$5 \times 3 - 5 \times y = 20$

$15 - 5y = 20$ subtract 15 from both sides of the equation

$15 - 5y - 15 = 20 - 15$

$^{-}5y = 5$ divide both sides by $^{-}5$

$\frac{^{-}5y}{^{-}5} = \frac{5}{^{-}5}$

$y = {^{-}1}$

3 $\frac{3(b+7)}{5} = 9$ multiply both sides of the equation by 5

$\frac{3(b+7)}{5} \times 5 = 9 \times 5$

$3(b+7) = 45$ expand the bracket

$3 \times b + 3 \times 7 = 45$

$3b + 21 = 45$ subtract 21 from both sides of the equation

$3b + 21 - 21 = 45 - 21$

$3b = 24$ divide both sides by 3

$\frac{3b}{3} = \frac{24}{3}$

$b = 8$

4 $3(x-2) - 2(x-5) + 4x = 26$ expand the brackets

$3x - 6 - 2x + 10 + 4x = 26$

$5x + 4 = 26$ subtract 4 from both sides subtract 4 from both sides

$5x + 4 - 4 = 26 - 4$

$5x = 22$ divide both sides by 5

$x = 4\frac{2}{5}$

EXERCISE

1 Solve for the pronumerals, showing a check for each answer.

a $5 + d = 13$
b $^-4m = 9$
c $^-x - 5 = 1$
d $\frac{f}{11} = 4$
e $3m - 2 = 10$
f $4 - 7m = 3$
g $5n - 7n + 11 = 17$
h $2y + 30 - 3y + 8 = 15$
i $4(3m - 2) = 32$
j $^-3(15 - 2n) = 15$
k $12 - 3(a - 1) = 8$
l $7 - (2 - d) = 4$
m $\frac{x}{4} - 8 = 4$
n $\frac{x-8}{4} = 4$
o $\frac{2(d+1)}{3} = 6$
p $\frac{6(2x-1)}{4} = 3$
q $3(t+5) + 2(2t-3) = 7$
r $7(2d-4) - (3d+3) = 4$
s $3(f-2) + 4f + 5(2f+7) = {}^-5$
t $2x - 3(x-10) + 2(x+4) = 61$

Solving equations with the unknown on both sides

To solve an equation with the unknown on both sides the first step is to 'remove' the pronumeral from one of the sides, using an inverse operation.

For example, if we are going to solve the equation $6x - 5 = 4 - 3x$ then the first step will be to add $3x$ to both sides of the equation. This will remove the term $3x$ from the left side of the equation: $6x - 5 + 3x = 4 - 3x + 3x$.

Adding like terms gives $9x - 5 = 4$.

The equation is now simple to solve by adding 5 to both sides so $9x = 9$ and dividing both sides by 9. Now $x = 1$.

Examples

1 Solve the equation $3(m+1) = 6 + 2(m-2)$.

2 Solve for x: $\frac{4x+11}{2} = x - 6$.

3 Solve $\frac{3x-5}{4} = \frac{2(x+7)}{9}$.

Answers

1

$3(m+1) = 6 + 2(m-2)$	expand the brackets on both sides of the equation
$3m + 3 \times 1 = 6 + 2m - 2 \times 2$	collect like terms
$3m + 3 = 2 + 2m$	subtract $2m$ from both sides
$3m + 3 - \mathbf{2m} = 2 + 2m - \mathbf{2m}$	collect like terms
$m + 3 = 2$	subtract 3 from both sides
$m + 3 - \mathbf{3} = 2 - \mathbf{3}$	collect like terms
$m = ^{-}1$	

2

$\frac{4x+11}{2} = x - 6$	multiply both sides by 2
$\frac{4x+11}{2} \times \mathbf{2} = \mathbf{2} \times (x-6)$	cancel and expand bracket
$4x + 11 = 2x - 2 \times 6$	
$4x + 11 = 2x - 12$	subtract $2x$ from both sides
$4x + 11 - \mathbf{2x} = 2x - 12 - \mathbf{2x}$	collect like terms
$2x + 11 = ^{-}12$	subtract 11 from both sides
$2x + 11 - \mathbf{11} = ^{-}12 - \mathbf{11}$	collect like terms
$2x = ^{-}23$	divide both sides by 2
$\frac{2x}{\mathbf{2}} = \frac{^{-}23}{\mathbf{2}}$	
$x = ^{-}11.5$	

3

$\frac{3x-5}{4} = \frac{2(x+7)}{9}$	expand the bracket
$\frac{3x-5}{4} = \frac{2 \times x + 2 \times 7}{9}$	
$\frac{3x-5}{4} = \frac{2x+14}{9}$	multiply both sides of the equation by 4
$\frac{3x-5}{4} \times \mathbf{4} = \frac{2x+14}{9} \times \mathbf{4}$	
$3x - 5 = \frac{4(2x+14)}{9}$	multiply both sides of the equation by 9
$\mathbf{9} \times (3x-5) = \frac{4(2x+14)}{9} \times \mathbf{9}$	
$9(3x-5) = 4(2x+14)$	expand the brackets
$9 \times 3x - 9 \times 5 = 4 \times 2x + 4 \times 14$	
$27x - 45 = 8x + 56$	subtract $8x$ from both sides
$27x - 45 - \mathbf{8x} = 8x + 56 - \mathbf{8x}$	
$19x - 45 = 56$	add 45 to both sides
$19x - 45 + \mathbf{45} = 56 + \mathbf{45}$	
$19x = 101$	divide both sides by 19
$\frac{19x}{\mathbf{19}} = \frac{101}{\mathbf{19}}$	
$x = 5\frac{6}{19}$	

EXERCISE

1 Find the value of the unknown that makes each of the following equations true.

a $5m = 15 - 8m$

b $5x = {}^{-}2 - 7x$

c $7m + 24 = 3m$

d $\frac{2a + 7}{3} = a$

e $4c - 5 = c + 16$

f $5 - y = 3(6 - 2y)$

g $7y - 11 = 3y + 5$

h $10x + 3 = 27 - 18x$

i $2(z + 5) = 21 - 13z$

j $9x + 7 - 4x = 3x + x + 11$

k $3(x - 1) = 2(x + 3) - 4$

l $2(f - 1) = 3(2f + 1) - 7f$

m $6 - 2(2m - 1) = 2(m + 5) - 5m$

n $6t - 5(4 - t) = 3t + 2$

o $3(x - 1) = 5 - (7 - 2x)$

p $4n + 3(n - 9) = 3n + 2(n - 1)$

q $4m^2 - 3(m - 1) = 4(m^2 - 1)$

r $2x(x + 5) - x + 5 = 2x(x - 1) - 3$

2 Solve these equations.

a $1 - g = \frac{2g - 7}{3}$

b $y + 4 = \frac{11 - 2y}{3}$

c $\frac{6x - 3}{5} = x + 1$

d $5n + 6 = \frac{7n + 9}{2}$

e $\frac{5x + 6}{2} = 2x + 5$

f $\frac{8p + 17}{3} = p + 4$

g $4y - 5 = \frac{2y - 5}{3}$

h $5x - 3 = \frac{48 - 4x}{3}$

Substituting in formulae

Scientists and mathematicians have studied and found the connection between different quantities, and they have developed rules that are used in everyday life. For example, Pythagoras discovered the rule that connects the length of the sides of a right-angled triangle. In Pythagoras's time this was expressed in words: 'the square on the hypotenuse is equal to the sum of the squares on the other two sides'. From the time that algebra, and the use of pronumerals, was developed this could be expressed as $c^2 = a^2 + b^2$.

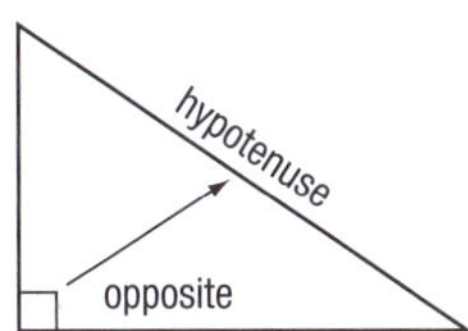

Already there will be many formulae that you have used in science and mathematics. What are some of these formulae?

In your working life you may use formulae in building, banking, engineering, medicine, cooking … in fact, in almost all areas we use formulae.

Substitution in formulae

If we know the value of all but one of the variables that are connected by a formula then, by substitution, we can find the value of the remaining variable.

Example

The formula for the volume, V, of a cone is $V = \frac{1}{3}\pi r^2 h$, where r is the radius of the base of the cone and h is the height of the cone.

Find the volume of a cone that has a diameter of 14 cm and a height of 12 cm.

Answer

There are three variables in this formula, V, r and h, so we can substitute the known values for r and h to find the value of V.

Note: r = half of the diameter = $\frac{1}{2}$ of 14 cm = 7 cm.

$$V = \frac{1}{3}\pi r^2 h$$
$$= \frac{1}{3} \times \frac{22}{7} \times 7^2 \times 12$$
$$= \frac{1}{\cancel{3}} \times \frac{22}{\cancel{7}} \times \cancel{7^2} \times \cancel{12}^4$$
$$= 22 \times 7 \times 4$$
$$= 616 \text{ cubic centimetres}$$

EXERCISE

1 Find the value of y when $x = 4$ in each of the following.

a $y = 4x - 7$

b $y = x^2 - 3x - 2$

c $y = 5\sqrt{x} - 3$

d $y = 2x^3 - 4x^2 - 8x + 9$

e $y = \frac{8}{(x-2)^2}$

f $y = \sqrt{x+5} - \frac{3}{\sqrt{x-3}}$

2 Evaluate each of the following formulae:

a the circumference of a circle $C = \pi d$ when $d = 42$ cm. Use $\pi = \frac{22}{7}$.

b the volume of a cube $V = l^3$ when the length of the edge of the cube is 6 cm.

c the formula to convert temperature degrees in centigrade to degrees in Fahrenheit $F = \frac{9}{5}C + 32$ when $C = 40°$.

d the formula $F = ma$ connecting force, F newton; mass, m kg; and acceleration, a m/s^2; where the mass is 60 kg and the acceleration is 3 m/s^2.

e the formula to calculate the simple interest amount earned, I, when the principle, P; interest rate, r; and time T years are known: $I = \frac{PRT}{100}$; when P = K100, R = 5% and T = 4 years.

f the surface area, A, of a solid cylinder, $A = 2\pi r^2 + 2\pi rh$ where the radius of the cylinder $r = 3.5$ cm, and $h = 12$ cm is the height of the cylinder. (Use $\pi = \frac{22}{7}$.)

3 The illumination, I lumens, from a light source can be measured at various distances, d m, from the light source. It is found that the relationship between the variables I and d is given by the formula $I = \frac{45}{d^2}$. Find the illumination in lumens from the light source at distances:

a 2 m

b 5 m

c 10 m.

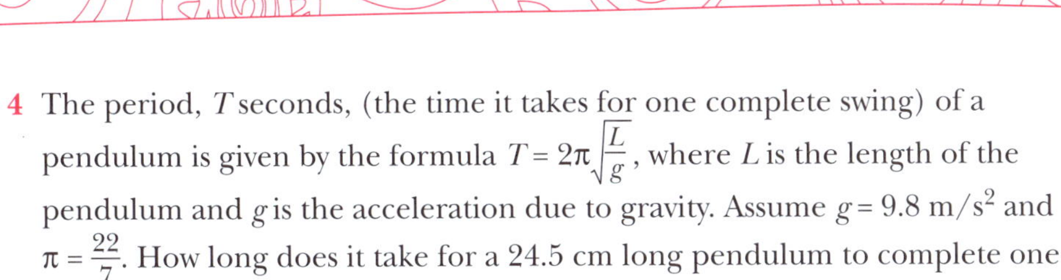

4 The period, T seconds, (the time it takes for one complete swing) of a pendulum is given by the formula $T = 2\pi\sqrt{\frac{L}{g}}$, where L is the length of the pendulum and g is the acceleration due to gravity. Assume $g = 9.8$ m/s^2 and $\pi = \frac{22}{7}$. How long does it take for a 24.5 cm long pendulum to complete one swing?

5 The area of cardboard A_s, needed to form a cone which can hold popcorn is given by the formula $A_s = \pi rs$, where r equals the radius of the base of the cone, and s equals the length of the slant edge of the cone. Find the area of cardboard needed to make the popcorn holder shown. (Use $\pi = \frac{22}{7}$.)

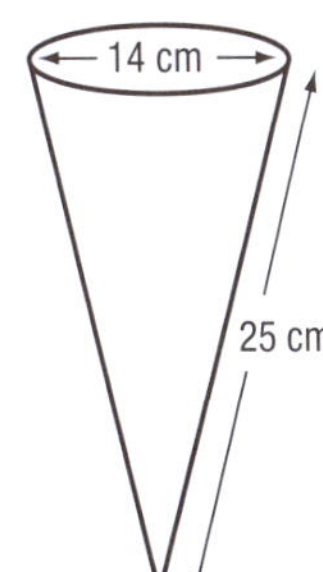

6 Students in a motor workshop class have to calculate the amount of electrical current I amps, passing through different parts of the car. To do this they use the equation:

$I = \frac{V}{R}$ where V = voltage and R = resistance.

Find the amount of current passing through the:

a tail light if $V = 12$ and $R = 0.5$

b head light if $V = 12$ and $R = 2.40$.

7 A nurse needs to give a specific dosage of a drug to a child. The formula used to calculate the child's dosage, d, in millilitres, when the child's age a, in years, is known is $d = \frac{12.6a}{a + 12}$. What is the correct dosage for an 8-year-old child?

8 Conrad the computer whizz bought a desktop computer for K2500 (A_0). The computer depreciates (loses value) at a rate, r, of 10% per year. Calculate the current value, A, of the computer after a period of time, t, of 3 years using the formula $A = A_0(1 - \frac{r}{100})^t$ where A_0 is the purchase price.

Lesson 11

Transposing formulae

When using formulae, we are sometimes required to find the value of a variable (pronumeral) that is not the subject of the formula. So we need to transpose the equation using inverse operations.

Example

The formula for the area, A, of a circle is $A = \pi r^2$ where r is the radius of the circle. Calculate the radius of a circle that encloses an area of 200 cm^2.

Answer

There are two ways we can approach this question.

i Substitute then transpose to find A.

Substituting $A = 200$ and $\pi = \frac{22}{7}$ in $A = \pi r^2$.

$200 = \frac{22}{7} \times r^2$ multiply both sides of the equation by 7

$1400 = 22r^2$ divide both sides of the equation by 22

$63.6363... = r^2$ take the square root of both sides of the equation

$r = 7.977$ cm

A circle of radius 7.98 cm has an area of approximately 200 cm^2.

ii Transpose the formula to make r the subject first and then substitute.

$A = \pi r^2$ divide both sides of the equation by π

$\frac{A}{\pi} = r^2$ take the square root of both sides of the equation

So $r = \sqrt{\frac{A}{\pi}}$

Now we can substitute $A = 200$ to find r.

$r = \sqrt{\frac{200}{\frac{22}{7}}} = 7.977\ldots \approx 7.98$ cm

EXERCISE

1 For each of the following:

i substitute the given values into the formula

ii solve the equation to find the value of the variable in the brackets.

a $C = 2\pi r \quad (r); \qquad C = 440$ cm^2, $\pi = \frac{22}{7}$

b $V = l^3 \quad (l); \qquad V = 1000$ cm^3

c $F = \frac{9}{5}C + 32; \qquad (C); \qquad F = 100°$

2 For each of the following:

i transpose the formula so that the variable in brackets is the subject

ii evaluate the transposed formula from part **i** using the given values.

a $F = ma \quad (a); \qquad F = 250$ N, $m = 20$ kg

b $A = 2\pi r^2 + 2\pi rh \quad (h); \qquad A = 2500$ cm^2, $r = 10$ cm, $\pi = \frac{22}{7}$

c $I = \frac{PRT}{100} \quad (T); \qquad I =$ K120, $P =$ K2000, $R = 4.8\%$

d $A = P(1 + \frac{r}{100})^n \quad (P); \qquad A =$ K15 000, $r = 5$, $n = 2$

e $V = \frac{1}{3}\pi r^2 h \quad (r); \qquad V = 1000$ cm^3, $h = 11.2$ cm, $\pi = \frac{22}{7}$

3 The time, t, (in minutes) for roasting a chicken at 180°C is 45 minutes per kilogram weight, w, of the chicken plus an additional thirty minutes, is $t = 45w + 30$.

a Make w the subject of the formula.

b What is the weight of a chicken that needs to be cooked for 80 minutes at 180°C?

4 The intensity of the light given out by the candle, I lumens, depends on the distance from the candle, d, in metres. The formula for the candle's light intensity is $I = \frac{k}{d^2}$.

a Transpose the equation to make d the subject.

b Calculate the distance from the candle, d centimetres, where the light intensity I, is 0.4 lumens and k, a constant, has the value of 0.096.

5 The formula $l = \frac{^{-}5t}{4} + 200$ gives the length, l, of a candle (in millimetres) after it has been burning for t minutes.

a What will be the candle's length after 16 minutes?

b Make t the subject of the formula.

6 The pendulum on a grandfather clock controls the speed of the clock. Adjusting the length of the pendulum (in metres), l, affects the timing of the clock. The time, t, (in seconds) for a complete swing of the pendulum back to its starting position, is given by the formula $t = 2\pi\sqrt{\frac{l}{g}}$ where g (which equals 9.8 m/s^2) is the gravitational acceleration.

a Does the weight on the end of the pendulum effect the timing of its swing? Explain your answer.

b Calculate the time for a complete swing if the length of the penulum is 0.882 m. (Use $\pi = \frac{22}{7}$.)

c Express l in terms of the other pronumerals then find how long the pendulum must be if the time for a complete swing is adjusted to 1.65 seconds. (Use $\pi = \frac{22}{7}$.)

GRAPHS

You will have already encountered many graphs in your schoolwork, and you will have noticed that they are used in newspapers, by the government, and used or consulted frequently in almost all occupations. Graphs are used to illustrate the relationship between variables. A properly constructed graph gives us a 'picture' of this relationship that is easy to read.

Graphs can take many forms; you will already have studied statistical graphs and some line graphs. In this topic we are looking at the construction and interpretation of linear graphs, in particular.

Lesson 12

Practical application and interpretation of graphs

Interpreting a graph means to explain features of the graph in terms that apply to the situation. To **interpret** a graph, the first thing you need to identify is 'what is the graph telling us?' To find the answer to this we can look at the title (if there is one) and at the labels on the axes.

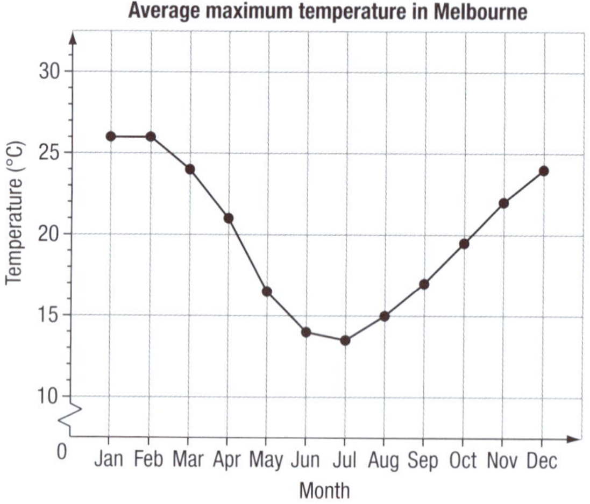

For example, this graph might appear in a travel brochure. We have a graph titled the 'Average maximum temperature in Melbourne' (Australia). The labels on the axes are 'Month' and 'Temperature' so we can read the average maximum temperature in Melbourne for each month of the year.

This type of graph is important if you are travelling and perhaps want to visit Melbourne in one of the warmer months.

To interpret this graph we could make a general statement: 'The average maximum temperature in Melbourne varies from a high in January of about 26°C to a low in July of about 13°C. The average maximum temperature is above 20°C for six months of the year, November to April'.

Otherwise we could select some important features that can be read from the graph:

- The warmest month of the year is January, closely followed by February.
- The coldest month of the year is July.
- For the months November to April, the average maximum temperature is above 20°C.
- For the months June, July and August, the average maximum temperature is below 15°C.

EXERCISE

1 This graph shows the variation in Helen's pulse rate during a one-hour exercise session.

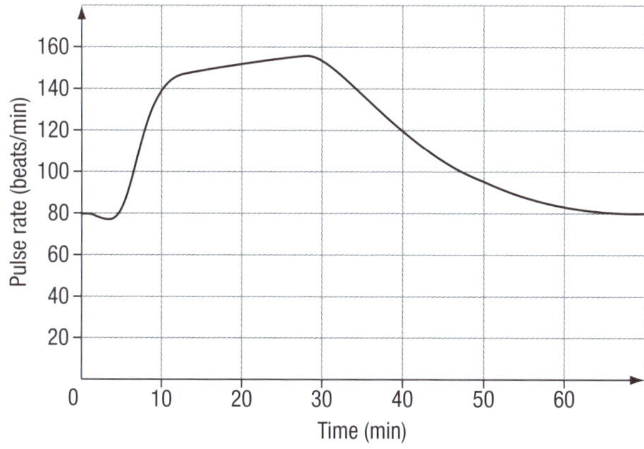

a What is Helen's maximum pulse rate reached during the exercise session?

b For how long was her pulse rate above 100 beats per minute?

c What is Helen's 'resting' (when she is not exercising) pulse rate?

d How long did it take for Helen's pulse rate to reach at least 140 beats/minute?

e For what period of time was Helen's pulse rate decreasing?

2 The time taken for a stone to fall various distances was observed and recorded in the table below.

Distance (m)	10	20	30	40
Time (s)	1.43	2.02	2.47	2.86

The points have been plotted on the set of axes at right, and a line graph drawn through them to illustrate the type of graph that might apply in this case.

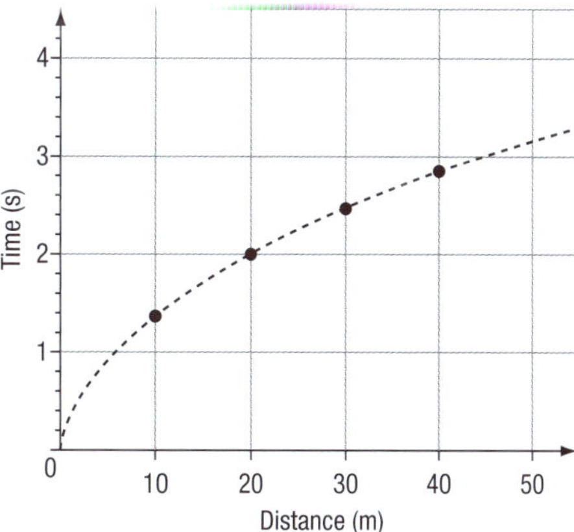

a Use the graph to estimate the time taken for the stone to fall:
   i 15 m   ii 35 m.

b Use the graph to **predict** the time taken for the stone to fall:
   i 5 m   ii 50 m.

c Why do you think the line drawn through the points starts at the point (0, 0)?

d Could the line that goes through the points have been a straight line? Explain.

3 The speed, in kilometres/hour, of a car travelling a lap of a racing circuit is graphed.

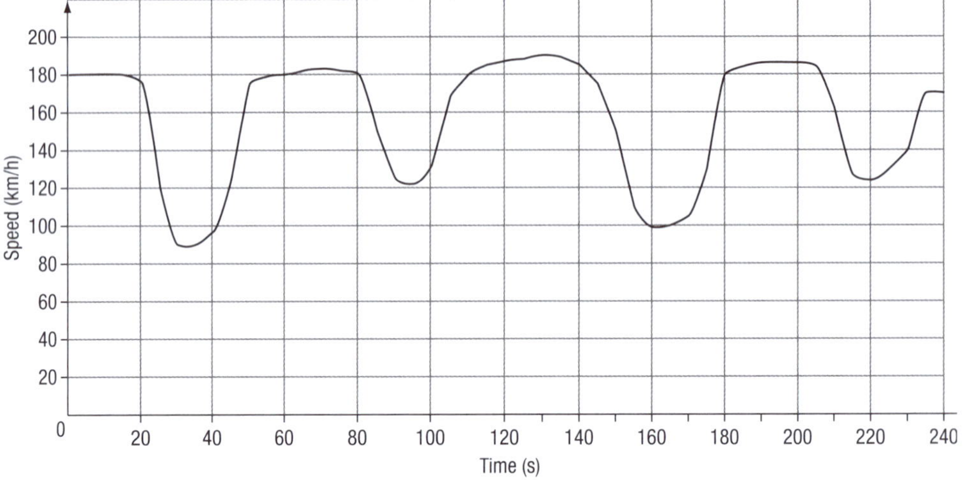

a What is the maximum speed reached on this lap of the circuit?

b How long did it take to complete this circuit of the racecourse?

c How many corners are there on this circuit? Explain your answer.

d What was the minimum speed on this lap of the circuit?

e For how long, on this circuit, was the speed below 120 km/h?

f Find the percentage of the time for this circuit when the speed was 180 km/h or more.

4 The graph below shows the incidence of HIV/AIDS in PNG for the years 1987 to 2005. Use the graph to answer the following.

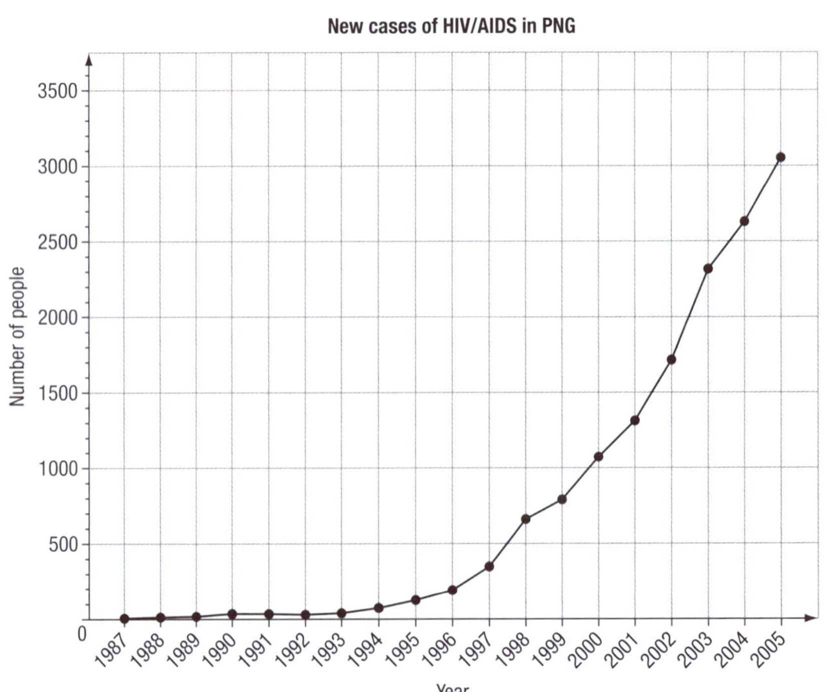

a How many new cases of HIV/AIDS were diagnosed in 2005?

b Describe the trend for the years 1994 to 2005.

c How many more new cases of HIV/AIDS were diagnosed in 2005 compared to 2001?

d Assuming the trend continues, use the graph to predict the number of new cases of HIV/AIDS in 2006.

5 Kevin is riding his bike from home to the village of his friend and then home again. This graph illustrates Kevin's journey.

a How far was it to Kevin's friend's place?

b How long did the complete journey take?

c If Kevin left home at 8.00 am, what time did he arrive back home?

d How long did he stay at his friend's place?

e Did it take longer to get to his friend's place or to get home from his friend's place? Explain.

f Did Kevin stop on his way home?

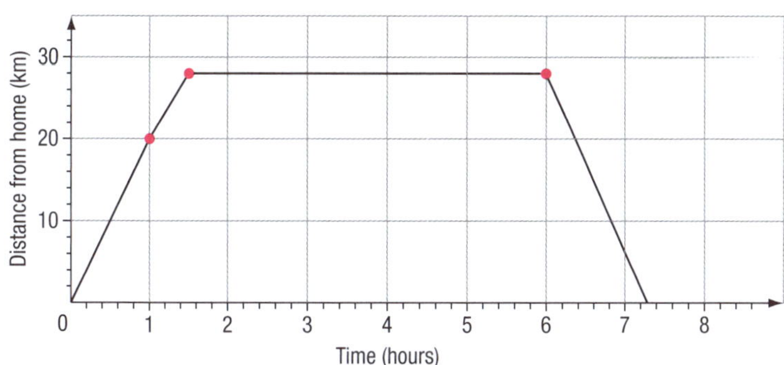

6 A delivery company has charges that are based on the distance required to deliver a parcel. Within a radius of 100 km of the depot, charges can be determined from a graph like that shown at right.

The open circle point means that this point is not included, and the filled-in circle point means that this point is included.

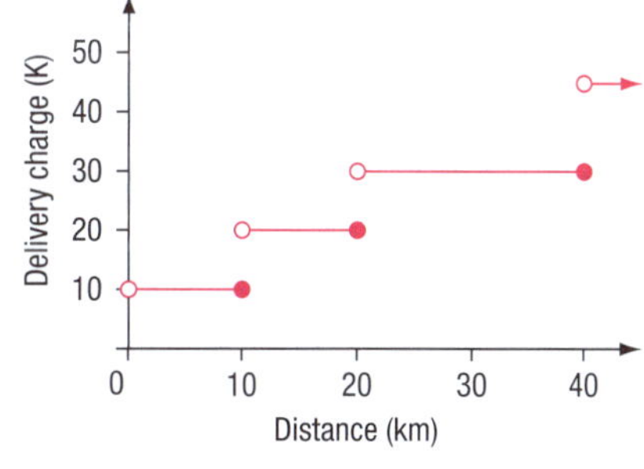

a What is the delivery charge if the article has to be delivered a distance of:

i 30 km ii 20 km iii 15 km

iv 10 km v 43 km?

b Ulato is charged K20 for the delivery of a parcel. What is the maximum distance that the parcel could be delivered for this cost?

c On his delivery round Elijah has to deliver parcels to places that are 12 km, 15 km, 18 km, 20 km, 25 km and 50 km from the depot. How much will he collect on this delivery round?

7 Lawrence has taken out a loan for K180 000 from a bank to buy a house. The reducing balance loan is charged 6.5% p.a. interest for a term of 20 years. Lawrence is given the graph below showing the amount still owing on the loan after each of the twenty years.

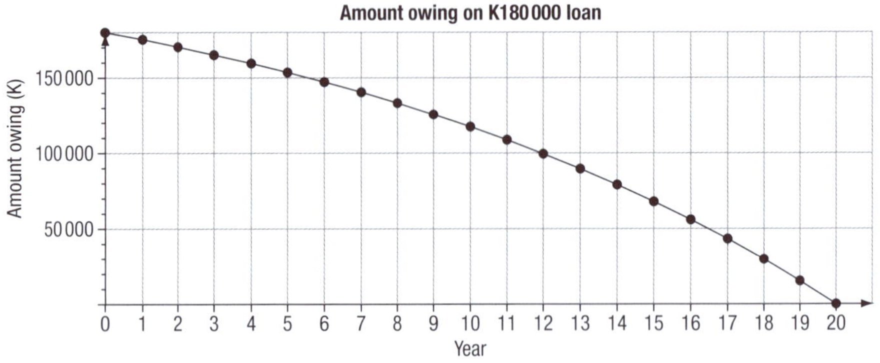

a Describe what the graph is showing about the amount still owing on the loan.
b Approximately how much is still owing on the loan after 7 years?
c After how many years will there first be less than K100 000 owing on the loan?
d In which year will approximately half the loan be repaid?
e In which year is the greatest amount paid off the loan?
f In which year is the least amount paid off the loan?

Investigation

8 Locate a graph from a newspaper, something you have recently worked on at school or at your place.
 a Write a paragraph explaining what the graph is meant to illustrate.
 b Locate some important points or features on the graph and interpret these points.

Lesson 13

Graphing on a Cartesian plane

A **Cartesian plane** contains of a **set of axes**, where the axes are number lines positioned at right angles to each other and intersecting at zero.

In the seventeenth century, the French mathematician **Rene Descartes** (hence the namc **Cartcsian** plane) suggested that a point on a plane could be identified by its position from a fixed point, *O*, called the **origin**—this is the point where the axes intersect.

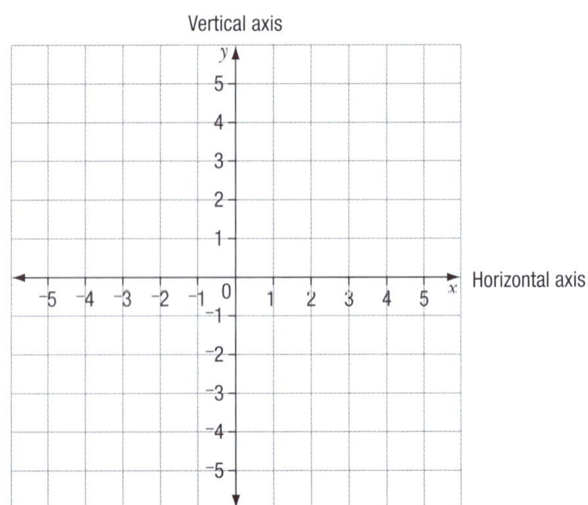

The axis going from left to right on a page is called the **horizontal axis** and the axis from bottom to top of a page is referred to as the **vertical axis**.

It is important that the axes are scaled accurately. On each of the axes the length representing a unit must be the same all along the axis. However the length representing a unit does not have to be the same on the vertical as it is on the horizontal axis.

The axes divide the Cartesian plane into four **quadrants**. The quadrants are commonly labelled first, second, third and fourth, as shown.

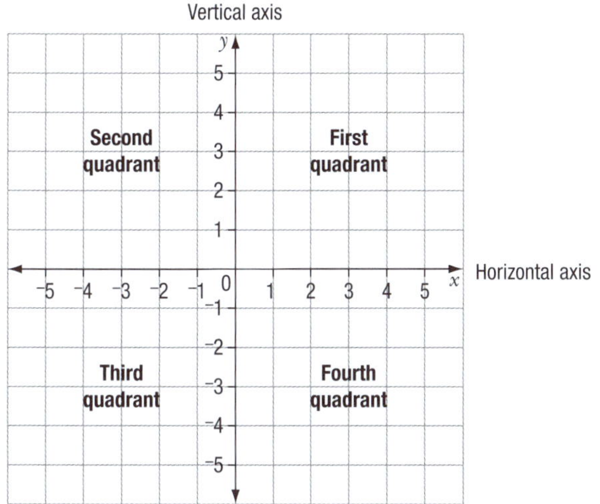

When identifying a point on a Cartesian plane, the **coordinates** relating to the positions on the axes are quoted in brackets with the coordinate that relates to the horizontal axis given first.

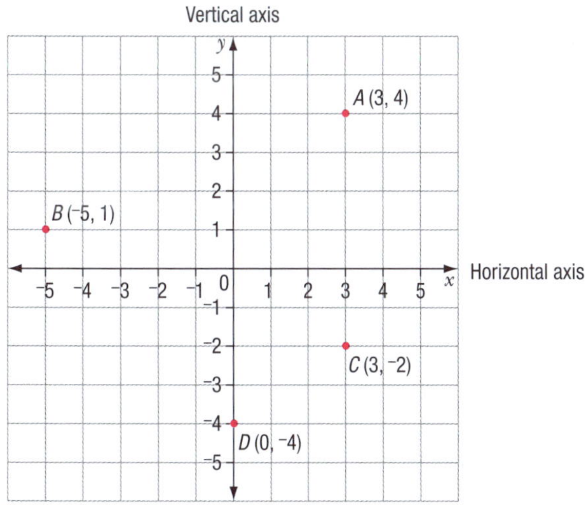

The points $A(3, 4)$, $B(^{-}5, 1)$, $C(3, ^{-}2)$ and $D(0, ^{-}4)$ are plotted on the axes above. A point such as (3, 4) is called an **ordered pair**.

Graphing linear relations

A **linear relation** is a rule connecting two variables which, when graphed on a Cartesian plane, produces points that are in a straight line (called **collinear** points). For example, the points $(^-2, ^-4)$, $(^-1, ^-3)$, $(0, ^-2)$, $(1, ^-1)$, $(2, 0)$, $(3, 1)$ and $(4, 2)$ are plotted on the Cartesian plane below. The points appear to be in a straight line.

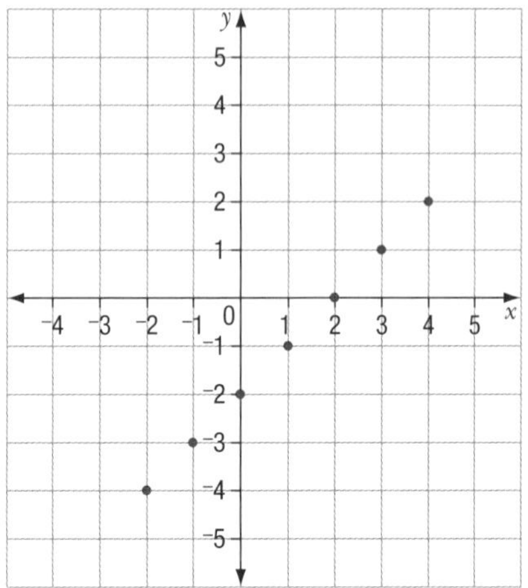

The two variables in this example are x (horizontal axis) and y (vertical axis) and the points are plotted as **ordered pairs** (x, y). For the point $(^-2, 4)$, the value $^-2$ is the ***x*-coordinate** and the value 4 is the ***y*-coordinate**.

If we analyse the given points, we can see they all are governed by the rule 'the y-coordinate equals the x-coordinate minus 2' or $y = x - 2$.

This rule is the **equation of the straight line** that goes through all the given points; there will be an infinite number of points that satisfy this equation. If they are all plotted then they will form a continuous straight line, as shown below.

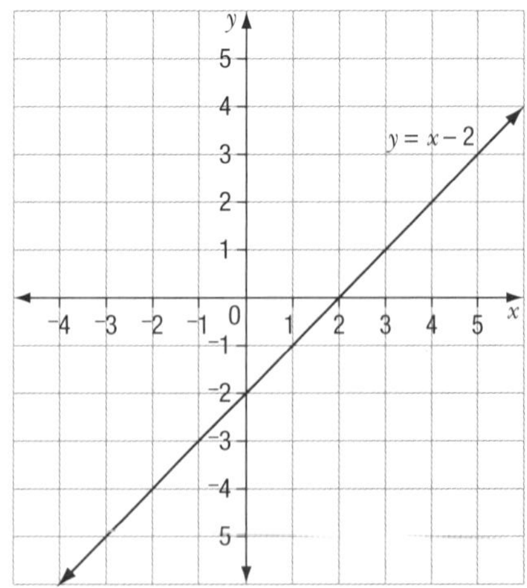

The point where a straight line crosses the x-axis is called the ***x*-intercept**, and the point where it crosses the y-axis is called the ***y*-intercept**. In the example at right, the x-intercept is 2 (coordinates $(2, 0)$) and the y-intercept is $^-2$ (coordinates $(0, ^-2)$). The point where two lines cross each other is called **the point of intersection** of the two lines. At this point the lines have the **same coordinates**.

EXERCISE

1 State, as ordered pairs, the coordinates of the points *A* to *G* that are graphed on this set of axes.

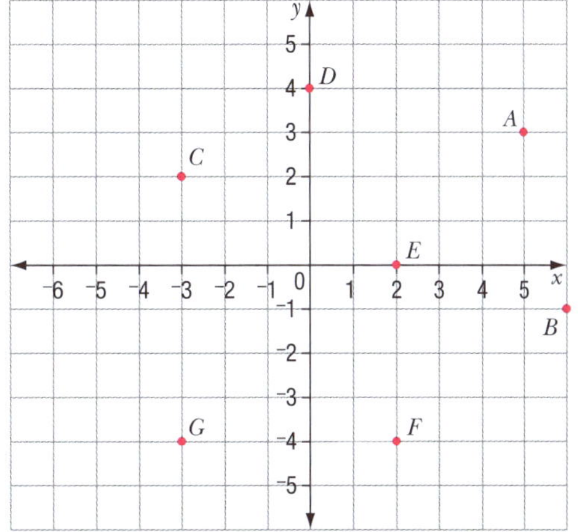

> **Remember**
> When scaling an axis, the length representing a unit must be the same all along the axis.

> **Remember**
> The first value in the brackets of the ordered pair is the *x*-coordinate.

2 Rule-up a set of axes, carefully scaling the axes from $^{-}5$ to $^{+}5$ (or use grid paper). Label the axes *x* (horizontal) and *y* (vertical). On the set of axes plot the points *A*(1, 4), *B*($^{-}3$, $^{-}3$), *C*($^{-}2$, 4), *D*($^{-}3$, 0), *E*(0, 2), *F*(3, $^{-}1$).

3 How many different straight lines can you draw through:
  i one point **ii** two points?

4 **a** Carefully construct a set of axes, scaling each axis from $^{-}6$ to $^{+}6$ and labelling the axes *x* and *y*.
  b For each of *A* to *D* [*A* ($^{-}1$, 2), (1, 6); *B* (0, 0), (2, 4); *C* ($^{-}5$, 6), (2, $^{-}1$); *D* ($^{-}3$, $^{-}4$), (3, $^{-}4$):
    i Plot the two points.
    ii Rule a straight line going through the two points. Label these lines *A* to *D*.
    iii State the coordinates of two other points that are on each line.
  c State the coordinates of the point where the lines *A* and *C* intersect. Which quadrant is this point of intersection in?
  d State the coordinates of the point where the line *A* crosses:
    i the *x* axis **ii** the *y* axis.
  e Find the:
    i *x*-intercept **ii** *y*-intercept
  of the line *C*.

5 On a set of axes:
  a Plot the points (3, 6) and ($^{-}1$, 2) and rule a straight line through these points. Call this line *A*.
  b Plot the points ($^{-}4$, 3) and (2, $^{-}3$) and rule a straight line through these points. Call this line *B*.
  c State the coordinates of the point of intersection of the lines *A* and *B*.
  d Write, as ordered pairs, the coordinates of the *x*- and *y*-intercepts of line *A*.
  e Write, as ordered pairs, the coordinates of the *x*- and *y*-intercepts of line *B*.

Lesson 14

The gradient of a straight-line graph

The gradient (or slope) of a straight-line graph refers to its **inclination** to the positive direction of the x-axis.

The straight line is said to have a **positive gradient** if the y-values increase when the x-values increase. A linear relation that has a positive gradient has this appearance.

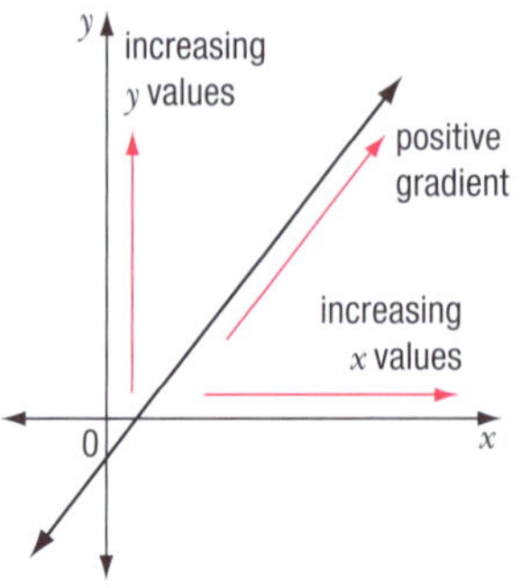

The y-values of a straight line with a negative gradient decrease as the x-values increase and they look like this.

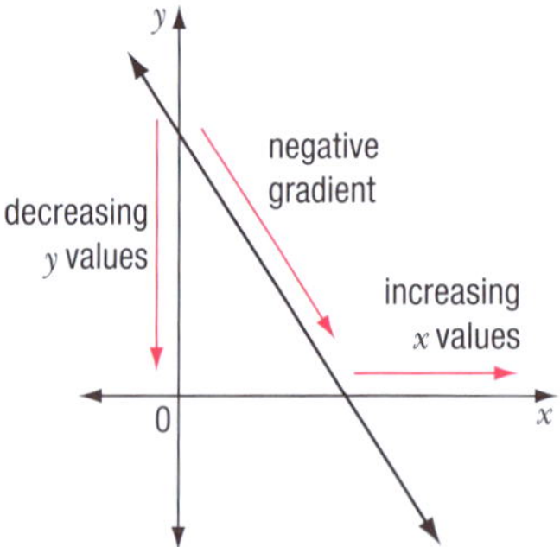

Finding the gradient of a straight line from a graph

Choose any two convenient points on the line and use gradient = $\frac{\text{rise}}{\text{run}}$.

Example 1

Find the gradient of the line graphed here.

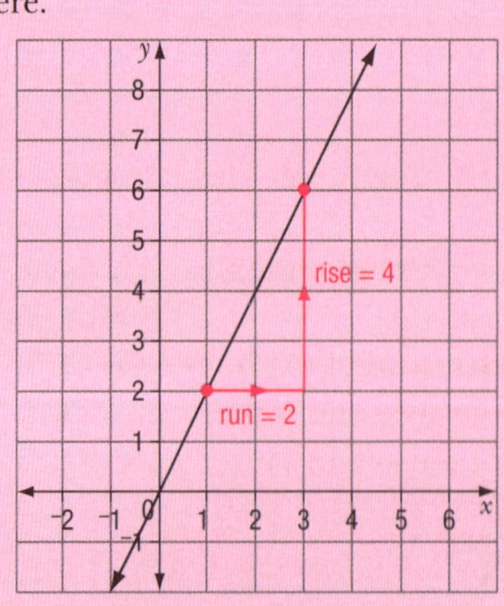

Answer

Two points that the straight line goes through are (1, 2) and (3, 6).

Moving from left to right, both rise and run are positive.

The gradient is found using $\frac{\text{rise}}{\text{run}} = \frac{6-2}{3-1}$

$= \frac{4}{2}$

$= 2$

So the gradient of this line is 2.

Example 2

Find the gradient of the line of this graph.

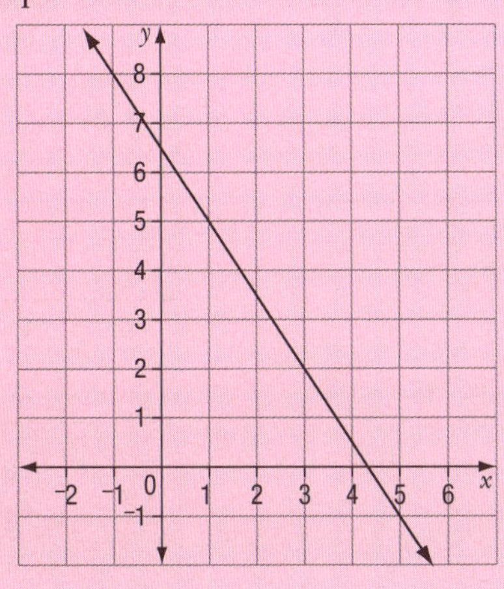

Answer

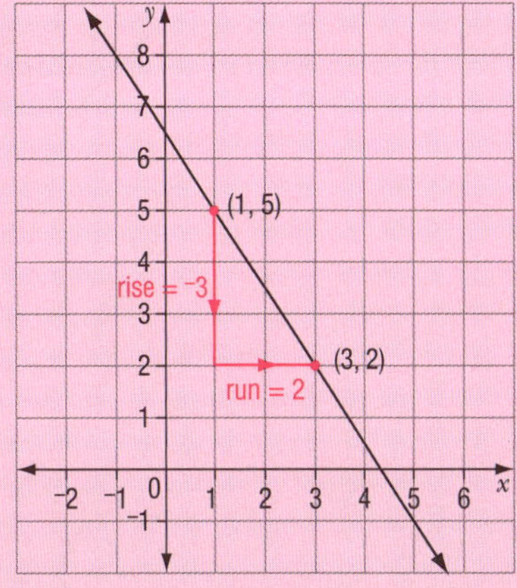

Using points (1, 5) and (3, 2) that are on the line and moving from left to right, the rise will be negative.

In this case, gradient = $\frac{\text{rise}}{\text{run}} = \frac{^-3}{2} = -\frac{3}{2}$.

Example 3

Here are two special cases.

a The straight line graphed at right is **parallel to the *x*-axis** and so has zero inclination to the *x*-axis and hence **zero gradient**.

Alternatively, using two points on the line (1, 2) and (4, 2), and gradient = $\frac{\text{rise}}{\text{run}}$, we can see that gradient = $\frac{0}{3} = 0$.

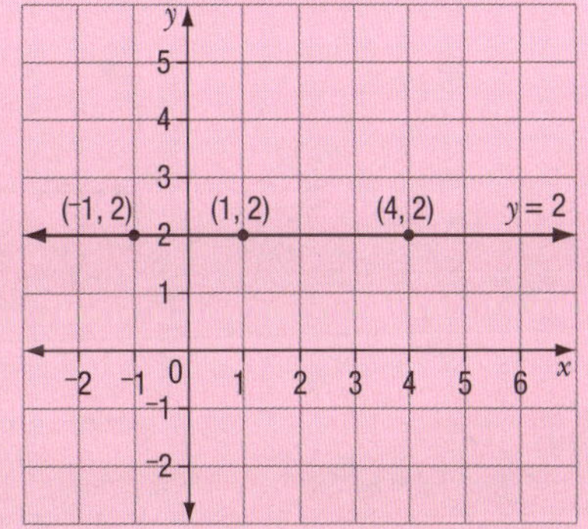

b Straight lines that are **parallel to the *y*-axis** have an **indeterminate gradient**. This means that the gradient is so large it cannot be determined.

If we use two points on the line (2, 3) and (2, ⁻3), and gradient = $\frac{\text{rise}}{\text{run}}$, we can see that gradient = $\frac{3-(^-3)}{0} = \frac{6}{0}$.

A fraction with zero in the denominator is indeterminate.

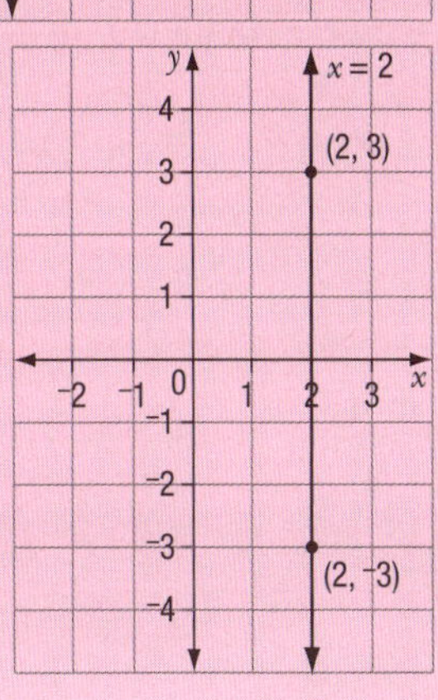

EXERCISE

1 Find the gradient of each of the following straight lines.

a 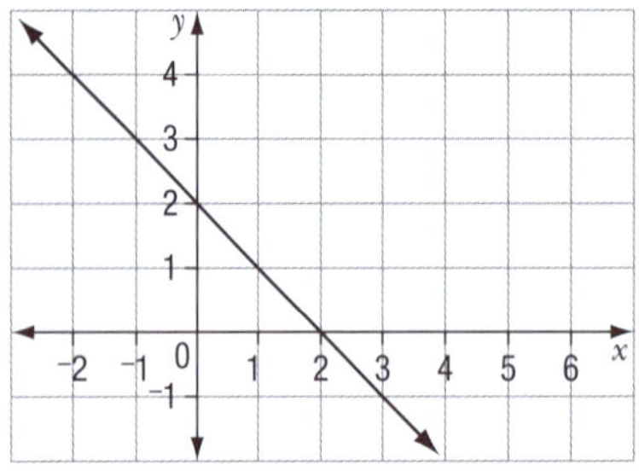

b 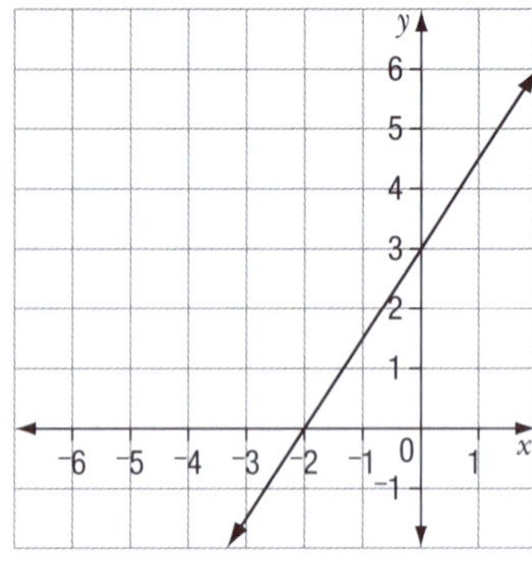

c 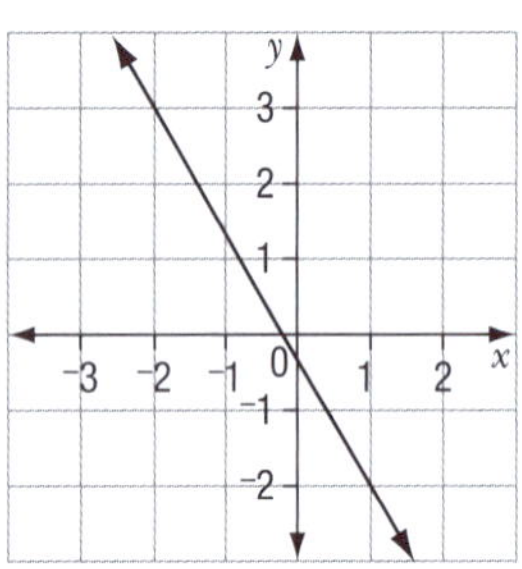

d 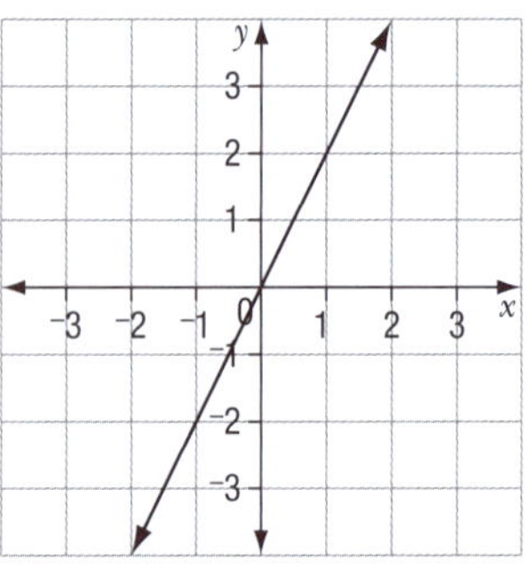

e 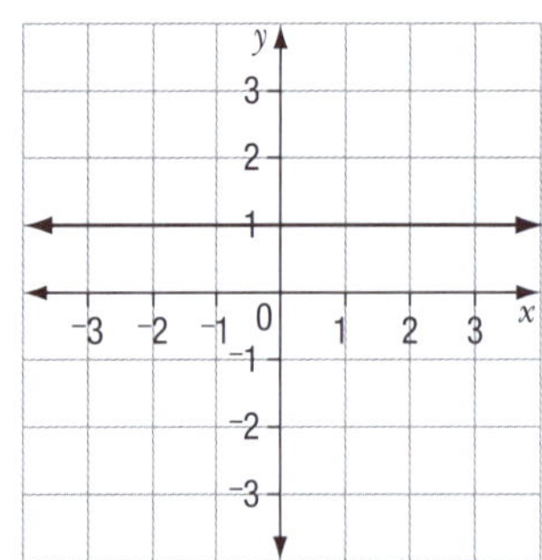

f 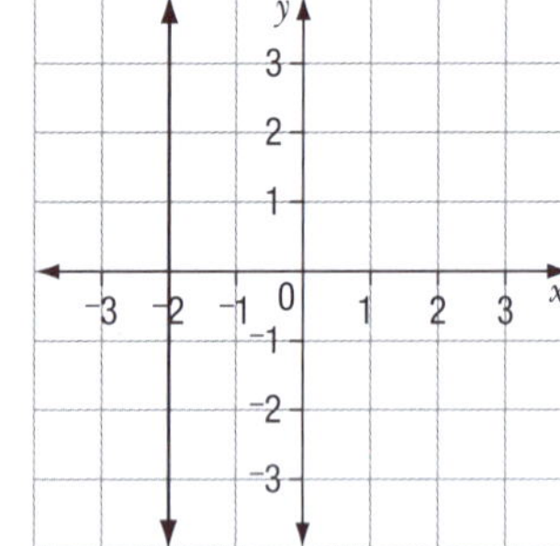

2 a Which of the straight lines, A to F, drawn on the axes below, have a positive gradient?

b Match the gradients of the straight lines, A to F, to the gradients given below.

i 3

ii $-\frac{1}{3}$

iii 0

iv -3

v indeterminate

vi $\frac{1}{3}$

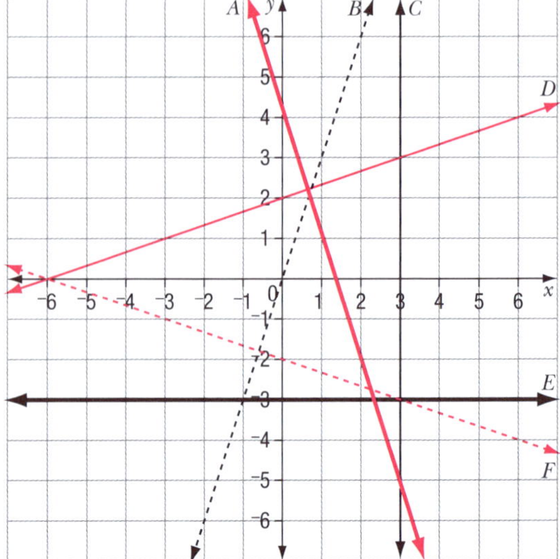

3 Plot the following pairs of points on a set of axes and find the gradient of the line that joins the points.

a (0, 0) and (2, 6)
b (1, 0) and (2, 5)
c (0, ⁻1) and (2, 5)
d (0, 0) and (⁻2, ⁻4)
e (⁻6, ⁻1) and (⁻1, ⁻6)
f (⁻2, ⁻3) and (4, ⁻3)
g (4, 0) and (⁻4, 0)
h (2, ⁻2) and (2, 3)

4 On a set of axes, draw a straight line with a gradient of:

a 5 b ⁻2 c $\frac{2}{3}$ d $-\frac{1}{3}$ e $\frac{5}{3}$ f 0

Finding the gradient from two points on a straight line

Lesson 15

In general, if two points, (x_1, y_1) and (x_2, y_2), on the straight line are known then the **gradient**, ***m***, is given by

$$m = \frac{y_2 - y_1}{x_2 - x_1}$$

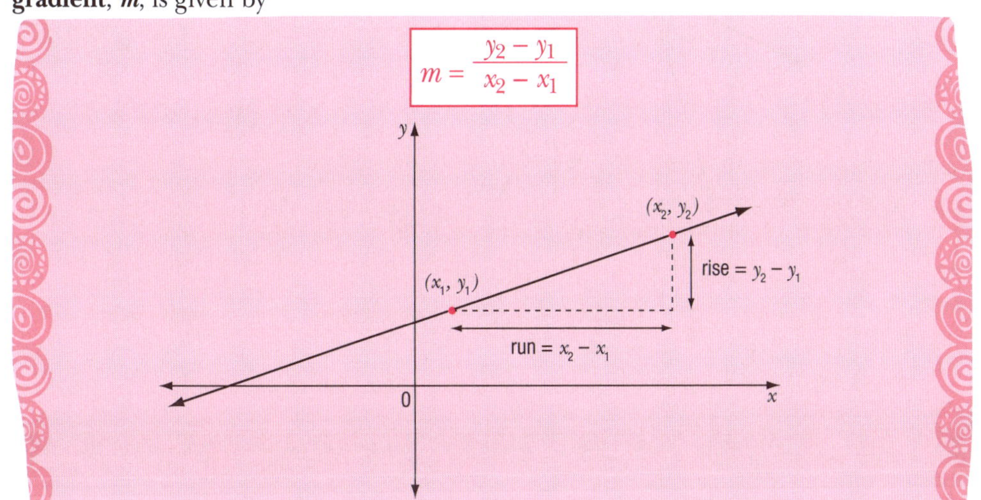

Example

Find the gradient of the straight line going through the points (⁻2, 5) and (3, ⁻7).

Answer

Using (⁻2, 5) as (x_1, y_1) and (3, ⁻7) as (x_2, y_2)

Substituting in $m = \frac{y_2 - y_1}{x_2 - x_1}$

gives $m = \frac{-7 - 5}{3 - (-2)}$

$= -\frac{12}{5}$

EXERCISE

1 Use the rule for gradient $m = \frac{y_2 - y_1}{x_2 - x_1}$ to find the gradient of the straight line joining the following pairs of points:

a (0, 0) and (2, 6)
b (0, ⁻1) and (2, 5)
c (1, 0) and (⁻3, 8)
d (5, 2) and (7, 9)
e (⁻2, ⁻7) and (⁻4, 3)
f (6, ⁻3) and (⁻3, 0)
g (⁻4, 6) and (0, 0)
h (⁻3, 5) and (5, 2)
i (23, 102) and (28, 187)

Investigation

2 Five straight lines have been plotted on the set of axes below and they are each labelled with their equation.

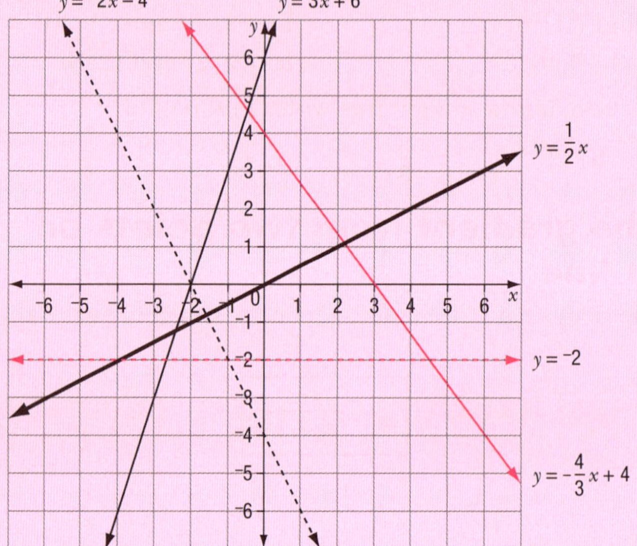

Remember

The equation of a straight line is the rule that connects the y-values to the x-values of all the points on the line.

a For each of the straight lines, calculate the gradient and write down their x- and y-intercepts. Copy the table below and record your answers.

Equation	Gradient	y-intercept	x-intercept
$y = {}^{-}2x - 4$			
$y = 3x + 6$			
$y = \frac{1}{2}x$			
$y = {}^{-}2$			
$y = -\frac{4}{3}x + 4$			

b Can you see a relationship between the gradient and the equation? Explain this in words.

c Can you see a relationship between the y-intercept and the equation? Explain this in words.

d Can you see a relationship between the x-intercept and the equation? Explain this in words.

e If the equation of a straight-line is $y = {}^{-}4x + 5$, can you identify:

i the gradient **ii** the y-intercept **iii** the x-intercept?

f Copy and complete the following statement:
If the equation of a straight line is given in the form $y = mx + c$, then the co-efficient of x (the number in front of x), m, is the ………… of the straight line and c has the value of the ……………..

Identifying the gradient and y-intercept from an equation

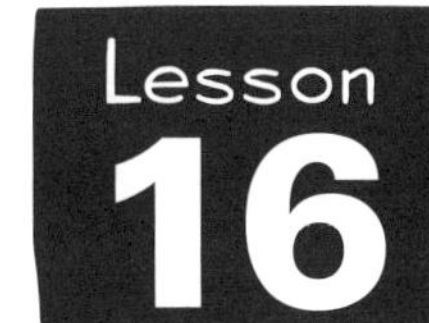

In the previous exercise you will have learnt that if the equation of a straight line is $y = mx + c$ then the number, m, the coefficient of the variable x, is the gradient of the straight line, and the constant term in the equation, c, is the y-intercept of the straight line.

Examples

1 Find: **i** the gradient **ii** the y-intercept of each of the following:

a $y = 5x - 3$ **b** $y = 2 - 7x$ **c** $y = \frac{2x + 5}{3}$ **d** $4x - 3y = 6$

2 Write the equation for each of the following linear relations given the information:

a gradient $^-3$, y-intercept 4

b gradient $\frac{2}{3}$, point $(0, {}^-2)$ on the line.

Answers

1 **a** **i** The gradient is the co-efficient (number in front of) of the variable x so the gradient, $m = 5$.

ii The y-intercept is the constant term in the equation so $c = {}^-3$.

b **i** The equation $y = 2 - 7x$ can be changed to the form $y = mx + c$ by re-arranging the terms to $y = {}^-7x + 2$. The gradient is the co-efficient of the variable x so the gradient $m = {}^-7$.

ii The y-intercept is the constant term in the equation so $c = 2$.

c We will need to separate the terms in the numerator by dividing each by the denominator so that the equation is in the form $y = mx + c$:

$$y = \frac{2x + 5}{3} = \frac{2x}{3} + \frac{5}{3} = \frac{2}{3}x + \frac{5}{3}$$

i The gradient is the co-efficient of the variable x so the gradient $m = \frac{2}{3}$.

ii The y-intercept is the constant term in the equation so $c = \frac{5}{3}$.

d We will need to re-arrange the equation to make y the subject so that we can identify the gradient and the y-intercept.

$4x - 3y = 6$	subtract $4x$ from both sides of the equation
$4x - 3y - 4x = 6 - 4x$	collect like terms
$^-3y = 6 - 4x$	divide both sides of the equation by $^-3$
$\frac{^-3y}{^-3} = \frac{6 - 4x}{^-3}$	
$y = \frac{6}{^-3} + \frac{^-4x}{^-3}$	simplify fractions
$y = {}^-2 + \frac{4}{3}x$	write in the form $y = mx + c$
$y = \frac{4}{3}x - 2$	

i The gradient is the co-efficient of the variable x so the gradient $m = \frac{4}{3}$.

ii The y-intercept is the constant term in the equation so $c = {}^-2$.

2 **a** We are given the values for the gradient, $m = {}^-3$, and the y-intercept, $c = 4$ so substituting in $y = mx + c$ gives the equation $y = {}^-3x + 4$.

b We are given the value for the gradient, $m = \frac{2}{3}$, and the point $(0, {}^-2)$ is the y-intercept, so $c = {}^-2$. Substituting in $y = mx + c$ gives the equation $y = \frac{2}{3}x - 2$.

EXERCISE

1 Determine:

i the gradient, and

ii the y-intercept,

of each of the following straight lines with equations:

a $y = 2x - 3$

b $y = 4 - 3x$

c $y = \frac{5}{6}x + 4$

d $y = \frac{x}{3} - 2$

e $y = 3(4x - 1)$

f $y = \frac{5x - 4}{2}$

g $y = 7 - 3x$

h $y = \frac{2(3x - 1)}{3}$

i $y = \frac{4x}{3}$

2 Write the equations of the straight lines given the information:

a gradient 4, y-intercept 6

b gradient $^-2$, point on line $(0, {^-3})$

c gradient $\frac{2}{3}$, points on line $(3, 0)$, $(0, {^-2})$

d gradient $\frac{1}{3}$, y-intercept 0

e gradient 0, point on line $(0, 4)$

3 Determine:

i the gradient,

ii the y-intercept,

of each of the following straight lines with equations:

a $x + y = 5$

b $2x + y = 3$

c $y - 3x = 5$

d $2x - 3y = 6$

e $2y + x - 8 = 0$

f $8 - 4y + 5x = 0$

g $3x + 8y = 24$

h $5y = 3$

Finding the equation of a straight line from a graph

If a graph of a straight line is available, then it is usually possible to find the equation of the line by calculating the gradient, m, from two points on the graph, reading the y-intercept, c, from the graph and substituting these values into the general equation, $y = mx + c$.

Example

Find the equation of each of the straight lines graphed below.

a

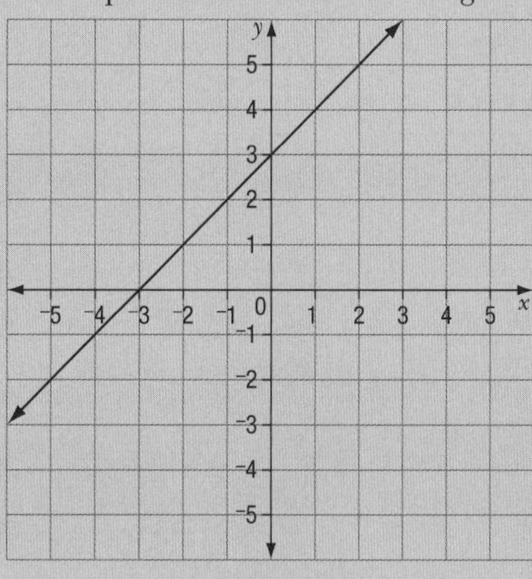

b

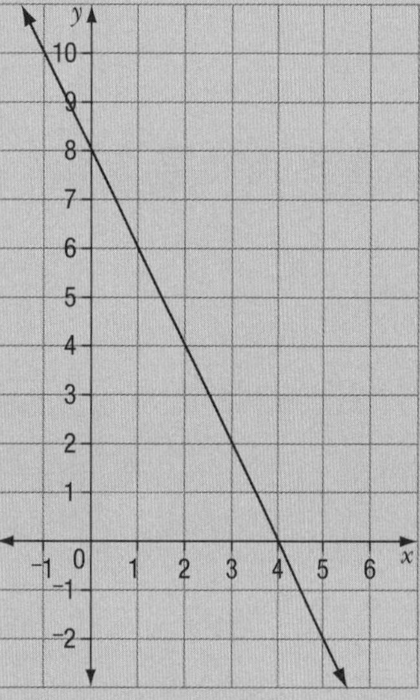

c

Answer

a The gradient is positive, and using the points (⁻3, 0) and (0, 3) the gradient, $m = \frac{\text{rise}}{\text{run}} = \frac{3}{3} = 1$. The y-intercept is $c = 3$ so, substituting in $y = mx + c$, the equation is $y = x + 3$.

b The gradient is negative, and using the points (4, 0) and (0, 8) the gradient, $m = \frac{8}{-4} = -2$. The y-intercept is $c = 8$ so, substituting in $y = mx + c$, the equation is $y = -2x + 8$.

c The gradient is positive, and using the points (⁻3, ⁻4) and (0, 0) the gradient, $m = \frac{4}{3}$. The y-intercept is $c = 0$ so, substituting in $y = mx + c$, the equation is $y = \frac{4}{3}x$.

EXERCISE

1 Find the equation of each of the following straight lines graphed below.

a

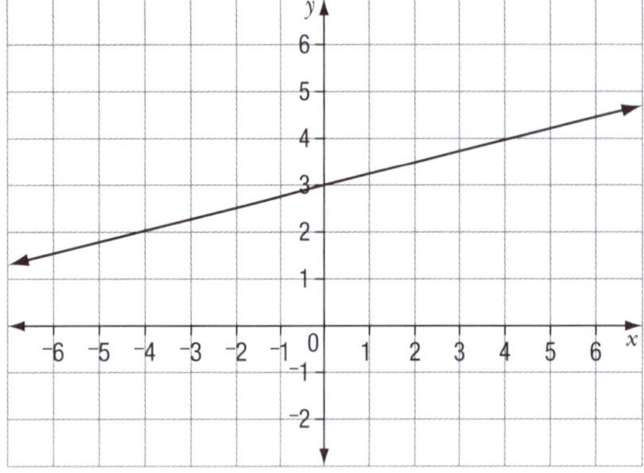

b

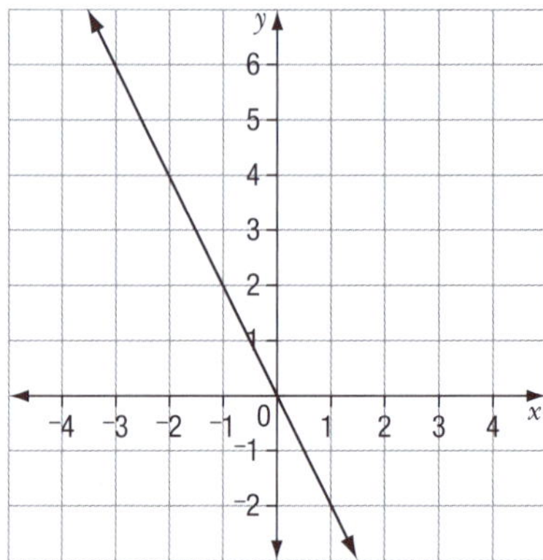

c

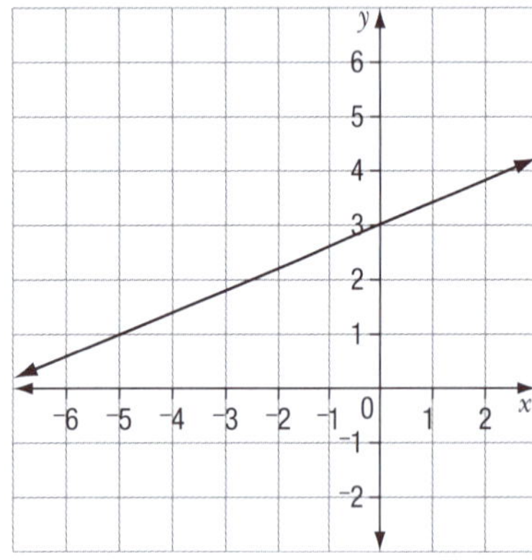

d

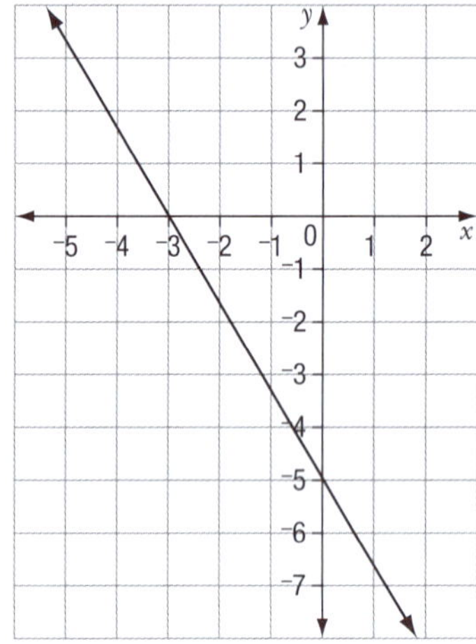

e

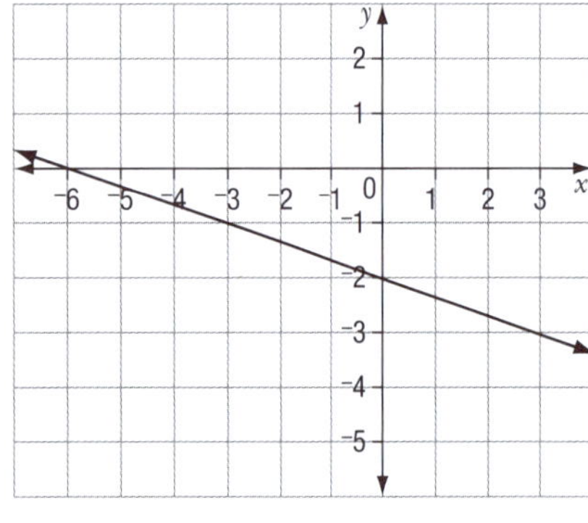

f

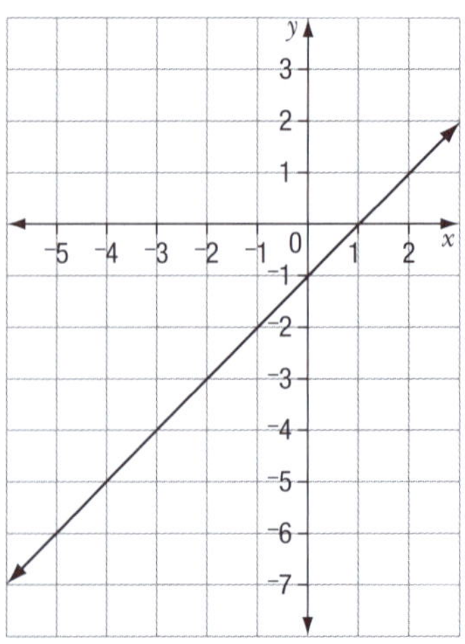

g

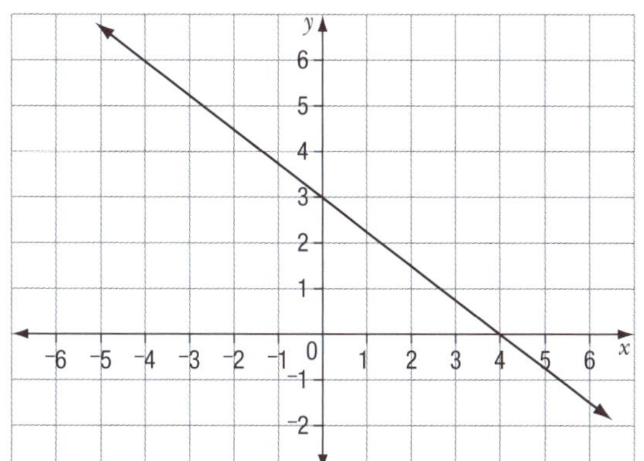

h

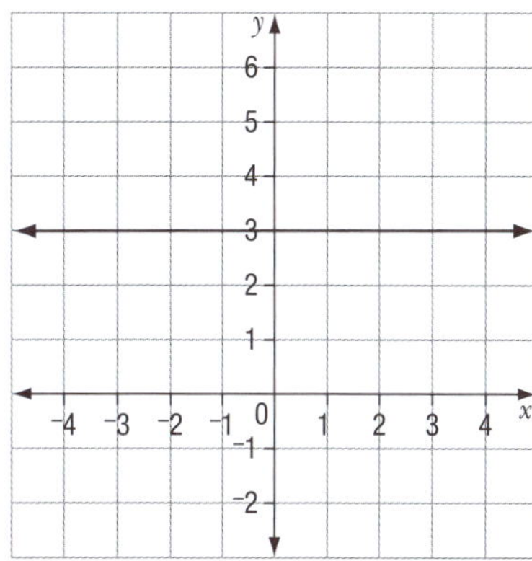

i

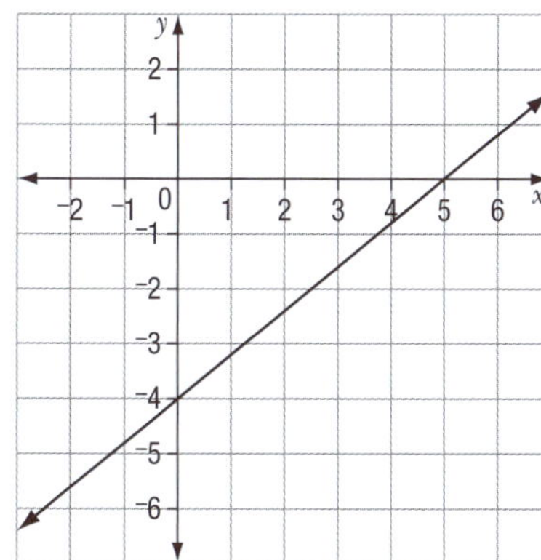

2 Match the equations **i** to **vii** to the straight lines, A to G, graphed opposite.

i $y = 3$

ii $y = 3 - x$

iii $3y = 3x$

iv $x = 3$

v $y = 3x + 3$

vi $y = 3 - 3x$

vii $y = \frac{1}{3}x$

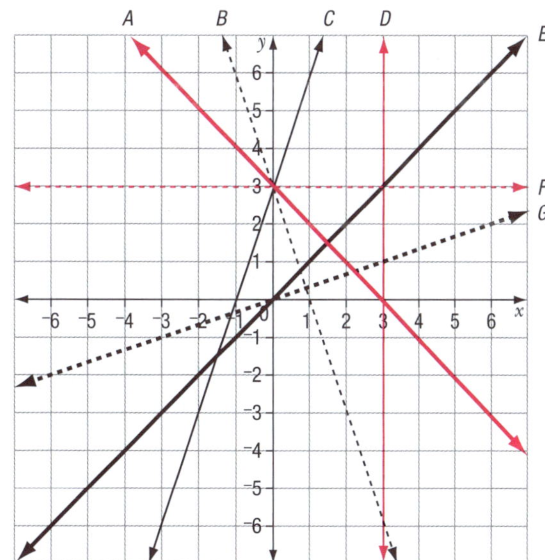

Lesson 18 Finding the equation from the coordinates of two points on a straight line

If we know the coordinates of two points, (x_1, y_1) and (x_2, y_2), on a straight line then we can find the equation of the straight line. To find the equation:

i calculate the gradient using $m = \frac{y_2 - y_1}{x_2 - x_1}$

ii substitute this m value into the general equation for the straight line $y = mx + c$

iii substitute the x and y values from one of the points into $y = mx + c$ to find the value of c.

Example

Find the equation of the straight line that goes through the points (1, 2) and (5, 4).

Answer

i Find the gradient using $m = \frac{y_2 - y_1}{x_2 - x_1}$: $m = \frac{4 - 2}{5 - 1} = \frac{2}{4} = \frac{1}{2}$.

ii Substitute this value into $y = mx + c$: $y = \frac{1}{2}x + c$.

iii Substitute the x and y values from one of the points into $y = \frac{1}{2}x + c$ to find the value of c.

Using the point (1, 2) we will substitute $x = 1$ and $y = 2$ into $y = \frac{1}{2}x + c$ and solve for c.

$2 = \frac{1}{2} \times 1 + c$

$2 = \frac{1}{2} + c$ subtract $\frac{1}{2}$ from both sides of the equation

$2 - \frac{1}{2} = \frac{1}{2} + c - \frac{1}{2}$ collect like terms

$\frac{3}{2} = c$

So the equation is $y = \frac{1}{2}x + \frac{3}{2}$.

Note: We would get the same equation if we substitute (5, 4) into $y = \frac{1}{2}x + c$ to find the value of c.

$4 = \frac{1}{2} \times 5 + c$

$4 = \frac{5}{2} + c$ subtract $\frac{5}{2}$ from both sides of the equation

$\frac{3}{2} = c$

EXERCISE

1 For each of the following, find the equation of the straight line that goes through the two given points.

- **a** (2, 3) and (⁻1, ⁻3)
- **b** (3, 0) and (⁻2, ⁻5)
- **c** (⁻3, 6) and (2, ⁻4)
- **d** (3, ⁻3) and (6, ⁻1)
- **e** (1, 3) and (⁻3, 1)
- **f** (4, ⁻8) and (⁻2, 1)
- **g** (⁻6, ⁻3) and (2, ⁻3)
- **h** (20, 800) and (50, 1100)

Graphing a straight line from an equation

To graph a straight line we need to identify two points on the line, graph these points and rule the line through them.

The equations of straight lines (linear relations) are commonly given in one of the following forms:

i The **gradient–intercept form**: $y = mx + c$ where m is the gradient and c is the y-intercept. For example, $y = 2x - 3$.

To graph a straight line given in this form we can substitute a convenient value for x into the equation to find the y-value (one point) and use the y-intercept $(0, c)$ as the other point.

Example 1

On a set of Cartesian axes, graph the straight line whose equation is $y = 2x - 3$.

Answer

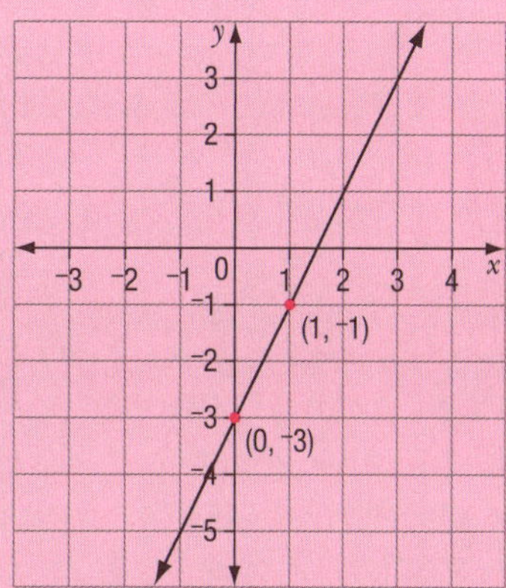

From the equation we can identify the point $(0, {}^{-}3)$ as the y-intercept.

To find another point on the line we need to choose a value for x, say $x = 1$, and substitute this in the equation to find its corresponding y-coordinate:

$y = 2x - 3 = 2 \times 1 - 3 = 2 - 3 = {}^{-}1$.

The point $(1, {}^{-}1)$ will be on the line.

Plot the points $(0, {}^{-}3)$ and $(1, {}^{-}1)$ and rule a line through them..

ii The **intercept form**: $px + qy = r$ (p, q and r are all constant numbers). For example, $3x + 4y = 12$.

If a linear relation is expressed as $px + qy = r$ then it is relatively easy to find the two points that are the intercepts with the axes. Graph these and rule a line through the points.

If we substitute $y = 0$ into the equation and solve for x, then we will have the x-intercept.

If we substitute $x = 0$ into the equation and solve for y, then we will have the y-intercept.

Example 2

On a set of Cartesian axes, graph the straight line whose equation is $3x + 4y = 12$.

Answer

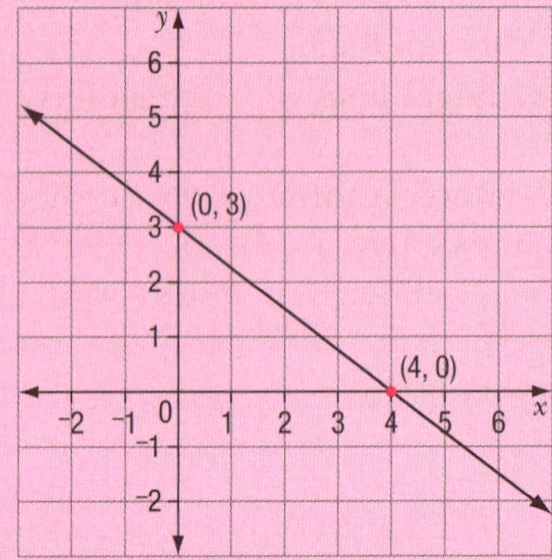

To find the x-intercept: substitute $y = 0$ into the equation.

$3x + 4y = 12$

$3x + 4 \times 0 = 12$

$3x = 12$

$\frac{3x}{3} = \frac{12}{3}$

$x = 4$

The x-intercept is the point (4, 0).

To find the y-intercept: substitute $x = 0$ into the equation.

$3x + 4y = 12$

$3 \times 0 + 4y = 12$

$4y = 12$

$\frac{4y}{4} = \frac{12}{4}$

$y = 3$

The y-intercept is the point (0, 3).

Plot the points (4, 0) and (0, 3) and rule a line through them as shown on the set of axes above.

EXERCISE

1 On a set of Cartesian axes, graph each of the following labelling the two points that you have plotted with their coordinates.

- a $y = 3x + 5$
- b $y = {}^{-}2x + 4$
- c $y = \frac{3}{2}x - 1$
- d $y = {}^{-}3 - \frac{1}{2}x$

2 On a set of Cartesian axes, graph each of the following labelling the two points that you have plotted with their coordinates.

- a $2x - 3y = 6$
- b $x + 2y = 4$
- c $4x + y = 6$
- d $5x - 2y = 8$

3 For each of the following straight lines:

  i Find the coordinates of two points that are on the line.
  ii Graph the straight line on a set of axes.
  iii Label the x- and y-intercepts with their coordinates.
  iv Determine the gradient of the line.

a $2y = x - 5$
b $6 + 3y = x$
c $y = -\frac{x}{3}$
d $2y + x - 4 = 0$

4 To **investigate parallel lines**:

a On a set of axes construct the graph of $y = 2x + 3$.
b Write down the gradient of the graph $y = 2x + 3$.
c On the same set of axes, rule a line that is parallel to the graph of $y = 2x + 3$ that also goes through the point $(0, {}^{-}2)$.
d Find the equation of this new line and compare the equation to $y = 2x + 3$. Which values are the same? Which values are different?
e Graph the straight lines

  i $y = {}^{-}3x - 1$ and ii $y = 4 - 3x$

  on the same set of axes.
  What do you notice about the two straight lines and their equations?
f What can you conclude about the parallel lines? Copy and complete the following:
  Straight lines that are parallel have the same ....

5 The converse statement to the one in Question 4 part f above is also true. Copy and complete the converse statement:
If straight lines have the same gradient then they are ....

6 For each of the following, choose the pair of equations of the straight lines that are parallel.

a i $y = 2x - 3$ ii $y = 5 - 2x$ iii $y - 2x = 5$
b i $3y = x + 6$ ii $y = 3x - 7$ iii $y = \frac{1}{3}x + 2$
c i $2x + 5y = 10$ ii $5x + 2y = 5$ iii $y = {}^{-}2.5x - 3$
d i $y = \frac{4x - 3}{2}$ ii $2x + y = 5$ iii $y = 5 + 2x$
e i $3(x + 2y) = 15$ ii $y = \frac{x - 3}{2}$ iii $4y - 2x = 6$

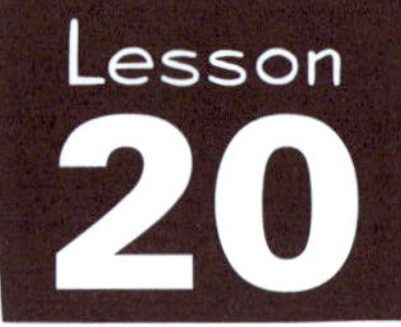

Lesson 20 Modelling everyday problems with linear relations

There are many situations in real life when two quantities (variables) are related in some way, and some of these can be modelled by a linear relation. For example, *weight* and *height*, *distance travelled* and *amount of fuel used*, *cooking time* and *weight of food*.

In previous lessons we have used the **variables** x (horizontal axis) and y (vertical axis) but in practical situations we will use more appropriate labels for the variables.

On the horizontal axis we place the **independent variable** and on the vertical axis we place the **dependent variable**.

For example, if we were representing the relationship between the variables *amount of petrol used* and *distance travelled* we could use P for the *amount of petrol used*, in litres, and d for the *distance travelled*, in kilometres. The variable d is the independent variable (horizontal axis) in this case and the variable P is the dependent variable (vertical axis) because the *amount of petrol used* **depends** on the *distance travelled*.

A graph that could represent the relationship between P and d could look like this.

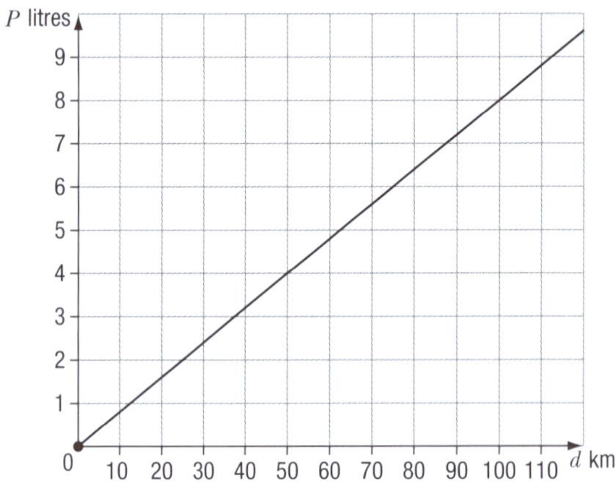

From this graph we can:

i read and interpret a point

Interpret means to explain in terms that apply to the situation. For example, the point (50, 4) means that it takes 4 litres of petrol to travel 50 kilometres.

ii calculate and interpret the gradient

For example, using the points (0, 0) and (100, 8) we can calculate the gradient as $m = \frac{\text{rise}}{\text{run}} = \frac{8}{100} = 0.08$. We could say that the gradient is 0.08 but to interpret the gradient we could say 'the petrol is used at a rate of 0.08 litres per kilometre' or, more usefully, '8 litres per 100 kilometres'.

iii write the equation of the straight line in terms of the variables

For example, the gradient $m = 0.08$ and the intercept with the vertical axis is $c = 0$, so the equation is $P = 0.08d$.

iv find the values of the independent variable that the situation applies to

For example, the graph does not extend into the negative numbers; there is a large filled-in point at (0, 0). This means that this situation only applies to values of d greater than or equal to zero, written as $d \geqslant 0$.

EXERCISE

1 A car is travelling along a highway at a constant speed. The distance, d kilometres, the car has travelled after t minutes is graphed at right.

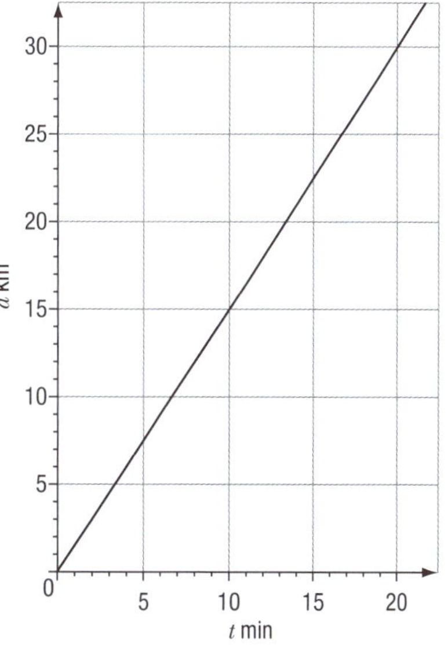

a How far has the car travelled after 10 minutes?

b How long does it take the car to travel 24 kilometres?

c Interpret the point (6, 9) in terms of distance travelled and time.

d If the gradient of the graph represents the speed of the car, what is the speed of the car in:

i kilometres per minute (km/min)

ii kilometres per hour (km/h)?

e Which of the variables, *distance travelled* (d) or *time* (t), is the dependent variable in this situation? Why?

f Find the equation of the graph, writing your equation in the form $d = mt + c$ where m is the gradient and c is the intercept with the vertical axis.

> **Remember**
> Interpret means to explain in terms that apply to the situation.

2 The cooking time for a roast chicken cooked at 180°C is given as '20 minutes plus 5 minutes for each 100 grams of weight'. The variables in this situation are C, the *cooking time* in minutes, and w, the *weight of the chicken* in kilograms. Here is a graph for this situation.

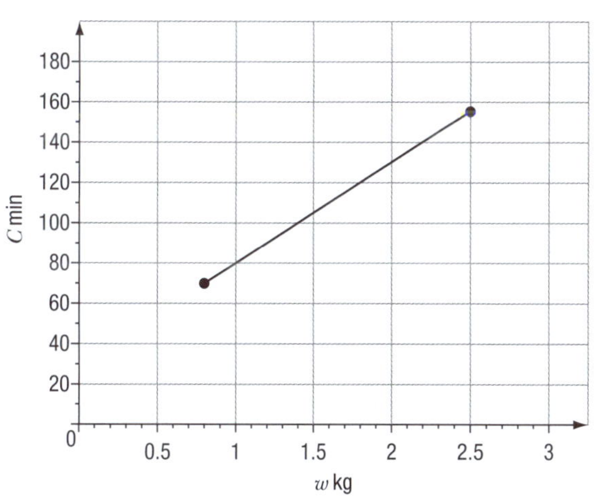

a Which of the variables is the dependent variable in this situation?

b How long will it take to cook a chicken that weighs 1.4 kg?

c A chicken took 80 min to cook. What was the weight of the chicken?

d Why do you think the graph starts at (0.8, 70) and finishes at (2.5, 155)?

e What is the gradient of the graph above? What does this mean in terms of cooking time and weight of chicken?

f Find the equation of the graph, writing it in terms of the variables C and w.

3 A 30-centimetre high candle burns down at a constant rate. A graph showing the height of the candle, H centimetres, after a certain time, t minutes, is given at right.

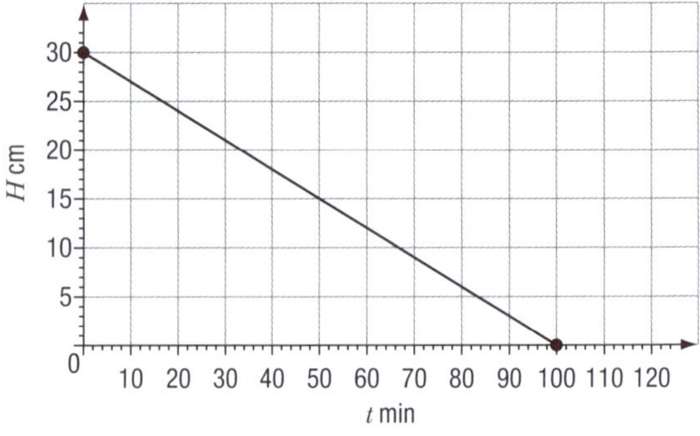

a Interpret the point (0, 30).

b Interpret the point (50, 15).

c How long does it take for the candle to burn down to one-third of its original height?

d What percentage of the candle's original height remains after 20 minutes?

e Find and interpret the gradient of the graph.

f Write the equation of the graph in terms of H and t.

g What values of t apply in this case? Write this as .... $\leqslant t \leqslant$ ....

4 Anton sells mobile phones, and he is paid a base salary of K200 plus K20 for each phone he sells per week. A graph showing the salary, S kina, he will receive per week if he sells n mobile phones is given at right.

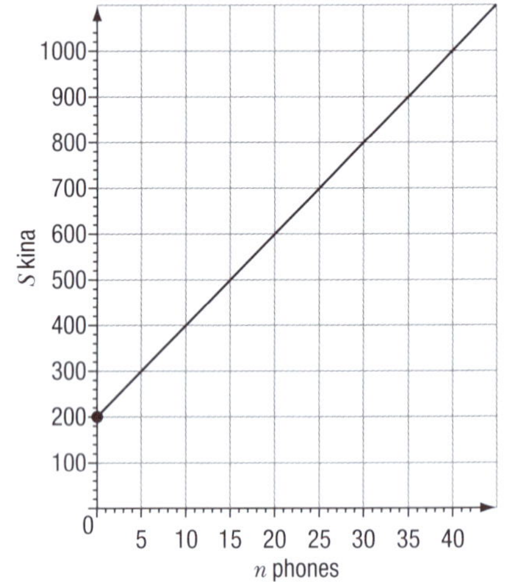

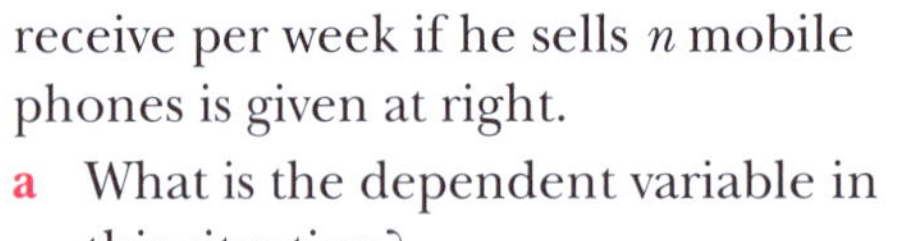

a What is the dependent variable in this situation?

b How much will Anton earn if he sells 16 mobile phones in a week?

c Interpret the point (22, 640).

d Interpret the point (0, 200).

e Find, and interpret, the gradient of the graph.

f State the equation of the graph in terms of the variables S and n.

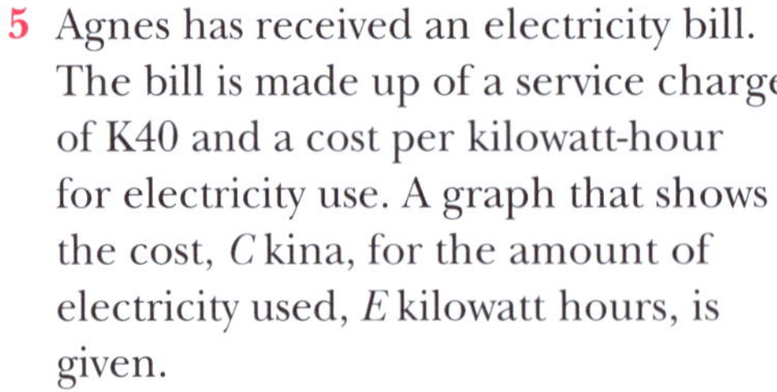

5 Agnes has received an electricity bill. The bill is made up of a service charge of K40 and a cost per kilowatt-hour for electricity use. A graph that shows the cost, C kina, for the amount of electricity used, E kilowatt hours, is given.

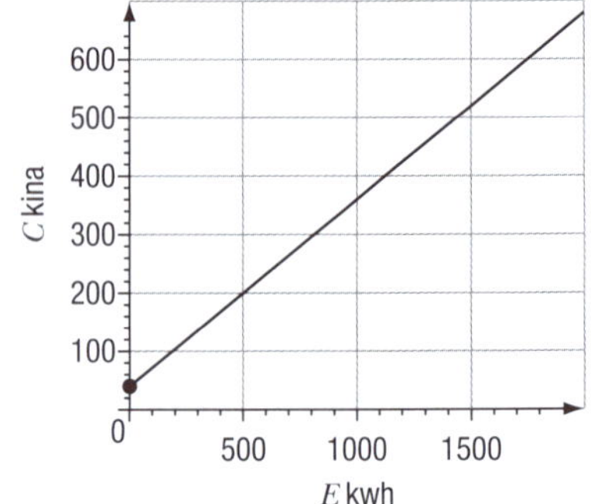

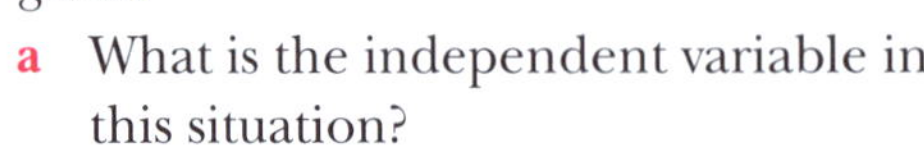

a What is the independent variable in this situation?

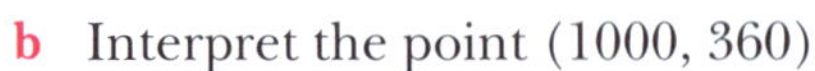

b Interpret the point (1000, 360).

c Calculate the gradient of the graph and interpret this value.

d Write the equation for the graph in terms of the variables C and E.

6 The value of a machine bought for K2400 decreases by 20% of the purchase price each year. A graph that shows the value, V kina, of the machine after n years is given.

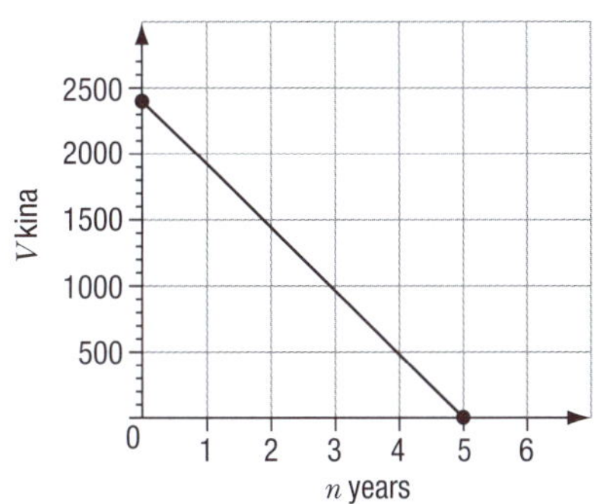

- **a** Interpret the point (2, 1440).
- **b** Interpret the point (5, 0).
- **c** Find, and interpret, the gradient of the graph.
- **d** Write the equation of the graph in terms of the variables.
- **e** Write the restrictions on the independent variable: .... $\leqslant n \leqslant$ ....

7 Maureen has an investment that pays 8% simple interest for the term of the investment. She was given the following graph that shows the growth of her investment over five years.

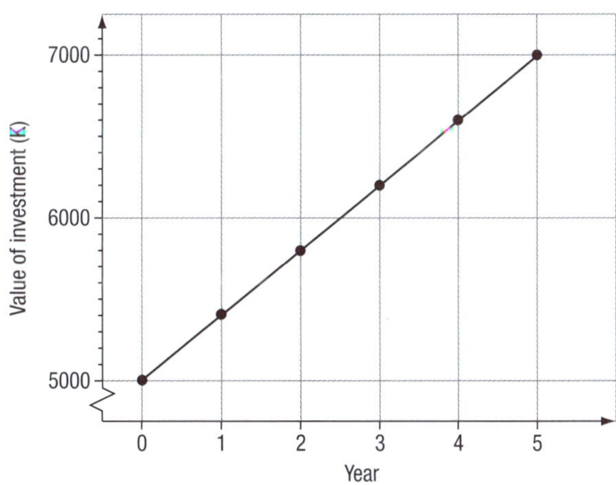

- **a** How much did Maureen invest?
- **b** What is the value of Maureen's investment after 5 years?
- **c** Interpret the point (1, 5400).
- **d** Write the equation of the straight line in terms of the variables, V (value of the investment) and t (time in years).

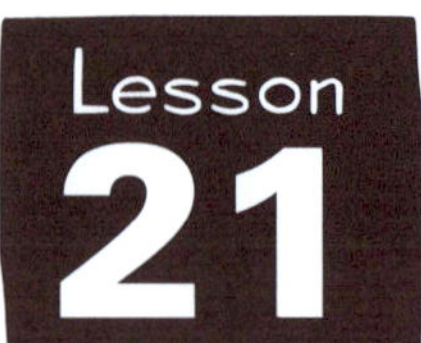

Lesson 21 Graphing linear relations that model everyday problems

Sometimes applying a mathematical model to an everyday problem provides an easier path to the answer. A graph can also be helpful in many situations.

Example

It costs Segana a set amount, K54 per week, plus K6 per bilum, to produce bilums for the market. She sells each of the bilums for K24 at the market. Segana needs to know how many bilums she should make so that she makes a profit.

a If the cost equation for the bilum is $C = mx + 54$ to produce x bilum, what is the value of m?
The revenue equation for bilum sales is $R = 24x$ when Segana sells x bilum.

b On a set of axes, graph the two equations placing x on the horizontal axis and C and R on the vertical axis.

c Find the point of intersection of the two graphs. This point is called the 'break-even' point. How many bilums does Segana need to make, and sell, before she is making a profit?

d Using the fact that Profit = Revenue – Costs, write an expression for Segana's profit, P, in terms of x.

e How much profit will Segana make if she makes and sells 10 bilums in a week?

Answer

a m represents the material costs per bilum, so $m = 6$. The cost equation is $C = 6x + 54$ and the revenue (amount Segana receives when she sells the items) equation is $R = 24x$.

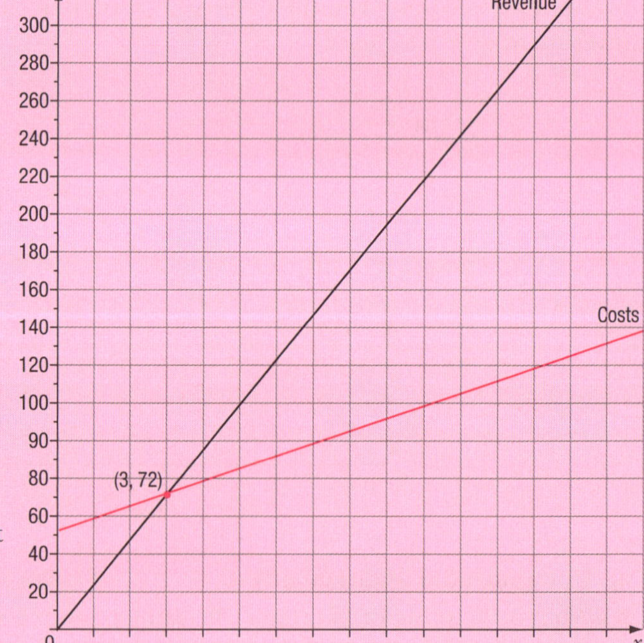

b See graph at right.

c The point of intersection of the two graphs is (3, 72) from the graph. This can be confirmed by solving $R = C$.
$24x = 54 + 6x$: subtract $6x$ from both sides
$18x = 54$: divide both sides by 3
$x = 3$
When $x = 3$; $R = 24 \times 3 = 72$.
The **break-even point** gives us the number of bilums that Segana needs to make and sell so that her costs are covered. Segana will need to make and sell 3 bilums to cover her costs for the week; if she makes and sells more than 3 bilums then she will be making a profit.

d Profit = Revenue − Costs

so $P = 24x - (54 + 6x)$

$= 24x - 54 - 6x$

$= 18x - 54$

e If $P = 18x - 54$ then substituting $x = 10$ will give a profit of $P = 18 \times 10 - 54 = 126$. Segana will make a profit of K126 if she makes and sells 10 bilums in a week.

EXERCISE

1 A taxi has a flag-fall (the set cost that is charged) of K2.20 and charges K1.50 per kilometre.

- a Write an equation of the form $C = mx + 2.2$ to represent the cost of a journey of x kilometres in the taxi.
- b Graph your equation for x values from 0 to 30 kilometres.
- c State the gradient of the straight line and interpret this value.
- d Use your graph to find the cost of a journey of 20 kilometres.
- e Approximately how far could you travel in the taxi for K15?

2 A car uniformly changes its speed from zero to 40 kilometres/hour in 15 seconds and then travels at a constant speed of 40 kilometres/hour for a further 20 seconds. The car then slows down uniformly to zero in 10 seconds.

- a What are the two variables and which is the independent variable?
- b What does 'uniformly' mean?
- c Construct a graph that shows the information above.
- d Find the gradient of each of the lines on your graph and interpret these values.

3 A furniture removal company charges K100 plus K80 per hour for the hire of a van and two men.

- a How much would it cost to hire the van and two men for three hours?
- b How much would it cost to hire the van and two men for six hours?
- c Write an equation for the cost, C kina, of hiring the van and two men for x hours.
- d Graph your equation on a suitably labelled set of axes for x values from 0 to 10 hours.
- e Martha is charged $200 to hire the van and two men. For how long did she hire the van and two men?

4 A candle has burnt down from 30 centimetres to 9 centimetres in 3 hours 30 minutes.

- a Assuming that the candle has burnt down uniformly, construct a graph for this situation choosing appropriate labels and units for the variables.
- b After how long was the candle half of its original size?
- c Find the gradient of the graph and interpret it.
- d How long would it take for the candle to completely burn out?

5 Luther's weekly pay is made up of a retainer (the amount he receives even when he does not sell any computers) and a set amount for each computer that he sells. For the last two weeks his weekly pay has been K428 when he sold 3 computers and K608 when he has sold 8 computers.

a What are the two variables in this situation, and which is the dependent variable?

b Using the information given, write down two ordered pairs.

c Plot the ordered pairs and construct a graph for the situation.

d Find the gradient of the straight line joining the two points and interpret this value.

e How much would Luther earn in a week if he sold 15 computers?

f Could Luther earn exactly K1000 in a week? Explain.

6 Konio has two recent electricity bills. One was for K162.44 for the use of 262 kWh (kilowatt hours) of electricity and the other was for K176.08 for the use of 284 kWh of electricity. Konio plans to graph the electricity use (kWh) against the cost (K).

a Which of the two variables, *electricity use* and *cost*, is the dependent variable?

b Write the information from each of the two bills as an ordered pair. Each of these ordered pairs will represent a point on a set of axes.

Remember

The dependent variable is placed on the vertical axis.

c Plot the two points on a carefully constructed set of axes and rule a straight line through the points. What assumption have we made when we rule a straight line through the two points?

d Find the equation of the straight line and write this in terms of the variables.

e State the gradient of the line and interpret this value.

f Use the equation to estimate the cost of using 300 kWh of electricity.

g Peter has an electricity bill for K201.50. How much electricity has Peter used in that billing period?

7 Angela produces doughnuts to sell at the weekly market. Angela has overhead costs of K22 and it costs 15 toea for ingredients for each doughnut. She sells each doughnut for 70 toea at the market.

a If the cost equation, C toea, for the doughnuts is $C = mx + 2200$ to produce x doughnuts, what is the value of m?

b The revenue equation for doughnut sales is $R = px$ when she sells x doughnuts. What is the value of p?

c On a set of axes, graph the two equations placing x on the horizontal axis and C and R on the vertical axis.

d Find the point of intersection of the two graphs. This point is called the **break-even** point. How many doughnuts does Angela need to make, and sell, before she is making a profit?

e Using the fact that Profit = Revenue – Costs, write an expression for Angela's profit, P, in terms of x.

f How much profit will Angela make if she makes and sells 80 doughnuts in a week?

REVISION AND ASSESSMENT

Directed numbers, indices and basic algebra

Multiple-choice questions

1 If $^-4 + 14 + ^-6 - ^-5$ is evaluated the answer will be:

A $^-1$ **B** 9 **C** 13 **D** $^-19$

2 If $\frac{24}{^-3} - \frac{5}{6} \times ^-54$ is evaluated the answer is:

A 37 **B** $^-53$ **C** 27 **D** 53

3 When $5^0 + 7^1$ is evaluated the answer is:

A 7 **B** 8 **C** 12 **D** 2

4 When $(2m)^2 \times (m^3)^{-2}$ is simplified the answer is:

A 1 **B** $2m^{-4}$ **C** $4m^{-2}$ **D** $\frac{4}{m^4}$

5 When 0.000 000 1 is multiplied by 1 100 000 the answer, in standard form, is:

A 1.1×10^2 **B** 1.1 **C** 1.1×10^{-1} **D** 1.1×10^{-2}

6 $4\sqrt{2}$ written as a power of 2 is:

A $\sqrt{8}$ **B** $2^{\frac{1}{2}}$ **C** $2^{\frac{3}{2}}$ **D** $2^{\frac{5}{2}}$

7 Which one of the following statements is **not** true for the expression $4p^3q + 4p^2q + 3pq - 3p - 4$?

A There are no factors (except 1) that are common to all terms in this expression.

B The constant term is $^-4$.

C The terms $3pq$ and ^-3p are like terms.

D The co-efficient of the third term is 3.

8 When expanded and simplified, the expression $2(x - 4) - 3(2 - x)$ is equal to:

A $5x - 14$ **B** $^-x - 10$ **C** $5x + 2$ **D** $3x - 10$

9 The solution to the equation $5x - 4 = x$ is:

A $\frac{4}{5}$ **B** $\frac{2}{3}$ **C** 1 **D** $^-1$

10 When the formula for the surface area of a sphere, $A = 4\pi r^2$, is transposed to make r the subject, the formula will be:

A $r = \sqrt{\frac{4A}{\pi}}$ **B** $r = \sqrt{\frac{A}{4\pi}}$ **C** $r = \frac{1}{4}\sqrt{\frac{A}{\pi}}$ **D** $r = \frac{A}{8\pi}$

Graphs

Multiple-choice questions

1 On a Cartesian set of axes, in which quadrant is the point (⁻2, 6)?

A 1st B 2nd C 3rd D 4th

Questions 2 and 3 refer to the graph below.

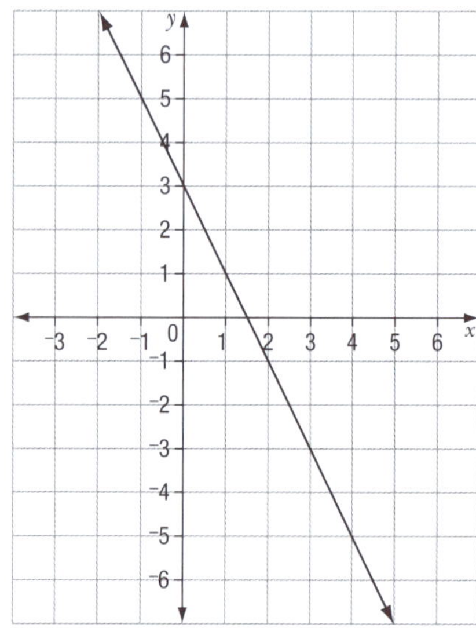

2 The gradient of the straight line graphed above is:

A ⁻2 B 2 C ⁻3 D ⁻1.5

3 Which one of the following points is **not** on the line?

A (2, ⁻1) B (4, ⁻5) C (3, 0) D (⁻1, 5)

4 The gradient of the straight line going through the points (4, ⁻1) and (⁻2, ⁻3) is:

A $\frac{1}{3}$ B $\frac{2}{3}$ C 3 D $\frac{-3}{2}$

5 The equation of the linear function that has a gradient of ⁻2 and y-intercept of 7 is:

A $y = 7x - 2$ B $y + 2x + 7 = 0$ C $y = 7x - 2$ D $2x + y = 7$

6 The line $3x - 4y = 24$ has x-intercept:

A (8, 0) B (3, 0) C (⁻8, 0) D (⁻6, 0)

7 The linear function with equation $8x + 3y = 12$ has a gradient of:

A 8 B 4 C $-\frac{8}{3}$ D $\frac{-3}{4}$

8 Which one of the following is the graph of the straight line with equation $x + y = 3$?

A

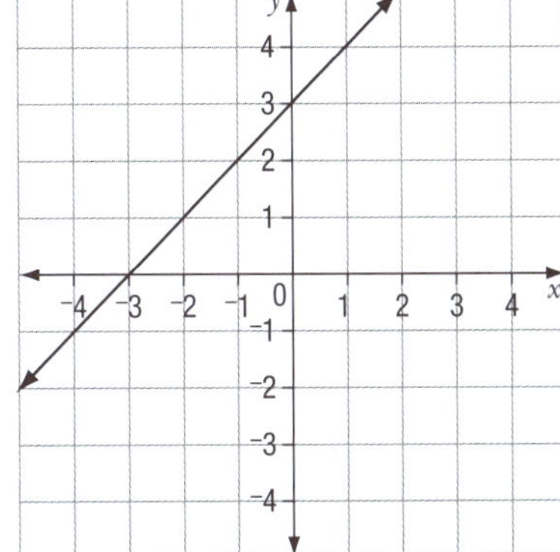

B

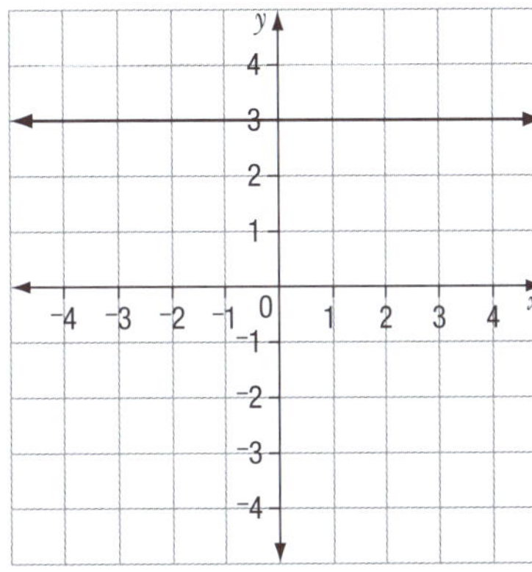

C

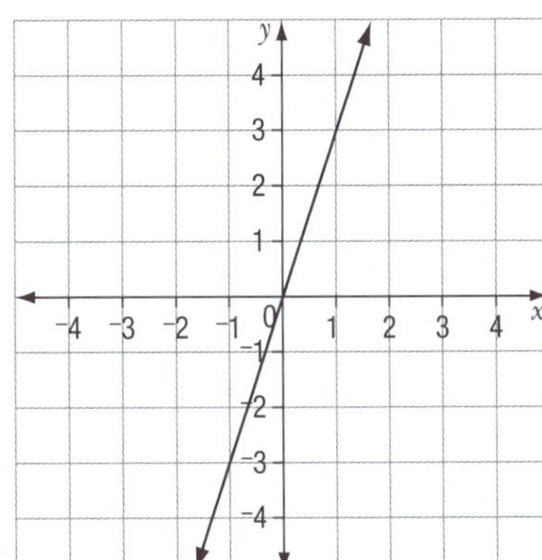

D

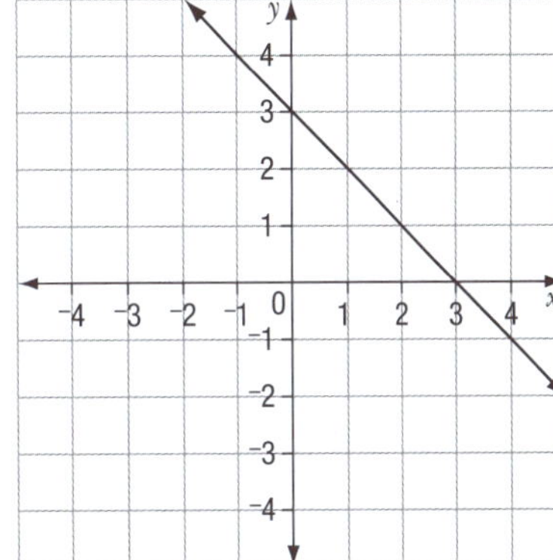

9 Which one of the following linear functions will be parallel to the linear function $2y = 5 - 8x$ when they are graphed?

A $y = 3 - 8x$ **B** $x + y = 8$ **C** $y = 4x + \frac{5}{2}$ **D** $y = 6 - 4x$

10 The equation of a straight line that passes through the point (3, 1) and is parallel to the y-axis is:

A $y = 1$ **B** $y = 3x$ **C** $x = 3$ **D** $y = \frac{1}{3}x$

Directed numbers, indices and basic algebra

Short answer questions

1 Evaluate each of the following.

a $9 - 5 + {}^-4 - ({}^-6)$

b $\frac{4 \times {}^-15}{{}^-6}$

c $3 \times {}^-4 + 5(3 - 5)$

d $\frac{6 \times {}^-7}{3} - 5(4 - 7)$

2 Evaluate each of the following.

a $({}^-3)^0$

b 4^{-2}

c $9^{\frac{1}{2}}$

3 Write in standard form:

a 0.000 005 68 b 399 500 000 000

4 Evaluate $(1.2 \times 10^6) \times (9 \times 10^{-3})$ writing your answer in standard form.

5 Use the index laws to simplify each of the following.

a $\frac{3x^4 \times 2x^2}{x^3}$ b $(3a^2b)^3$ c $4m^5 \div (2m)^{-3}$

6 a Write $\sqrt{8}$ as a power of 2.

b Write $\frac{6^{-2}}{3^{-3}}$ as a fraction.

c Write 27 as a power of 3, and hence evaluate $27^{\frac{2}{3}}$.

7 Expand and simplify:

a $3(a + 3) + 4(2 - a)$ b $2(3m - 5) - 5(12 - 7m)$

8 Factorise fully:

a $6a^2 + 9a$ b $^-10p^2q - 2p$

9 Solve the following equations.

a $2x - 4 = 6$ b $19 - 4(y + 1) = 3$

c $\frac{x-3}{6} - 5 = 0$ d $5y - 4 = 3y + 7$

e $4x - 3(x + 1) = 9x - 35$ f $\frac{4y-3}{3} = 5 - y$

10 Using the formula for $s = ut + \frac{1}{2}at^2$:

a find the value of s if $u = 5$, $t = 3$ and $a = 4$

b transpose the formula to make a the subject.

Graphs

Short answer questions

1 Two linear functions are graphed on the set of axes at right. State the coordinates of:

a the x-intercept of line A

b the y-intercept of line B

c the point of intersection of lines A and B

d a point in the 2nd quadrant that is on line A

e a point in the 4th quadrant that is on line B

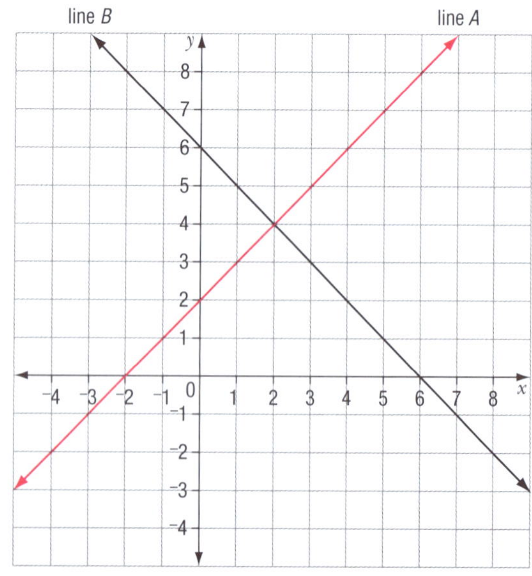

2 Find the gradient of the following linear functions.

a

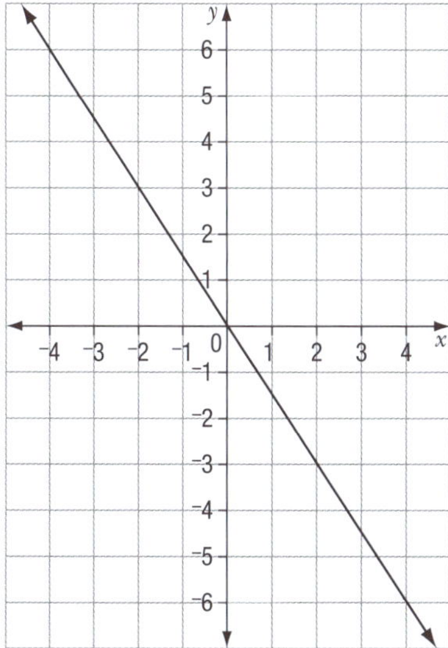

b

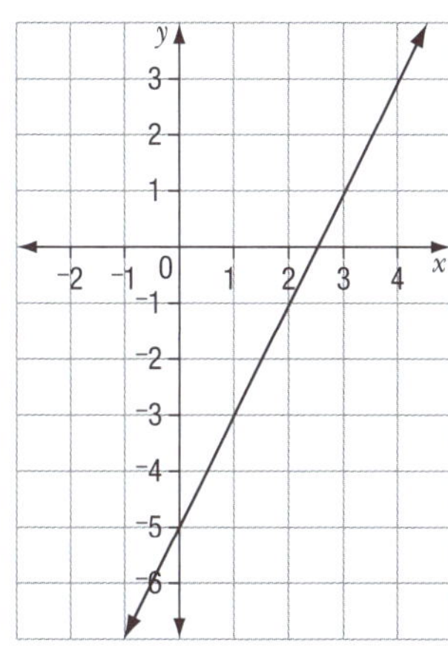

c

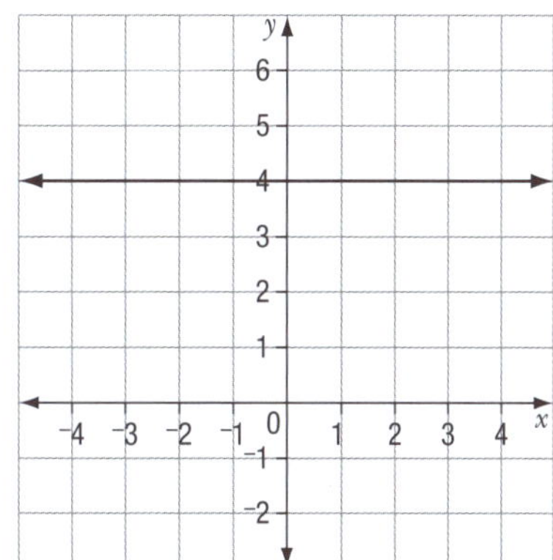

3 On the set of axes at right, construct a straight line with a gradient of $\frac{2}{3}$ that goes through the point $(0, ^-2)$.

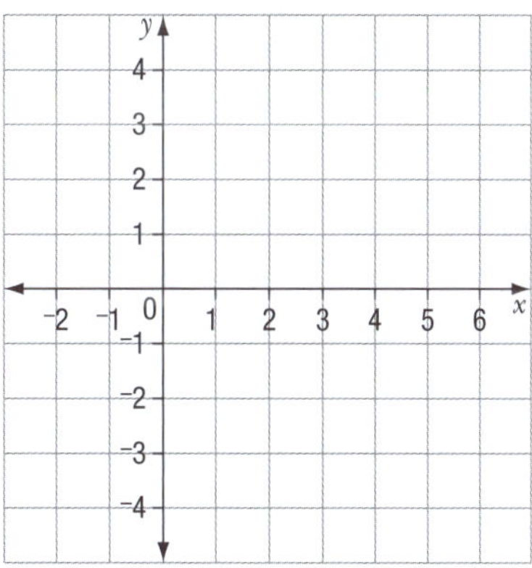

4 Find the gradient of the straight line that goes through the points:

a $(^-5, 3)$ and $(1, 1)$ b $(6, 2)$ and $(6, 7)$

5 For each of the following linear equations, find:

i the gradient ii the y-intercept

a $y = 7x - 4$ b $2x - y = 13$ c $8y + 5x = 16$

6 For each of the following:

i State two points that are on the line.

ii Plot the two points on a set of axes and rule a straight line through them.

Use the same set of axes for both lines.

a $y = 1.5x - 3$ b $5x + 4y = 10$

7 State the equation of the line that goes through the origin and that also is parallel to the straight line $y = 1.5x - 3$.

8 Find the equations of the straight lines A, B, C and D graphed on the set of axes.

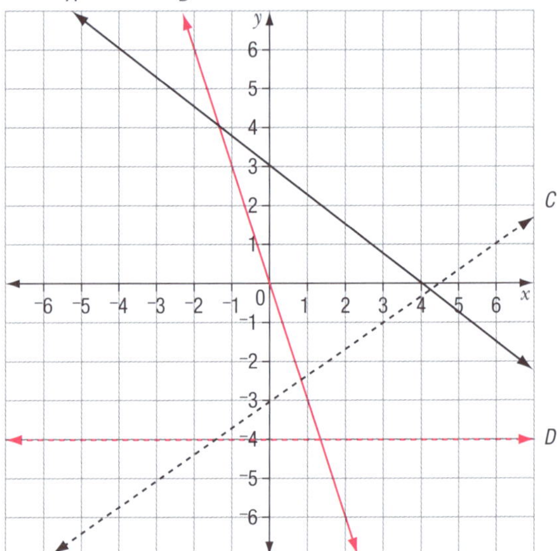

9 For each of the following, find the equation of the straight line that goes through the given points.

a (6, 5) and (⁻2, 1) b (0, 4) and (7, 0) c (2, 5) and (⁻3, 5)

10 A pier is built on a tidal river. The graph below shows the depth of water below the pier, measured over a 24-hour period.

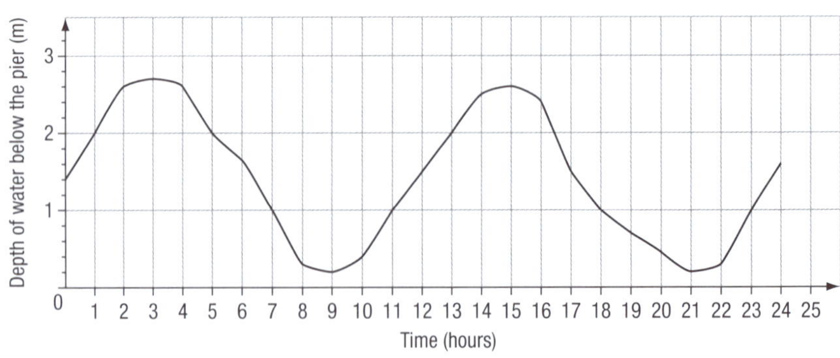

a What is the greatest depth of the water below the pier?

b How far below the pier is the water at high tide?

c For how long in the 24-hour period is the water less than 1 metre below the pier?

d If time 0 hours is 8 a.m., what time of the day is the first high tide?

OPTION A: ALGEBRA AND GRAPHS

If this option is chosen then it will take the remaining five weeks of the Functions and graphs unit, and will be assessed by an individual directed investigation involving algebra and graphs.

Topics covered in this option include:

- finding the line-of-best-fit on a graph drawn from experimental data
- constructing and interpreting scatter graphs
- rates of change
- calculating and interpreting quantities where the rate of change is constant
- comparing different rates of change
- constructing and comparing different graphs
- producing and interpreting story graphs.

Finding the line-of-best-fit for a graph drawn from experimental data

Lesson 1

Many important relationships between two quantities have been discovered by recording and graphing the results from an experiment.

For example, in 1660 Robert Boyle did experiments involving the compression of gases that led to what is today called **Boyle's law**.

Boyle carried out an experiment in which he applied pressure to a closed volume of gas at a constant temperature. The pressure compressed the gas, and he measured the resulting volume. He would have obtained results similar to those in this table.

Pressure (kilopascals)	1	1.2	1.4	1.6	1.8	2
Volume (cubic metres)	4	3.3	2.86	2.5	2.2	2

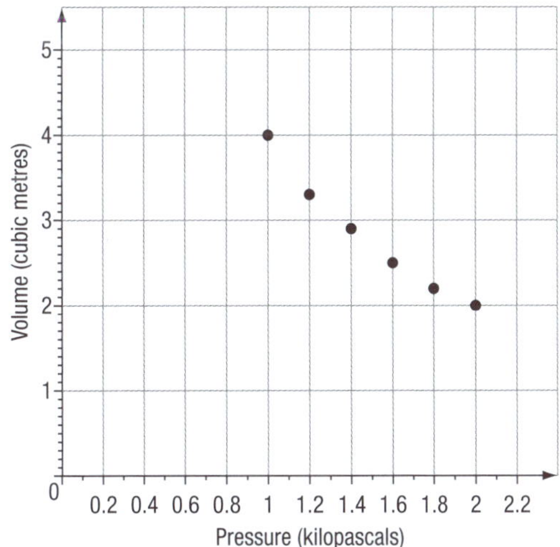

For the graph of these results we have used the variable *volume* as the **dependent variable** because for this experiment the *volume* depends on the *pressure* applied.

The variable *pressure* is placed on the horizontal axis as this is the **independent variable**.

The points on the graph below are clearly not in a straight line (not collinear).

A **line-of-best-fit** is a line (straight or curved) that appears to fit the general pattern of the points plotted. If the points show a definite pattern then the line-of-best-fit would be an accurate indication of the relationship between the two variables, and a rule could be found for this relationship.

Predictions can be made from the line or rule and these can be used to further test the relationship between the variables.

If a **line-of-best-fit** is applied to the points on the graph on the right we could have used this to predict that, for example, the volume of the gas would be 3.2 cubic metres when the pressure applied is 1.25 kilopascals.

From an experiment like this the rule for the relationship between the volume and the pressure of a gas was found and stated as:

For a gas in a closed container at a constant temperature, 'the product of the pressure and volume is a constant'. This is Boyle's law.

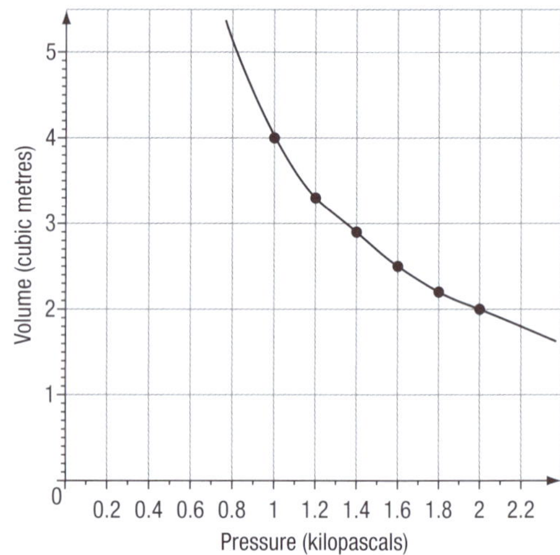

EXERCISE

1 The volume of a cylinder of height 10 cm is investigated, and the results for varying diameters are recorded and graphed.

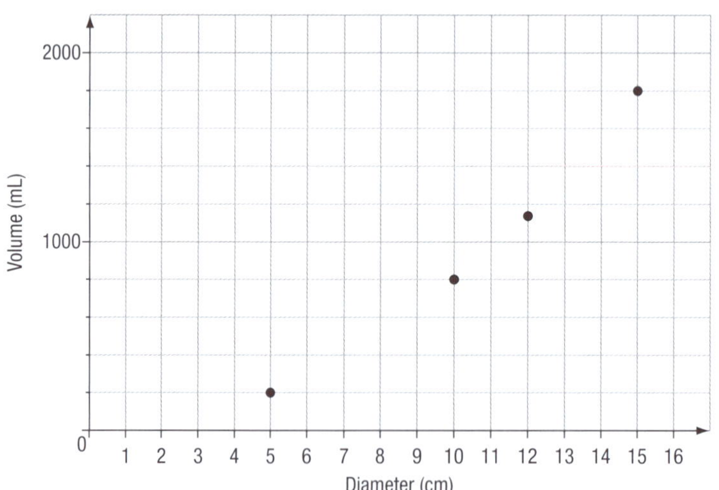

a What is the dependent variable in this experiment?

b Would the graph go through the point (0, 0)?

c Do the points on the graph appear to be linear?

d Draw a line-of-best-fit for the points and use it to predict the volume of a cylinder of diameter:

i 7 cm ii 14 cm.

Remember

The dependent variable (vertical axis) is determined when a particular value of the independent variable (horizontal axis) is known.

2 For a truncated cone the change in volume as the depth increases is investigated. The results are recorded and graphed.

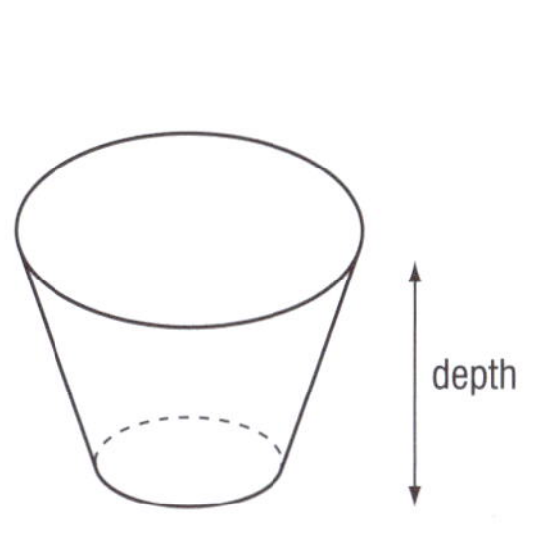

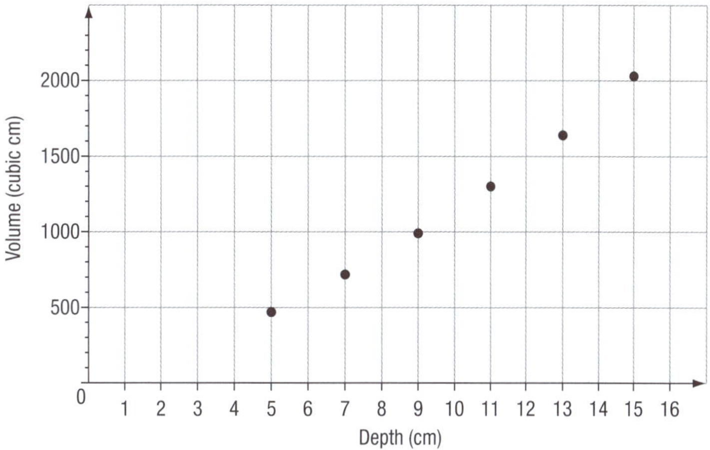

a What is the independent variable in this experiment?

b Would the graph go through the point (0, 0)?

c Do the points appear to be linear?

d Draw a line-of-best-fit through the points and use it to predict the volume of a truncated cone of depth:

i 7 cm   ii 10 cm.

3 This experiment involves timing the period (time to swing from one side to the other and then back again) of pendulums of different lengths. The time was taken for five periods and then the average per period was recorded. The results are given below.

Length (cm)	5	10	15	20	25	30	35	40
Period (s)	0.45	0.63	0.78	0.90	1.00	1.10	1.18	1.27

a What are the two variables in this experiment?

b Which of the variables is the dependent variable?

c Graph the points on a carefully constructed set of axes.

d Should the point (0, 0) be included in this graph?

e Do the points appear to be linear?

f Draw a line-of-best-fit for the points and use it to predict the period of a pendulum of length:

i 3 cm   ii 28 cm.

g What would be the length of a pendulum if it had a period of 0.7 s?

4 The braking distance (the time from the first reaction to complete stop) for a particular model of car has been recorded for the car travelling at different speeds. The results are given in the table below.

Speed (km/h)	50	60	80	100	110
Braking distance (m)	24	32	52	76	90

a What are the two variables in this experiment?

b Which of the variables is the dependent variable?

c Graph the points on a carefully constructed set of axes.

d Should the point (0, 0) be included in this graph?

e Do the points appear to be linear?

f Draw a line-of-best-fit through the points and use it to predict the braking distance of a car travelling at:
   i 90 km/h   ii 30 km/h.

g At what speed would the car be travelling if it took 60 m to stop?

5 Kala has asked his bank the amount that will still be owing on his ten-year, K10 000 reducing balance loan after 1, 3, 4 and 6 years. He was given the figures below.

Year	1	3	4	6
Amount owing (K) at the end of the year	9259	7629	6734	4765

a What are the two variables in this information?

b Which of the variables is the dependent variable?

c Graph the points on a carefully constructed set of axes.

d What are the end points on this graph?

e Do the points appear to be linear?

f Draw a line-of-best-fit through the points and use it to predict the amount still owing after:
   i 2 years   ii 8 years.

g After approximately what time will Kala still owe K5000 on the loan?

6 Robert Hooke was a scientist who, in 1660, discovered a relationship between the extension of an elastic body (a spring is an elastic body) and the force applied to it. An experiment involves the application of varying weights to a spring and measurement of the length that the spring stretches, called the extension. The results are tabled below.

Weight (g)	100	200	300	400	500
Extension (mm)	20	41	60	81	102

a What are the two variables in this experiment?

b Which of the variables is the dependent variable?

c Graph the points on a carefully constructed set of axes.

d Should the point (0, 0) be included in this graph?

e Do the points appear to be linear?

f Draw a line-of-best-fit through the points and use it to predict the extension of the spring for an applied weight of 250 g.

g What weight has been applied if the spring has been extended 5 cm?

h Do you think this relationship would apply for much larger weights? What would happen to the spring if, for example, a 5 kg weight was applied?

7 Pauline's baby girl, Rose, was born on 1 March and she has weighed Rose on the first of the month for the first 8 months of her life. The weights are recorded below.

Month	0	1	2	3	4	5	6	7	8
Weight (g)	3200	4120	4800	5410	6005	6800	7100	7500	8000

a What are the two variables in this experiment?

b Which of the variables is the independent variable?

c Graph the points on a carefully constructed set of axes.

d What does the point (0, 3200) mean?

e Do the points appear to be linear?

f Draw a line-of-best-fit through the points and use it to predict Rose's weight at 10 months.

g In which month does Rose double her birth weight?

h Do you think that you would be able to use your line-of-best-fit to predict Rose's weight at 36 months? Explain.

Constructing and interpreting scatter graphs

Lesson 2

A scatter graph (sometimes called a scatter plot) is used to compare two numerical variables. The form of the points plotted on the scatter graph will indicate whether there is a relationship between the two variables.

For example, the relationship between *weight* and *height* of a football team is being investigated. We would expect there to be a fairly strong association between the variables *weight* and *height* as it is generally thought that the taller the person, the more they will weigh.

Both *weight* and *height* are numerical variables as a number is recorded when each of the variables is measured.

The height, in centimetres, and weight, in kilograms, of each of the 15 players in the team is recorded and these values form a coordinate pair for each of the players.

Player	Height	Weight	Player	Height	Weight	Player	Height	Weight
1	173	87	6	171	89	11	190	115
2	173	90	7	177	104	12	187	106
3	190	114	8	189	111	13	180	96
4	172	85	9	178	95	14	182	105
5	171	93	10	179	97	15	185	98

Before a scatter graph is constructed you need to establish which of the variables is the **independent variable** and which is the **dependent variable**.

In this case we would assume that weight *depends* on height and so **weight** is the dependent variable (vertical axis) and **height** is the independent variable (horizontal axis).

The points are plotted as coordinate pairs (height, weight) for each of the individuals in the investigation. The 15 points are plotted below to form a **scatter graph**.

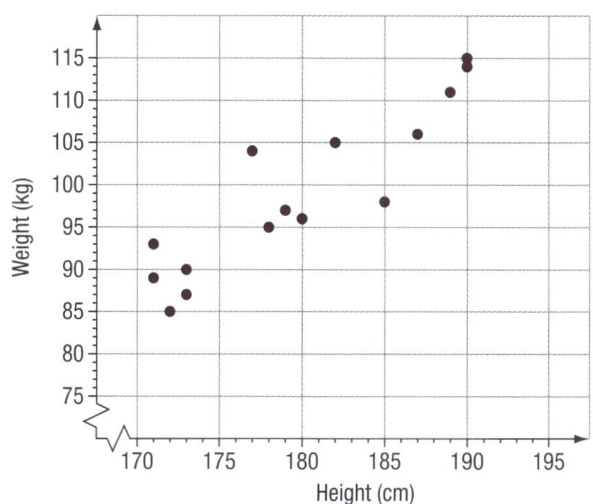

Interpretation of a scatter graph

There are four aspects we consider when discussing the association between two numerical variables on a scatter graph.

1 Direction

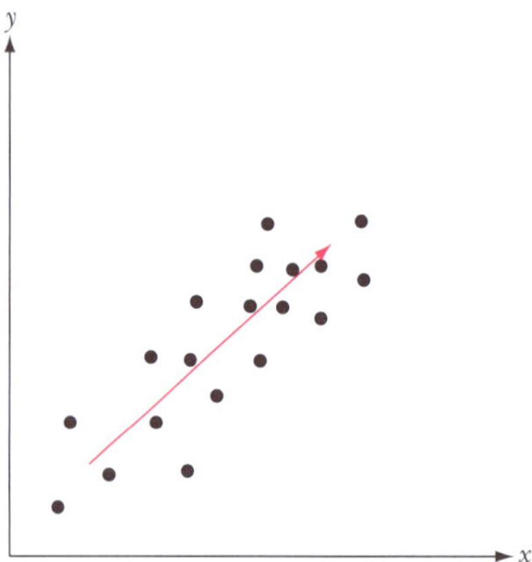

Positive correlation: the points generally go up as *x* increases, so it is similar to a positive gradient on a straight line. The interpretation is:
'As the **independent variable** (x) increases the **dependent variable** (y) also increases'.

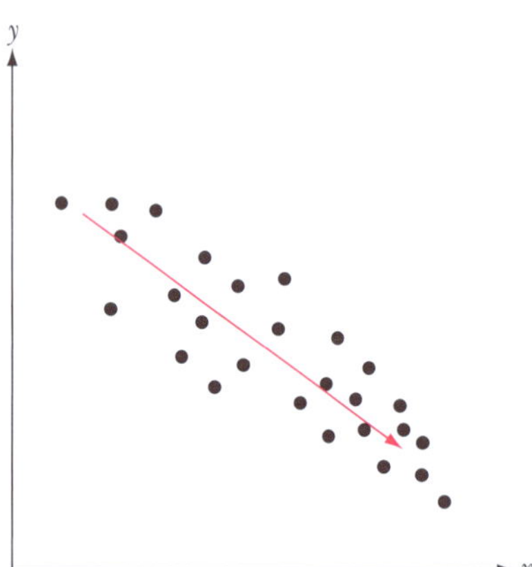

Negative correlation: the points generally go down as *x* increases, so it is similar to a negative gradient on a straight line. The interpretation is:
'As the **independent variable** (x) increases the **dependent variable** (y) decreases'.

2 **Form: linear or not linear?**

The two scatter graphs above appear to be linear as the points are generally in a straight line. The three scatter graphs below would not be of linear form.

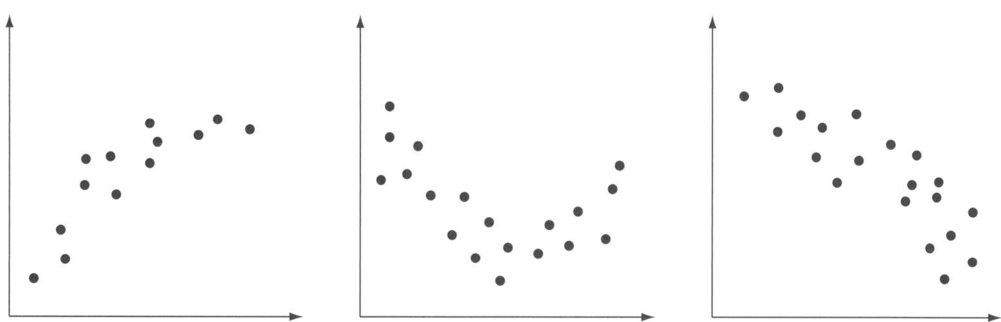

3 **Strength**

If the points are almost in a straight line then the association is said to be **strong**. For example, the graph on the left is **strong positive** while the one to the right is **strong negative**.

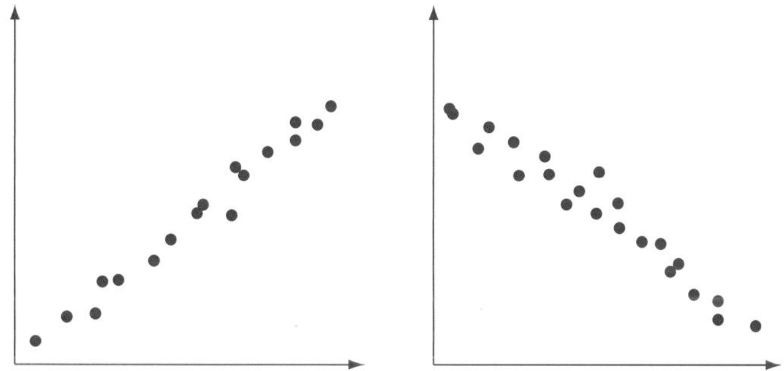

If the points are still obviously in a straight line but more scattered, then the strength is said to be **moderate**. For example, the graph on the left is **moderate positive** while the one to the right is **moderate negative**.

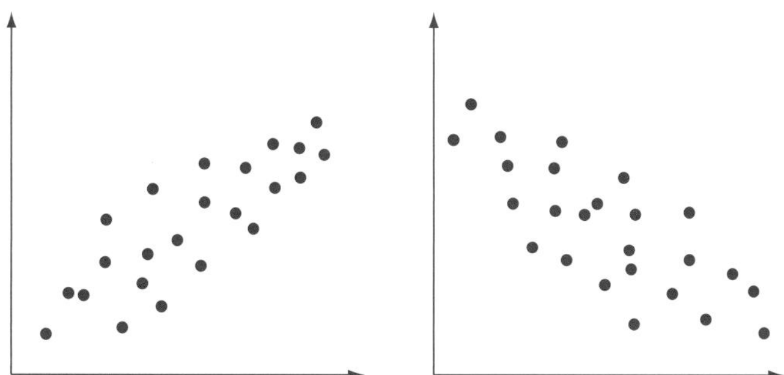

If the points are scattered but a general direction is still discernable then the association is said to be **weak**. For example, the graph on the left is **weak positive** while the one to the right is **weak negative**.

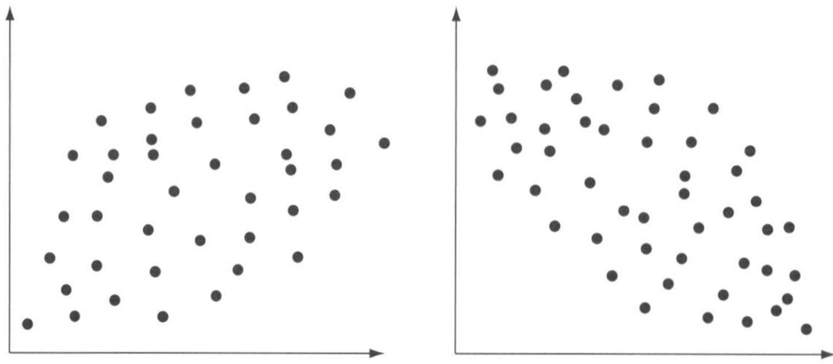

If the points appear to be randomly scattered then there is **no association** between the variables. As an example see the graph below.

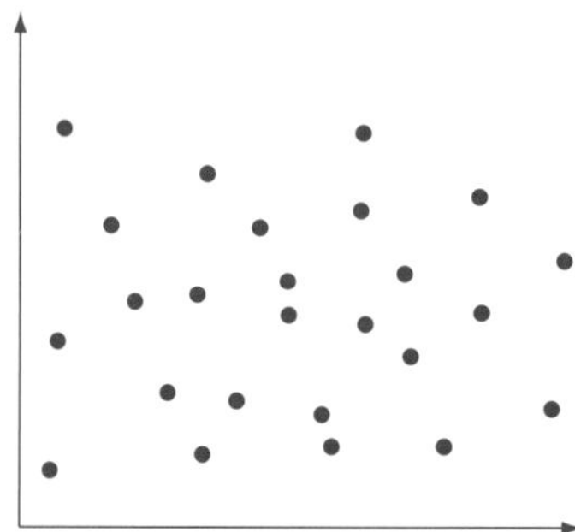

4 **Outliers**

Outliers stand out from the general body of the data. For example, this graph is described as moderate positive correlation with one outlier.

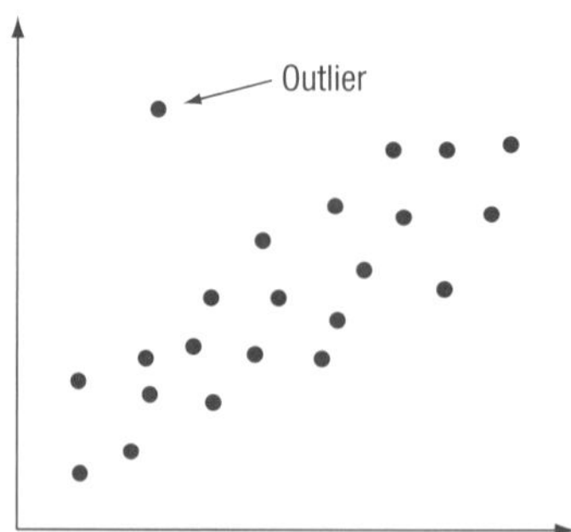

Outliers should be investigated to establish whether they are genuine outstanding data, errors in the data or errors in plotting. A decision can be made to ignore them as they influence models (lines) fitted to the data; however this should only be done after careful consideration.

From what we have learnt of scatter graphs, we can **interpret** the *weight* versus *height* example from earlier in this lesson:

'There appears to be **strong positive** association between the variables *height* and *weight*. This means that as height increases, weight increases. The relationship appears to be **linear** and there are no obvious **outliers**.'

Causation

A strong association between two variables does not necessarily mean that a change in one variable *causes* a change in the other variable. For example:

1 The heights and reading speeds of children were measured and a strong positive correlation was found between the variables *height* and *reading speed*. Does this mean that the taller children can read faster or that an increase in reading speed will cause you to grow? These suggestions are obviously not sensible.

   The strong correlation can be attributed to the fact that both variables are closely associated with a third variable *age*. As *age* increases both the variables *height* and *reading speed* increase.

2 The number of television sets sold in Port Moresby and the number of stray dogs collected in Mt Hagen were recorded over several years, and a strong positive association was found between the variables. Obviously the number of television sets sold in Port Moresby was not influencing the number of stray dogs collected in Mt Hagen. Both variables have simply been increasing over the period of time that their numbers were recorded.

When variables that are related such that a change in one variable does cause a change in the other variable, then we say that a **causal relationship** exists between the variables.

For example, if the height and weight of a group of children is measured then there will be a strong positive correlation between the variables *height* and *weight*. This will be a causal relationship because an increase in height will, in general, cause an increase in weight.

EXERCISE

1 For each of the following:
   - i State whether you would expect to find a positive, negative or no relationship between the variables.
   - ii Indicate what you would estimate as the strength (none, weak, moderate or strong) of the relationship.

   a Speed and the time taken for a journey
   b Number of occupants in a household and the water consumption of the household
   c Age and hearing ability
   d Height and shoe size

2 For each of the scatter graphs on the next page state:
   - i whether there is positive, negative or no association between the variables
   - ii whether the relationship between the variables appears to be linear or not
   - iii the strength of the association between the variables (zero, weak, moderate or strong).

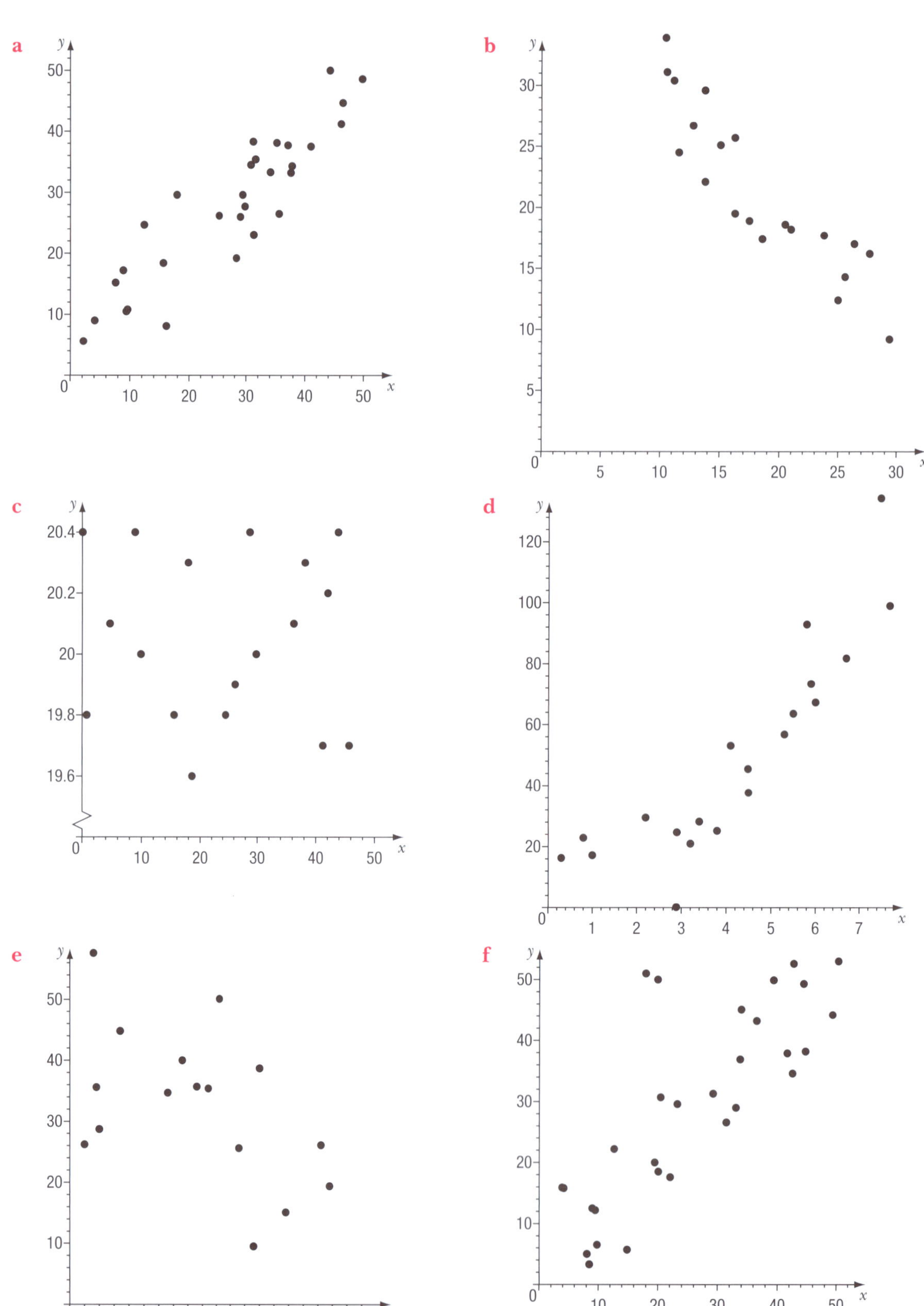
a
y
50
40
30
20
10
0
10
20
30
40
50
x
b
y
30
25
20
15
10
5
0
5
10
15
20
25
30
x
c
y
20.4
20.2
20
19.8
19.6
0
10
20
30
40
50
x
d
y
120
100
80
60
40
20
0
1
2
3
4
5
6
7
x
e
y
50
40
30
20
10
0
2
4
6
8
10
x
f
y
50
40
30
20
10
0
10
20
30
40
50
x

3 Copy and complete the following.

a If the variables x and y are positively associated, then as x increases, y ....

b If there is negative correlation between the variables m and n, then as m increases, n ....

c If there is no association between two variables then the points on the scatter graph appear to be ....

4

x	1	2	3	4	5	6	7	8	9	10
y	2	1	4	3	5	6	5	5	7	8

a Construct a scatter graph for the data in this table. *You will need the scatter graph from this question for exercises in Lessons 3 and 4.

b State whether the association between the variables appears to be:

i linear or not

ii positive, negative or no association

iii weak, moderate or strong.

5 The following data was collected by a store owner over fifteen consecutive days.

Maximum daily temperature (°C)	29	38	35	30	34	34	27	27	23	37	24	23	25	36	30
Number of drinks sold	119	164	161	152	206	169	122	143	63	208	155	96	125	248	139

a Which of the two variables is the dependent variable?

b Construct a scatter graph of the data. *You will need the scatter graph from this question for exercises in Lessons 3 and 4.

c Interpret the scatter graph in terms of the variables mentioning direction, strength, linearity and outliers.

d Do you think there is a causal relationship between the variables? Explain.

6 A class of 25 students were asked to record the time (in minutes) spent preparing for a topic test. The table below gives the score out of a possible 50 that they achieved on the test and the recorded preparation time.

Score	25	31	30	38	55	20	39	47	35	45	32	33	34	38	17
Time spent preparing (min)	75	30	35	65	110	60	40	80	56	70	50	110	18	80	22

Score	38	17	17	26	41	50	30	45	36	23
Time spent preparing (min)	30	15	10	85	100	60	55	80	50	75

a Which of the two variables is the independent variable?

b Construct a scatter graph of the data. *You will need the scatter graph from this question for exercises in Lessons 3 and 4.

c Interpret the scatter graph in terms of the variables mentioning direction, strength, linearity and outliers.

Lesson **3**

Fitting a straight line-of-best-fit 'by eye' to a scatter graph and finding its equation

'By eye' means that you will use your best judgement to draw a line through the points on a scatter graph in a manner that captures the general direction of the data. Everybody will have a different line, so this is not a very accurate way to find a rule connecting two variables.

Example

The data below gives the fat content (grams) and the energy (kilojoules) of 17 different foods.

a Fit a line-of-best-fit by eye (linear) to the data and find its equation.

b Predict (using your equation) the energy content of a food containing 40 grams of fat.

c Check your answer on the graph.

Fat (g)	15	55	18	45	17	24	30	30	30	16	11	9	30	24	24	30	32
Energy (kJ)	1255	3555	1800	1880	1670	2520	2300	2300	2090	1340	1130	1150	2300	1670	1670	2510	1460

Answer

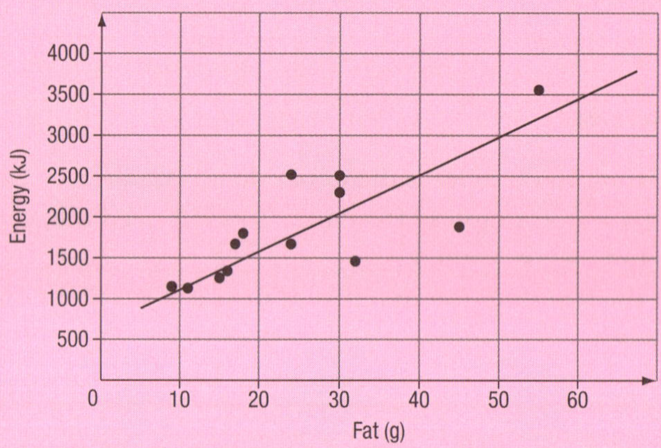

a A scatter graph of this data is drawn above. The data shows moderate positive correlation between the variables *energy* and *fat content*. Also, there appears to be a **linear** relationship between the variables.

Using a ruler I will draw the line that I think will best fit the data, my line-of-best-fit by eye (see the line drawn on the scatter graph above). This is the line that I think best fits the flow of the data.

b **Finding the equation of the regression line**

The equation will be in the form $y = mx + c$ which in terms of the variables will be:

$$\text{Energy} = m \times \text{Fat} + c$$

where m is the gradient (slope) of the line and

c is the y-intercept or in this case the energy-intercept

i Find two points on the line.

On my line I will choose the points (17, 1500) and (50, 3000) as shown on the graph.

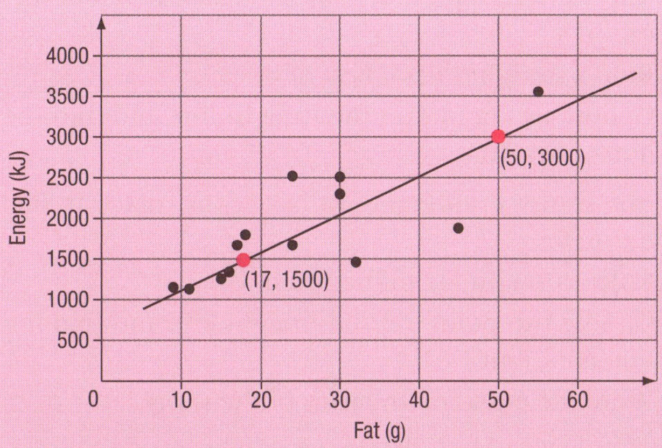

ii Calculate the gradient of the line using the formula $m = \frac{y_2 - y_1}{x_2 - x_1}$.

gradient $= \frac{3000 - 1500}{50 - 17} = 45.45$, rounded to two decimal places.

At this point the equation is Energy = 45.45 × Fat + c

iii Find the 'energy-intercept' c.

Note: Often the intercept can be read from the graph; it is the point where the line touches the y-axis. But it must be the point that corresponds to the value 0 on the horizontal axis, that is $(0, c)$. If the horizontal axis does not start at zero then the value of c cannot be read from the graph.

As my line does not extend to the axis I will use one of the points to find c.

Using the point (50, 3000) and substituting the values into the equation:

Energy = 45.45 × Fat content + c

$3000 = 45.45 \times 50 + c$

$3000 = 2272.5 + c$

Subtract 2272.5 from both sides of the equation:

$727.5 = c$

The equation is Energy = 45.45 × Fat + 727.5.

c Substituting Fat = 40 into the equation:

Energy = 45.45 × 40 + 727.5 = 2545.5 kilojoules

Checking this on the graph it looks the correct value.

EXERCISE

1 For each of the following sets of data:

i Construct a scatter graph.

ii Rule a straight line-of-best-fit by eye.

iii Choose two points on the straight line and find the equation of the line-of-best-fit.

a

x	1	2	3	4	5	6	7
y	11	8	7	5	4	2	1

b

Time (s)	1	2.5	5	8	12.5	16	18	20
Distance (m)	14	28	53	82	127	157	181	200

2 On the scatter graph that you have constructed in the exercise in Lesson 2 Question 4:

a Rule a straight line-of-best-fit by eye.

b Choose two points on the straight line and find the equation of the line-of-best-fit.

3 On the scatter graph that you have constructed in the exercise in Lesson 2 Question 5:

a Rule a straight line-of-best-fit by eye.

b Choose two points on the straight line and find the equation of the line-of-best-fit.

c Write the equation in terms of the variables.

4 On the scatter graph that you have constructed in the exercise in Lesson 2 Question 6:

a Rule a straight line-of-best-fit by eye.

b Choose two points on the straight line and find the equation of the line-of-best-fit.

c Write the equation in terms of the variables.

Lesson 4

Interpreting the gradient and intercept of a linear line-of-best-fit and using the line to predict

For the example in Lesson 3, the regression equation is Energy = 45.45 × Fat + 727.5 in which the gradient is the value 45.45.

The gradient can be interpreted as:

'for every increase of one gram of fat there is an increase of 45.45 kilojoules of energy'.

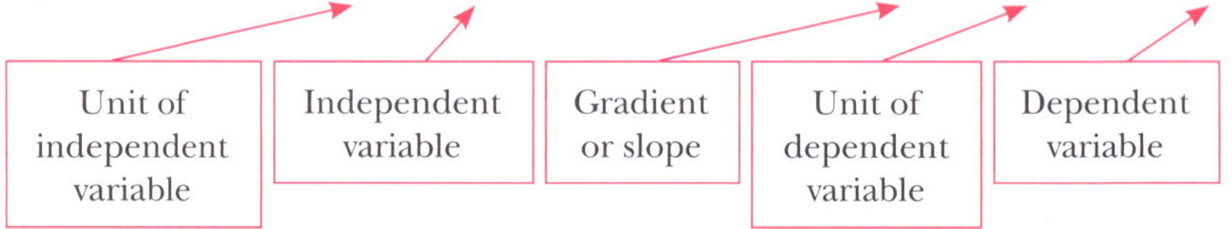

The intercept has the value 727.5. This can be interpreted as 'when the fat content of a food is **zero** then the energy provided by the food is 727.5 kilojoules'. This interpretation seems sensible as zero fat is a reasonable concept and zero is not too far removed from the data set (*fat* values of 9 to 55 grams).

Interpolation and extrapolation

Interpolation means predicting values from the equation for values from **within** the range of data from which the regression equation was based.

For example, in the Example in Lesson 3 the Fat data ranged from 9 to 55. Using a value within this range to predict an *Energy* value would be a case of interpolation. When we predicted, using the equation, that the energy content of a food containing 40 grams of fat was 2545.5 kilojoules this was a case of interpolation.

Extrapolation means predicting values from an equation for values from **outside** the range of data from which the regression equation was based.

For example, in the Example in Lesson 3 the Fat data ranged from 9 to 55. Using a value outside this range to predict an *Energy* value would be a case of extrapolation. Predicting, using the equation, the energy content of a food containing, say, 60 grams of fat or 5 grams of fat would both be cases of extrapolation.
Note: Extrapolation too far beyond the data on which the equation was based may not be reliable.

EXERCISE

1 Which one of the regression lines fitted by eye to the scatter graph below would best capture the flow of the data?

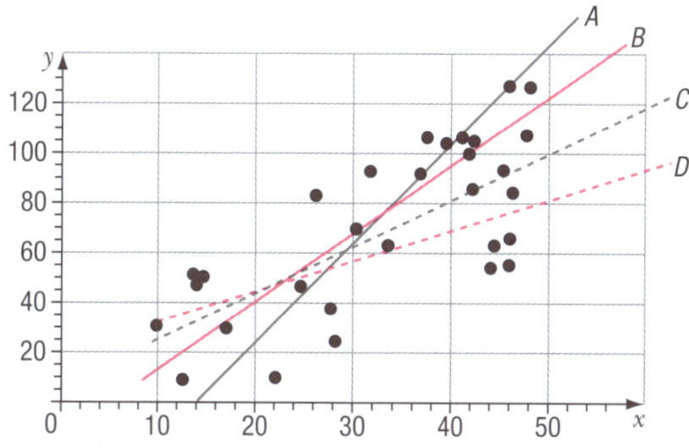

2 For each of the following:

i Draw a line-of-best-fit by eye on the scatter graph.

ii Find the equation of your line.

iii Interpret the gradient and intercept of the equation.

a

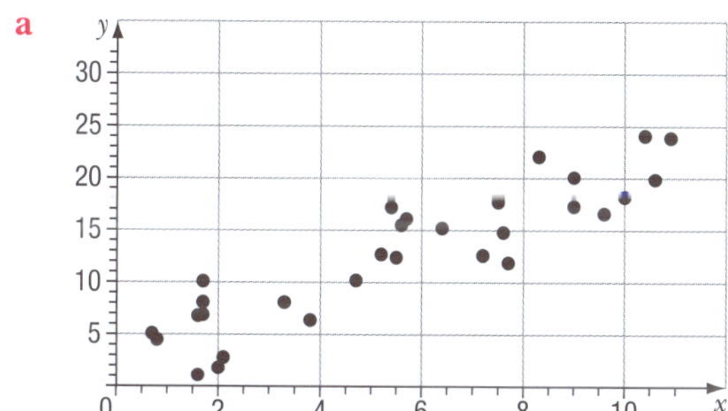

b

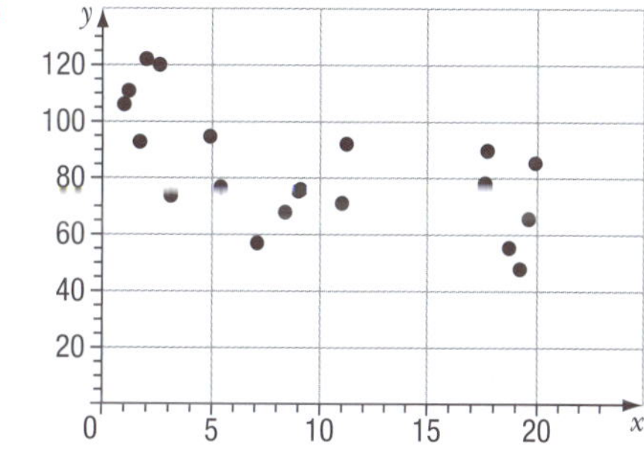

c

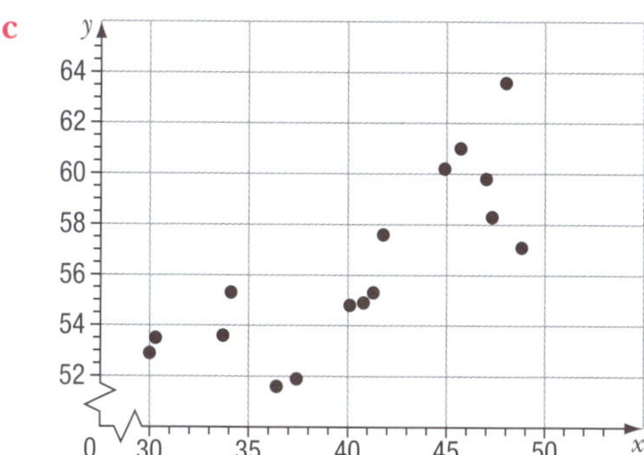

3 The number of houses sold and the distance travelled in a month has been recorded for a group of real estate agents. The data is graphed below.

a Draw a line-of-best-fit by eye on the scatter plot.

b Find the equation of your line expressing it in terms of the variables.

c Interpret the gradient and intercept of your equation in terms of the variables.

d Use your line to predict the number of houses sold by an agent who travels 2000 km in a month. Is this a case of interpolation or extrapolation?

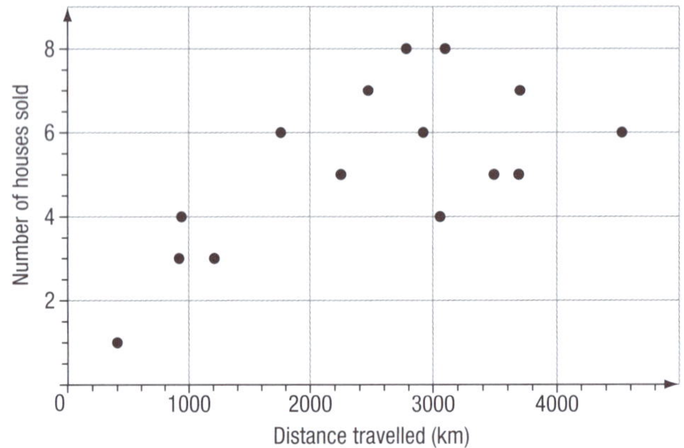

4 Susan is investigating the association between age and body mass index ($\text{BMI} = \frac{\text{Weight in kg}}{\text{Height in metres}^2}$) and her data is plotted on the scatter plot below. She has fitted a line by eye to the scatter plot.

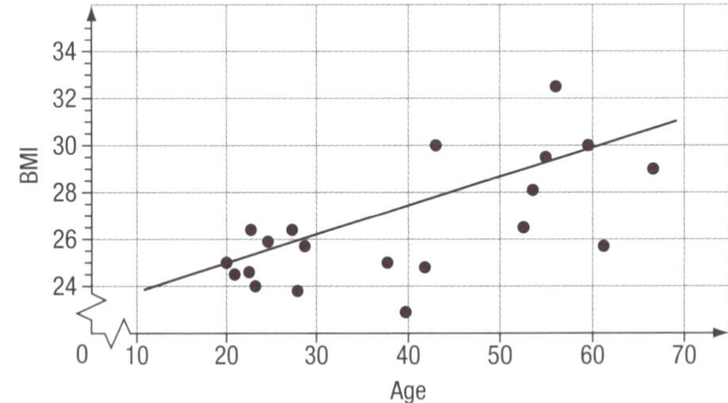

a Susan has found the gradient of this line to be 0.1. If the line goes through the point (50, 28) then the equation of the regression line is:

A BMI = 0.1 × Age + 23

B BMI = 0.1 × Age + 24

C BMI = 23 × Age + 0.1

D Age = 0.1 × BMI + 24

E Age = 0.1 × BMI + 23

b Which one of the following would **not** be a true interpretation of Susan's linear model?

A The BMI increases by 0.1 units per year increase in age.

B As age increases, BMI appears to increase.

C An increase in age causes an increase in BMI.

D The line would be overpredicting for a person aged 40 whose BMI is 25.

E Using this line to predict the BMI of a person aged 80 would be a case of extrapolation.

5 A class of 25 students was asked to record the time (in minutes) spent preparing for a topic test. Using the scatter graph and equation you have found for Question 4 in Lesson 3, interpret the gradient and intercept of your equation in terms of the variables.

Constant rates of change

A straight-line graph with the equation $y = 5x - 4$ is shown at right. For this graph the independent variable is x, the dependent variable is y, the gradient is 5 and the y-intercept $^-4$.

The **rate of change** of y, with respect to x, is another description of the **gradient**. So for this graph the rate of change of y with respect to x is 5, a **constant** value.

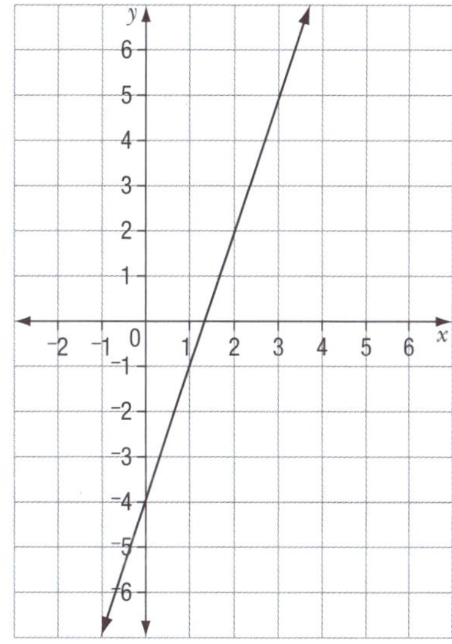

This next graph shows the *distance travelled* graphed against *time* for a car trip. The **rate of change** of *distance travelled* with respect to *time* is the gradient of the graph, so the **rate of change** of distance with respect to time is $\frac{\text{rise}}{\text{run}} = \frac{300 - 0}{5 - 0} = \frac{300}{5} = 60$.

If we were to interpret this rate we would say that 'the car was travelling at a **constant** speed of 60 km/h for the trip'.

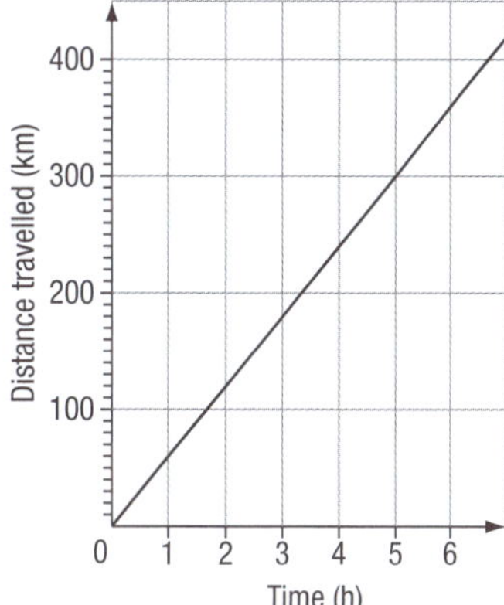

A **constant rate of change** means that the rate of change (**gradient**) does not change. If the rate of change between two variables is constant, then their graph will be a **straight-line graph**.

Example 1

The graph below shows the volume, in litres, in a water tank over a period of 24 hours from midnight to midnight. Describe the changes in volume for this tank over the 24 hours.

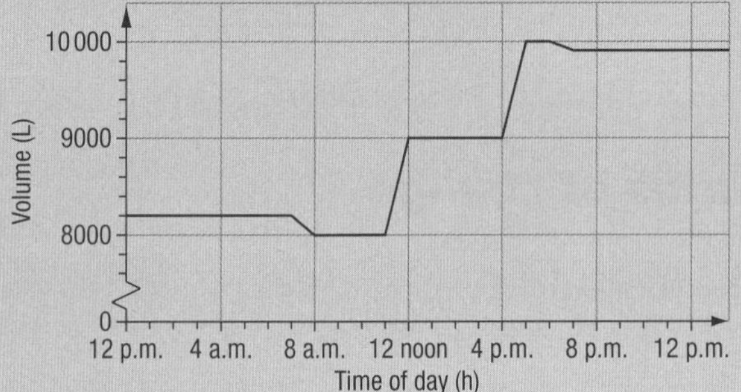

Answer

For the first 7 hours of the day the volume of water in the tank remained constant at 8200 L. Between 7 a.m. and 8 a.m. 200 L of water was used at a constant rate of ⁻200 L/h. The volume of water remained constant at 8000 L for the next 3 hours, but from 11 a.m. to 12 noon the volume in the tank increased at a constant rate of 1000 L/h to 9000 L (it probably rained). There was no change in the volume of water in the tank from 12 noon to 4 p.m. and then the volume in the tank increased at a constant rate of 1000 L/h to 10 000 L (it probably rained again). There was no change in the volume in the tank from 5 p.m. to 6 p.m., but 100 L was used between 6 p.m. and 7 p.m. at a constant rate of ⁻100 L/h. From 7 p.m. to midnight the volume of water in the tank remained constant at 9900 L.

Example 2

The velocity–time graph below shows the journey of a car on a short trip. Describe what is happening on this car trip.

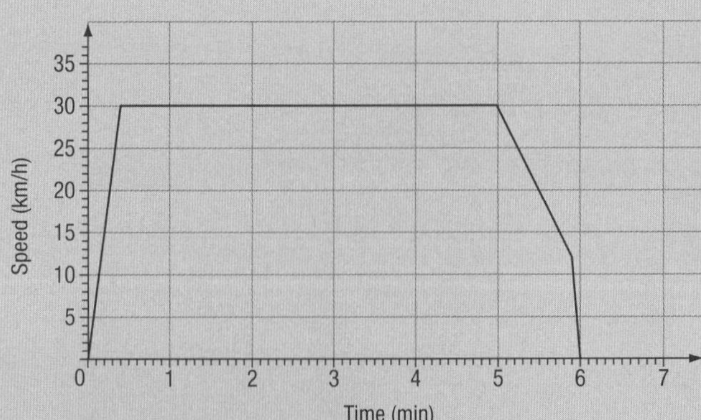

Answer

We can make the following observations from the graph:

- The rate of change of speed with respect to time (the acceleration) is positive and constant for the first 0.4 minute. The speed changes from 0 to 30 km/h in the first 0.4 minute (24 seconds).
- The rate of change of speed with respect to time is zero from 0.4 minute to 5 minutes (a total of 5 minutes, 36 seconds).
- The rate of change of speed with respect to time is constant and negative (deceleration) from 5 minutes to 5.9 minutes (54 seconds). The speed changes from 30 km/h to 12 km/h in this time.
- The rate of change of speed with respect to time is constant and negative from 5.9 minutes to 6 minutes (6 seconds). The speed changes from 12 km/h to 0 km/h in this time.

Describing the 'story' of the trip in everyday language we could say:

'The car accelerates at a constant rate from zero to 30 km/h in the first 24 seconds. It then travels at 30 km/h for the next 5 minutes 36 seconds. The car slows down (decelerates) over the next minute; the speed decreasing at a constant rate from 30 km/h to 12 km/h over 54 seconds and then decreasing at a constant rate from 12 km/h to zero over the final 6 seconds.'

EXERCISE

1 The graph below shows Jacob's trip to the local store and home again. He stopped and talked to a friend on his way to the store.

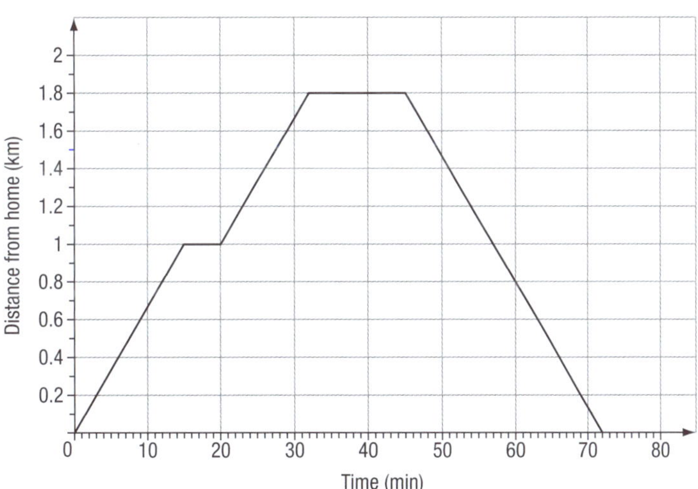

a At what speed, in km/h, did Jacob walk to the store?

b How long did it take Jacob to get to the store?

c How long did Jacob talk to his friend on the way to the store?

d How much time did Jacob spend at the store?

e What distance is Jacob's home from the store?

f How long did it take Jacob to get home from the store?

g At what speed, in km/h, did Jacob walk on his way home from the store?

2 The charges for delivery for two different companies (Company A and Company B) are graphed on this set of axes. Each company's charge is made up of a set charge and a charge per kilometre.

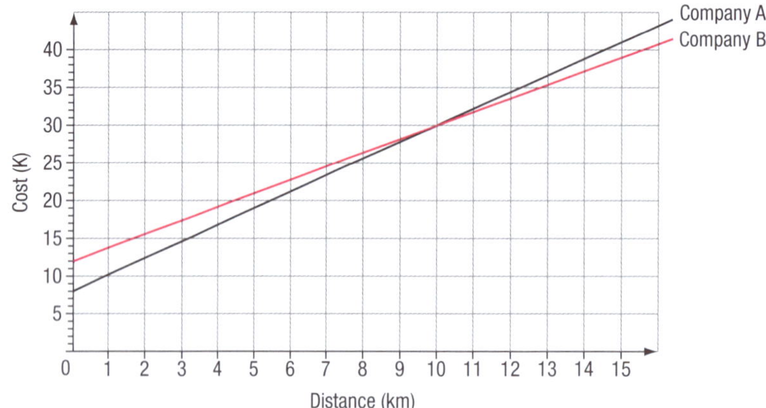

a Which company has the higher set charge for delivery?
b What is the rate per kilometre charge for Company A?
c What is the rate per kilometre charge for Company B?
d For what distances is it cheaper to use Company A for delivery?
e What is the difference in charge for delivering an article at a distance of 5 kilometres?

3 On the same set of axes, graph and label the speed–time graphs of three trucks:
i Truck A travelling at a steady speed of 50 km/h
ii Truck B starting from rest and uniformly reaching a speed of 40 km/h in 30 s, which it then maintains
iii Truck C travelling at 30 km/h for 10 s, accelerating uniformly to 55 km/h over another 10 s and then slowing to a stop uniformly over the next 20 s.
a For which of the trucks was the rate of change of speed with respect to time:
i the greatest? ii negative?
b Find the rate of change of speed with respect to time, in (km/h)/second of:
i Truck B for the time 0 to 30 s
ii Truck C for the time from 20 to 40 s.

4 Betty took 20 min to walk the 1.5 km to the store. She stayed at the store for 15 min and then walked at a rate of 5 km/h to her friend Jan's place, taking 24 min to get there. She stayed at Jan's place for 1 h 20 min and then walked straight home in 40 min. If all places are in a straight line and the store is on the way to Jan's place:
a What was the furthest distance that Betty was from her home?
b Construct a graph to show Betty's distance, d km, from home at time, t hours.
c What was the rate, in km/h, at which Betty walked on her way to the store?
d What was the rate, in km/h, at which Betty walked on her way home from Jan's place?
e If Betty left home at 10.30 a.m., what time did she arrive back home?

5 Thelma has planted a corn seed and measured the height of the plant on the same day each week. The results for the beginning of the first five weeks are given in the table.

Beginning of week	1	2	3	4	5
Height (cm)	0	8	14.8	20.1	24.7

a Plot the points on a set of axes. Join the points with straight-line segments.

b Calculate the rate of growth for each week, completing the table below.

Week	1	2	3	4	5
Rate of growth (cm/week)					

c Describe what is happening to the rate of growth as the weeks increase?

6 Tepi leaves his village at 8 a.m. to walk the 10 km to town. He arrives in town at 10 a.m. and stays in town until 12.30 p.m. After that he returns to his village, arriving at 2.30 p.m.

Didya leaves the same village at 8.30 a.m. but catches the bus to town; the trip takes 30 min. Didya stays in town until 1 p.m. then catches the bus home to the village with the trip taking 30 min again.

a On the same set of axes, construct a graph for both Tepi and Didya showing their distance from the village from 8 a.m. until 3 p.m. Label the graphs appropriately.

b What is the average speed of Tepi on his trip from the village to town?

c What is the average speed of the bus on the trip from the village to town?

d What do the points of intersection of the two graphs represent?

e The bus (with Didya on board) passes Tepi on the road on two occasions. Read from your graph the approximate times that the bus passes Tepi.

7 The graph at right shows the journeys of two women, Railala and Eleni, along the road from their village to town on a particular day.

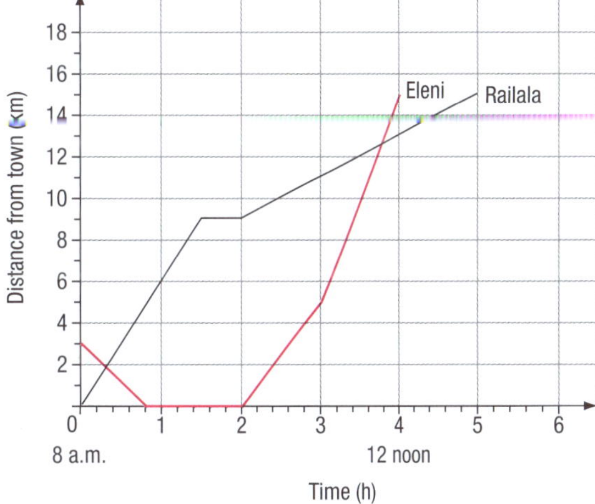

a How far was Eleni from the village at 8 a.m.?

b How far was the town from the village?

c At what time did:

  i Railala

  ii Eleni leave the village on their journey to the town?

d At what time did Railala stop to rest and for how long was her rest?

e At what rate did Railala walk before and after the rest?

f Calculate the rates that Eleni walked on all sections of her journey.

g At approximately what time did Eleni pass Railala on the road?

h How far were Railala and Eleni from the town when Eleni passed Railala on the road?

i At what times were Railala and Eleni 4 km from town?

j How much sooner did Eleni arrive in the town?

Lesson 6

The rate of change for non-linear graphs

Graphs that do not have a constant gradient (rate of change) will appear as non-linear curves.

To find the gradient of a graph that does not have a constant rate of change (gradient) we need to look at the gradient of straight lines that touch the graph at one point only.

For example, the graph at right has a gradient that is not constant. The gradient of a curve such as this one is different at every point on the curve.

The gradient of the graph at this point is equal to the gradient of the straight line that touches the graph at that point.

The gradient of the graph at a particular point is equal to the gradient of the straight line that touches the graph at that point (called a tangent). For all the points on the graph, except (0, 0), the gradient is positive because all the straight lines (tangents) drawn will have a positive gradient. The gradient at the point (0, 0) will be zero as the straight line drawn at this point will be horizontal.

For this graph we could say that 'the rate of change (gradient) is **positive and increasing** as x increases' because the gradients of the tangents are increasing as x increases.

A graph with **a positive but decreasing rate of change** would have a shape like this.

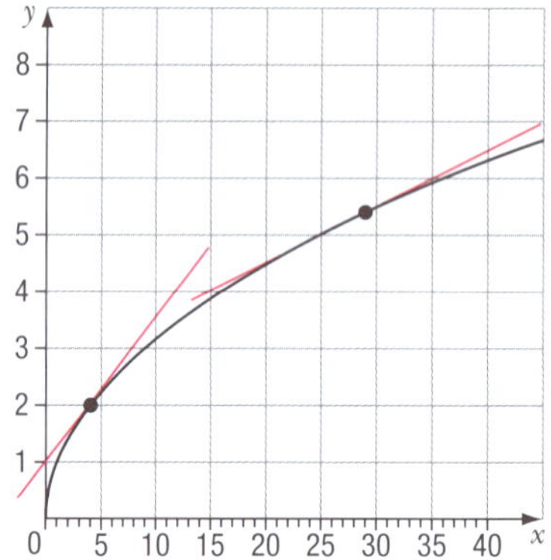

A graph with a **negative and decreasing rate of change** (gradient) would appear in this form.

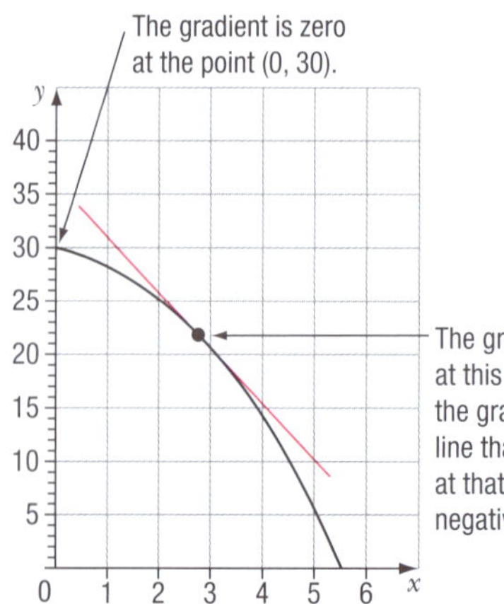

A graph with a **negative and increasing rate of change** (gradient) would appear like the one at right.

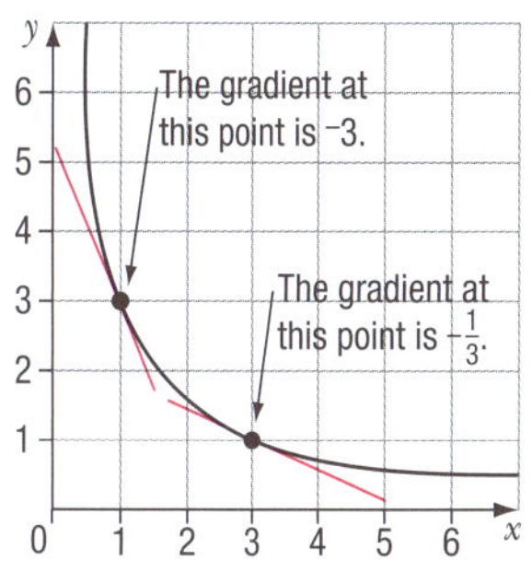

EXERCISE

1 A graph that demonstrates Boyle's law is shown at right.

  a Describe the gradient of this graph.

  b Describe what is happening to the rate of change of volume with respect to pressure as the pressure increases.

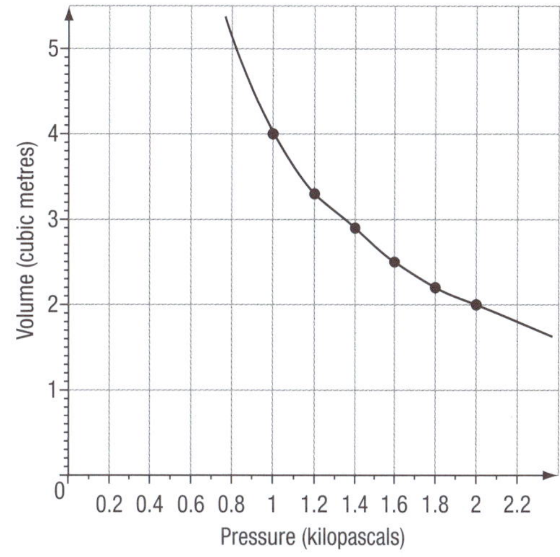

2 This graph shows the variation in the pulse rate of a person during a 1-hour exercise session.

  a Describe the rate of change of pulse rate with respect to time for the time interval 5 min to 20 min.

  b Describe the rate of change of pulse rate with respect to time for the time interval 30 min to 60 min.

  c When is the rate of change of pulse rate with respect to time zero?

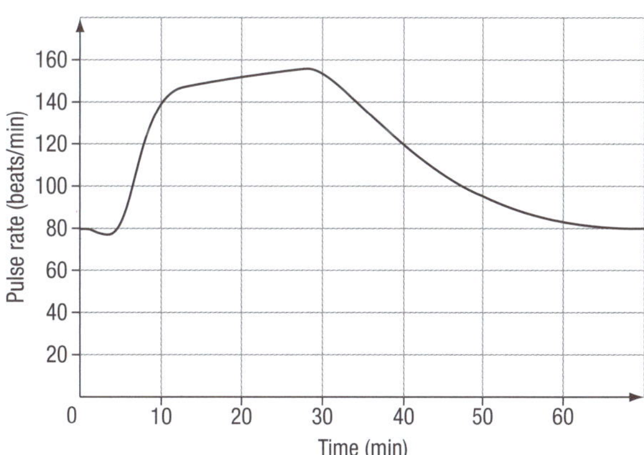

3 The graph at right shows the incidence of HIV/AIDS in PNG for the years 1987 to 2005. Describe the rate of change of new cases of HIV/AIDS with respect to time over these years.

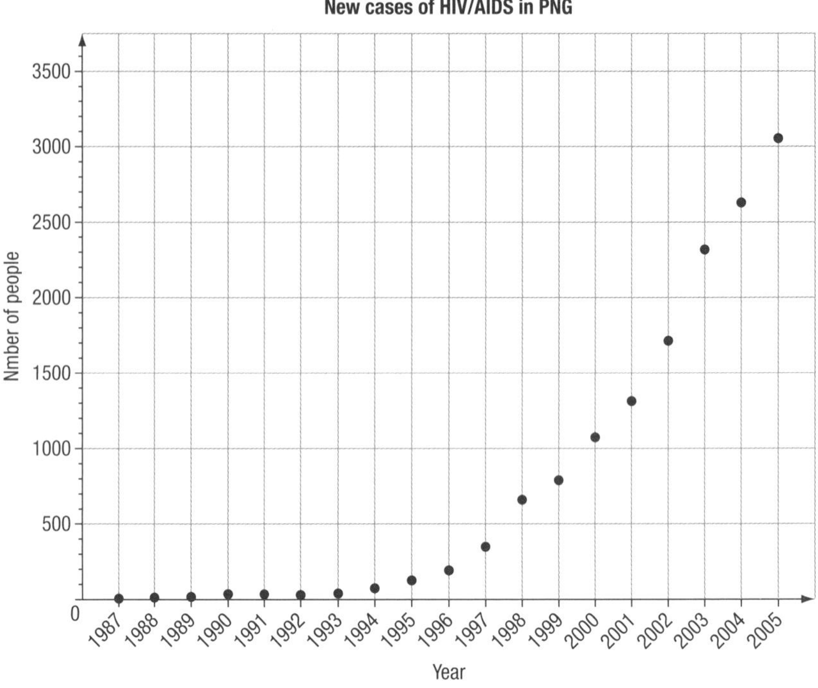

4 The weight of a piglet has been recorded for the first eight weeks of its life. The results are graphed at right. Describe the rate of change of the weight of the piglet with respect to time as time increases.

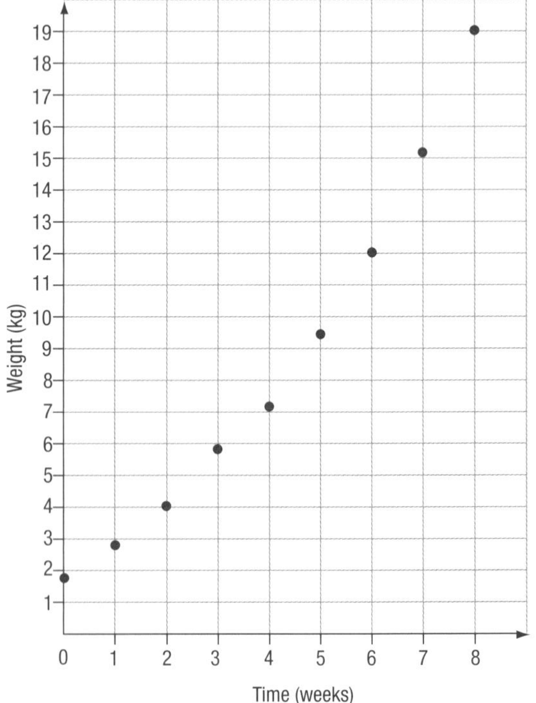

5 The amount still owing (debt) on a car loan is graphed at right for the ten years of the loan. Describe the rate of change of debt with respect to time as the time increases.

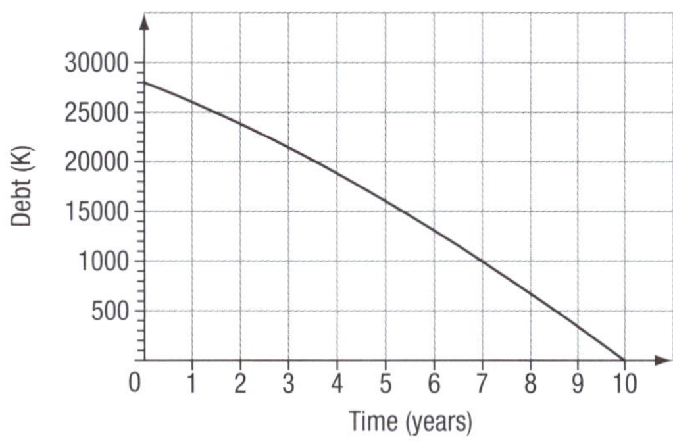

6 This graph shows the radius of a circle for a given area. Describe the rate of change of radius with respect to area, as area increases.

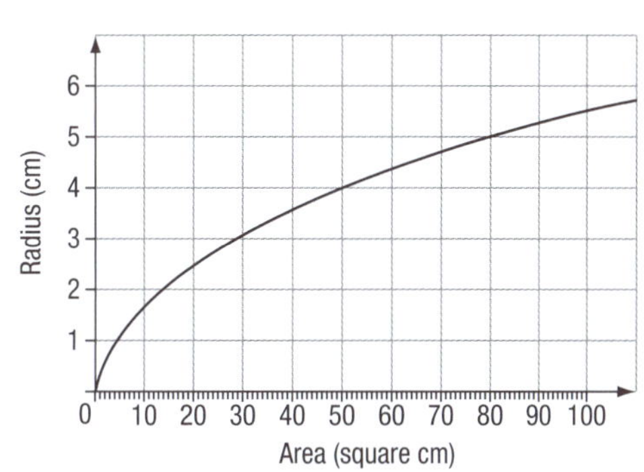

Producing and interpreting story graphs

For assessment of this option, you are required to draw and interpret a story graph from your local context.

To produce a story graph you will need to spend some time:

a deciding the context of your graph
b collecting information for your graph
c planning the best way you can present the information as a graph and then constructing the graph
d interpreting the graph in terms of the variables
e presenting the graph and its interpretation in a manner that is easily understood.

Your assessment task will be assessed on the extent to which you can:

- demonstrate appropriate investigation skills
- choose and apply relevant mathematical techniques
- make an effective communication of the project results.

A story graph can be obtained from experimental data, from historical data or something that you have observed.

- If you are using experimental data then you will have to take care that you observe and note the conditions in which you are conducting the experiment.
- If you are collecting sample data, you must make sure that it is not a biased sample.
- If your data involves measurement of a variable then you must be consistent in the way that you do this.

It is important that a graph is correctly constructed and labelled so that anybody is able to 'read the story' represented by the graph. This means that the features of the graph should be easy to interpret in terms of the variables involved.

Here are some suggestions that may give you an idea for your story graph. You could investigate:

1 The temperature of a glass of water after an ice-cube has been placed into it. You could compare what happens if you use water at room temperature or tap water (usually colder).
2 How long it takes for a cup of boiling water to cool to room temperature. You could vary the volume and/or the container.
3 The weight of an ice-cube as it melts.
4 The growth of a bamboo shoot or another plant or animal. Height or weight can be measured.
5 The evaporation of a particular volume of water. You could vary the area exposed to the air.
6 Comparison of delivery rates for goods, furniture or people by different companies.
7 Break-even analysis as shown in the example in Lesson **21** on page 132.

OPTION B: QUADRATIC EQUATIONS

If this option is chosen then it will take the remaining five weeks of the Functions and graphs unit, and will be assessed by an individual directed investigation involving quadratic functions and their graphs.

Topics covered in this option include:

- expanding factors that produce quadratic expressions
- factorising quadratic expressions
- solving quadratic equations
- graphing quadratic functions
- using simultaneous equations to find the equations of quadratic functions
- practical problems involving quadratic functions.

Lessons and questions in the exercises involving advanced concepts are marked with an asterisk (*).

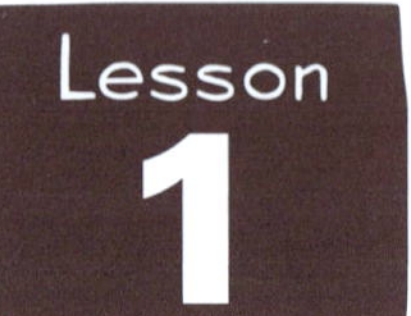

Quadratic expressions and expansions that produce quadratic expressions

A **quadratic expression** has the form $\boldsymbol{ax^2 + bx + c}$ where x is the variable and a, b and c are constants (numbers).

a, b and c can be any real numbers except that a cannot be zero ($a \neq 0$). This means that a quadratic expression must have an 'x^2' term.

A **quadratic equation** has the form $\boldsymbol{ax^2 + bx + c = 0}$ where $a \neq 0$ and a, b and c are real numbers.

Quadratic expressions model many everyday situations. For example, if an object is projected vertically upwards then the height of the object at a particular time can be modelled by a quadratic expression.

Expansions that produce quadratic expressions

If two linear expressions are multiplied then the result will be a quadratic expression.

Example 1

Expand the following products of linear expressions:

a $x(x+2)$ **b** $(x+2)(x+3)$ **c** $(2x+1)(x-2)$ **d** $(x-4)(2x-3)$

Answer

a $x(x+2) = x \times x + x \times 2$
$= x^2 + 2x$

b $(x+2)(x+3) = (x+2) \times x + (x+2) \times 3$ using $(x+2)$ as the 'number outside the bracket'
$= x(x+2) + 3(x+2)$
$= x \times x + x \times 2 + 3 \times x + 3 \times 2$ expanding each of the brackets
$= x^2 + 2x + 3x + 6$
$= x^2 + 5x + 6$ collecting like terms

c $(2x+1)(x-2) = (2x+1) \times x - (2x+1) \times 2$
$= x(2x+1) - 2(2x+1)$
$= x \times 2x + x \times 1 - [2 \times 2x + 2 \times 1]$ expanding each of the brackets
$= 2x^2 + x - [4x+2]$
$= 2x^2 + x - 4x - 2$
$= 2x^2 - 3x - 2$ collecting like terms

d $(x-4)(2x-3) = (x-4) \times 2x - (x-4) \times 3$
$= 2x(x-4) - 3(x-4)$
$= 2x \times x - 2x \times 4 - [3 \times x - 3 \times 4]$ expanding each of the brackets
$= 2x^2 - 8x - [3x - 12]$
$= 2x^2 - 8x - 3x + 12$
$= 2x^2 - 11x - 12$ collecting like terms

Inspection of the second last line of each of the answers above reveals that the answers can be obtained by multiplying each term in the first bracket by each term in the second bracket, which gives a total of four terms. We will use this method for Example **1b**, **c** and **d** above.

Example 2

a $(x+2)(x+3)$
b $(2x+1)(x-2)$
c $(x-4)(2x-3)$

Answer

a $(x+2)(x+3) = x \times x + x \times 3 + 2 \times x + 2 \times 3$
$= x^2 + 3x + 2x + 6$
$= x^2 + 5x + 6$

b $(2x+1)(x-2) = (2x+1)(x+(^-2))$
$= 2x \times x + 2x \times {}^-2 + 1 \times x + 1 \times {}^-2$
$= 2x^2 - 4x + x - 2$
$= 2x^2 - 3x - 2$

c $(x-4)(2x-3) = (x+(^-4))(2x+(^-3))$
$= x \times 2x + x \times {}^-3 + (^-4) \times 2x + (^-4) \times (^-3)$
$= 2x^2 - 3x - 8x + 12$
$= 2x^2 - 11x + 12$

EXERCISE

1 Which of the following expressions are quadratic expressions?

a $x^2 + 3$ b $5x - x^2 + x^3$ c $^-x^2$ d $8 + x - x^2 + 2 + 5x^2$

e $\frac{1}{x^2 + 3}$ f $9 - x - x$ g $\frac{x^3 + 5}{x}$

2 Expand and simplify each of the following.

a $(x + 2)(x + 4)$ b $(m + 5)(m + 3)$ c $(n + 7)(n + 1)$ d $x(x + 6)$

e $(2x + 5)(3x + 2)$ f $(7 + 3x)(x + 3)$ g $(4y + 5)(3y + 4)$

3 Expand and simplify each of the following.

a $(x + 3)(x - 5)$ b $(4x + 3)(x - 2)$ c $(y - 4)(y + 7)$ d $(5y - 3)(2y + 1)$

e $3x(x - 8)$ f $(3 - y)(y + 4)$ g $(4m - 1)(3 + 2m)$

4 Expand and simplify each of the following.

a $(x - 8)(x - 2)$ b $(2x - 3)(x - 4)$ c $(5 - x)(x - 3)$ d $^-y(2y - 3)$

e $(5n - 1)(3n - 7)$ f $(6 - 7x)(x - 5)$ g $(5 - x)(6 - x)$

5 Expand and simplify each of the following.

a $(3x + 5)(x - 4)$ b $(2x - 1)(2x + 1)$ c $(3 - 5x)(6 - 5x)$

d $x(x + 3) + (x - 4)(x - 5)$ e $(y + 4)(y + 4)$ f $(x + 4)(2x + 3) + (x - 4)(2x + 3)$

g $y(y - 9) - y(3 - y)$ h $(4 - 3y)(2 - 5y) - 2y(8 - 3y)$

Lesson 2

Expansion of perfect squares

A **perfect square** is an expression of the form $(a + b)^2$.

$(a + b)^2$ can also be written as $(a + b)(a + b)$ and expanded as follows:

$$\begin{aligned}(a + b)(a + b) &= a \times a + a \times b + b \times a + b \times b \\ &= a^2 + ab + ab + b^2 \\ &= a^2 + 2ab + b^2\end{aligned}$$

Similarly $(a - b)^2$ is a perfect square and this can be expanded to:

$$\begin{aligned}(a - b)(a - b) &= a \times a + a \times {^-b} + {^-b} \times a + {^-b} \times {^-b} \\ &= a^2 - ab - ab + b^2 \\ &= a^2 - 2ab + b^2\end{aligned}$$

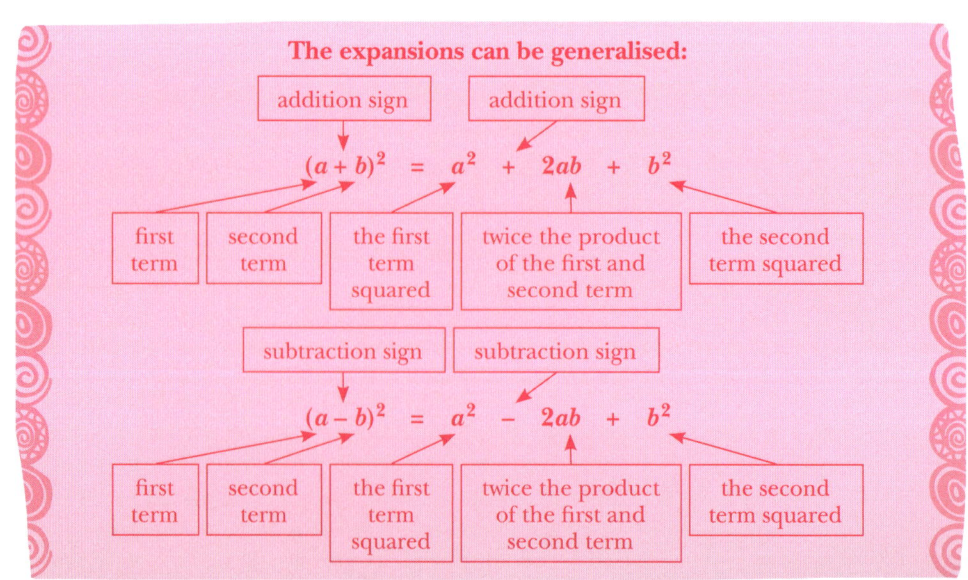

Example 1

Use the rule
$(a+b)^2 = a^2 + 2ab + b^2$
to expand:

a $(x+5)^2$

b $(2x+7)^2$

Answer

a The first term is x and the second term is 5.

$$(x+5)^2 = x^2 + 2 \times x \times 5 + 5^2$$
$$= x^2 + 10x + 25$$

b The first term is $2x$ and the second term is 7.

$$(2x+7)^2 = (2x)^2 + 2 \times 2x \times 7 + 7^2$$
$$= 4x^2 + 28x + 49$$

Example 2

Use the rule
$(a-b)^2 = a^2 - 2ab + b^2$
to expand:

a $(x-6)^2$

b $(5y-3)^2$

Answer

a The first term is x and the second term is 6.

$$(x-6)^2 = x^2 - 2 \times x \times 6 + 6^2$$
$$= x^2 - 12x + 36$$

b The first term is $5y$ and the second term is 3.

$$(5y-3)^2 = (5y^2)^2 - 2 \times 5y \times 3 + 3^2$$
$$= 25y^2 - 30y + 9$$

EXERCISE

1 Use the rule $(a+b)^2 = a^2 + 2ab + b^2$ to expand:

a $(x+3)^2$ **b** $(2x+1)^2$ **c** $(p+q)^2$ **d** $(4m+7)^2$

e $(6x+1)^2$ **f** $(3y+11)^2$ **g** $(8n+3)^2$ **h** $(3p+4q)^2$

i $(5+3y)^2$ **j** $(0.2+5x)^2$

2 Use the rule $(a-b)^2 = a^2 - 2ab + b^2$ to expand:

a $(x-4)^2$ **b** $(2x-3)^2$ **c** $(m-n)^2$ **d** $(6y-1)^2$

e $(5x-12)^2$ **f** $(2p-5q)^2$ **g** $(5-x)^2$ **h** $(1-2y)^2$

i $(6-3x)^2$ **j** $(x-0.8)^2$

3 Expand using one of the perfect square rules, then simplify.

a $(0.8x+0.3)^2$ **b** $(ax-1)^2$ **c** $(100-6x)^2$ **d** $(x^2-2)^2$

e $(25+24y)^2$ **f** $(\frac{5}{2}+x)^2$ **g** $(\frac{x}{2}-5)^2$ **h** $(0.5x-0.8)^2$

i $(5+1.4y)^2$ **j** $(ax-b)^2$

Activity

Finding the square of a number using the expansion of a perfect square

For example, find the square of 71.

$71^2 = (70+1)^2$

$= 70^2 + 2 \times 70 \times 1 + 1^2$ using 70 as the first term, 1 as the second

$= 4900 + 140 + 1$ term and the perfect square expansion

$= 5041$

4 Use the expansion of a perfect square rule to evaluate the following.

a 41^2 **b** 39^2 **c** 52^2 **d** 1001^2

e 998^2 **f** 85^2 **g** 123^2 **h** 950^2

i 6.05^2 **j** 4.8^2

Lesson 3

Expansions that produce a difference of two squares

Consider the expansion of an expression of the form $(a + b)(a - b)$.

$(a + b)(a - b) = a \times a + a \times {}^{-}b + b \times a + b \times {}^{-}b$

$= a^2 - ab + ab - b^2$: the two middle terms cancel each other out

$= a^2 - b^2$

The expression $a^2 - b^2$ is called '**a difference of two squares**' and this is abbreviated to **DOTS**.

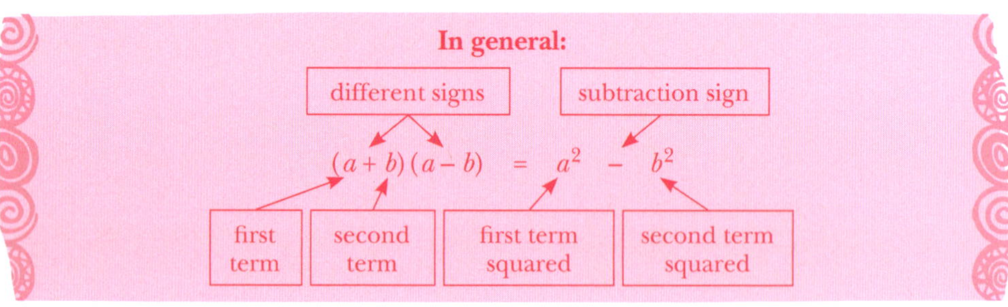

Example

Expand the following using the DOTS rule.

a $(x + 2)(x - 2)$

b $(3x - 5)(3x + 5)$

c $(1 + 5y)(1 - 5y)$

Answer

a $(x + 2)(x - 2)$ — the first term is x and the second term is 2

$(x + 2)(x - 2) = x^2 - 2^2$

$= x^2 - 4$

b $(3x - 5)(3x + 5)$ — the first term is $3x$ and the second term is 5

$(3x - 5)(3x + 5) = (3x)^2 - 5^2$

$= 9x^2 - 25$

c $(1 + 5y)(1 - 5y)$ — the first term is 1 and the second term is $5y$

$(1 + 5y)(1 - 5y) = 1^2 - (5y)^2$

$= 1 - 25y^2$

EXERCISE

1 Expand the following using the DOTS rule.

a	$(x + 3)(x - 3)$	**b**	$(m + n)(m - n)$
c	$(2x + 3)(2x - 3)$	**d**	$(7 - x)(7 + x)$
e	$(5x + 3)(5x - 3)$	**f**	$(1 - y)(1 + y)$
g	$(4x + 3)(3 - 4x)$	**h**	$(4 - 5y)(4 + 5y)$
i	$(7x + 3y)(7x - 3y)$	**j**	$3(x - 5)(5 + x)$

2 The following questions contain expansions of all the types covered so far. Expand then simplify, if possible.

a	$(x + 3)(x - 8)$	**b**	$(2m + n)(m - n)$
c	$(x + 3)(2x - 13)$	**d**	$(y - 9)^2$
e	$(5x + 12)^2$	**f**	$(1 - 3x)(2 + 5x)$
g	$(0.7x + 0.8)(0.7x - 0.8)$	**h**	$(5y + 6)(5y + 6)$
i	$(1 + 4x)(2 - 3x)$	**j**	$(5 + 6y)^2$
k	$(x + 0.5)(4x + 7)$	**l**	$(ab - c)(ab + c)$

Factorising quadratic expressions

Lesson 4

Factorisation is the reverse process to expanding, and for quadratics it is generally a more difficult task.

Why do we need to factorise quadratic expressions?

The answer is we need to be able to factorise quadratic expressions so that we can **solve a quadratic equation**.

A quadratic equation cannot be 'unravelled' like a linear equation by applying reverse operations: this process comes 'unstuck' because of the x^2 term. We will explore solving quadratic equations in later lessons.

Factorisation of quadratic expressions

Quadratic expressions fall into three types:

1 quadratic expressions that are a **difference of two squares**, e.g. $x^2 - 9 = x^2 - 3^2$, in which both terms can be written as perfect squares and they are separated by a subtraction sign.

2 quadratic expressions that are **perfect squares**, e.g. $x^2 + 6x + 9 = x^2 + 2 \times x \times 3 + 3^2$, where the first and last terms are perfect squares of two quantities and the middle term is twice the product of the quantities.

3 **quadratic trinomials**, e.g. $x^2 + 7x + 12 = x^2 + 3x + 4x + 3 \times 4 = (x + 3)(x + 4)$.

To factorise a quadratic expression we categorise the expression by the following list so that we can use the most appropriate method.

A checklist for factorisation of quadratic expressions

- **Is there a factor common to all the terms?**
  Common factors should be taken out before any other factorisation technique is applied.
- **Does the expression have two terms?**
  It could be a **difference of two squares** where both terms are squares and one is subtracted from the other:
  $$a^2 - b^2 = (a + b)(a - b)$$
- **Does the expression have three terms?**
  - i It could factorise to a **perfect square** where two terms are squares and the remaining term is twice the product of the terms in the bracket:
    $$a^2 + 2ab + b^2 = (a + b)^2 \text{ or } a^2 - 2ab + b^2 = (a - b)^2$$
  - ii It could be a **quadratic trinomial**:
    $$x^2 + (a + b)x + ab = (x + a)(x + b)$$

In this lesson we will concentrate on recognising the type of quadratic expression and on factorising DOTS and perfect squares. Factorising quadratic trinomials will be covered in Lessons 5 and 6.

Factorising an expression that is the difference of two squares

The **expanded form** of an expression that is the difference of two squares is $\boldsymbol{a^2 - b^2}$.

The **factorised form** of this difference of two squares is $\boldsymbol{(a + b)(a - b)}$.

Note: This factorised form can be written as $(a - b)(a + b)$; the order of the factors is not important.

To factorise an expression that is the difference of two squares (DOTS):

1 Establish that the expression contains two terms that can both be written as squares. Write the two terms as squares.

2 Check that the two terms are separated by a subtraction sign.

3 Use the rule $a^2 - b^2 = (a + b)(a - b)$ to factorise the expression.

Example 1

Factorise the following DOTS expressions:

a $m^2 - n^2$

b $x^2 - 16$

c $9a^2 - 49$

d $8x^2 - 18$

Answer

a $m^2 - n^2$
So $m^2 - n^2 = (m + n)(m - n)$

Both terms are squares; the terms are separated by a subtraction sign.

b $x^2 - 16$

$x^2 - 16 = x^2 - 4^2$
$= (x + 4)(x - 4)$

Both terms are squares, $16 = 4^2$; the terms are separated by a subtraction sign.

c $9a^2 - 49$

$9a^2 - 49 = (3a)^2 - 7^2$
$= (3a + 7)(3a - 7)$

Both terms are squares, $9a^2 = (3a)^2$, $49 = 7^2$; the terms are separated by a subtraction sign.

d $8x^2 - 18$

$8x^2 - 18 = 2[4x^2 - 9]$
$= 2[(2x)^2 - 3^2]$
$= 2(2x + 3)(2x - 3)$

There is a common factor of 2 in both the terms. Both terms in the bracket are squares.

Factorising expressions that produce a perfect square

The **expanded version** of a perfect square is $a^2 + 2ab + b^2$ or $a^2 - 2ab + b^2$.
The **factorised version** of a perfect square is $(a + b)^2$ or $(a - b)^2$ respectively.

To factorise a quadratic expression that is a perfect square:

1 Establish that the first and last terms are perfect squares of two quantities and the middle term is twice the product of the quantities.

2 Write the first and last term as a square of a quantity and the middle term as twice the product of the two quantities.

3 Write the expression as a perfect square checking that the sign in the bracket is the same as the sign in front of the middle term.

Example 2

Establish that the following quadratic expressions are perfect squares and then factorise:

a $x^2 + 8x + 16$ **b** $2x^2 - 20x + 50$

Answer

a $x^2 + 8x + 16$

The first term is x squared and the last term is 4 squared, so these terms are squares. The middle term has to be twice the product of the quantities that are squared: $8x = 2 \times x \times 4$ so this expression is a perfect square.

Hence $x^2 + 8x + 16 = x^2 + 2 \times x \times 4 + 4^2$

$= (x + 4)^2$

Checking that the sign in the bracket is the same as the sign in front of the middle term.

b $2x^2 - 20x + 50$

There is a factor of 2 common to all the terms in this expression so this will be dealt with first.

$2x^2 - 20x + 50 = 2[x^2 - 10x + 25]$

Now the first term in the bracket is x squared and the last term is 5 squared, so these terms are squares. The middle term has to be twice the product of the quantities that are squared: $10x = 2 \times x \times 5$ so this expression is a perfect square, noting that the sign in front of the middle term is a subtraction sign.

$2[x^2 - 10x + 25] = 2[x^2 - 2 \times x \times 5 + 5^2]$

$= 2(x - 5)^2$

EXERCISE

1 Which of the following expressions are 'difference of two squares'?

a $c^2 - d^2$ **b** $9 - a$ **c** $1 - x^2$ **d** $x^2 - x$

e $36 - y^2$ **f** $(4x)^2 + y^2$ **g** $16y^2 - 9$ **h** $25 - 4x$

2 Copy and complete the following factorisations of DOTS.

a $a^2 - 4b^2 = a^2 - (\ldots.)^2 = (a + \ldots.)(a - \ldots.)$

b $x^2 - 36 = x^2 - (\ldots.)^2 = (x - \ldots.)(x + \ldots.)$

c $9 - 4a^2 = \ldots.^2 - (2a)^2 = (\ldots. - \ldots.)(\ldots. + 2a)$

d $25x^2 - 4y^2 = (5x)^2 - [\ldots.]^2 = (5x - \ldots)(5x + \ldots)$

3 Factorise the following DOTS expressions using the rule $a^2 - b^2 = (a + b)(a - b)$.

a $p^2 - q^2$ **b** $x^2 - 4$ **c** $9m^2 - 1$ **d** $y^2 - 36$

e $16 - 9y^2$ **f** $4x^2 - 1$ **g** $25x^2 - 9y^2$ **h** $64p^2 - 25q^2$

i $50x^2 - 2$ **j** $50 - 18k^2$ **k** $100x^2 - 25$ **l** $100x^2 - 121$

m $48 - 75a^2$ **n** $x^4 - x^2$ **o** $4x - x^3$ **p** $5p^2 - 5p^2q^2$

4 Which of the following quadratic expressions will factorise to a perfect square?

a $x^2 + 4x + 4$	b $y^2 - 18x + 9$	c $x^2 - 36$
d $x^2 + 16 + 8x$	e $x^2 - 14x - 49$	f $m^2 - 50m + 25$
g $4x^2 + 8x + 4$	h $3p^2 - 12p - 12$	

5 Factorise the following expressions to perfect squares (check for common factors first).

a $x^2 + 4x + 4$	b $y^2 - 6x + 9$	c $x^2 - 12x + 36$
d $x^2 + 16 + 8x$	e $x^2 - 14x + 49$	f $m^2 - 10m + 25$
g $4x^2 + 8x + 4$	h $3p^2 - 12p + 12$	i $2x^2 + 18 + 12x$
j $3y^2 - 24y + 48$	k $72 - 24x + 2x^2$	l $ax^2 + 2ax + a$

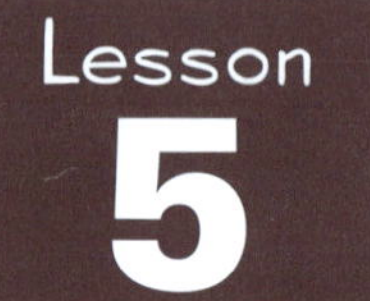

Factorisation of quadratic trinomials of the form $ax^2 + bx + c$

A **quadratic trinomial of the form** $ax^2 + bx + c$ is a quadratic expression with **three terms**. The variable in this expression is x and the pronumerals, a, b and c, represent numbers.

In this expression a is called the coefficient of x^2, b is the coefficient of x and c is the constant term.

Quadratic trinomials are produced when an expression such as $(x + 2)(x + 5)$ is expanded:

$$\begin{aligned}(x+2)(x+5) &= x \times x + x \times 5 + 2 \times x + 2 \times 5 \\ &= x^2 + 5x + 2x + 10\text{: collect like terms} \\ &= x^2 + 7x + 10\end{aligned}$$

Although four terms are produced in the expansion, two of the terms are like terms and hence the expression is reduced to a trinomial.

First we are considering factorisations of quadratic trinomials where the **coefficient** of x^2 is one.

Consider the expansion of $(x + 2)(x + 5) = x \times x + x \times 5 + 2 \times x + 2 \times 5$

$$\begin{aligned} &= x^2 + 5x + 2x + 10 \\ &= x^2 + 7x + 10\end{aligned}$$

Considered in detail:

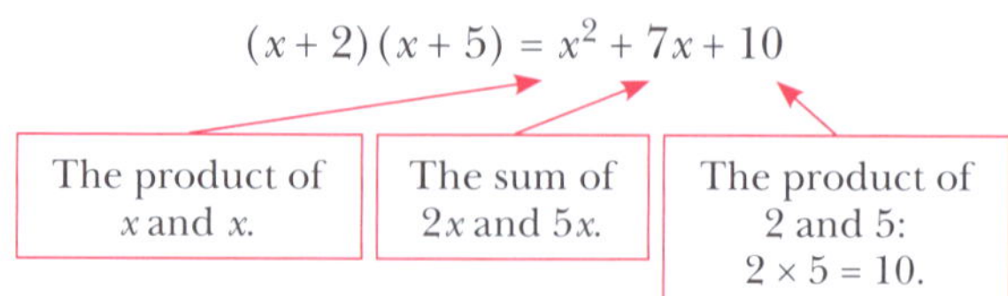

A general term that will produce a quadratic trinomial where all the terms are positive has the form $(x + a)(x + b)$, where the pronumerals a and b represent positive numbers.

Expanding this:

$$\begin{aligned}(x+a)(x+b) &= x \times x + x \times b + a \times x + a \times b \\ &= x^2 + bx + ax + ab \\ &= x^2 + (a+b)x + ab\end{aligned}$$

$(a + b)$ is the coefficient of x.

ab is the constant term.

Factorising a quadratic trinomial requires us to go from an expression in the form $x^2 + (a + b)x + ab$ to the factorised form $(x + a)(x + b)$.

$$x^2 + (a+b)x + ab = (x+a)(x+b)$$

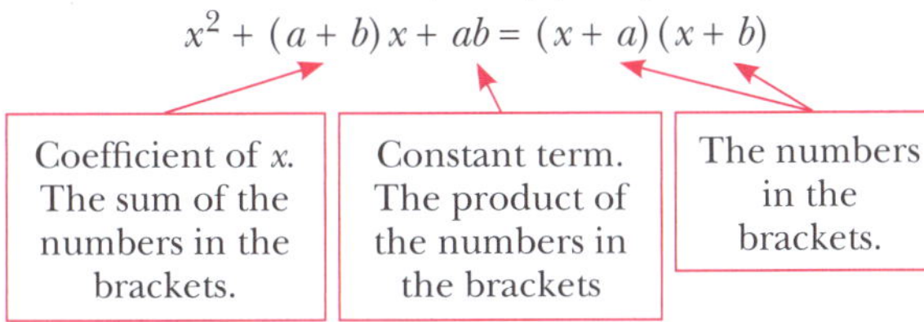

In summary, we need to find two numbers that are a factor pair of the constant term that also sum to the coefficient of x. These two numbers will be the 'numbers in the brackets' of the factorised expression.

Here is an example.

Example 1

Factorise $x^2 + 11x + 24$.

Answer

The first step is to consider the factor pairs of the constant term 24; we need to find a pair that sums to the coefficient of x, which is 11.

Factor pairs:	24×1	12×2	8×3	6×4
Sum of factors:	$24 + 1 = 25$	$12 + 2 = 14$	$8 + 3 = 11$	$6 + 4 = 10$

This is the pair of factors we are looking for. ($8 + 3 = 11$)

Hence $x^2 + 11x + 24 = x^2 + 8x + 3x + 24$

$= x(x + 8) + 3(x + 8)$ Grouping the terms and factorising.

$= (x + 8)(x + 3)$ Common factor of $(x + 8)$

Example 2

Factorise $x^2 + 9x + 18$.

Answer

Factor pairs of 18:	18×1	9×2	6×3
Sum of factors:	$18 + 1 = 19$	$9 + 2 = 11$	$6 + 3 = 9$

The factors we are looking for are 6 and 3.

$$\begin{aligned}\text{Hence } x^2 + 9x + 18 &= x^2 + 6x + 3x + 18 \\ &= x(x+6) + 3(x+6) \\ &= (x+6)(x+3)\end{aligned}$$

When you are sure of the technique then these two lines can be left out.

Factorising quadratic trinomials of the form $x^2 - bx + c$

What types of factors produce a quadratic trinomial where the coefficient of x (the middle term) is negative but the constant term is positive?

Consider the expansion of $(x-5)(x-3)$.

$$\begin{aligned}(x-5)(x-3) &= x \times x + x \times {}^{-}3 + {}^{-}5 \times x + {}^{-}5 \times {}^{-}3\\ &= x^2 - 3x - 5x + 15\\ &= x^2 - 8x + 15\end{aligned}$$

$^{-}8$ is the sum of $^{-}5$ and $^{-}3$.

15 is the product of $^{-}5$ and $^{-}3$.

This type of expansion, where both the numbers in the brackets are subtracted, produces a quadratic trinomial of the form $ax^2 - bx + c$.

Coefficient of x Negative because we are adding two negative numbers.

Constant term Positive because we are multiplying two negative numbers.

In summary, to factorise an expression of the form $ax^2 - bx + c$ we need to find factors of c (they will both be negative) that add to ^{-}b. If these numbers are ^{-}m and ^{-}n then the expression is factorised as $(x-m)(x-n)$.

Example 3

Factorise:

a $x^2 - 14x + 24$ **b** $x^2 - 9x + 8$

Answer

a $x^2 - 14x + 24$

We need to find a pair of negative factors of 24 that sum to $^{-}14$.

Factor pairs:	$^{-}24 \times {}^{-}1$	$^{-}12 \times {}^{-}2$	$^{-}8 \times {}^{-}3$	$^{-}6 \times {}^{-}4$
Sum of factors:	$^{-}25$	$^{-}14$	$^{-}11$	-10

These are the factors we are looking for.

$$\begin{aligned}\text{Hence } x^2 - 14x + 24 &= x^2 - 12x - 2x + 24\\ &= x(x-12) - 2(x-12) && \text{grouping and factorising}\\ &= (x-12)(x-2) && (x-12) \text{ is a common factor}\end{aligned}$$

b $x^2 - 9x + 8$

We need to find a pair of negative factors of 8 that sum to $^{-}9$.

Factor pairs:	$^{-}8 \times {}^{-}1$	$^{-}4 \times {}^{-}2$
Sum of factors:	$^{-}9$	$^{-}6$

These are the factors we are looking for.

When you are sure of the technique then these two lines can be left out.

$$\begin{aligned}\text{Hence } x^2 - 9x + 8 &= x^2 - 8x - x + 8\\ &= x(x-8) - (x-8) && \text{grouping and factorising}\\ &= (x-8)(x-1) && (x-8) \text{ is a common factor}\end{aligned}$$

EXERCISE

1 Find two numbers that:

a multiply to 16 and add to 10
b multiply to 32 and add to 18
c multiply to 28 and add to $^-16$
d multiply to 48 and add to $^-14$
e multiply to 64 and add to 65
f multiply to 64 and add to 34
g multiply to 72 and add to $^-17$
h multiply to 54 and add to $^-15$

2 Factorise the following quadratic trinomial expressions where all the terms are positive.

a $x^2 + 5x + 6$
b $m^2 + 5m + 4$
c $p^2 + 11p + 24$
d $q^2 + 14q + 24$
e $x^2 + 12x + 36$
f $x^2 + 11x + 10$
g $x^2 + 10x + 21$
h $a^2 + 15a + 36$
i $b^2 + 16b + 48$
j $x^2 + 16x + 63$
k $t^2 + 100t + 2400$
l $s^2 + 18s + 81$

3 Factorise the following quadratic trinomial expressions where the constant term is positive and the 'middle' term is negative.

a $x^2 - 3x + 2$
b $m^2 - 7m + 12$
c $p^2 - 13p + 36$
d $q^2 - 8q + 12$
e $x^2 - 14x + 45$
f $x^2 - 13x + 12$
g $t^2 - 16t + 64$
h $a^2 - 15a + 36$
i $b^2 - 23b + 132$
j $x^2 - 24x + 63$
k $x^2 - 14x + 49$
l $x^2 - 10x + 21$

Factorisation of quadratic trinomials of the form $x^2 + bx - c$ or $x^2 - bx - c$

What types of factors produce a quadratic trinomial where the coefficient of x is positive but the constant term is negative?

Consider the expansion of $(x + 5)(x - 3)$.

$$
\begin{aligned}
(x + 5)(x - 3) &= x \times x + x \times {}^-3 + 5 \times x + 5 \times {}^-3 \\
&= x^2 - 3x + 5x - 15 \\
&= x^2 + 2x - 15
\end{aligned}
$$

2 is the sum of 5 and $^-3$.

$^-15$ is the product of 5 and $^-3$.

If we had to factorise the expression $x^2 + 2x - 15$ we would be looking for a factor pair of $^-15$ where the factors add to 2. One number would have to be positive and the other negative because their product is negative.

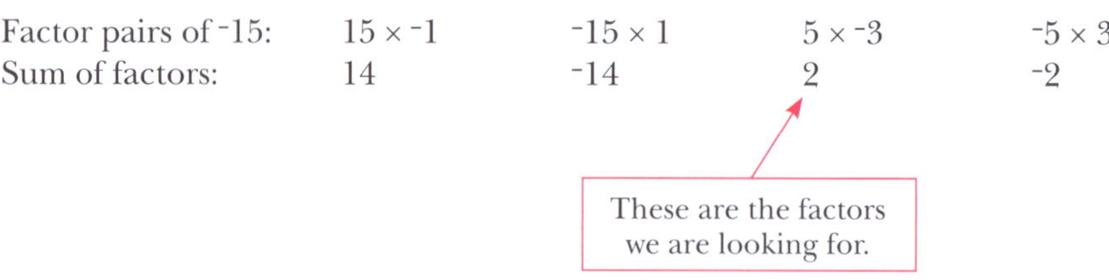

Factor pairs of -15:	15×-1	-15×1	5×-3	-5×3
Sum of factors:	14	-14	2	-2

The factors are 5 and -3, hence the factorisation of $x^2 + 2x - 15$ is $(x + 5)(x - 3)$.

What types of factors produce a quadratic trinomial where the coefficient of x is negative and the constant term is negative?
Consider the expansion of $(x - 6)(x + 4)$.

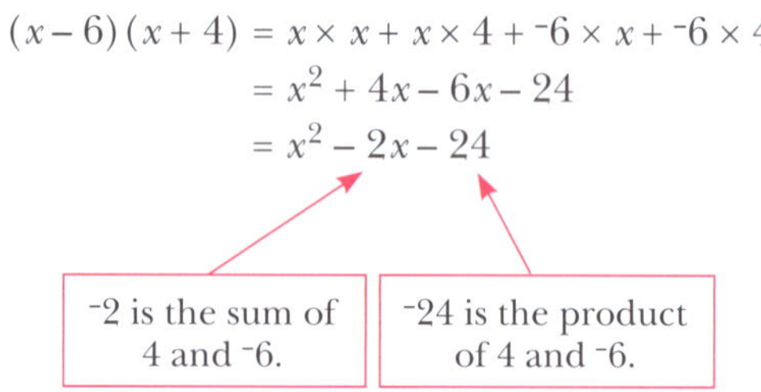

$$\begin{aligned}(x - 6)(x + 4) &= x \times x + x \times 4 + {}^{-}6 \times x + {}^{-}6 \times 4 \\ &= x^2 + 4x - 6x - 24 \\ &= x^2 - 2x - 24\end{aligned}$$

The constant term is negative because one number in the brackets is positive and the other is negative. The coefficient of the x term (the middle term) is negative in this case because $4 + {}^{-}6 = {}^{-}2$; the negative factor has greater magnitude than the positive factor.
If we had to factorise the expression $x^2 - 2x - 24$ we would be looking for a factor pair of -24 where the factors add to -2.
One number would have to be positive and the other negative because their product is negative.

Factor pairs of -24:	24×-1	-24×1	12×-2	-12×2	8×-3	-8×3	6×-4	-6×4
Sum of factors:	23	-23	10	-10	5	-5	2	-2

These are the factors we are looking for.

The factors are -6 and 4, hence the factorisation of $x^2 + 2x - 15$ is $(x - 6)(x + 4)$.
In summary, if the constant term is negative then the factor pair will consist of one positive number and one negative number. The coefficient of x will be positive or negative depending on the relative magnitude of the factors.

Example

Factorise these quadratic trinomials with a negative constant term.

a $x^2 + 3x - 10$ **b** $m^2 - 5m - 14$ **c** $p^2 - p - 56$

Answer

a $x^2 + 3x - 10$

We are looking for a factor pair of $^{-}10$ where the factors add to 3.

Factor pairs:	$10 \times {}^{-}1$	$^{-}10 \times 1$	$5 \times {}^{-}2$	$^{-}5 \times 2$
Sum of factors:	9	$^{-}9$	3	$^{-}3$

These are the factors we are looking for.

Hence $x^2 + 3x - 10 = x^2 + 5x - 2x - 10$

$= x(x + 5) - 2(x + 5)$

$= (x + 5)(x - 2)$

b $m^2 - 5m - 14$

We are looking for a factor pair of $^{-}14$ where the factors add to $^{-}5$.

Factor pairs:	$14 \times {}^{-}1$	$^{-}14 \times 1$	$7 \times {}^{-}2$	$^{-}7 \times 2$
Sum of factors:	13	$^{-}13$	5	$^{-}5$

These are the factors we are looking for.

Hence $m^2 - 5m - 14 = m^2 - 7m + 2m - 14$

$= m(m - 7) + 2(m - 7)$

$= (m - 7)(m + 2)$

c $p^2 - p - 56$

We are looking for a factor pair of $^{-}56$ where the factors add to $^{-}1$.

Factor pairs:	$56 \times {}^{-}1$	$^{-}56 \times 1$	$26 \times {}^{-}2$	$^{-}26 \times 2$	$14 \times {}^{-}4$	$^{-}14 \times 4$	$8 \times {}^{-}7$	$^{-}8 \times 7$
Sum of factors:	55	$^{-}55$	24	$^{-}24$	10	$^{-}10$	1	$^{-}1$

These are the factors we are looking for.

Hence $p^2 - p - 56 = p^2 - 8p + 7p - 56$

$= p(p - 8) + 7(p - 8)$

$= (p - 8)(p + 7)$

EXERCISE

1 Factorise the following quadratic trinomial expressions where the constant term is negative.

a $x^2 - 2x - 8$
b $m^2 + 2m - 15$
c $p^2 - 4p - 32$
d $q^2 - q - 20$
e $x^2 + 4x - 21$
f $x^2 - x - 42$
g $x^2 + 2x - 99$
h $a^2 - 5a - 36$
i $b^2 + 2b - 35$
j $x^2 + 5x - 24$
k $m^2 - 5m - 66$
l $p^2 + 9p - 22$
m $a^2 - 4a - 32$
n $b^2 + 12b - 45$
o $x^2 + x - 56$
p $t^2 + 4t - 60$

2 Factorise the following 'mixed' quadratic trinomials.

a $x^2 - 9x + 8$
b $m^2 - 2m - 15$
c $p^2 + 14p - 32$
d $q^2 + 12q + 36$
e $x^2 - 4x - 21$
f $x^2 + 11x - 42$
g $x^2 + 10x - 39$
h $a^2 - 9a - 36$
i $b^2 + 36b + 35$
j $x^2 - 5x + 4$
k $m^2 + 5m - 84$
l $p^2 + 7p - 60$
m $a^2 - 26a + 120$
n $b^2 + 18b + 45$
o $x^2 + x - 90$
p $t^2 - 8t - 240$
q $s^2 - 30s - 400$
r $x^2 - 24x + 144$
s $b^2 - b - 56$
t $x^2 - 22x + 120$
u $m^2 + 2m - 63$
v $p^2 - 21p - 72$
w $a^2 + 37a + 70$
x $b^2 + b - 72$
y $x^2 - 13x - 48$
z $t^2 - 17t + 42$

Factorisation of quadratic trinomials of the form $ax^2 + bx + c$ $(a \neq 1)$

Factorisation of quadratic trinomials where the **coefficient of x^2 is not one** is more complex because the factors of this coefficient affect the middle term.

For example, consider the expansion of $(2x + 1)(x + 3)$.

$$\begin{aligned}(2x + 1)(x + 3) &= 2x \times x + 2x \times 3 + 1 \times x + 1 \times 3 \\ &= 2x^2 + 6x + x + 3 \\ &= 2x^2 + 7x + 3 \quad \text{a quadratic trinomial with 2 as the coefficient of } x^2\end{aligned}$$

The 'middle' term is $7x = 6x + x$; the sum of the 'outers' and the 'inners' when expanding. Considering only the factors of the constant term when factorising will not lead to the middle term. So the middle term is a combination of the factors of the x^2 term and the factors of the constant term.

To factorise $2x^2 + 7x + 3$:

Factor pairs of the x^2 term: $2x$ and x
Factor pairs of the constant term: 3 and 1

There is a special way these can be set out to find the correct arrangement of these factors.

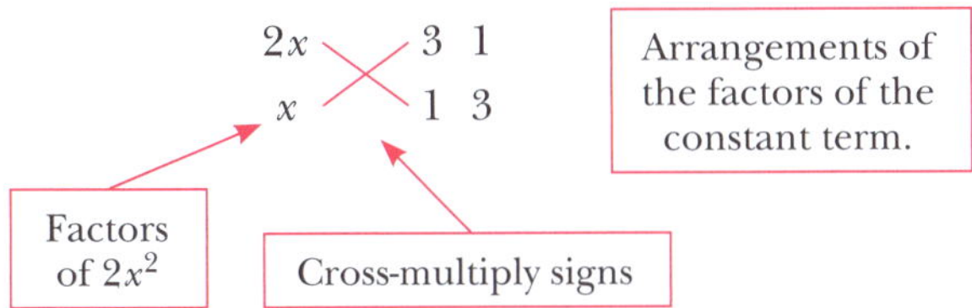

Having set up the factors we cross-multiply to find which arrangement gives the middle term.

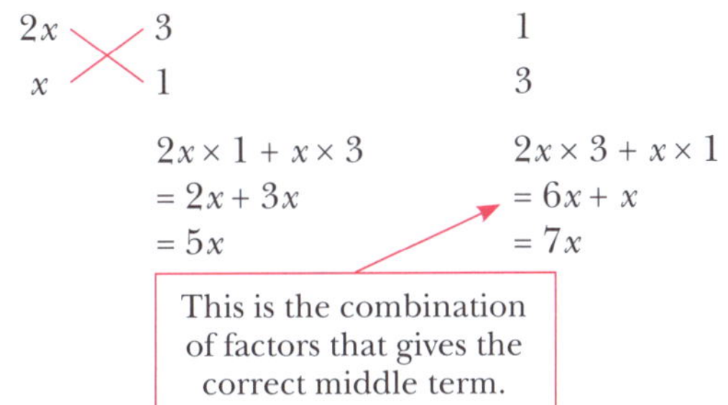

Note: The brackets of the factorised form can be seen in the cross-multiplication setting-out:

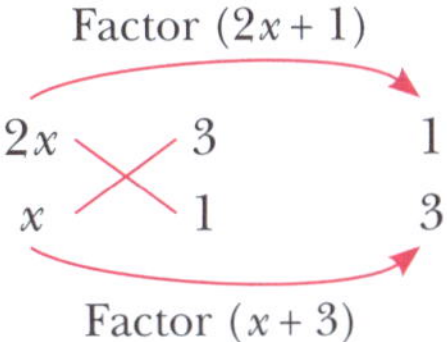

Factorisation by grouping confirms this:

$$
\begin{aligned}
2x^2 + 7x + 3 &= 2x^2 + 6x + x + 3 \\
&= 2x(x + 3) + (x + 3) \\
&= (x + 3)(2x + 1)
\end{aligned}
$$

Example 1

Use the method outlined above to factorise the quadratic trinomial $3x^2 + 8x + 4$.

Answer

We need to find factor pairs of $3x^2$ and 4:

Factors of $3x^2$	Factors of 4		
$3x$	4	1	2
x	1	4	2
Cross-multiplying to find middle term	$3x \times 1 + x \times 4$ $= 3x + 4x$ $= 7x$	$3x \times 4 + x \times 1$ $= 12x + x$ $= 13x$	$3x \times 2 + x \times 2$ $= 6x + 2x$ $= 8x$

This is the combination of factors that gives the correct middle term.

The factors of $3x^2 + 8x + 4$ are therefore $(3x + 2)(x + 2)$.

Confirming this:

$$
\begin{aligned}
3x^2 + 8x + 4 &= 3x^2 + 6x + 2x + 4 \\
&= 3x(x + 2) + 2(x + 2) \\
&= (x + 2)(3x + 2)
\end{aligned}
$$

Example 2

Factorise $2x^2 - 7x + 6$.

Answer

The middle term of the expression is negative but the constant term is positive, so we are looking at two negative factors of the constant term.

Factors of $2x^2$	Negative factors of 6			
$2x$	$^-6$	$^-1$	$^-3$	$^-2$
x	$^-1$	$^-6$	$^-2$	$^-3$
Cross-multiplying to find middle term	$2x \times {}^-1 + x \times {}^-6$ $= {}^-2x + {}^-6x$ $= {}^-8x$	$2x \times {}^-6 + x \times {}^-1$ $= {}^-12x + {}^-x$ $= {}^-13x$	$2x \times {}^-2 + x \times {}^-3$ $= {}^-4x + {}^-3x$ $= {}^-7x$	$2x \times {}^-3 + x \times {}^-2$ $= {}^-6x + {}^-2x$ $= {}^-8x$

This is the combination of factors that gives the correct middle term.

The factors of $2x^2 - 7x + 6$ are therefore $(2x - 3)(x - 2)$.

Confirming this: $2x^2 - 7x + 6 = 2x^2 - 4x - 3x + 6$

$= 2x(x - 2) + {}^-3(x - 2)$

$= (x - 2)(2x - 3)$

Example 3

Factorise $5x^2 + 2x - 3$.

Answer

The middle term of the expression is positive but the constant term is negative, so we are looking at one negative factor and one positive factor of the constant term.

Factors of $5x^2$	Factors of $^-3$			
$5x$	3	$^-1$	$^-3$	1
x	$^-1$	3	1	$^-3$
Cross-multiplying to find middle term	$5x \times {}^-1 + x \times 3$ $= {}^-5x + 3x$ $= {}^-2x$	$5x \times 3 + x \times {}^-1$ $= 15x + {}^-x$ $= 14x$	$5x \times 1 + x \times {}^-3$ $= 5x + {}^-3x$ $= 2x$	$5x \times {}^-3 + x \times 1$ $= {}^-15x + x$ $= {}^-14x$

This is the combination of factors that gives the correct middle term.

The factors of $5x^2 + 2x - 3$ are $(5x - 3)(x + 1)$.

Confirming this: $5x^2 + 2x - 3 = 5x^2 + 5x - 3x - 3$

$= 5x(x + 1) - 3(x + 1)$

$= (x + 1)(5x - 3)$

If there is a choice of factors for the coefficient of the x^2 term then more combinations of factors need to be considered.

Example 4

Factorise $6x^2 + 25x + 4$.

Answer

All terms are positive so we are looking at positive factors only.

The factors of $6x^2$ are $3x$ and $2x$ or $6x$ and x.

The factors of 4 are 4 and 1, 1 and 4 or 2 and 2.

Factors of $6x^2$	**Factors of 4**		
$3x$ ╳ $2x$	4 1	1 4	2 2
Cross-multiplying to find middle term	$3x \times 1 + 2x \times 4$ $= 3x + 8x$ $= 11x$	$3x \times 4 + 2x \times 1$ $= 12x + 2x$ $= 14x$	$3x \times 2 + 2x \times 2$ $= 6x + 4x$ $= 10x$

Factors of $6x^2$	**Factors of 4**		
$6x$ ╳ x	4 1	1 4	2 2
Cross-multiplying to find middle term	$6x \times 1 + x \times 4$ $= 6x + 4x$ $= 10x$	$6x \times 4 + x \times 1$ $= 24x + x$ $= 25x$	$6x \times 2 + x \times 2$ $= 12x + 2x$ $= 14x$

Hence $6x^2 + 25x + 4 = (6x + 1)(x + 4)$.

This is the combination of factors that gives the correct middle term.

Example 5

Factorise $4x^2 + 5x - 6$.

Answer

The constant term is negative indicating that one number in the factor pair of the constant term will be positive and the other negative.

The factors of $4x^2$ are $2x$ and $2x$ or $4x$ and x.

The factors of ${}^{-}6$ are 6 and ${}^{-}1$, ${}^{-}6$ and 1, 3 and ${}^{-}2$ or ${}^{-}3$ and 2.

Factors of $4x^2$	**Factors of ${}^{-}6$**			
$2x$ ╳ $2x$	6 ${}^{-}1$	${}^{-}6$ 1	3 ${}^{-}2$	${}^{-}3$ 2
Cross-multiplying to find middle term	$2x \times {}^{-}1 + 2x \times 6$ $= {}^{-}2x + 12x$ $= 10x$	$2x \times 1 + 2x \times {}^{-}6$ $= 2x + {}^{-}12x$ $= {}^{-}10x$	$2x \times {}^{-}2 + 2x \times 3$ $= {}^{-}4x + 6x$ $= 2x$	$2x \times 2 + 2x \times {}^{-}3$ $4x + {}^{-}6x$ $= {}^{-}2x$

Factors of $4x^2$	**Factors of ${}^{-}6$**			
$4x$ ╳ x	6 ${}^{-}1$	${}^{-}6$ 1	3 ${}^{-}2$	${}^{-}3$ 2
Cross-multiplying to find middle term	$4x \times {}^{-}1 + x \times 6$ $= {}^{-}4x + 6x$ $= 2x$	$4x \times 1 + x \times {}^{-}6$ $= 4x + {}^{-}6x$ $= {}^{-}2x$	$4x \times {}^{-}2 + x \times 3$ $= {}^{-}8x + 3x$ $= {}^{-}5x$	$4x \times 2 + x \times {}^{-}3$ $= 8x + {}^{-}3x$ $= 5x$

Hence $4x^2 + 5x - 6 = (4x - 3)(x + 2)$.

This is the combination of factors that gives the correct middle term.

Note: The answers to the factorisations have been set out in full in the examples above. With practice you will be able to do most of the work mentally. Here is a typical abbreviated setting out of Example 5.

Factorise $4x^2 + 5x - 6$.

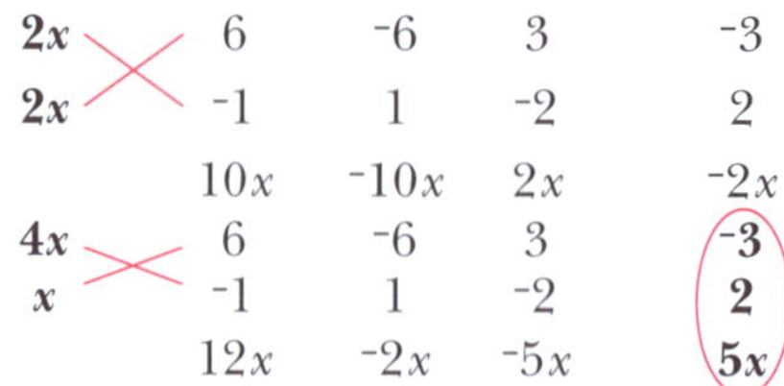

$4x^2 + 5x - 6 = (4x - 3)(x + 2)$ taking the brackets directly from the setting out.

EXERCISE*

1 Factorise the following quadratic trinomials, considering both factors of the 'squared' term and the constant term.

a $3x^2 + 11x + 6$
b $5c^2 + 19c + 12$
c $7m^2 + 29m + 4$
d $2x^2 + 9x + 10$
e $11p^2 + 15p + 4$

2 Factorise the following quadratic trinomials, considering both factors of the 'squared' term and the constant term.

a $2x^2 - 7x + 3$
b $5c^2 - 17c + 14$
c $7m^2 - 20m + 12$
d $13x^2 - 25x + 12$
e $3p^2 - 28p + 32$

3 Factorise the following quadratic trinomials, considering both factors of the 'squared' term and the constant term.

a $3x^2 - 10x - 8$
b $3c^2 + c - 10$
c $2m^2 - m - 21$
d $5x^2 + 48x - 20$
e $17p^2 + 3p - 14$
f $2x^2 + 3x - 9$
g $7x^2 - 20x - 32$
h $13b^2 + 6b - 64$

4 Factorise the following quadratic trinomials, considering both factors of the 'squared' term and the constant term.

a $6x^2 + 23x + 20$
b $4c^2 + 19c + 12$
c $10m^2 + 27m + 18$
d $12x^2 + 41x + 24$
e $12p^2 + 29p + 15$

5 Factorise the following quadratic trinomials considering both factors of the 'squared' term and the constant term.

a $15x^2 + 7x - 2$
b $14c^2 - 27c + 9$
c $12m^2 + m - 20$
d $12x^2 + 4x - 33$
e $12p^2 + 28p + 15$
f $8x^2 - 42x + 27$
g $9q^2 - 14q - 64$
h $16t^2 + 16t - 45$

Solving quadratic equations that can be factorised

A **quadratic equation** has the form $\mathbf{ax^2 + bx + c = 0}$ where a, b and c are constants and $a \neq 0$.

A quadratic equation cannot be 'unravelled' like a linear equation by applying reverse operations: this process does not work because of the x^2 term.

To solve a quadratic equation we need to first factorise the quadratic expression and then apply the **Null factor law** that states:

'If two factors that are multiplied have a result that is zero, then either or both of the factors must be zero'.

Example 1

Solve the quadratic equation $x^2 + x - 6 = 0$ and check your solutions.

Answer

First we factorise the left-hand side of the equation.

Looking at factors of $^-6$ that add to 1 gives 3 and $^-2$, so the equation in factorised form is $(x + 3)(x - 2) = 0$.

The Null factor law implies that either or both of the factors will be equal to zero, so we can say $x + 3 = 0$ or $x - 2 = 0$.

If $x + 3 = 0$ then $x = {}^-3$ and if $x - 2 = 0$ then $x = 2$.

This means that there are two possible solutions to this quadratic equation: $x = {}^-3$ and $x = 2$.

Checking the solutions by substituting:

$x = {}^-3$ into $x^2 + x - 6$ gives $({}^-3)^2 + ({}^-3) - 6 = 9 - 3 - 6 = 0$ ✓

$x = 2$ into $x^2 + x - 6$ gives $2^2 + 2 - 6 = 4 + 2 - 6 = 0$ ✓

Example 2

Solve the following quadratic equations:

a $(x + 6)(x - 2) = 0$ **b** $x^2 - 9 = 0$ **c** $x^2 - 8x + 16 = 0$

d $3x^2 - x - 4 = 0$ **e** $x^2 - 24 = 5x$

Answer

a $(x + 6)(x - 2) = 0$: The left-hand side of this equation is already factorised.
So using the Null factor law we can say $x + 6 = 0$ or $x - 2 = 0$.
If $x + 6 = 0$ then $x = {}^-6$.
If $x - 2 = 0$ then $x = 2$.
The two solutions are $x = {}^-6$ and $x = 2$.

b $x^2 - 9 = 0$: The left-hand side of this equation is a DOTS so factorising this gives $(x + 3)(x - 3) = 0$.
So using the Null factor law we can say $x + 3 = 0$ or $x - 3 = 0$.
If $x + 3 = 0$ then $x = {}^-3$.
If $x - 3 = 0$ then $x = 3$.
The two solutions are $x = {}^-3$ and $x = 3$.

c $x^2 - 8x + 16 = 0$: The left-hand side of this equation is a perfect square so factorising this gives $(x - 4)(x - 4) = (x - 4)^2 = 0$.
So using the Null factor law we can say $x - 4 = 0$ or $x - 4 = 0$.
If $x - 4 = 0$ then $x = 4$: the two solutions will be the same in this case so there is only one solution $x = 4$.

d $3x^2 - x - 4 = 0$: To factorise the left-hand side of this equation we need to look at factors of $3x^2$ and $^-4$ that give a middle term of $^-1$.
Factorised this equation will be $(3x - 4)(x + 1) = 0$.
So using the Null factor law we can say $3x - 4 = 0$ or $x + 1 = 0$.
If $3x - 4 = 0$ then $3x = 4$ and $x = \frac{4}{3}$.
If $x + 1 = 0$ then $x = {}^-1$.
The two solutions are $x = \frac{4}{3}$ and $x = {}^-1$.

e The equation $x^2 - 24 = 5x$ must be re-arranged so that all the terms are on one side and 0 is on the other side of the equation.
Subtracting $5x$ from both sides of the equation gives $x^2 - 24 - 5x = 5x - 5x$.
$x^2 - 5x - 24 = 0$: Now the equation is in the correct form to factorise.
Factors of $^-24$ that sum to $^-5$, are $^-8$ and 3, so $(x - 8)(x + 3) = 0$.
So using the Null factor law we can say $x - 8 = 0$ or $x + 3 = 0$.
If $x - 8 = 0$ then $x = 8$.
If $x + 3 = 0$ then $x = {}^-3$.
The two solutions are $x = 8$ and $x = {}^-3$.

EXERCISE

1 Solve the following quadratic equations that are in factorised form.

a $(x + 4)(x + 2) = 0$ **b** $x(x - 6) = 0$
c $(y - 4)(y - 1) = 0$ **d** $(x + 1)(2 - x) = 0$
e $(2x + 1)(x + 3) = 0$ **f** $(3m + 4)(2m - 5) = 0$

2 For each of the following quadratic equations:

i factorise the left-hand side of the equation
ii solve the equation
iii check your solutions.

a $x^2 + 8x + 15 = 0$ **b** $x^2 - x - 30 = 0$
c $x^2 + 4x - 45 = 0$ **d** $3x^2 - 27 = 0$

3 Solve the following quadratic equations.

a $x^2 - 18x + 32 = 0$ **b** $x^2 + 15 = 16x$
c $3x^2 - 39x + 36 = 0$ **d** $5x^2 + 5x = 60$
e $18 - x^2 = 7x$ **f** $x^2 - 5x = 24$

4* Solve the following quadratic equations.

a $2x^2 + 7x + 3 = 0$ **b** $3x^2 - 13x + 4 = 0$
c $3x^2 - 2x - 5 = 0$ **d** $2x^2 - 9x + 9 = 0$
e $24x^2 + 14x - 3 = 0$ **f** $12x^2 + 11x - 15 = 0$

Solving quadratic equations that cannot be factorised

If a quadratic equation cannot be factorised so that it can be solved, then we can use what is commonly referred to as the **quadratic formula** to give a solution.

If a quadratic equation has the form $ax^2 + bx + c = 0$ then the solution to this equation is given by

$$x = \frac{^{-}b \pm \sqrt{b^2 - 4ac}}{2a}$$

This quadratic formula will work for all quadratic equations that have a solution, and it gives **two different solutions** if they exist.

One solution is $x = \frac{^{-}b + \sqrt{b^2 - 4ac}}{2a}$ and the other is $x = \frac{^{-}b - \sqrt{b^2 - 4ac}}{2a}$; the difference is the sign in front of the square-root symbol.

There are two conditions that must apply for the formula to give a solution:

1. $a \neq 0$: if $a = 0$ then the equation is not a quadratic equation **and**
2. $b^2 - 4ac \geqslant 0$: if $b^2 - 4ac$ is negative then a solution cannot be found because we cannot find the square root of a negative number.

Example

Solve the quadratic equation $2x^2 - 5x - 1 = 0$ using the quadratic formula. Give your answer as an exact value and as a decimal approximation, correct to three decimal places.

Answer

For this equation $a = 2$, $b = {^{-}5}$ and $c = {^{-}1}$.

Substituting these values into $x = \frac{^{-}b \pm \sqrt{b^2 - 4ac}}{2a}$ gives $x = \frac{^{-}(^{-}5) \pm \sqrt{(^{-}5)^2 - 4 \times 2 \times {^{-}1}}}{2 \times 2}$

$$= \frac{5 \pm \sqrt{25 - (^{-}8)}}{4}$$

$$= \frac{5 \pm \sqrt{25 + 8}}{4}$$

So $x = \frac{5 + \sqrt{33}}{4}$ or $x = \frac{5 - \sqrt{33}}{4}$. These are the solutions as exact values.

Decimal approximations for these values are 2.686 and $^{-}0.186$ respectively.

How was the quadratic formula established?

The solution to the equation $2x^2 - 5x - 1 = 0$ can be found by the process of **completing the square**.

$2x^2 - 5x - 1 = 0$

$2[x^2 - \frac{5}{2}x - \frac{1}{2}] = 0$ Take the coefficient of the x^2 term out as a common factor.

$2[x^2 - \frac{5}{2}x + (\frac{5}{4})^2 - \frac{1}{2} - (\frac{5}{4})^2] = 0$ Complete the square by adding and subtracting the square of half of the coefficient of the x term. The first three terms in the bracket now form a perfect square.

$2[(x+\frac{5}{4})^2 - \frac{1}{2} - \frac{25}{16}] = 0$ Write the first three terms as a perfect square.

$2[(x+\frac{5}{4})^2 - \frac{8}{16} - \frac{25}{16}] = 0$ Change the remaining terms to the same denominator.

$2[(x+\frac{5}{4})^2 - \frac{33}{16}] = 0$ Combine the last two terms.

$(x+\frac{5}{4})^2 - \frac{33}{16} = 0$ Divide both sides of the equation by 2.

$(x+\frac{5}{4})^2 = \frac{33}{16}$ Add $\frac{33}{16}$ to both sides of the equation.

$x+\frac{5}{4} = \pm\sqrt{\frac{33}{16}}$ Take the square root of both sides of the equation. We have to include the fact that the square root could be positive or negative.

$x = \frac{^-5}{4} \pm \frac{\sqrt{33}}{4}$ Subtract $\frac{5}{4}$ from both sides and write the square root of 16 as 4.

$x = \frac{^-5 \pm \sqrt{33}}{4}$ This is the same solution we found using the quadratic formula.

As you can see, 'completing the square' is a long process and for this reason **the quadratic formula was established by the process of 'completing the square' on the general quadratic equation $ax^2 + bx + c = 0$.**

EXERCISE*

1 Use the quadratic formula $x = \frac{^-b \pm \sqrt{b^2 - 4ac}}{2a}$ to find the solution to each of these quadratic equations. Give your answers as exact solutions and as decimal approximations.

a $5x^2 + 2x - 4 = 0$
b $x^2 + 5x + 2 = 0$
c $^-2x^2 + 6x + 3 = 0$
d $4x^2 - 3x - 6 = 0$
e $6x - 7x^2 + 3 = 0$
f $3x^2 - 5 = 2x$

2 Each of the following quadratic equations is a perfect square.

i Factorise the left side of the equation and find the solution.

ii Use the quadratic formula $x = \frac{^-b \pm \sqrt{b^2 - 4ac}}{2a}$ to confirm the solution.

a $x^2 + 10x + 25 = 0$
b $x^2 - 6x + 9 = 0$

3 Referring to Question 2 above, answer the following.

a How many solutions are there for quadratic equations that are perfect squares?

b What do you notice about the value of $b^2 - 4ac$ for a quadratic equation that is a perfect square?

4 Use the quadratic formula $x = \frac{^-b \pm \sqrt{b^2 - 4ac}}{2a}$ to find the solution to each of the following quadratic equations.

a $x^2 + 3x + 5 = 0$
b $3x^2 - 4x + 4 = 0$
c $^-5x^2 + 3x - 2 = 0$

5 a How many solutions were there for each of the equations in Question 4?

b What do you notice about the value of $b^2 - 4ac$ for the quadratic equations in Question 4?

6 Copy and complete the following statements about the number of solutions to a quadratic equation.

a If the quantity $b^2 - 4ac$ in the quadratic formula is positive, then there will be .... solutions to the quadratic equation.

b If the quantity $b^2 - 4ac$ in the quadratic formula is zero, then there will be .... solutions to the quadratic equation.

c If the quantity $b^2 - 4ac$ in the quadratic formula is ...., then there will be no solutions to the quadratic equation.

Graphing quadratic functions

Lesson 10

Quadratic functions of the form $y = ax^2 + bx + c$ can be represented by a set of points (x, y) on the Cartesian plane.

A table of values and the graph for the simplest quadratic graph, $y = x^2$, is given below.

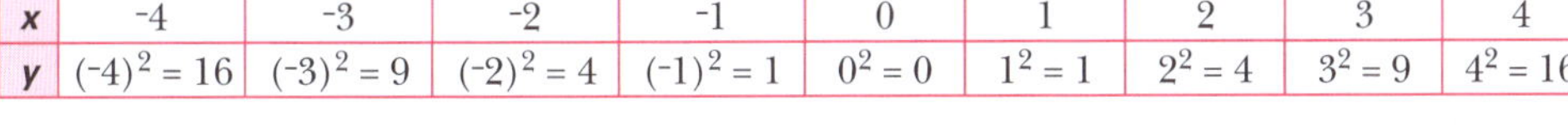

x	-4	-3	-2	-1	0	1	2	3	4
y	$(^-4)^2 = 16$	$(^-3)^2 = 9$	$(^-2)^2 = 4$	$(^-1)^2 = 1$	$0^2 = 0$	$1^2 = 1$	$2^2 = 4$	$3^2 = 9$	$4^2 = 16$

The graph of a quadratic function has a special name, the **parabola**.

The graph at left has a **minimum point** (the y-value is a minimum) at (0, 0). This point is also often called the **turning point**.

A parabola is **symmetrical** about the vertical line passing through its turning point. This line is called the **axis of symmetry**.

For the graph of $y = x^2$ the axis of symmetry is the y-axis which has equation $x = 0$.

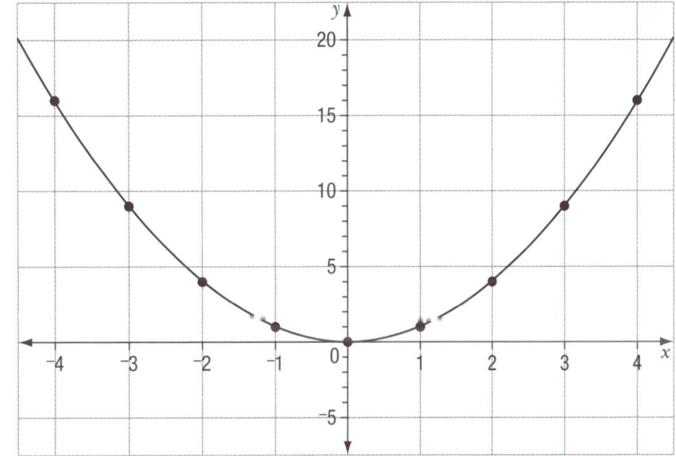

The graph of $y = {}^-x^2$ is given at right. This graph is a reflection of the graph of $y = x^2$ in the x-axis.

This graph has a **maximum point** (the y-value is a maximum) at (0, 0). This point too can be called the turning point.

For the graph of $y = {}^-x^2$ the axis of symmetry is the y-axis which has equation $x = 0$.

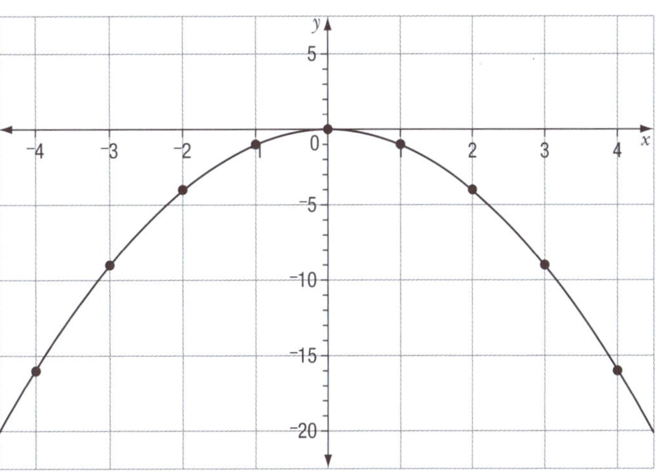

Example

For the quadratic function $y = x^2 - 4x + 3$:

a Construct a table of values for $^{-}3 \leq x \leq 5$.

b Graph this function for $^{-}3 \leq x \leq 5$.

c Write the coordinates of the turning point indicating whether this is a maximum or minimum point.

d Write the equation of the axis of symmetry.

e Write the coordinates of the x-intercepts.

f Write the coordinates of the y-intercept.

Answer

a To find the y-values we substitute the x-values into $y = x^2 - 4x + 3$.

x	-3	-2	-1	0	1	2	3	4	5
y	24	15	8	3	0	-1	0	3	8

$(^{-}3)^2 - 4 \times (^{-}3) + 3 = 24$

$2^2 - 4 \times 2 + 3 = {}^{-}1$

b We plot the points from the table and draw the parabola through them as shown in the graph.

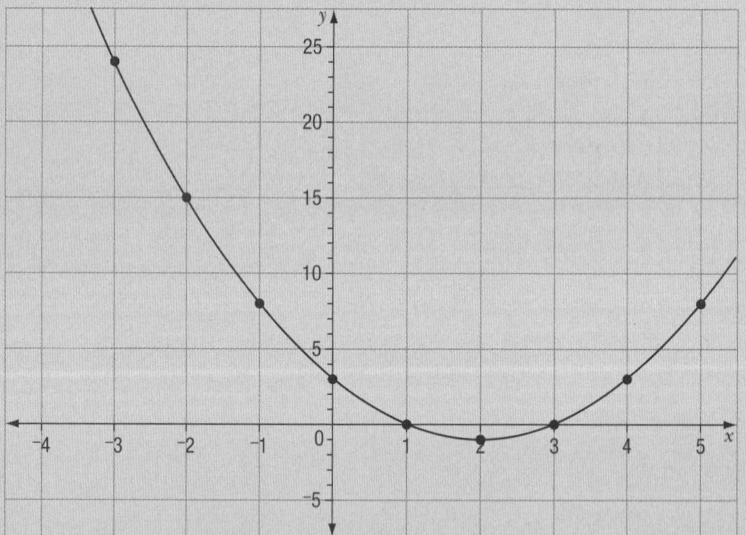

c The turning point for this graph is the point where the graph changes from decreasing to increasing. The turning point is the minimum point $(2, {}^{-}1)$.

d The axis of symmetry is the vertical line going through the turning (minimum) point. The equation is $x = 2$.

e The x-intercepts are the points where the graph cuts the x-axis. The x-intercepts are 1 and 3; coordinates $(1, 0)$ and $(3, 0)$.

f The y-intercept is the point where the graph cuts the y-axis. For this function the y-intercept is 3; coordinates $(0, 3)$.

EXERCISE

1 For each of the following quadratic functions:

  i Construct a table of values.
  ii Graph the function.
  iii Write the coordinates of the turning point indicating whether this is a maximum or minimum point.
  iv Write the equation of the axis of symmetry.
  v Write the coordinates of the *x*-intercepts.
  vi Write the coordinates of the *y*-intercept.

a $y = x^2 - 4x - 5$ b $y = 4 - x^2$ c $y = x^2 - 2x$
d $y = x^2 + 2x + 3$ e $y = 4x^2 - 1$ f* $y = 4x^2 - 4x - 3$

Using the intercepts with the axes and the turning point to sketch a quadratic function

- The ***y*-intercept** of a quadratic function can be found by substituting $x = 0$ into the equation. If the equation is in the form $y = ax^2 + bx + c$ then the ***y*-intercept is the point (0, *c*)**.
- The ***x*-intercepts** of a quadratic equation can be found by solving $y = 0$. If the equation is in the form $y = ax^2 + bx + c$ then this will involve **factorising** the left-hand side of $ax^2 + bx + c = 0$, or **using the quadratic formula**. For some quadratic functions there are no *x*-intercepts.
- The **turning point** of a quadratic function has an *x*-coordinate halfway between the *x*-intercepts. Once the *x*-coordinate of the turning point is known then this value can be substituted into the equation to find the *y*-coordinate of the turning point.

Example

Find the intercepts with the axes and the coordinates of the turning point. Use these points to sketch the quadratic function $y = x^2 - 2x - 8$.

Answer

The *y*-intercept is found by substituting $x = 0$ into the equation: $y = 0^2 - 2 \times 0 - 8 = {}^-8$.
The *y*-intercept is the point (0, ⁻8).
The *x*-intercepts can be found by solving $y = 0$. Solving $x^2 - 2x - 8 = 0$ we need to factorise the left side of this equation or use the quadratic formula to solve the equation.
Looking at factors of ⁻8 that sum to ⁻2 we can see that the values ⁻4 and ⁺2 satisfy this condition.
So $(x - 4)(x + 2) = 0$: factorising
and $x - 4 = 0$ or $x + 2 = 0$: using the Null factor rule
This gives solutions $x = 4$ and $x = {}^-2$.
The *x*-intercepts are the points (4, 0) and (⁻2, 0).
The *x*-coordinate of the turning point is halfway between the *x*-intercepts ($x = {}^-2$ and $x = 4$), which is $x = 1$.
Substituting $x = 1$ into $y = x^2 - 2x - 8$ gives $y = (1)^2 - 2 \times 1 - 8 = 1 - 2 - 8 = {}^-9$.
The coordinates of the turning point are (1, ⁻9).

By plotting the four points and drawing a line through them this graph is produced.

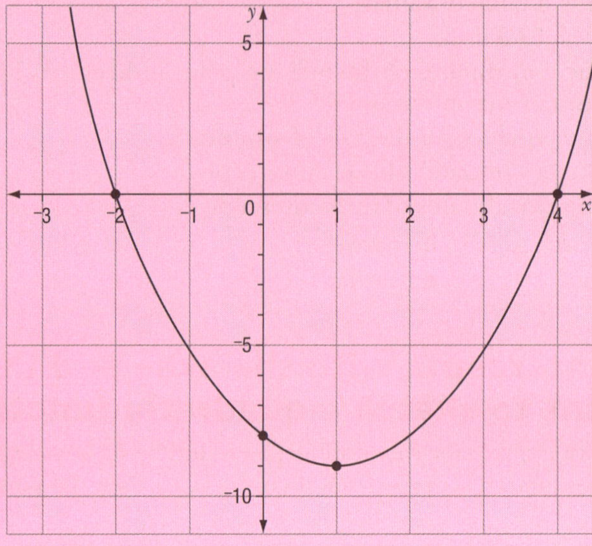

EXERCISE

1 For each of the following quadratic functions, find the intercepts with the axes and the coordinates of the turning point. Use these points to sketch the graph of the function.

a $y = x^2 + 2x - 3$

b $y = x^2 - 4$

c $y = {}^{-}(4 + x)(x - 2)$

d $y = x^2 - 3x$

e $y = (2x - 3)(2x + 1)$

2* For each of the following quadratic functions, find the intercepts with the axes and the coordinates of the turning point. Use these points to sketch the graph of the function.

a $y = 2x^2 + 3x - 2$

b $y = 3x^2 + 2x - 8$

c $y = 6x^2 - x - 15$

3* The following quadratic equations do not have *x*-intercepts. Find the *y*-intercept and the coordinates of the turning point. Use these points to sketch the graph of the function.

a $y = x^2 - 2x + 5$

b $y = 3x^2 + 6x + 5$

c $y = {}^{-}2x^2 + 8x - 9$

Solving simultaneous equations to find the equation of a quadratic function

If we are given some points on a graph then we can use a process called '**solving simultaneous equations**' to find the equation of the graph. Simultaneous means 'happening at the same time'.

We have already learnt that for a linear function we need to know two points to find the equation of the straight line. The general equation for a linear function is $y = mx + c$; there are two unknowns, m and c, so we need two points to find these values.

For the general equation of a quadratic function $y = ax^2 + bx + c$ there are three unknown values a, b and c, so we will need three points on the graph to find these values.

Sometimes the quadratic equation is given in a 'reduced' form and fewer points are needed. For example, if the quadratic function is of the form $y = ax^2 + c$ or $y = (x - p)(x - q)$ or $y = (x - h)^2 + k$ then only two points are needed to find the equation. The general rule is that the number of points needed must be equal to the number of unknown constants in the equation.

The process of finding the constant values by solving simultaneous equations is best shown by examples.

Example 1

A quadratic function is of the form $y = ax^2 + c$ and its graph goes through the points (2, 11) and ($^-1$, 2). Find the values of a and c.

Answer

We are going to substitute the x and y values from each of the points into the equation to obtain equations in a and c that we can solve simultaneously.

Substituting (2, 11) into $y = ax^2 + c$:

$11 = a \times 2^2 + c$ gives the equation $4a + c = 11$ ....(1)

Substituting ($^-1$, 2) into $y = ax^2 + c$:

$2 = a \times (^-1)^2 + c$ gives the equation $a + c = 2$ ....(2)

To solve the equations $4a + c = 11$ ....(1) and $a + c = 2$ ....(2) simultaneously, we will subtract equation (2) from equation (1) because the coefficients of c are the same (both 1) and this will eliminate c.

$(4a + c) - (a + c) = 11 - 2$

$4a + c - a - c = 9$ removing the brackets

$3a = 9$ collect like terms

$a = 3$ divide both sides of the equation by 3

If $a = 3$ then we can substitute this value into equation (2) to find c

$3 + c = 2$ subtract 3 from both sides of the equation

$c = {}^-1$

The constants are $a = 3$ and $c = {}^-1$.

The equation of the quadratic function is $y = 3x^2 - 1$.

Example 2

A quadratic function of the form $y = ax^2 + bx + c$ goes through the points $(0, ^-4)$, $(1, 1)$ and $(^-2, ^-2)$. Find the values of a, b and c and write the equation of the function.

Answer

There are three unknown to be found and we have three different points, so we can find the values of the unknowns.

Substituting $(0, ^-4)$ into $y = ax^2 + bx + c$:

$^-4 = a \times 0 + b \times 0 + c$ gives the equation $\mathbf{c = {}^-4}$.

You may have recognised this point as the y-intercept. We now know the value of c and can use this in the other equations.

Substituting $(1, 1)$ into $y = ax^2 + bx + c$:

$1 = a \times 1^2 + b \times 1 + {}^-4$ gives the equation $1 = a + b - 4$

adding 4 to both sides gives $\boldsymbol{a + b = 5}$ **.........(1)**

Substituting $(^-2, ^-2)$ into $y = ax^2 + bx + c$:

$^-2 = a \times (^-2)^2 + b \times {}^-2 + {}^-4$ gives the equation $^-2 = 4a - 2b - 4$

adding 4 to both sides $2 = 4a - 2b$:

divide both sides by the common factor 2 to give $\boldsymbol{2a - b = 1}$ **.........(2)**

To solve the equations $a + b = 5$ ....(1)

and $2a - b = 1$ ....(2) simultaneously, we will add equation (1) and equation (2) because the coefficients of b are the same value but different signs and this will eliminate b.

$a + b + (2a - b) = 5 + 1$ remove the brackets

$a + 2a + b - b = 6$ group like terms

$3a = 6$ divide both sides by 3

$\boldsymbol{a = 2}$

Substitute $a = 2$ into equation (1): $2 + b = 5$

subtract 2 from both sides: $\boldsymbol{b = 3}$

We have found the values of the constants so the equation of the function is $y = 2x^2 + 3x - 4$.

Example 3

A quadratic function of the form $y = ax^2 + bx + c$ goes through the points $(^-1, 0)$, $(2, 15)$ and $(3, 36)$. Find the values of a, b and c and write the equation of the function.

Answer

There are three unknown to be found and we have three different points, so we can find the values of the unknowns.

Substituting ($^-1$, 0) into $y = ax^2 + bx + c$:
$0 = a \times (^-1)^2 + b \times {^-1} + c$ gives the equation $\mathbf{0 = a - b + c}$ **.........(1)**
Substituting (2, 15) into $y = ax^2 + bx + c$:
$15 = a \times 2^2 + b \times 2 + c$ gives the equation $\mathbf{15 = 4a + 2b + c}$ **.........(2)**
Substituting (3, 36) into $y = ax^2 + bx + c$:
$36 = a \times 3^2 + b \times 3 + c$ gives the equation $\mathbf{36 = 9a + 3b + c}$ **.........(3)**
We now have three equations with three unknowns.
To eliminate c we can reduce these equations by subtracting equation (1) from equation (2) and also (1) from (3).

(2) – (1): $15 - 0 = 4a + 2b + c - (a - b + c)$
$15 = 4a + 2b + c - a + b - c$
$15 = 3a + 3b$: divide both sides by the common factor 3
$\mathbf{5 = a + b}$ **.......(4)**

(3) – (1): $36 - 0 = 9a + 3b + c - (a - b + c)$
$36 = 9a + 3b + c - a + b - c$
$36 = 8a + 4b$: divide both sides by the common factor 4
$\mathbf{9 = 2a + b}$ **........(5)**

We now have two equations (4) and (5) with the unknowns a and b.
To eliminate b we could subtract (4) from (5).

(5) – (4): $9 - 5 = 2a + b - (a + b)$
$4 = 2a + b - a - b$
$\mathbf{4 = a}$

To find b we can substitute $a = 4$ into **(4)** [or **(5)**].
$5 = 4 + b$: subtract 4 from both sides
$\mathbf{b = 1}$

To find c we can substitute $a = 4$ and $b = 1$ into equation **(1)**:
$0 = a - b + c$ ....**(1)**
$0 = 4 - 1 + c$
$0 = 3 + c$
$\mathbf{c = {^-3}}$

We have found the value of the three unknowns so the equation of the quadratic function is $y = 4x^2 + x - 3$.

EXERCISE*

1 A quadratic function is of the form $y = ax^2 + c$ and its graph goes through the points (2, $^-3$) and ($^-1$, 3). Find the values of a and c and write the equation of the function.

2 A quadratic function is of the form $y = ax^2 + c$ and its graph goes through the points ($^-2$, $^-17$) and (1, 7). Find the values of a and c and write the equation of the function.

3 A quadratic function has the form $y = a(x - p)(x - q)$. The x-intercepts for this function are 4 and 2 and the y-intercept is 16. Find the values of a, p and q and write the equation of the function.

4 A quadratic function is of the form $y = ax^2 + bx + c$ and its graph goes through the points $(^-2, 3)$ and $(1, ^-4)$ and it has a y-intercept of $^-3$. Find the values of a, b and c and write the equation of the function.

5 A quadratic function of the form $y = ax^2 + bx + c$ goes through the points $(1, 1)$, $(^-2, 4)$ and $(3, 29)$. Find the values of a, b and c and write the equation of the function.

6 A quadratic function of the form $y = ax^2 + bx + c$ goes through the points $(1, 4)$, $(2, 3)$ and $(^-3, ^-32)$. Find the values of a, b and c and write the equation of the function.

7 A quadratic function of the form $y = ax^2 + bx + c$ goes through the points $(2, ^-1)$, $(^-2, ^-5)$ and $(^-4, ^-13)$. Find the values of a, b and c and write the equation of the function.

8 A quadratic function is a perfect square of the form $y = a(x - h)^2$. Find the values of a and h if the x- and y-intercepts are 2 and 12 respectively. Write the equation of the function.

9 A quadratic function is a perfect square of the form $y = a(x - h)^2$. Find the values of a and h if the x- and y-intercepts are $^-3$ and $^-8$ respectively. Write the equation of the function.

10 A quadratic equation has x-intercepts of $^-4$ and $^+4$ and a y-intercept of $^-32$. Write the equation of the function.

Practical problems involving quadratic functions

After solving a practical problem involving a quadratic function, it is important to test the solutions for practicality. For example, if the solutions involve lengths then a negative answer is not a practical answer.

Example 1

The sum of two numbers is 15. Find the two numbers so that their product is a maximum. Find this maximum product.

Answer

There are many possibilities for the two numbers; too many to test all of them so we will use algebra and our knowledge of quadratic equations to solve the problem.

Let one of the numbers be x. This means that the other number is $15 - x$.

The product, P, of the two numbers is $P = x \times (15 - x) = x(15 - x) = 15x - x^2$; a quadratic function.

The quadratic function has a negative coefficient of the x^2 term so it has a maximum point.

The x-coordinate of the maximum value will be halfway between the x-intercepts so we need to find the x-intercepts.
$P = x(15 - x)$ is in factorised form.
The x-coordinates are where $y = 0$ so solving $x(x - 15) = 0$ gives $x = 0$ or $x - 15 = 0$.
The solutions are $x = 0$ or $x = 15$.
Halfway between $x = 0$ and $x = 15$ is $x = 7.5$.

Substituting $x = 7.5$ into $P = x(15 - x)$ gives $P = 7.5(15 - 7.5) = 7.5^2 = 56.25$.
The two numbers are 7.5 and 7.5 and the maximum product is 56.25.

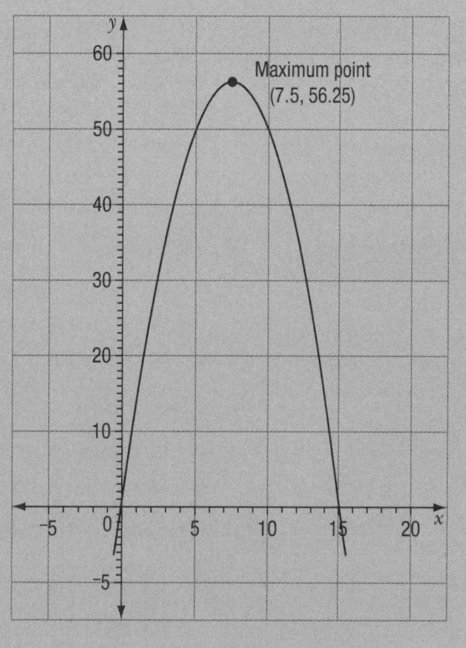

Example 2

Ilikis estimates that he needs 84 square metres of garden. The garden is to be rectangular and 8 metres longer than it is wide. What dimensions will he make the garden?

Answer

Let x be the width of the garden so the length of the garden will be $(x + 8)$ metres.
The area of the garden will be A = length × width

$= (x + 8) \times x$

$= x(x + 8)$ a quadratic function

If the area is to be 84 square metres then we can solve $84 = x(x + 8)$

$84 = x \times x + x \times 8$

$84 = x^2 + 8x$ subtract 84 from both sides of the equation

$0 = x^2 + 8x - 84$ to factorise we look for two numbers that multiply to $^-84$ and also add to 8. These numbers are 14 and $^-6$.

$0 = (x + 14)(x - 6)$

So $x + 14 = 0$ and $x - 6 = 0$

If $x + 14 = 0$ then $x = {}^-14$ it is not possible to have a negative length so we will discard this solution.

If $x - 6 = 0$ then $x = 6$

The width, x, of the garden will be 6 metres and the length will be $x + 8 = 14$ metres.

EXERCISE

Remember

Consecutive means 'one after the other'.

1 The sum of two numbers is 24.

a If x is one of the numbers, write the other number in terms of x.

b Write the product, P, of the number in terms of x.

c Graph the quadratic function, P.

d What is the maximum value that P can have, and what is the value of x when this occurs?

2 The product of two consecutive whole numbers is 272.

a If one of the numbers is x, what is the other number?

b Write the product of the two numbers in terms of x.

c Solve your quadratic equation to find the two numbers.

3 The sum of the first n natural numbers is $\frac{n(n+1)}{2}$.

a Use this expression to find the sum of the first 11 natural numbers.

b How many natural numbers were added together if their sum is 325?

4 Use DOTS to calculate the shaded areas in these shapes.

a

10 cm

15 cm

b

3 cm

25 cm

Write your answer in terms of π.

5 The path of a ball, hit at ground level, is given by the equation $h = x - \frac{x^2}{60}$ where x is the horizontal distance travelled by the ball and h is the vertical height of the ball, both in metres.

a What is the value of h when the ball hits the ground?

b Calculate the greatest horizontal distance the ball will travel before it hits the ground.

c Graph the path of the ball until it hits the ground.

d Calculate the maximum height that the ball reaches.

6 Lucas has 80 metres of fencing to enclose his rectangular garden. There is an existing fence along one side of the garden so this side will not need a fence. x is the width of the fence.

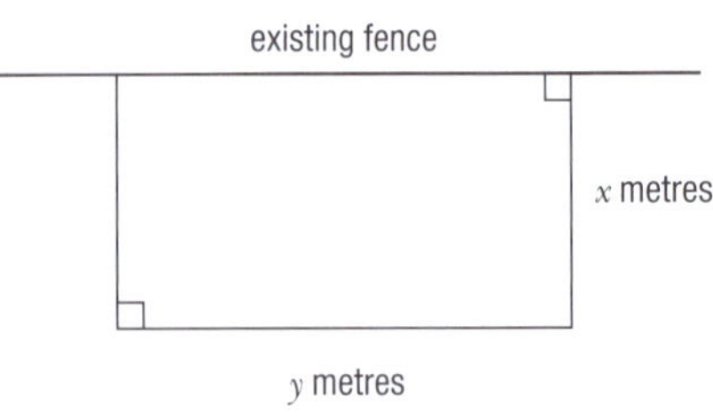

a Write an equation for the length of the fencing in terms of x and y and hence write y in terms of x.

b Write the area, A, of Lucas's land in terms of x.

c Find the maximum area that can be enclosed by the 80 metres of fencing.

d What are the dimensions of Lucas's garden that give this maximum area?

7 Kenime's garden is a rectangle and she wants to fence it and divide it into three equal sections with fencing, using the 96 metres of fencing she has available.

- **a** Using the fact that there are 96 metres of fencing available, show that $y = 48 - 2x$.
- **b** Write the area, A, of the whole garden in terms of x.
- **c** Sketch the graph of the quadratic function, A.
- **d** Find the coordinates of the turning point of the function and hence the maximum area enclosed by the whole garden.
- **e** What are the dimensions of each of the three smaller gardens when the whole area enclosed is a maximum?

8 Bernard is two years older than his sister Therese and four years younger than his sister Josephine. The product of his sisters' ages is the age of his mother, 40.

- **a** If x is Bernard's age, write his two sisters' ages in terms of x.
- **b** Write Bernard's mother's age in terms of x and solve this quadratic equation.
- **c** What is Bernard's age?

9 Edris wants to build a pottery kiln that is to have a parabolic cross section. He wants the kiln to have an available space of 1 cubic metre and has sketched the cross section on a set of Cartesian axes, as shown.

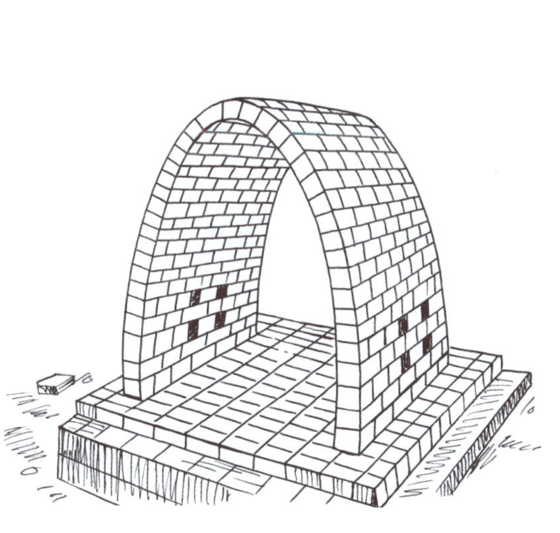

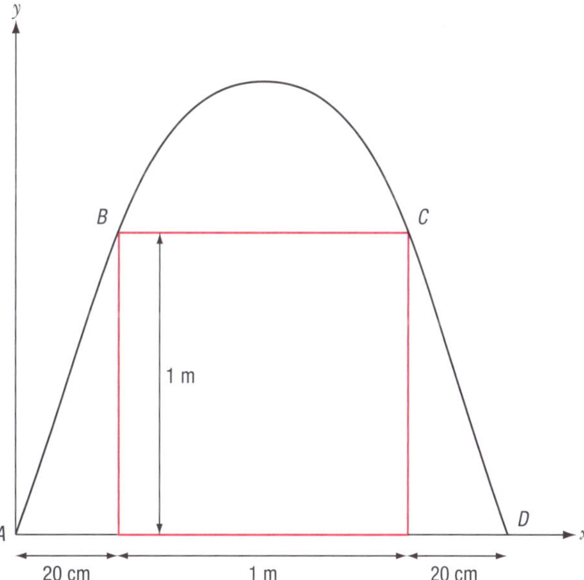

- **a** State the coordinates of the points A, B, C and D.
- **b** The equation of the arch is a quadratic function of the form $y = px(x - q)$ where p and q are constants. Write down the value of q.
- **c** Use the coordinates of either point B or C to calculate the value of p.
- **d** State the x-coordinate of the maximum point and then find the maximum height of this kiln.

10* Tracy makes woven baskets that she sells at the variable rate of K$(20 - 0.05x)$ per basket, where x is the number of baskets sold.

a If Tracy sells 10 baskets to a customer, how much does she charge per basket?

b If Tracy sells 10 baskets to a customer, how much does she receive in total revenue?

c If Tracy sells x baskets to a customer, how much will she receive in total revenue.

d If Tracy's yearly costs are K$(175 + 2x)$ to produce x baskets and she sells x baskets to a customer, show that Tracy's profit for the year is $P = {}^-0.05x^2 + 18x - 175$. (Profit = revenue – costs).

e Use the quadratic formula to solve the equation $0 = {}^-0.05x^2 + 18x - 175$ and find the x-intercepts of the quadratic function $P = {}^-0.05x^2 + 18x - 175$.

f Find the coordinates of the turning point of $P = {}^-0.05x^2 + 18x - 175$. What is the maximum profit that Tracy can make in a year and the number of baskets she needs to make and sell in a year to make this profit.

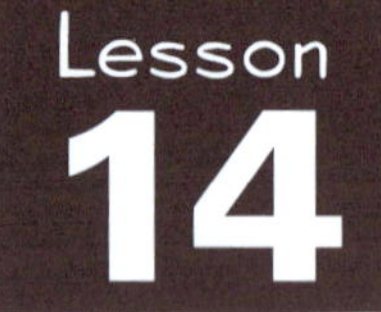

Practice for the assessment task

The assessment task for this option is an individual investigation where you will be expected to construct graphs for and interpret quadratic functions. You will be required to present your findings through a written report.

You will be assessed on the extent to which you can:

- demonstrate appropriate investigation skills
- choose and apply relevant mathematical techniques
- make an effective communication of the project results.

Your teacher will direct you as to what you are to investigate, however the exercise below contains some suggestions and exercises to prepare you for the assessment task.

EXERCISE

1 Investigate the graphs of quadratic functions of the form $y = ax^2$ for different values of a. Use both positive and negative values for a, and values between $^-1$ and 1. Construct graphs to illustrate the effect the value of a has on the graphs. Write a report on your findings.

2 Investigate the graphs of quadratic functions of the form $y = (x - h)^2$ for different values of h. Use both positive and negative values for h. Construct graphs to illustrate the effect the value of h has on the graphs. Write a report on your findings.

3 Investigate the graphs of quadratic functions of the form $y = x^2 + k$ for different values of k. Use both positive and negative values for k. Construct graphs to illustrate the effect the value of k has on the graphs. Write a report on your findings.

4 Investigate the development of the quadratic formula by completing the square on the general quadratic equation $ax^2 + bx + c = 0$.

5 Investigate the number and type (rational or irrational) of solutions of a quadratic equation and how this relates to $b^2 - 4ac$ in the quadratic formula. Write a report on your findings.

6 Investigate how the number of solutions of a quadratic equation relates to the graph of a quadratic function. Write a report on your findings.

7 For the quadratic function $y = x^2 + 3x - 10$ and the linear function $y = 2x - 8$:

  a Find the coordinates of the points of intersection by solving the equation $x^2 + 3x - 10 = 2x - 8$.

  b Sketch the graphs of the functions on the same set of axes to confirm your answers from part a.

8 A quadratic function of the form $y = ax^2 + bx + c$ goes through the points $(1, {}^-9)$, $({}^-1, {}^-7)$ and $(3, 5)$.

  a Use these points to find the values of a, b and c.

  b Graph the function giving the coordinates of the intercepts with the axes and the turning point.

9 a On the same set of axes, sketch the graphs of the quadratic functions $y = \frac{1}{3}x(x - 6)$ and $y = {}^-x^2 - 2x + 12$.

  b Find the coordinates of the points of intersection of these two functions from:

    i the graphs

    ii algebraically by solving $\frac{1}{3}x(x - 6) = {}^-x^2 - 2x + 12$.

10 A ball is thrown vertically upwards from a height of 2 metres. Its height, h metres, above the ground at time, t seconds, is given by the expression $h = {}^-5t^2 + 9t + 2$.

  a What is the height of the ball after 1 second?

  b How long does it take for the ball to hit the ground?

  c Draw a graph of quadratic function $h = {}^-5t^2 + 9t + 2$ for the time it is in the air.

  d Determine the maximum height that the ball reaches.

Unit 3 Trigonometric applications

Unit summary

This unit focuses on the geometry and trigonometry used in a wide range of activities in modern society. It emphasises the development of the skills of mathematically modelling environmental situations, such as land areas and navigation routes.

The core part of the unit requires the students to demonstrate an understanding of:

- similar triangles
- Pythagoras's rule
- trigonometric ratios and
- situations where they can be used.

The core part of the unit will take five weeks and will be assessed with tests. The tests will require the students to demonstrate understanding of mathematical concepts and correctly choose and apply mathematical techniques.

The remaining five weeks of the unit will be spent on one of the options:

- Option A: Surveying
- Option B: Navigation.

Assessment for the option will be a group project with the students presenting their findings in written form.

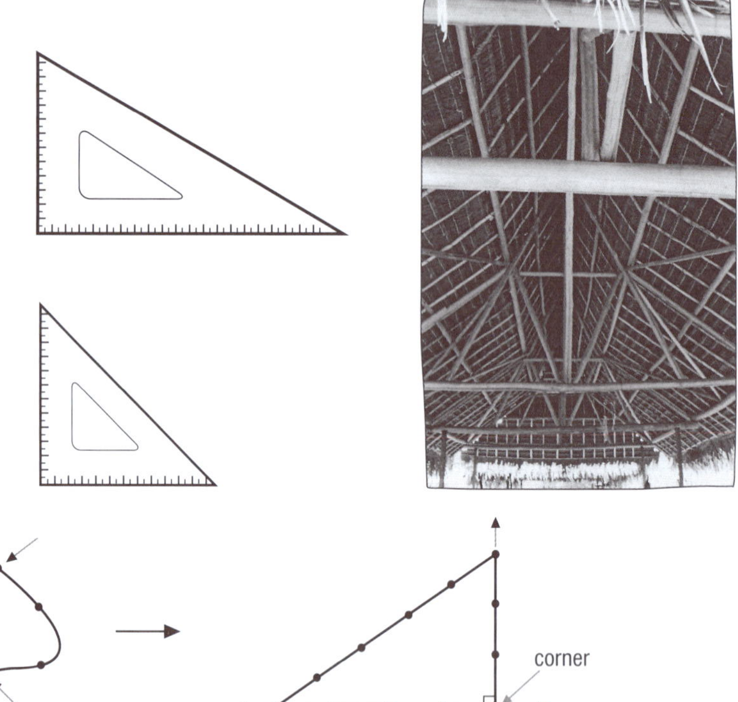

Syllabus references

10.3.1 Demonstrate understanding of the basic concepts of similar figures and trigonometric ratios

10.3.2 Apply Pythagoras's rule and trigonometric ratios to solve right-angle triangles and find lengths and angles in simple real problems

10.3.3 Identify and apply calculations and be aware of whether or not the result is reasonable

10.3.4 Communicate mathematical processes and results

10.3.5 Undertake investigations individually and cooperatively in which mathematics can be applied to solve problems

Trigonometric applications: Topics

TRIANGLE PROPERTIES AND TERMINOLOGY

Triangle terminology and properties

A triangle is a closed, plane figure with three straight sides.

Terminology

The triangle, at right, has three **vertices** (points of intersection of the sides), marked *A*, *B* and *C*.

The side (or the length of the side) going from vertex *A* to vertex *B* is referred to as 'side *AB*', *AB* or $\overline{AB}$.

The **interior angle** at vertex *B* is written as *angle ABC* or $\angle ABC$.

$\angle ABC$ is described as the **included angle** for sides *AB* and *BC*.

The side *BC* is described as the **opposite side** to $\angle BAC$.

Triangle terminology associated with side lengths:

- A **scalene** triangle has sides that are all of different lengths.

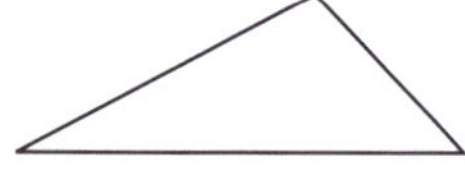

Did you know?

Sides that are the same length are marked in the same manner.

- An **isosceles** triangle has two sides that are of equal length.

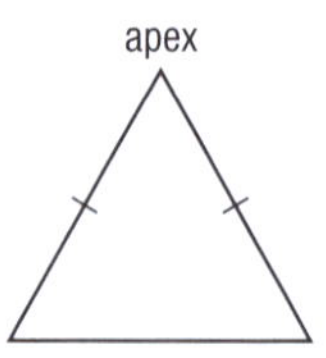

- An **equilateral** triangle has all sides of the same length.

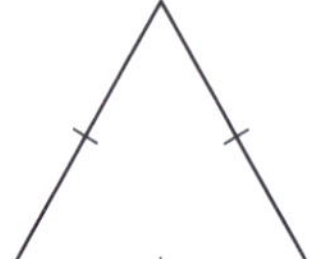

Triangle terminology associated with angles:

- An **obtuse**-angled triangle has one angle greater than 90°.
- An **acute**-angled triangle has all angles less than 90°.

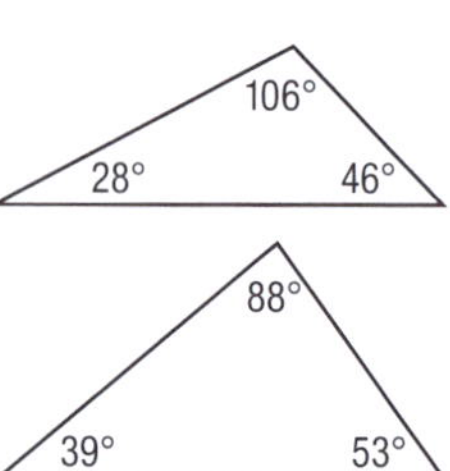

- A **right**-angled triangle has one angle of 90°. This angle is marked with the ∟ symbol.
- The side opposite the right angle is referred to as the **hypotenuse**.

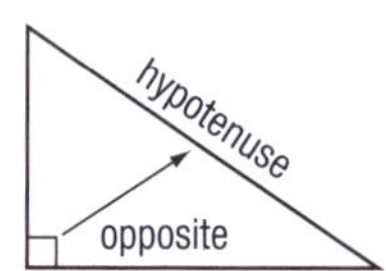

Properties

- The interior angles of a triangle sum to 180°. $a + b + c = 180$

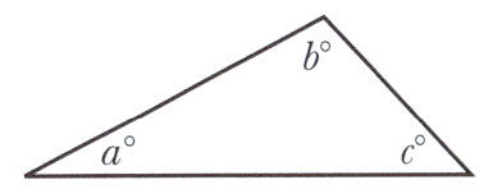

- The exterior angle of a triangle is equal to the sum of the two opposite interior angles. $z = x + y$

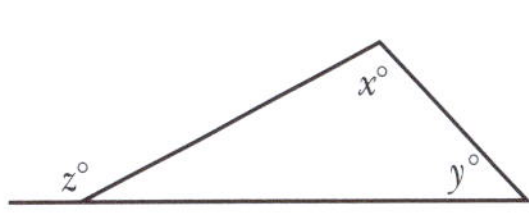

- The longest side of a triangle is opposite the largest angle.

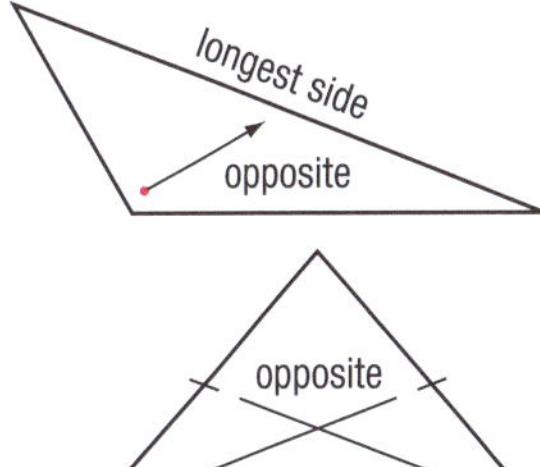

- The angles opposite the equal-length sides in an isosceles triangle are equal in size.

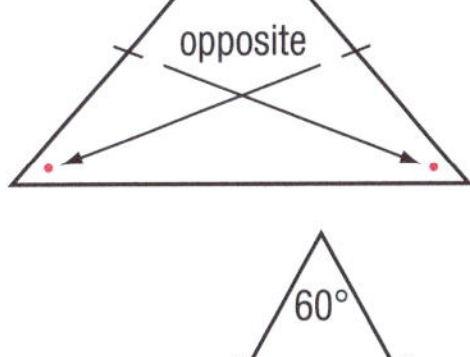

- The angles in an equilateral triangle are all of size 60°.

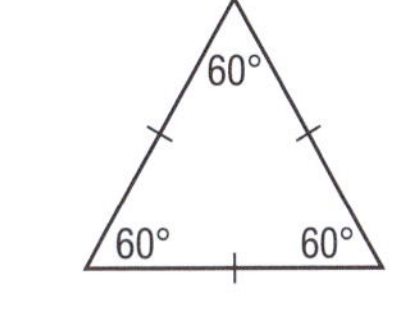

- For an isosceles triangle (this includes equilateral triangles), the line drawn from the **apex** to the midpoint of the opposite side **bisects** (cuts in half) the apex angle and is **perpendicular** (at right angles) to the opposite side.

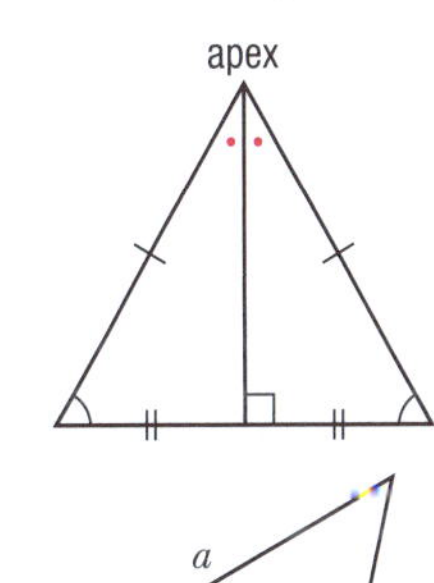

Did you know?

Angles that are the same size are marked in the same manner.

- To be able to construct a triangle, the length of the longest side of the triangle must be **less** than the sum of the lengths of the other two sides. $a < b + c$

EXERCISE

1 Classify each of these triangles using side length terminology.

a 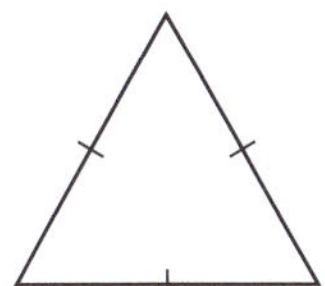

b 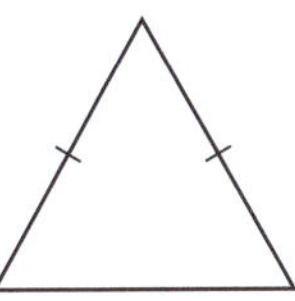

c 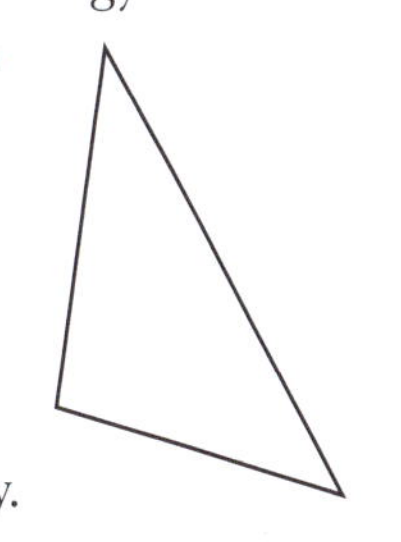

2 Classify each of these triangles using angle terminology.

a 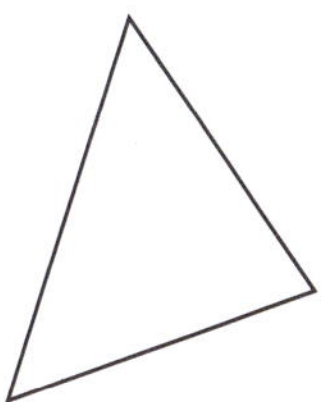

b 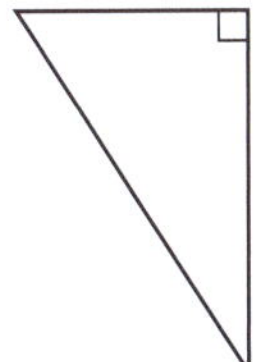

c

3 Describe each of these triangles and state two important properties that apply to the triangle.

a

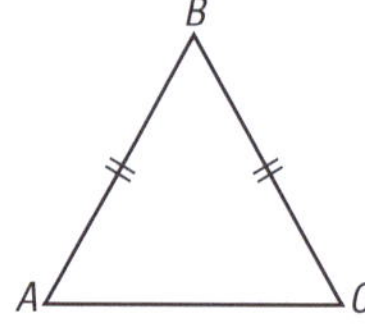

b

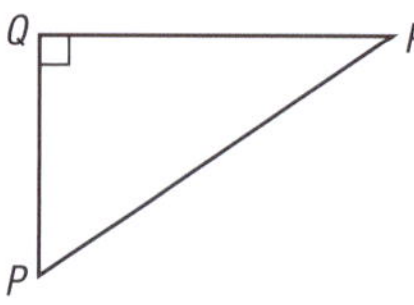

c

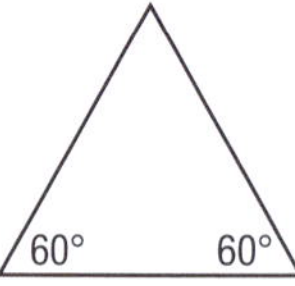

4 Draw each of the following triangles marking angles and sides that are equal in the same manner.

a A right-angled isosceles triangle.
b An obtuse, scalene triangle.
c An acute, isosceles triangle.
d An equilateral triangle with a perpendicular bisector drawn from one vertex to its opposite side.

5 Construct each of the following triangles using a ruler, compass or protractor where necessary. Describe each of the triangles using side length and angle terminology.

a A triangle with side lengths 3 cm, 2 cm and 4.5 cm.
b A triangle with angles 30°, 50° and 100°.
c A right-angled triangle with the two shorter sides of length 4 cm and 5 cm.
d A triangle PQR with $PQ = 3.5$ cm and $QR = 5$ cm and $\angle PQR = 65°$.
e An isosceles triangle with an angle of 40° at the apex and the side opposite this angle of length 5 cm.
f A right-angled triangle with one angle of size 60° and the hypotenuse of length 6 cm.
g A triangle with sides of length 2.5 cm and 4 cm with an included angle of 50°.
h $\triangle ABC$ with $\angle ABC = 50°$, $\angle BCA = 70°$ and $AB = 5$ cm.

6 **a** Complete the following for the triangle at right:

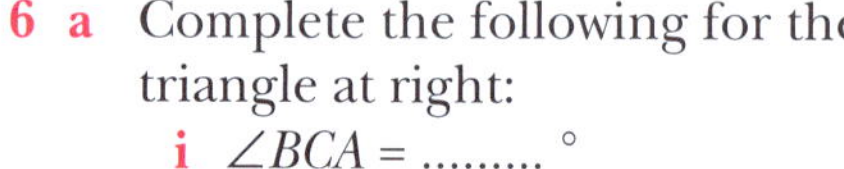

i $\angle BCA =$ .... °
ii $AB =$ .... cm
iii The side opposite the angle of size 93° is ....
iv The angle of size 51° is the included angle for the sides .... and ....
v The triangle can be described as a ...., ... (two words) triangle.

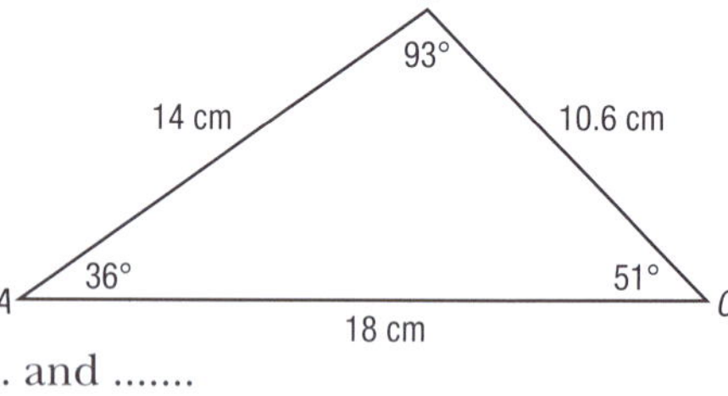

b Complete the following for the triangle at right.

i $QR =$ .... cm
ii $\angle QPR =$ .... °
iii $QP =$ .... cm
iv $\angle QPR = \angle$.........
v QM is .... to PR
vi $\angle QMP =$ .... °
vii $PM =$ ......cm
viii The triangle can be described as a ...., ... (two words) triangle.

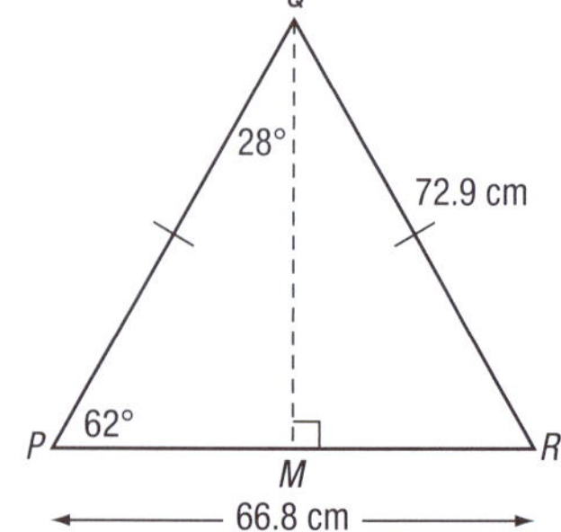

c Complete the following for the triangle at right.

i $\angle MLN =$ .... °

ii The side opposite the right angle is called the ....

iii LM is ....to side MN.

iv The angle included by the sides NL and NM is $\angle$......

v The triangle can be described as a ...., ... (two words) triangle.

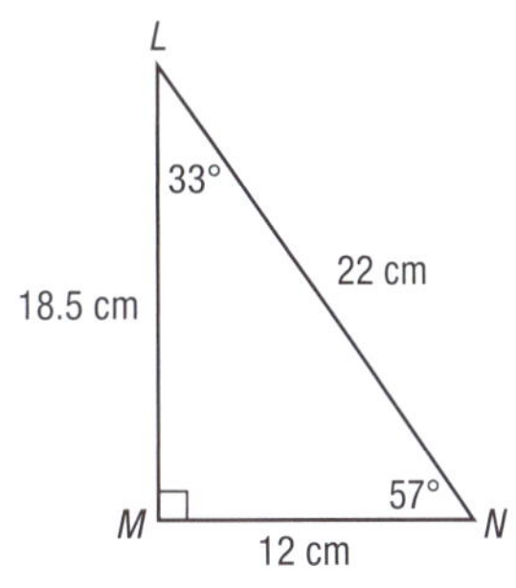

Working with the properties of triangles

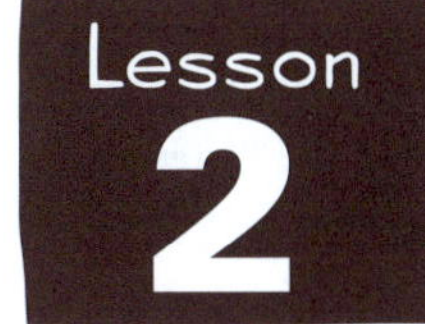

The properties of triangles are listed in Lesson 1 and below in the following Help boxes.

Example

Find the value of the pronumerals in each of the following.

a

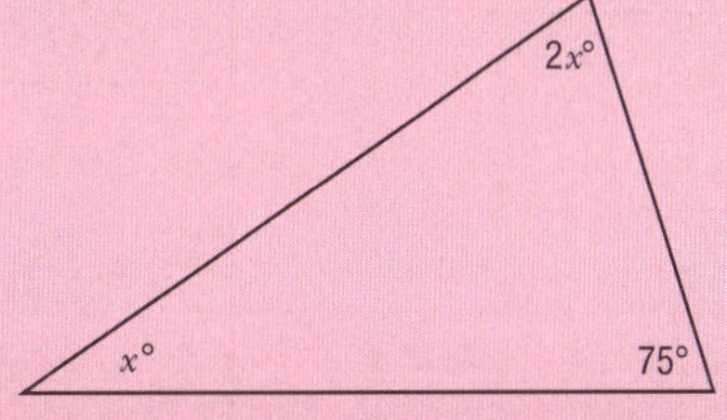

b

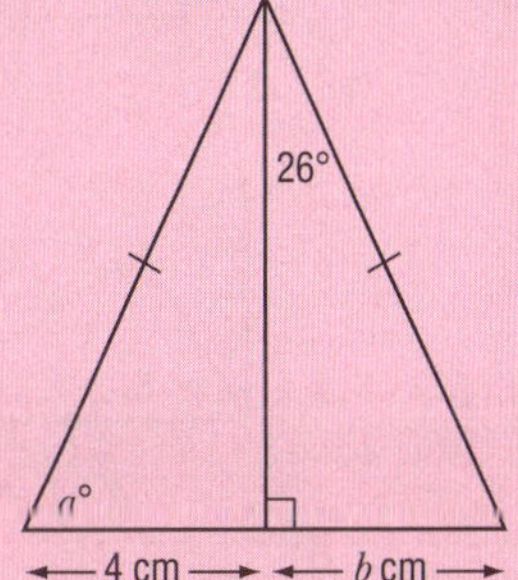

Answer

a Using the property that the interior angles of a triangle sum to 180°:

$x + 2x + 75 = 180$ add like terms

$3x + 75 = 180$ subtract 75 from both sides of the equation

$3x = 105$ divide both sides of the equation by 3

$x = 35$

b The triangle is isosceles with a perpendicular bisector drawn from the apex. Using the property that for an isosceles triangle, the line drawn from the apex to the midpoint of the opposite side bisects (cuts in half) the apex angle and is perpendicular (at right angles) to the opposite side.

The triangle at right shows angles and sides that are equal in size.

Using the angles in a triangle sum to 180° in one of the right-angled triangles:

$a + 26 + 90 = 180$ add like terms

$a + 116 = 180$ subtract 116 from both sides of the equation

$a = 64$

Sides marked in the same manner are equal in length so $b = 4$.

Properties of triangles

- To be able to construct a triangle the length of the longest side of the triangle must be less than the sum of the lengths of the other two sides. $a < b + c$
- The interior angles of a triangle sum to 180°.
- The exterior angle of a triangle is equal to the sum of the two opposite interior angles.

EXERCISE

1 The measurements given below are for the side lengths or angles of a triangle. Which of the triangles cannot be drawn?

a 25 cm, 30 cm, 12 cm

b 20 cm, 30 cm, 50 cm

c 34°, 58°, 92°

d 48 m, 23 m, 70 m

e 45°, 45°, 90°

2 Find the value of the pronumerals in each of the following diagrams.

a 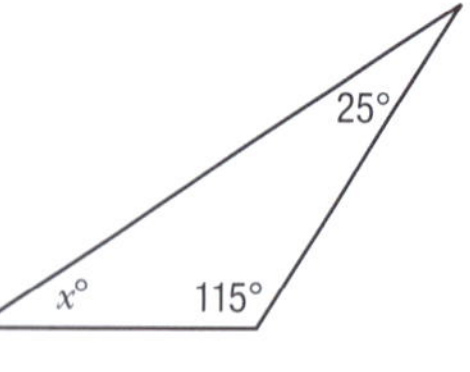

b 

c 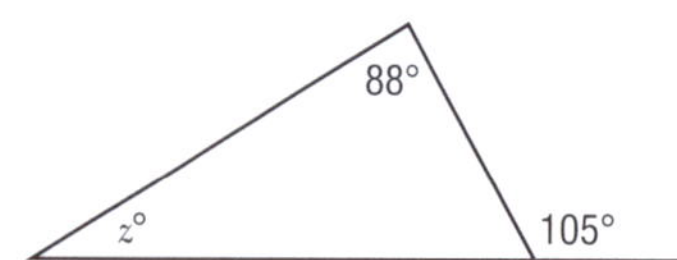

d 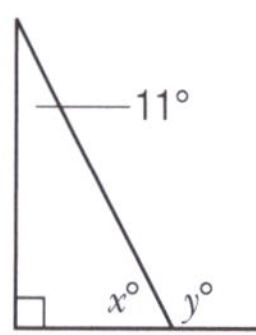

e 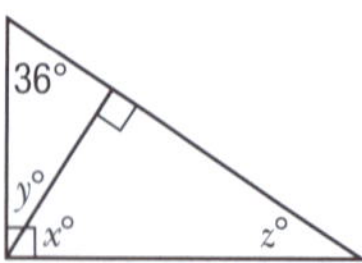

f 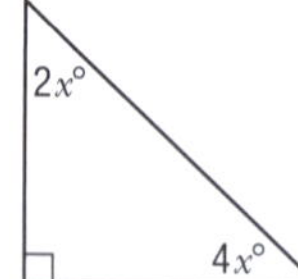

g 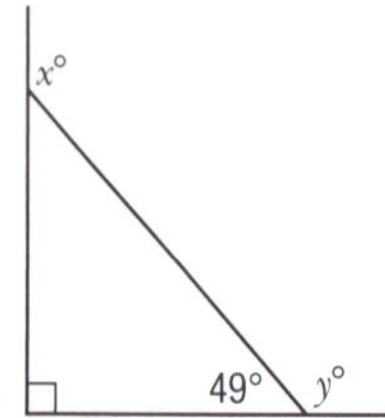

3 Find the value of the pronumerals in each of the following diagrams.

a 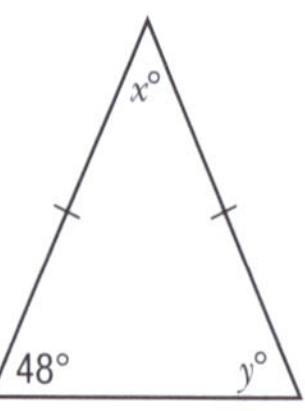

b 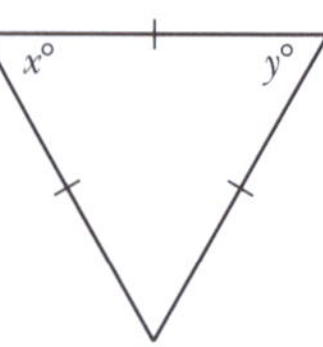

c 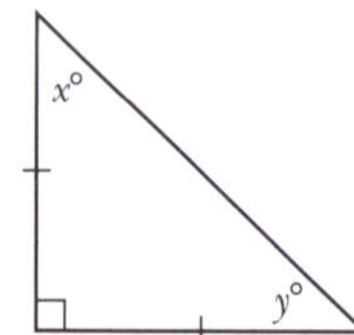

d 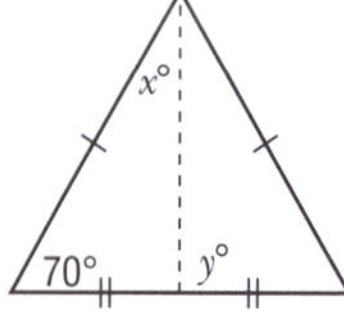

e

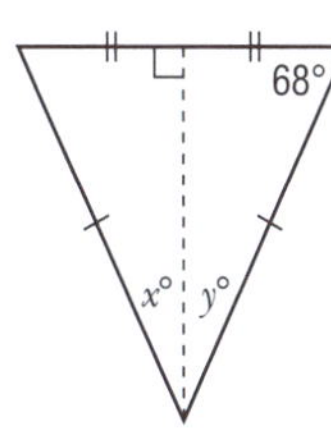

f

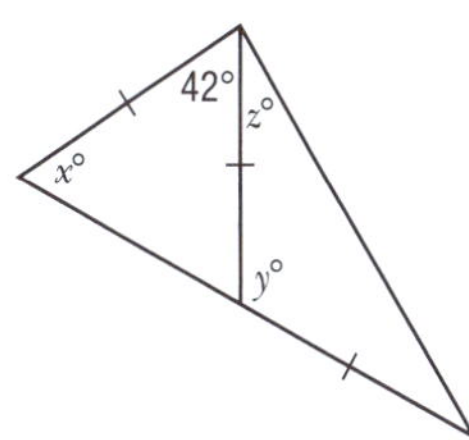

4 Find the value of the pronumerals in each of the following diagrams.

a

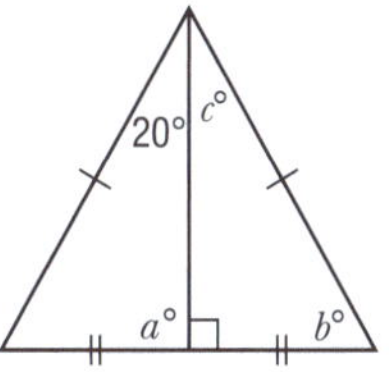

b

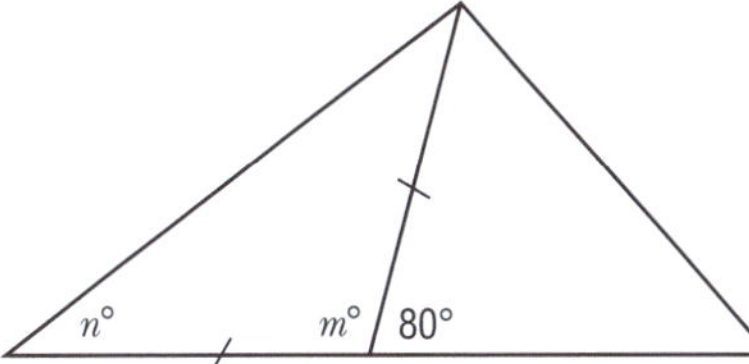

c

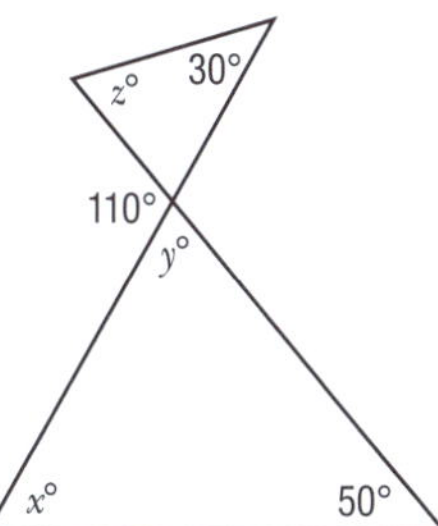

d

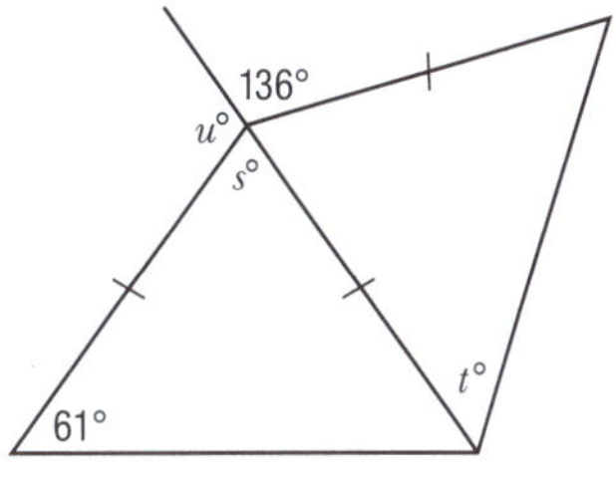

e

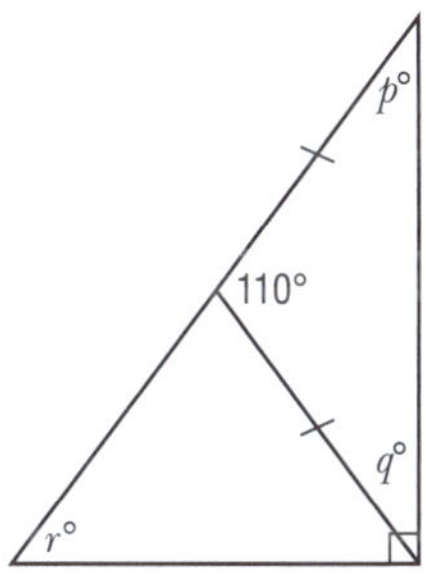

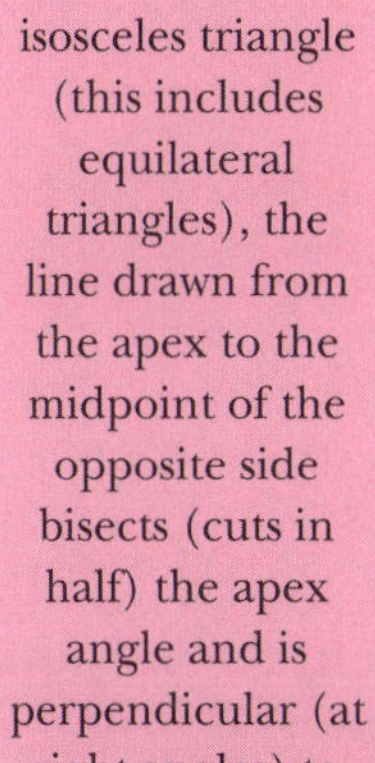

Properties of triangles

- The angles opposite the equal-length sides in an isosceles triangle are equal in size.
- The angles in an equilateral triangle are all of size 60°.
- For an isosceles triangle (this includes equilateral triangles), the line drawn from the apex to the midpoint of the opposite side bisects (cuts in half) the apex angle and is perpendicular (at right angles) to the opposite side.

5 In the diagram, JM bisects $\angle LJK$ and KM bisects $\angle JKL$. Find $\angle JMK$.

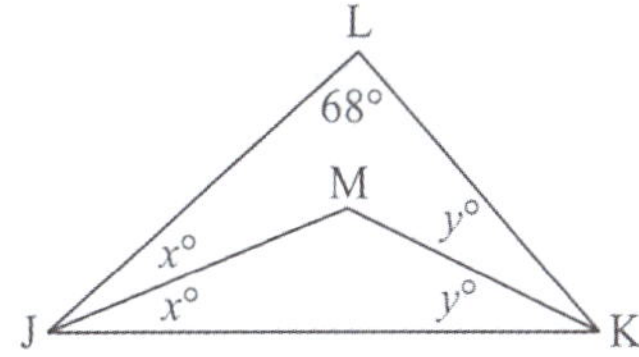

SIMILAR TRIANGLES

Similar triangles

Similar figures have the same shape but differ in size.

Similar figures have **corrresponding angles that are equal in size** and their **corresponding sides are in the same ratio**.

All squares are similar figures.

For these two squares:

- Corresponding angles are all 90°.
- Corresponding sides are in the ratio 1 : 2.

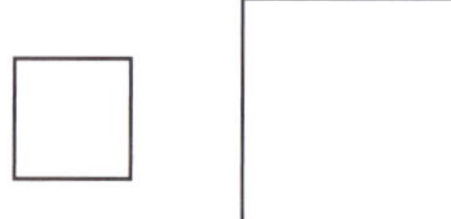

All circles are similar figures.

For these two circles:

- Corresponding diameters are in the ratio 1 : 3.

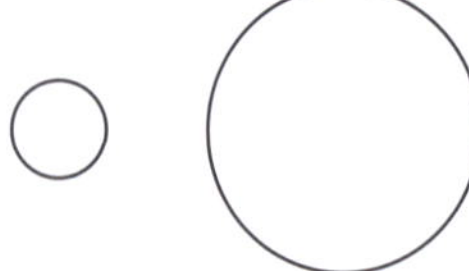

All equilateral triangles are similar figures.

For these two equilateral triangles

- Corresponding angles are all 60°.
- Corresponding sides are in the ratio 2 : 3.

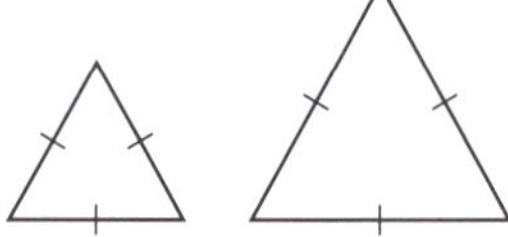

What other shapes that you have studied are similar figures?

Similar triangles

Not all triangles are similar figures. There are three tests you can apply to pairs of triangles to see if they are similar.

Two triangles are similar if:

i all their corresponding sides are in the same ratio

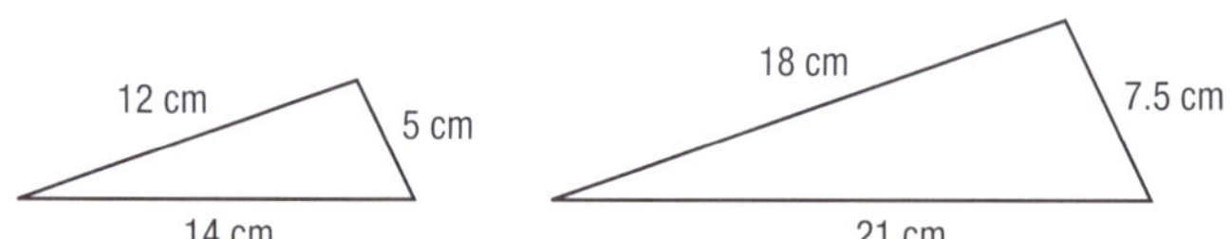

The corresponding sides in this pair of triangles are in the ratio 2 : 3.

OR

ii all the corresponding angles are of equal size

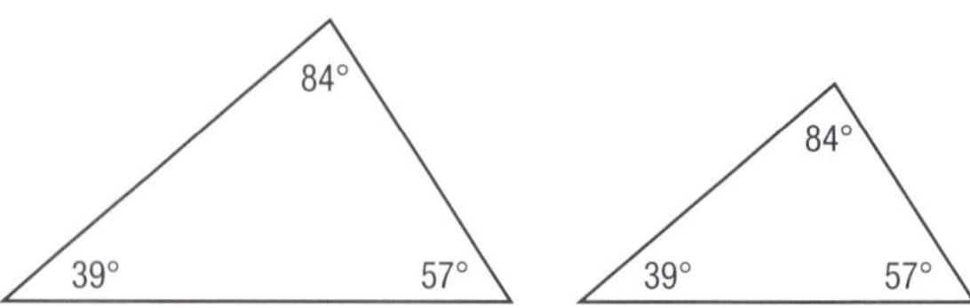

OR

iii a corresponding pair of adjacent sides in the triangles are in the same ratio and the angle between these adjacent sides (the included angle) is the same size.

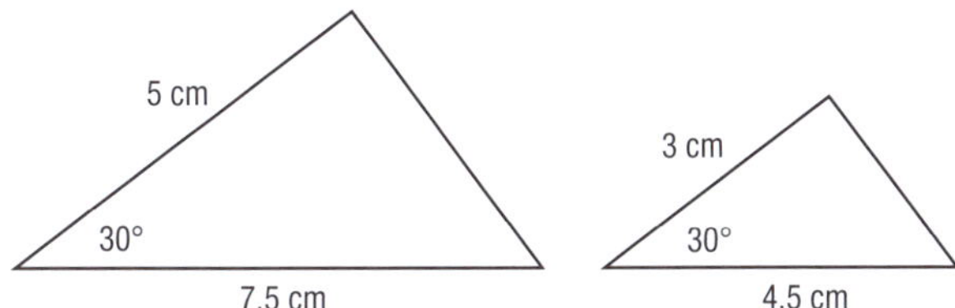

The ratio of the two adjacent sides in the example above is 5 : 3 and the included angle is 30°.

EXERCISE

1 Which of the following figures are similar.

a

b

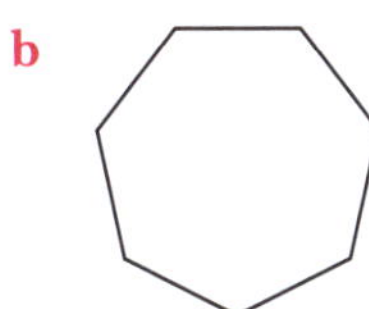

c

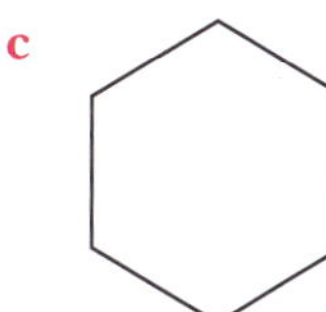

d

e

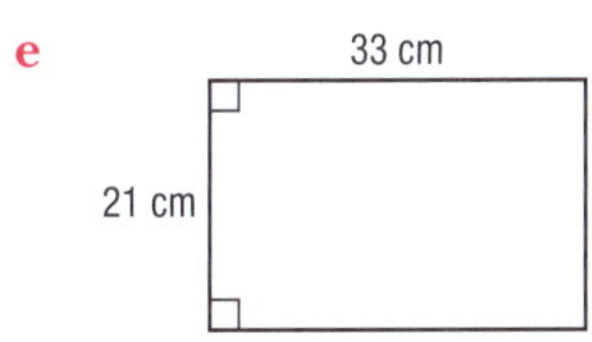

f

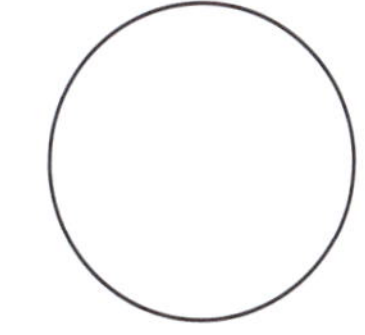

g

25°

h

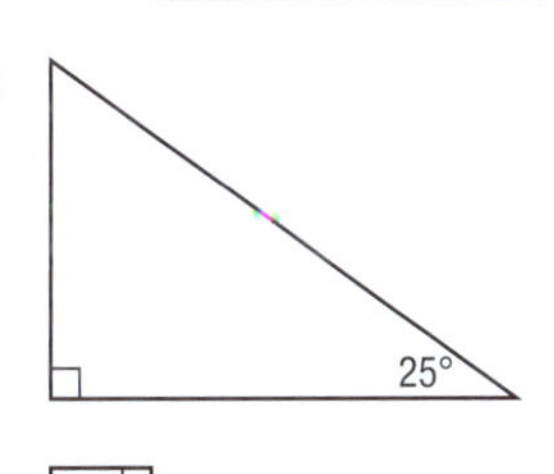

i

j

k

11 cm
7 cm

l

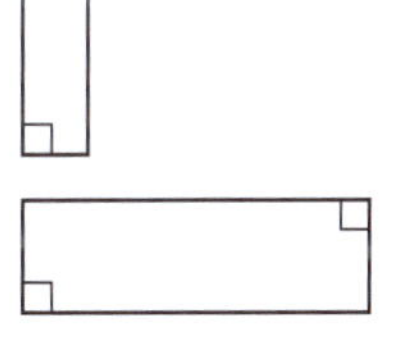

2 Identify the pair of similar triangles in each of the following, giving a reason for your answer. (Angles the same OR corresponding sides in the same ratio OR two sides in the same ratio and the included angle the same.)

a

b

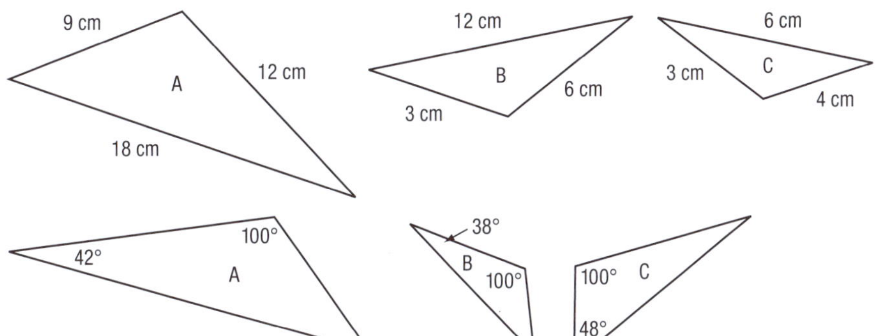

c

d

e

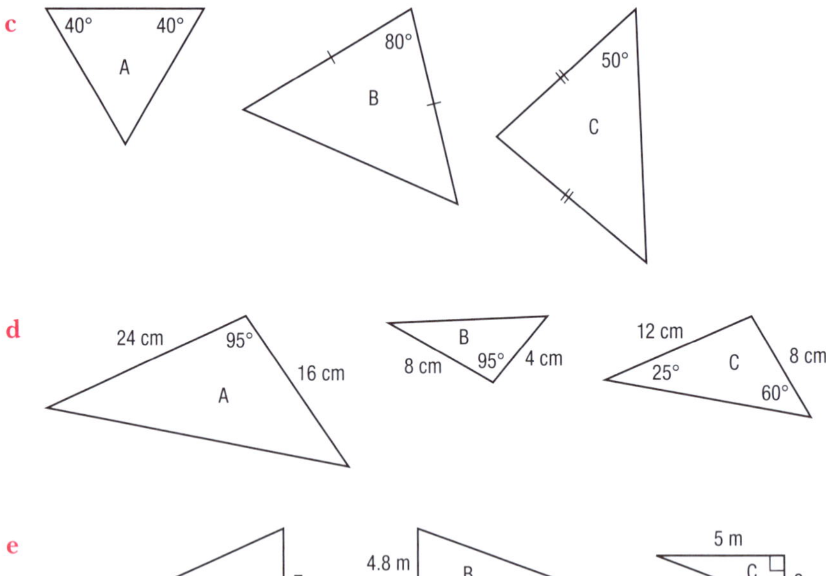

3 Construct a triangle similar to the triangles given below. Use a ruler, compass or protractor if necessary. Mark on your triangle the angles or lengths that show that it is similar to the given triangle.

a

b

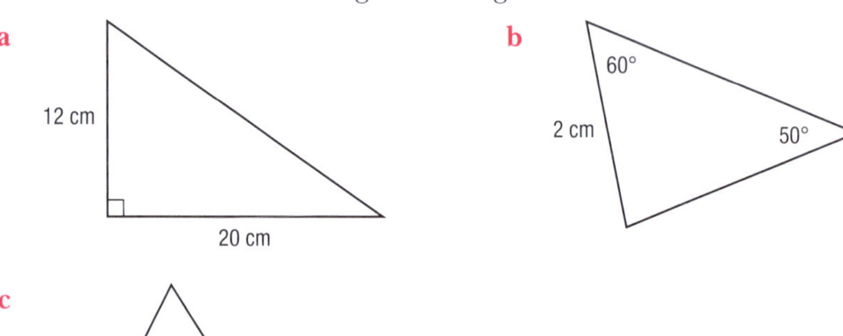

c

10 cm

65°

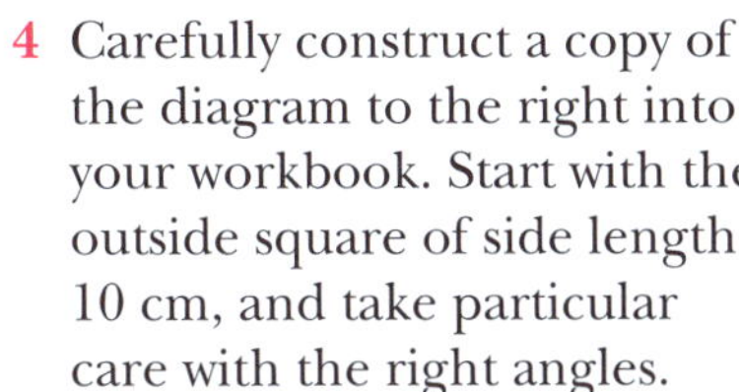

4 Carefully construct a copy of the diagram to the right into your workbook. Start with the outside square of side length 10 cm, and take particular care with the right angles.

a What similar shapes have you produced in your diagram? How do you know they are similar?

b How many different-sized similar triangles can you see in the diagram?

c If the largest triangle is the '1st triangle', what is the ratio between the corresponding sides of the:

i 1st and 3rd triangles?

ii 2nd and 4th triangles?

iii 1st and 5th triangles?

iv 1st and 2nd triangles?

Using scale factors for similar triangles

Consider this pair of triangles:

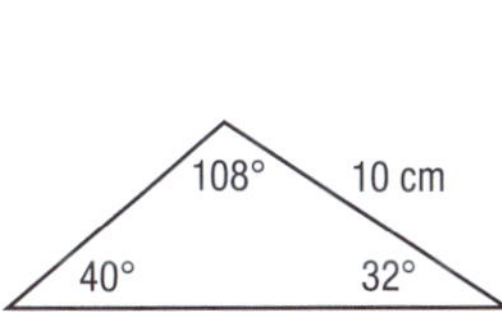

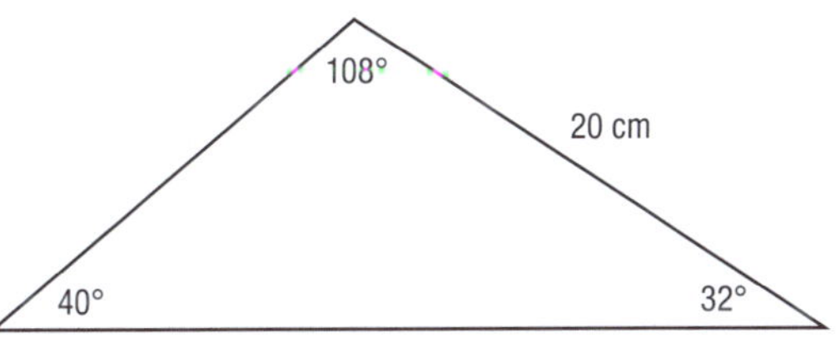

They have the same shape but differ in size. These two triangles are similar because the corresponding angles are equal. The sides are in the ratio 10 : 20 = 1 : 2.

The **scale factor** for this pair of similar triangles is $\frac{2}{1} = 2$; a side in the second figure is twice the length of the corresponding side in the first figure.

The second figure is an **enlargement** of the first figure, so the **scale factor is greater than one**.

Consider these two regular hexagons:

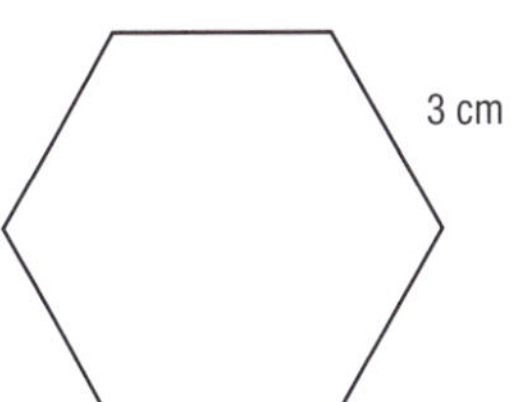

All regular hexagons are similar figures; their corresponding sides are in the same ratio and their interior angles are equal.

The scale factor in this case is $\frac{2}{3}$ so a side in the second figure is $\frac{2}{3}$ of the length of the corresponding side in the first figure.

The second figure is a **reduction** of the first figure and hence the **scale factor is less than one**.

$$\text{Scale factor} = \frac{\text{Length of the side in the second figure}}{\text{Corresponding length in the first figure}}$$

Drawing enlargements and reductions using scale factor

Example 1

From a point *O*, either inside or outside the figure *ABC*, the distance to each of the points *A*, *B* and *C* is measured then multiplied by the scale factor, 3 in this case. Now the distance from *O* to the image points *A′*, *B′* and *C′* is three times the distance from *O* to *A*, *B* and *C*, respectively. The image points are then connected to form the enlarged image *A′B′C′*.

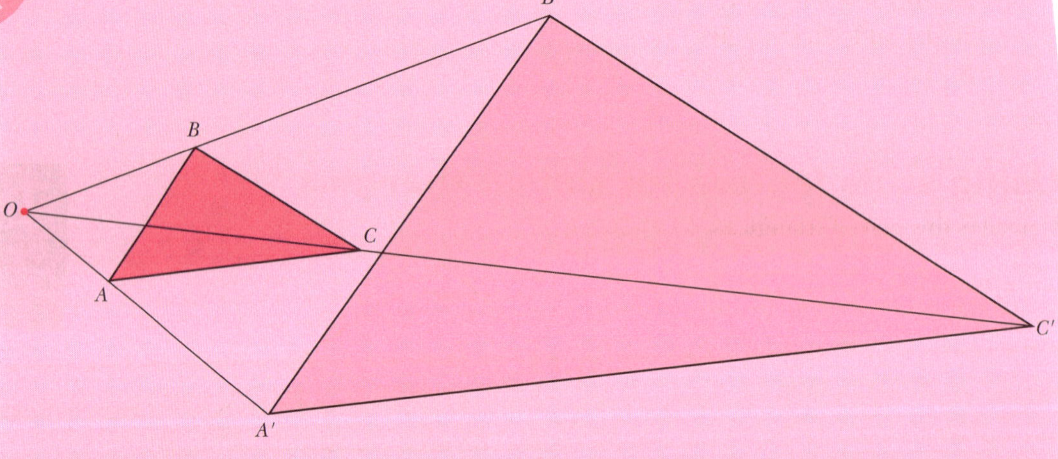

A similar process is done for a reduction.

Example 2

The scale factor is $\frac{1}{2}$ so $OA' = \frac{1}{2}OA$, $OB' = \frac{1}{2}OB$, $OC' = \frac{1}{2}OC$.

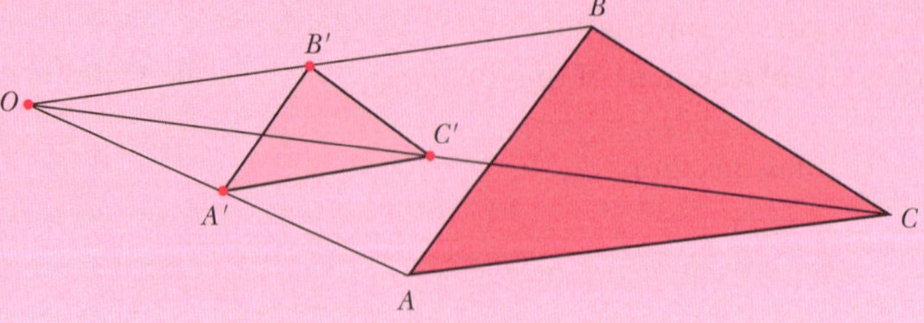

Finding the length of sides using scale factor

Example

The triangles below are similar and have a scale factor of 0.75. Find x and y.

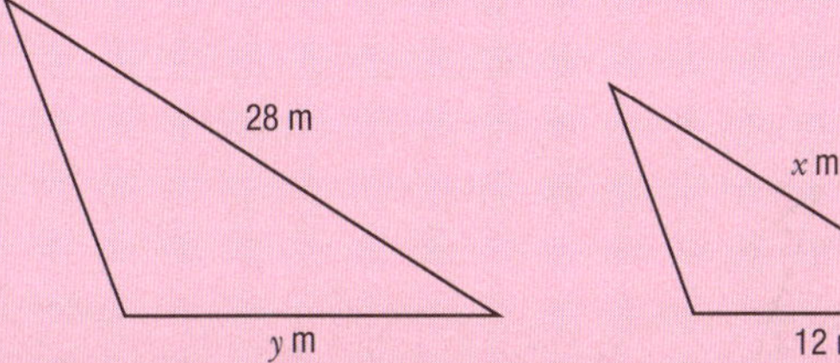

The side marked x in the second figure is a corresponding side to the side marked 28 m in the first figure. The scale factor is 0.75.

Substituting in the formula:

$$0.75 = \frac{x}{28} \quad \text{multiply both sides of the equation by 28}$$

$$28 \times 0.75 = x$$

$$x = 21$$

The side marked y in the first figure is a corresponding side to the side marked 12 m in the second figure. The scale factor is 0.75.

Substituting in the formula:

$$0.75 = \frac{12}{y} \quad \text{multiply both sides of the equation by } y$$

$$0.75y = 12 \quad \text{divide both sides of the equation by 0.75}$$

$$y = 16$$

EXERCISE

1 Estimate, and then find the scale factor for each of the following enlargements or reductions. Assume all pairs of triangles are similar.

a

b

c

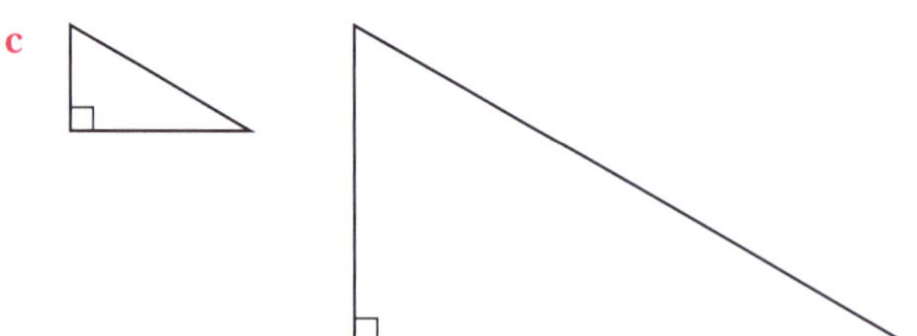

d

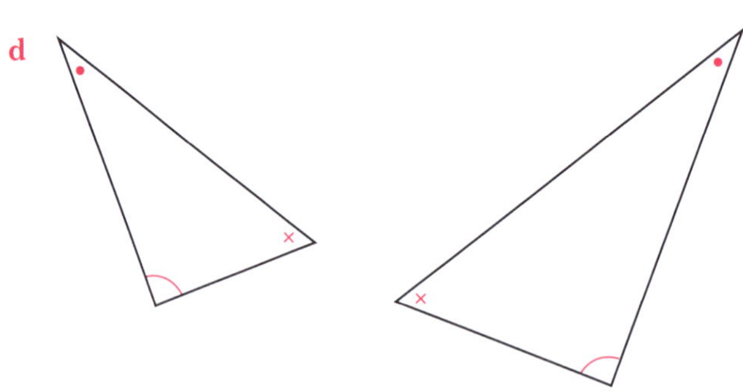

2 Find the value of the pronumeral in each of the following using the scale factor given. Assume the triangles are similar.

a Scale factor = 0.5

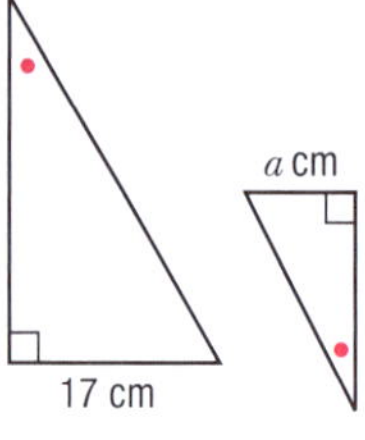

b Scale factor = 1.6

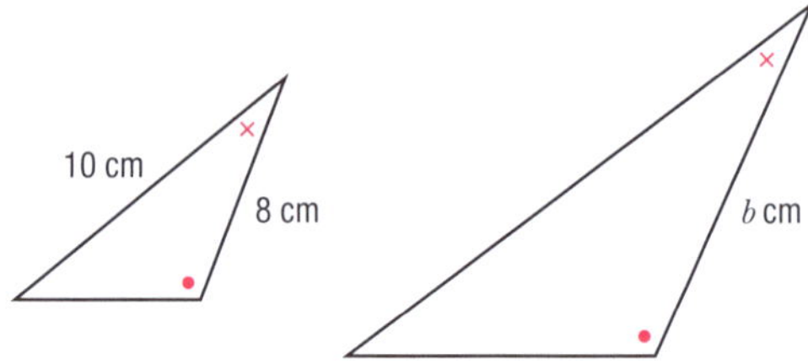

c Scale factor = $\frac{5}{3}$

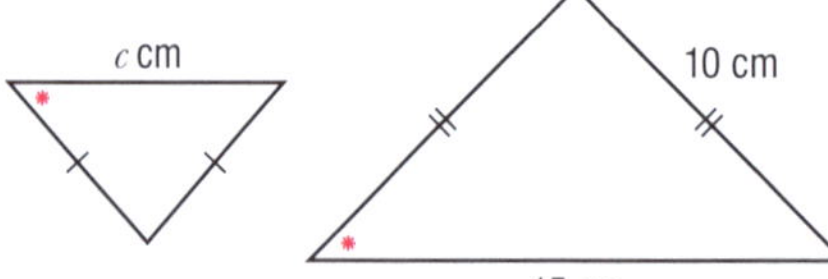

d Scale factor = $\frac{5}{8}$

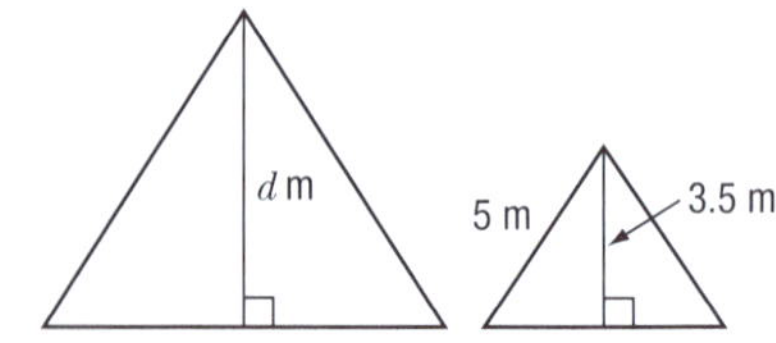

3 a Draw a scalene triangle and enlarge it by a scale factor of 2.

b Draw an isosceles triangle and reduce it by a scale factor of $\frac{2}{3}$.

c Draw a right-angled triangle and enlarge it by a scale factor of 1.6.

Applications of similar triangles

If we know that two triangles are similar then we can use the facts that:

i their corresponding angles are equal,

and

ii their corresponding sides are in the same ratio,

to find unknown sides and angles.

Example 1

Find the value of x.

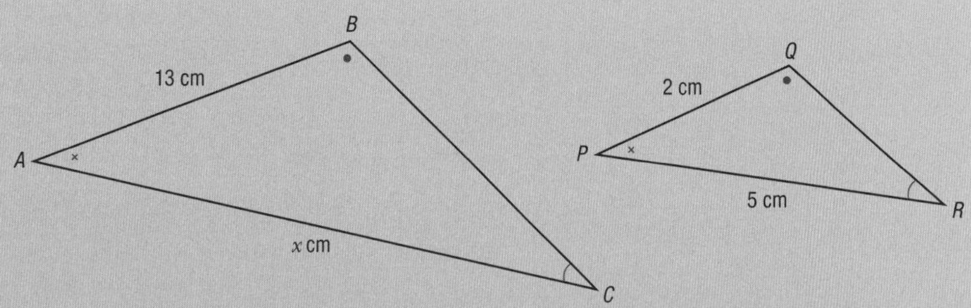

Triangles ABC and PQR are similar because their corresponding angles are equal. The sides AB and PQ are a pair of corresponding sides and AC and PR are also corresponding sides, so we know that these pairs of sides will be in the same ratio.

Answer

We can say:

$AC : PR = AB : PQ$ rewrite these ratios as fractions

$\frac{AC}{PR} = \frac{AB}{PQ}$ substitute the given side lengths

$\frac{x}{5} = \frac{13}{2}$ multiply both sides of the equation by 5

$x = \frac{13 \times 5}{2}$

$= 32.5$

Example 2

Identify the pair of similar triangles in the diagram, draw them separately and then find the value of the pronumeral.

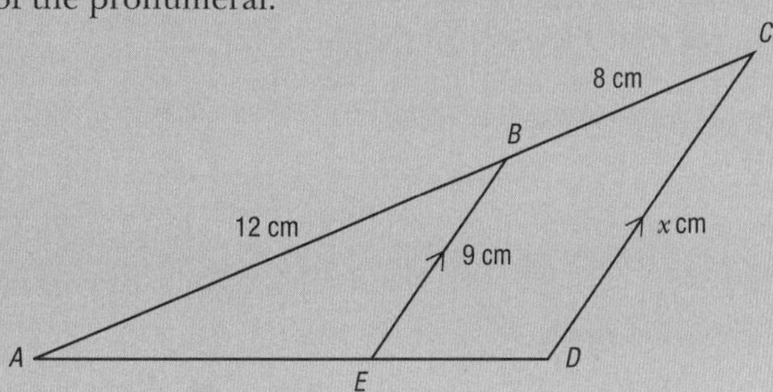

Answer

The two triangles, drawn separately below, are similar because their corresponding angles are equal:

$\angle ACD = \angle ABD$ and $\angle ADC = \angle AEB$ because these are corresponding angles formed when a transversal crosses a pair of parallel lines.

$\angle CAD = \angle BAE$ because both triangles have this angle in common.

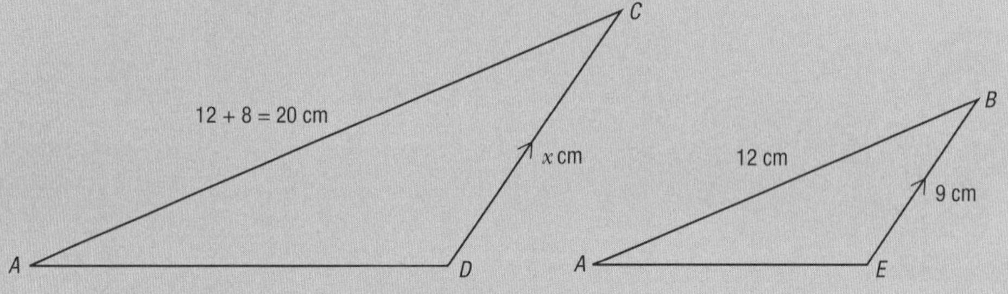

AC and AB are corresponding sides as are DC and EB so we can say:

$\frac{DC}{EB} = \frac{AC}{AE}$ — substitute the lengths that we know

$\frac{x}{9} = \frac{20}{12}$ — multiply both sides of the equation by 9

$x = \frac{20 \times 9}{12}$

$= 15$

Example 3

Find the value of x in the following diagram.

Answer

The triangles ABC and ADE are similar triangles as their corresponding angles are equal.
The triangles are drawn separately at right.
AC and AE are corresponding sides and BC and DE are corresponding sides, so they will be in the same ratio.

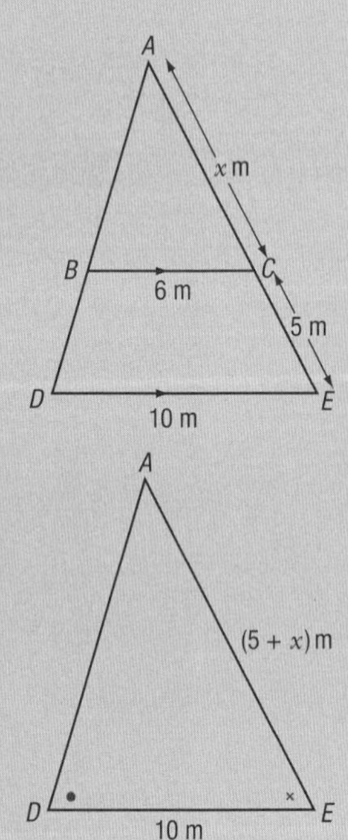

$\frac{AC}{AE} = \frac{BC}{DE}$

$\frac{x}{5 + x} = \frac{6}{10}$ — multiply both sides by $(5 + x)$

$x = \frac{6}{10} \times (5 + x)$ — multiply both sides by 10

$10x = 6(5 + x)$ — expand bracket

$10x = 30 + 6x$ — subtract $6x$ from both sides

$4x = 30$ — divide both sides by 4

$x = 7.5$

EXERCISE

1 The following pairs of triangles are similar. Find the value of the pronumeral.

a

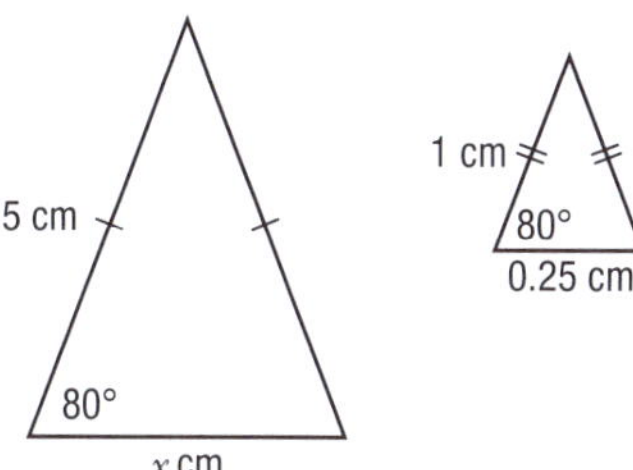

b

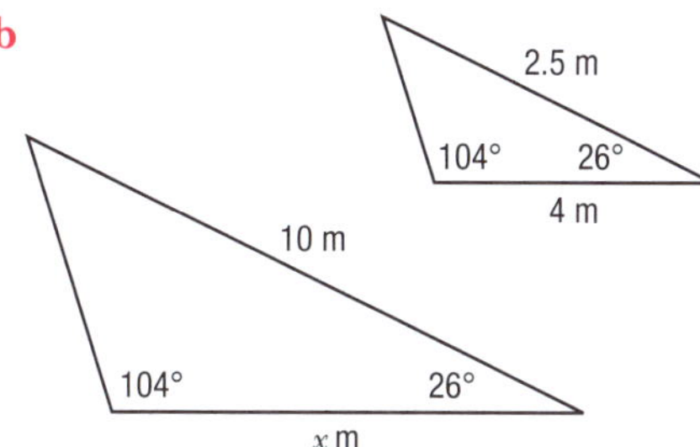

c

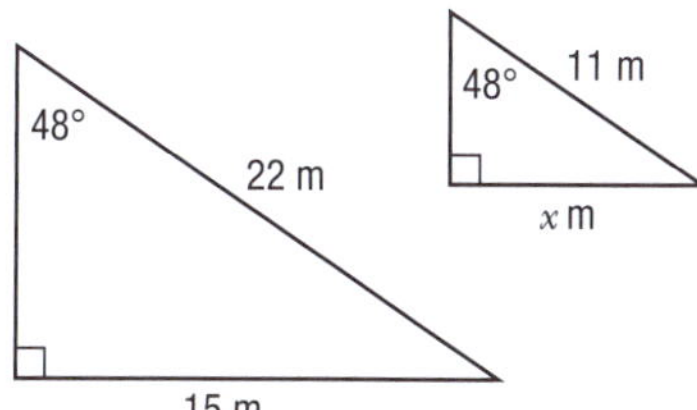

d

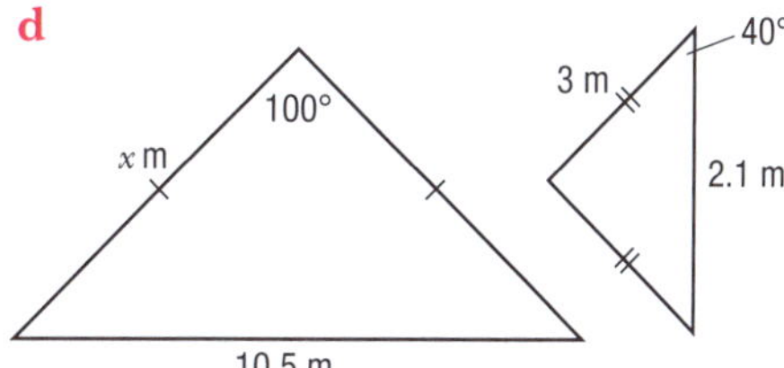

e

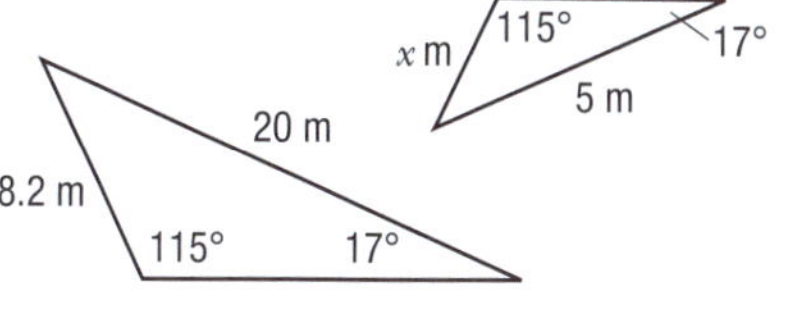

f

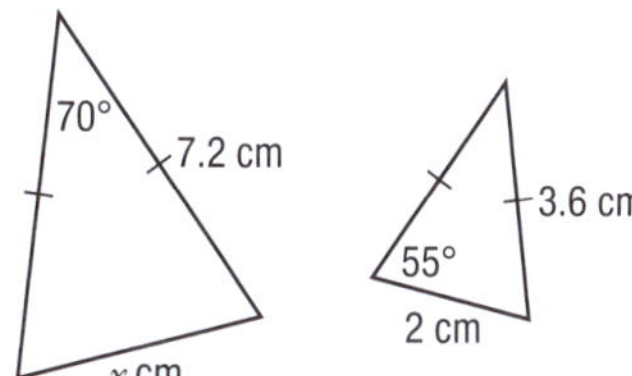

2 Identify the two similar triangles in each of the following, draw them separately and find the value of the pronumeral.

a

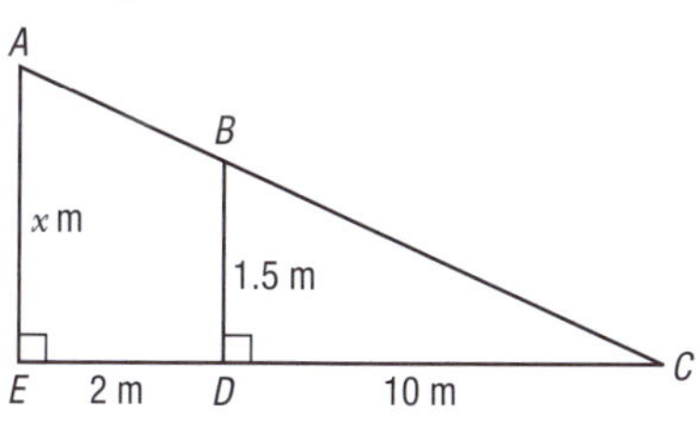

b

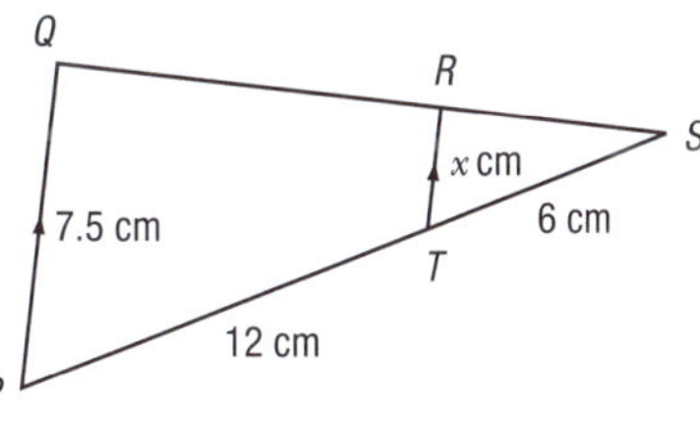

c

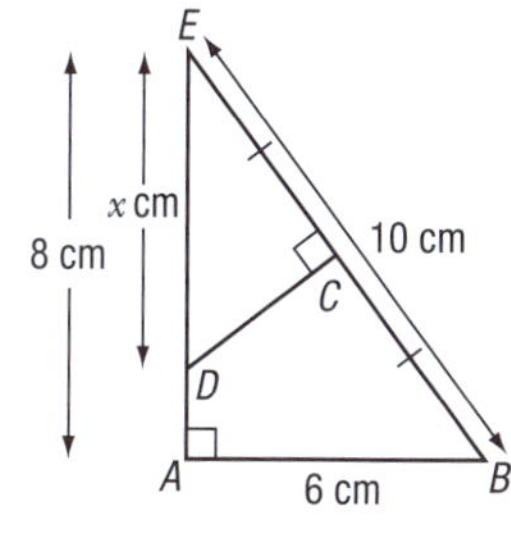

d

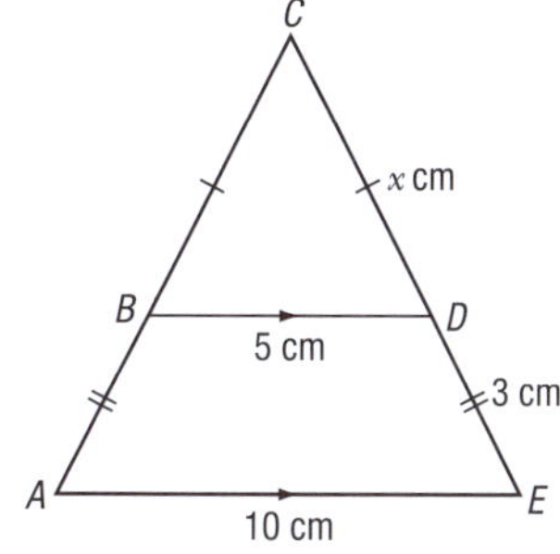

e

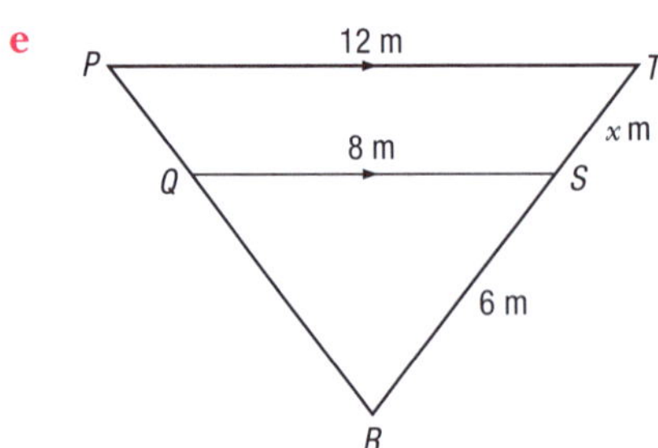

3 Show that the following pairs of triangles are similar and find the value of the pronumeral.

a

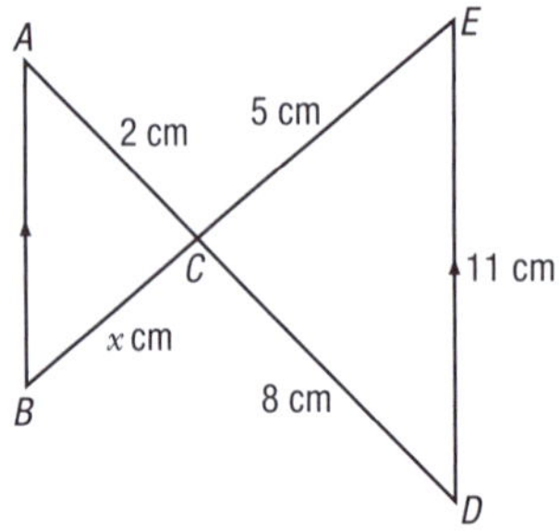

b

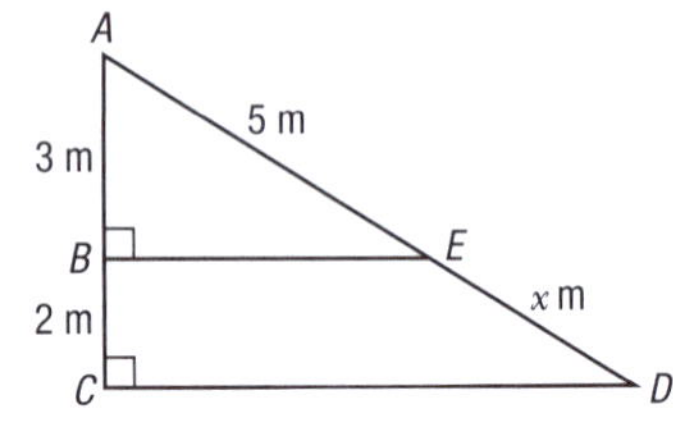

c

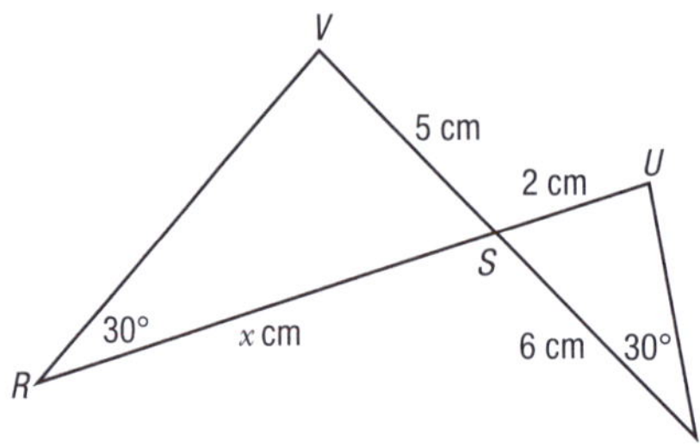

d

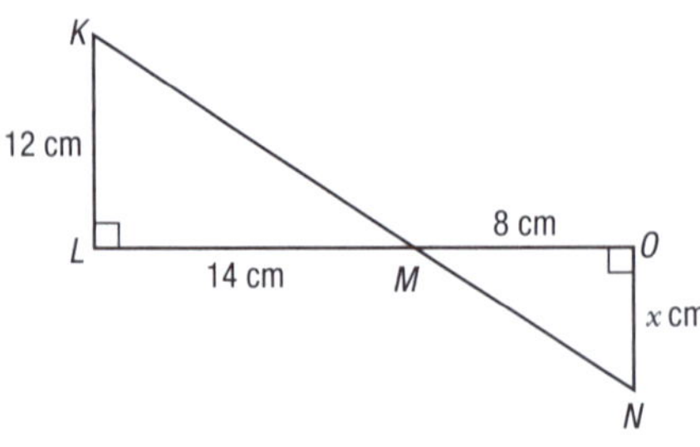

e

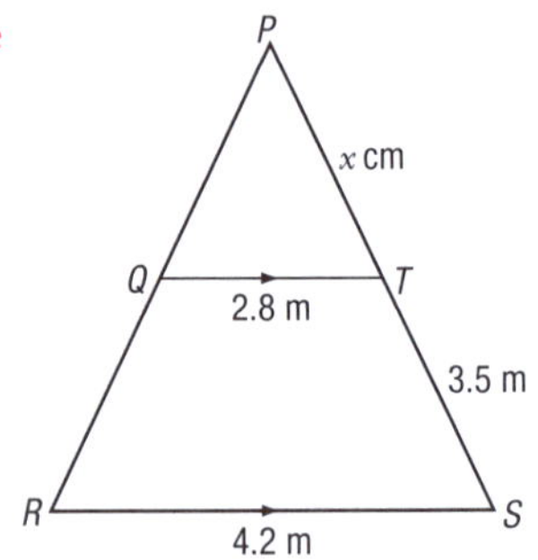

Practical applications of similar triangles

Similar triangles are often used in applications where it is not possible to measure a length.

Example 1

Veru wants to know the height of a large tree growing in her garden. It is too tall to measure.

Early one morning she takes a one-metre ruler, stands it vertically and then measures its shadow. She then measures the shadow of the tree and draws the following diagrams. Use the measurements to find the height of the tree.

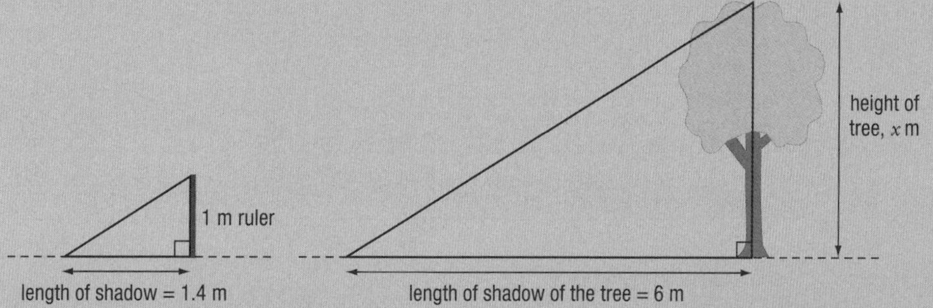

Answer

The right-angled triangles formed by the ruler and the tree will be similar because the measurements were taken at the same time of the day when the sun's rays were casting shadows at a particular angle.

The shadows will be corresponding sides and the heights will be corresponding sides.

So $\frac{\text{Height of tree}}{\text{Height of ruler}} = \frac{\text{Length of shadow of the tree}}{\text{Length of shadow of the ruler}}$

substitute the given lengths: $\frac{x}{1} = \frac{6}{1.4}$

$x = 4.286$

The tree is 4.286 metres high.

Example 2

Rufus wants to find the width of a river. He starts at the point A, directly opposite a tree on the other side of the river. He walks 20 paces along the riverbank, places a stake at the point B then walks another 8 paces along the riverbank, placing another stake at C. He then walks in a direction perpendicular to the riverbank until he reaches a point where he can line up the stake at B and the tree. This point D is 18 paces from C.

If Rufus's pace is 80 cm, use these measurements to find the width of the river.

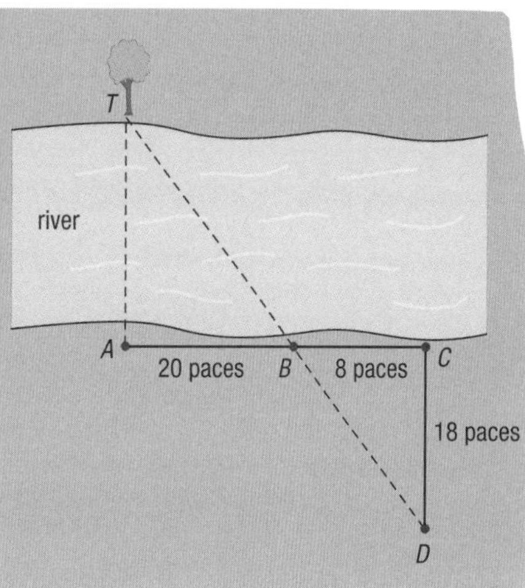

Answer

The two triangles are similar because their corresponding angles are equal:

$\angle TAB = \angle BCD = 90°$

$\angle TBA = \angle CBD$ because they are vertically opposite angles.

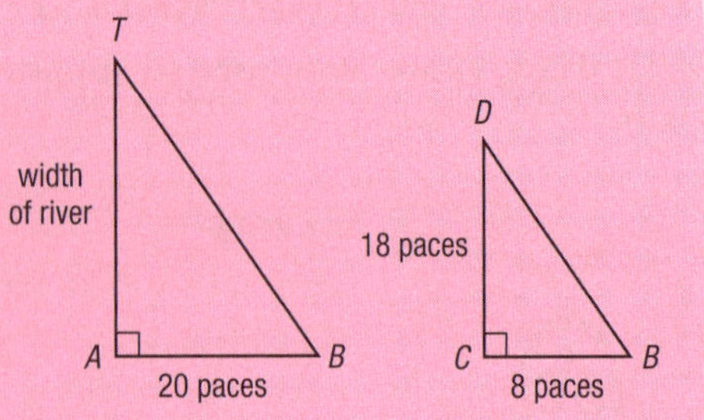

AB and *BC* are corresponding sides, as are *TA* and *DC*.

$\frac{TA}{DC} = \frac{AB}{BC}$ substitute the known lengths

$\frac{TA}{18} = \frac{20}{8}$ multiply both sides of the equation by 18

$TA = \frac{20 \times 18}{8} = 45$

The river is 45 paces wide. $45 \times 0.80 = 36$

The river is 36 metres wide.

EXERCISE

1 Milla wants to find the height of a flagpole using the 'shadow method'. She has a stick that she has measured as 80 cm long. At a particular time of the day the stick, held vertically, casts a shadow that is 64 cm long. At the same time the flagpole casts a shadow that is 3.4 metres long.

- a Draw diagrams that represent the stick, the flagpole and their shadows, labelling the appropriate sides with the lengths given.
- b Why are the triangles in your diagram similar triangles?
- c Use the similar triangles to calculate the height of the flagpole.

2 Kema wants to measure the distance across a lagoon, *XT*. He can see a large tree on the other side of the lagoon at the point *T* and uses this as a sight point. He places pegs at the points *X*, *A*, *B* and *C* as shown in the diagram and measures the distances *XA*, *AB* and *BC*.

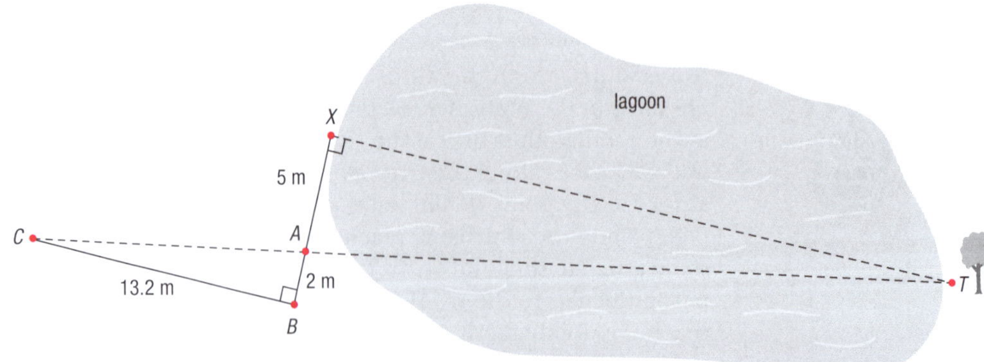

- a Why are the triangles *AXT* and *ABC* similar triangles?
- b Use the similar triangles to calculate the distance *XT*.

ACTIVITY 1

In groups of two or three, use the shadow method (see Example 1) to find the height of a tall object in your school grounds.

You will need a tape measure and a stick or long ruler to use as the shadow stick.

- Draw diagrams showing your measurements and show your calculations.
- Compare your calculation of the height with other groups.
- Discuss any problems that you found with this method. How could you improve the accuracy of your measurement?

ACTIVITY 2

In groups of two or three, use the method outlined in Example 2 to find the width across a sports ground or something similar in your school grounds.

You will need a tape measure and four pegs.

- Draw diagrams showing your measurements and show your calculations.
- Compare your calculation of the width with other groups.
- Discuss any problems that you found with this method. How could you improve the accuracy of your measurement?

PYTHAGORAS'S RULE

Introduction to Pythagoras's rule

Lesson 7

In ancient times it was commonly known that a triangle with sides of length 3, 4 and 5 units was a **right-angled triangle** and this fact was used to accurately measure an angle of 90°.

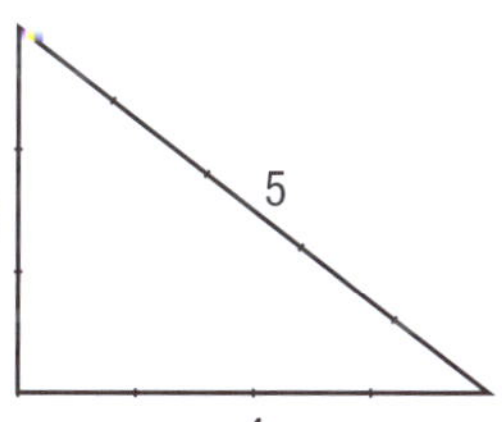

Use a protractor to measure the largest angle in the triangle on the previous page. It should be exactly 90° and therefore is a **right angle**.

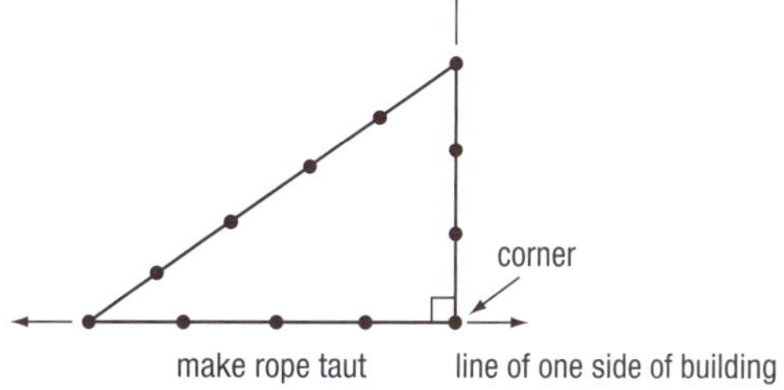

Although many cultures were aware that there was a relationship between the lengths of the sides of a right-angled triangle, **Pythagoras** the Greek philosopher and mathematician is credited with the **rule**, or **theorem**, that describes this relationship. Pythagoras lived in the years 580 to 500 BC.

Pythagoras's rule states:

For any right-angled triangle, the square on the hypotenuse is equal to the sum of the squares on the other two sides.

Remember

The **hypotenuse** is a special name for the longest side in a right-angled triangle. It is the side described as **opposite the right angle**.

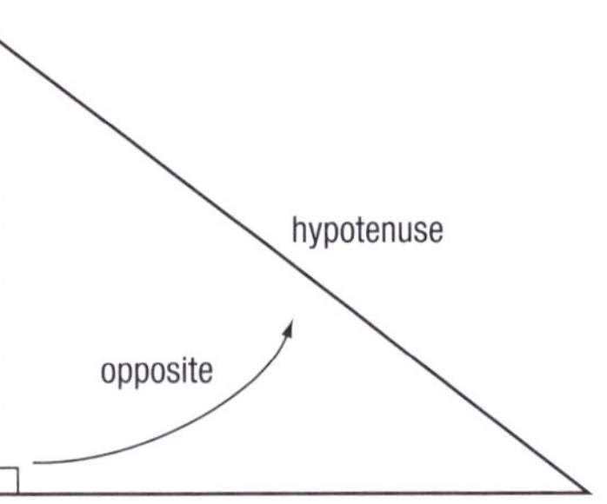

Pythagoras's rule is demonstrated below for a right-angled triangle with side lengths 3, 4 and 5 units. The hypotenuse in this triangle is 5 units long.

The square on the hypotenuse

$= 5^2$

$= 5 \times 5$

$= 25$ square units

the sum of the squares on the other two sides

$= 3^2 + 4^2$

$= 3 \times 3 + 4 \times 4$

$= 9 + 16$

$= 25$ square units

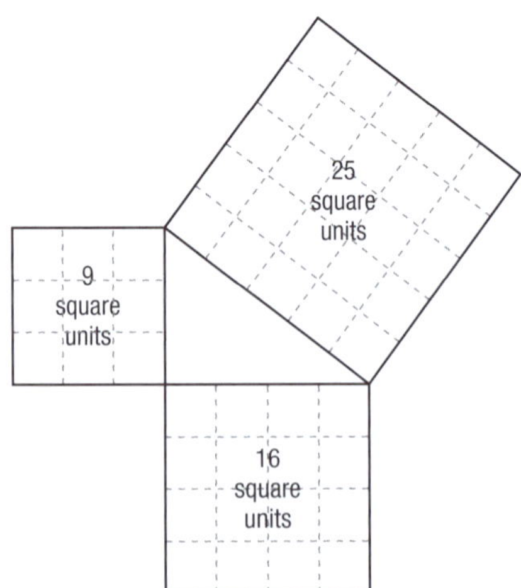

Hence we can say the **square on the hypotenuse** (25 square units) is equal to the **sum of the squares on the other two sides** (25 square units).

Algebra and pronumerals were not in use in Pythagoras's time, so the rule was stated in words only.

Today, using the pronumeral c to represent the length of the hypotenuse and a and b to represent the lengths of the other two sides, Pythagoras's rule can be stated as:

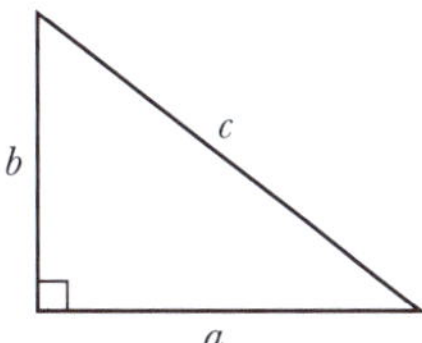

$$c^2 = a^2 + b^2$$

How would the formula change if the hypotenuse was z and the other two sides x and y?
The formula would be $z^2 = x^2 + y^2$.

EXERCISE

1 **a** Using a ruler and a protractor to measure the right angle, carefully draw the two sides of a right-angled triangle with the measurements shown.

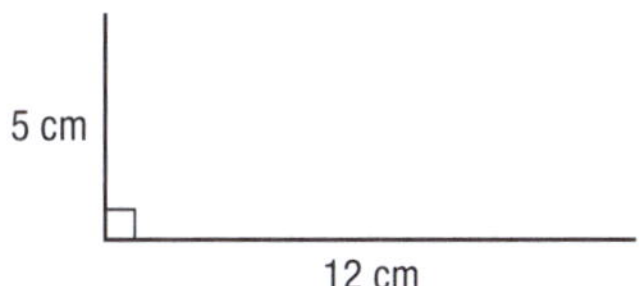

b Draw in the hypotenuse for this triangle and carefully measure its length.

c Copy and complete the following to check Pythagoras's rule for the given triangle:

Length of hypotenuse = ... centimetres
The square on the hypotenuse = ... ^{2}
= .... square centimetres
The sum of the squares on the other two sides = ... ^{2} + ... ^{2}
= .... + ...
= .... square centimetres

2 There are numerous triangles whose side lengths are called **Pythagorean triads**.

The special property of these triangles is that their side lengths are all whole numbers. Some of these are given in the table below. Show that Pythagoras's rule applies to these triangles. Copy and complete the following table.

Pythagorean triads

Side lengths	(Hypotenuse)^2	Sum of squares of other two sides
3, 4, 5	$5^2 = 25$	$3^2 + 4^2 = 9 + 16 = 25$
7, 24, 25		
9, 40, 41		
11, 60, 61		

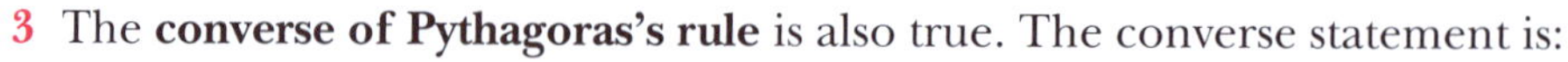

3 The **converse of Pythagoras's rule** is also true. The converse statement is:

> If the square of the longest side of a triangle is equal to the sum of the squares of the other two sides then the triangle is a right-angled triangle.

Which of the following triangles is a right-angled triangle?

a For this triangle copy and complete:
(Hypotenuse)^2 = ...
Sum of squares and the other two sides = ...
Right-angled triangle? Yes/No

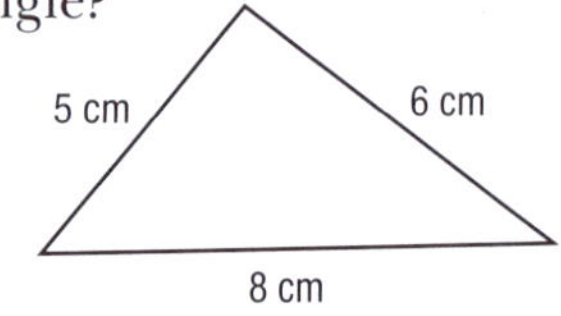

b Show similar calculations for these triangles.

i

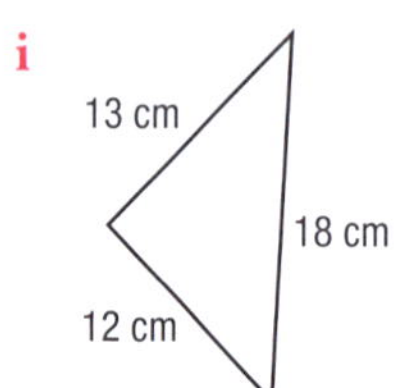

ii

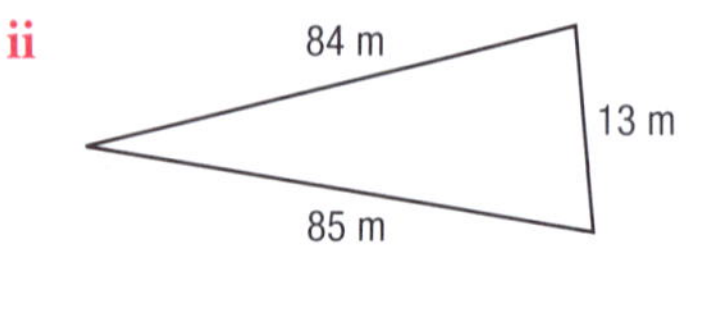

4 Write down Pythagoras's rule, in terms of the pronumerals given, for each of the following triangles.

a

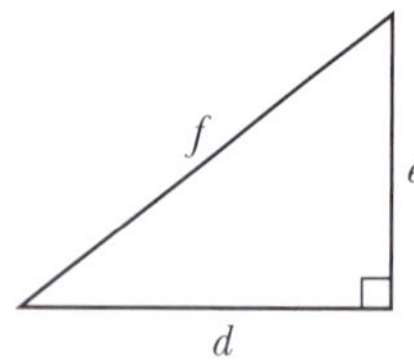

$f^2 = \ldots + \ldots$

b

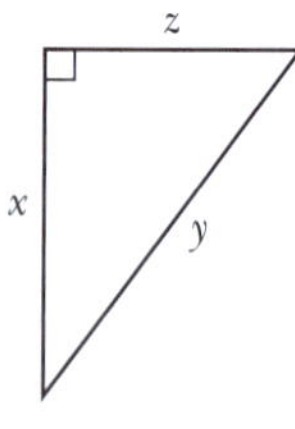

c

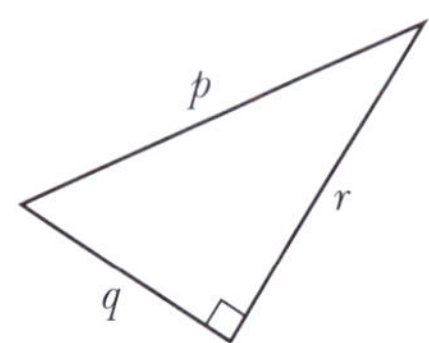

ACTIVITY A demonstration of Pythagoras's rule

For this activity to be successful it is important that you accurately measure the side lengths and use a protractor to draw the right angles.

Equipment required: a page in your workbook, a fine pen or pencil, a ruler and protractor, tracing paper, coloured pencils or highlighters, scissors, glue.

Step 1: Draw a right-angled triangle in the centre of your page using the protractor for the right angle. Make sure that you leave enough room around it to draw the squares on the three sides.

Step 2: Using the ruler and the protractor carefully draw the squares on the three sides of your triangle.

Step 3: Extend the sides of the square on the hypotenuse into the squares on the other two sides, as shown by the dotted lines in the diagram. Label the five areas produced 1 to 5, and colour each a different colour.

Step 4: Using the tracing paper copy the squares on the two shorter sides including the lines and numbers, colouring the copies the same colours as above. Carefully cut out and separate these five areas.

Step 5: Fit these five areas into the square on the hypotenuse. They should fit exactly! When you are sure that they are in the correct position glue them in place.

You have shown that 'the square on the hypotenuse is equal to the sum of the squares on the other two sides'.

Irrational numbers and surds

Pythagoras belonged to a powerful brotherhood whose motto was 'All is number'. They believed that all things in the universe could be explained in terms of whole numbers or ratios. These are the numbers that we now call **rational numbers**.

Rational numbers can be positive or negative, and have the special property that they can all be written as a **ratio**, that is, as one whole number divided by another.

Notice that the first five letters of the word **ratio**nal spell the word **ratio**.

Many of the numbers that you deal with are rational numbers. Examples of rational numbers are:

- Whole numbers, for example, the number 3 can be written as $\frac{3}{1}$ which is a ratio of the whole numbers 3 and 1.
- Decimals, for example, the number 0.345 can be written as $\frac{345}{1000}$ which is a ratio of whole numbers.
- Recurring decimals, for example, $0.1\dot{2} = 0.122\,222\,2\ldots$ can be written as $\frac{9}{11}$ which is a ratio of whole numbers.

For Pythagoras's brotherhood these rational numbers appeared to be the only type of numbers until they considered the length of the hypotenuse of certain right-angled triangles. They probably considered examples like this.

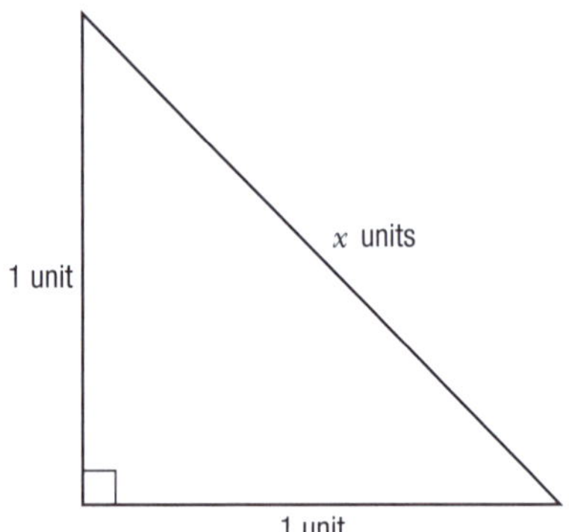

What is the length of the hypotenuse in an isosceles right-angled triangle with the shorter sides of length 1 unit?

Today we could accurately draw the triangle with shorter sides 1 m (or 1000 mm) long.

Measuring the hypotenuse gives a value of approximately 1414 mm corresponding to 1.414 units.

Using Pythagoras's rule to find the length of the hypotenuse, marked x, gives:

$$x^2 = 1^2 + 1^2$$
$$x^2 = 1 + 1$$

So, $x^2 = 2$.

The solution to this equation is the number which, when multiplied by itself, gives the answer 2.

This is the number we write as $\sqrt{2}$, and say as 'the square root of 2' or 'root 2'.

$\sqrt{2}$ is the number such that $\sqrt{2} \times \sqrt{2} = 2$

Your calculator or tables will give a decimal approximation to the number $\sqrt{2}$.

$\sqrt{2} = 1.414$ rounded to three decimal places

A mathematical dilemma!

The existence of numbers such as $\sqrt{2}$ would not have been a problem to Pythagoras's brotherhood except that they proved that this number could not be a rational number.

Numbers like $\sqrt{2}$ appeared to be **infinite**, **non-recurring decimals**—the decimal parts go on forever, not repeating the pattern of decimals and so cannot be written as a ratio.

This caused great consternation for Pythagoras's brotherhood as it ruined their 'All is number' theory. They decided to keep their knowledge of these numbers secret and threatened death by drowning to anyone who revealed their secret.

The existence of these numbers was finally revealed 400 years after their discovery, and they are the numbers we call **irrational numbers** today. (The word 'irrational' means 'not rational'.)

Surds are square roots that are **irrational** numbers.
The **exact value of a surd** is written as
$\sqrt{2}, \sqrt{3}, \sqrt{5}, \sqrt{6.85}, \sqrt{7.69}, \sqrt{1256}$ ..... and so on
although approximations can be found for all of them.

Note:

- One of the answers to an equation such as $x^2 = 7$ is $x = \sqrt{7}$ because $\sqrt{7} \times \sqrt{7} = 7$.
  The other answer is $x = -\sqrt{7}$ because $-\sqrt{7} \times -\sqrt{7} = 7$.
  However when using Pythagoras's rule we do not consider the negative answer because we are dealing with lengths of sides and these are always positive.
- There are some square roots that are not irrational numbers, that is they are rational numbers!
  We know that $2 \times 2 = 4$ so $\sqrt{4} = 2$ which is a rational number. $3 \times 3 = 9$ so $\sqrt{9} = 3$.
  What are the values of $\sqrt{16}, \sqrt{25}, \sqrt{36}, \sqrt{49}, \sqrt{1.21}, \sqrt{6.25}, \sqrt{900}, \sqrt{1\,000\,000}$?
  The answers are respectively: 4, 5, 6, 7, 1.1, 2.5, 30, 1000; all rational numbers.
  The numbers 16, 25, 36, 49, 1.21 6.25, 900 and 1 000 000 and numbers with the same properties are called **perfect squares** because they have **rational** square roots.

EXERCISE

1 Which of the following numbers are surds?
$\sqrt{101}$, 81, $\sqrt{36}$, $\sqrt{17}$, 1.732, 9.7, $\sqrt{8}$, $\sqrt{83.4}$, $\sqrt{1.44}$, $\sqrt{4900}$

2 **a** Use your calculator or the square root tables to find the square roots of the following numbers, rounding to two decimal places where necessary.
64, 99, 144, 40, 81, 400, 50, 1.69, 900

b Which of the numbers in part a are perfect squares?

3 Use your calculator or tables to find decimal values, rounded to three decimal places where necessary, for each of the following.

a $\sqrt{243}$ **b** $\sqrt{121}$ **c** $\sqrt{0.65}$ **d** $\sqrt{1000}$ **e** $\sqrt{67}$

4 Find a rational answer for each of the following.

a $\sqrt{11} \times \sqrt{11}$ **b** $\sqrt{81}$ **c** $\sqrt{(7 \times 7)}$ Brackets first!
d $\sqrt{(36 + 64)}$ **e** $\sqrt{(4 \times 9)}$

5 Find a positive solution to each of the following giving your answer as:
i a surd value
ii a decimal approximation rounded to three decimal places.

a $a^2 = 10$ **b** $b^2 = 79$ **c** $c^2 = 3054$
d $d^2 = 16 + 49$ **e** $x^2 = 8^2 + 3^2$

6 The lengths of the sides of these triangles are all rational values. Use Pythagoras's rule (and your calculator or tables) to show that they are right-angled triangles and to find the simplest ratio of the sides.

Copy and complete:

a (Longest side)^2 = $(.....)^2$ = ...

Sum of squares on other two sides

= .....^{2} + ...^{2}

= .....

The triangle ..... right angled.

Ratio of sides, smallest to largest

= 1.2 : 1.6 : 2

= … : … : …   multiplying by 10

= … : … : …   divide by the HCF, which is ……

2
1.2
1.6

b (Longest side)^2 = $(.....)^2$ = ...

Sum of squares on other two sides

= .....^{2} + ...^{2}

= .....

The triangle ..... right angled.

Ratio of sides, smallest to largest

= 2.7 : 12 : 12.3

= … : … : …

= … : … : …   divide by the HCF, which is ……

2.7 cm
12.3 cm
12 cm

c (Longest side)^2 = $(.....)^2$ = ...

Sum of squares on other two sides

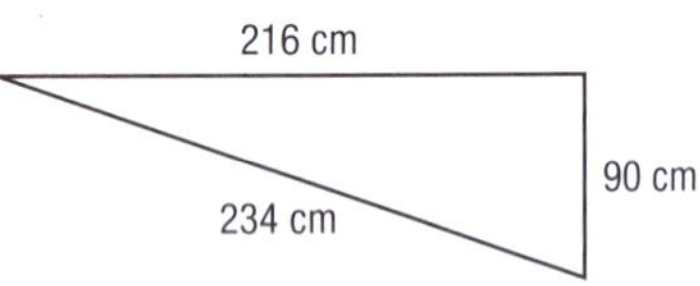

= .....^{2} + ...^{2}

= .....

The triangle ..... right angled.

Ratio of sides, smallest to largest

= 90 : 216 : 234

= … : … : …

= … : … : …   divide by the HCF, which is ……

ACTIVITY Generating Pythagorean triads

A Pythagorean triad is a set of three rational numbers that satisfy Pythagoras's rule. Choose a whole positive number for n in each of the following expressions to produce a Pythagorean triad.

Number 1: $2n + 1$

Number 2: $2n^2 + 2n$

Number 3: $2n^2 + 2n + 1$

For example, if $n = 3$

Number 1: $2n + 1 = 2 \times 3 + 1 = 7$

Number 2: $2n^2 + 2n = 2 \times 3^2 + 2 \times 3 = 24$

Number 3: $2n^2 + 2n + 1 = 2 \times 3^2 + 2 \times 3 + 1 = 25$

We have produced the Pythagorean triad 7, 24, 25.

Checking that these numbers satisfy Pythagoras's rule:

(Hypotenuse)$^2 = 25^2 = 625$

Sum of squares of other two sides = $7^2 + 24^2 = 49 + 576 = 625$.

Choose five other values for n to produce five other Pythagorean triads, checking that each triad satisfies Pythagoras's rule.

Calculating the length of the hypotenuse using Pythagoras's rule

Example 1

Find the length of the hypotenuse, marked x, in the given right-angled triangle. Give your answer as a surd value and also as a decimal approximation rounded to two decimal places.

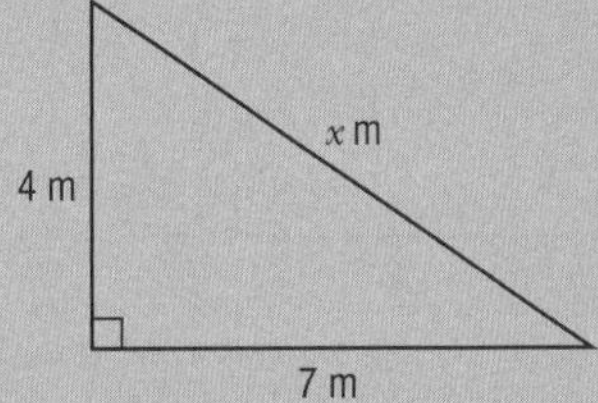

Answer

Pythagoras's rule for this triangle gives:

$x^2 = 4^2 + 7^2$

$x^2 = 16 + 49$

$x^2 = 65$

Hence, $x = \sqrt{65}$ surd answer

i.e., $x \approx 8.06$ decimal approximation

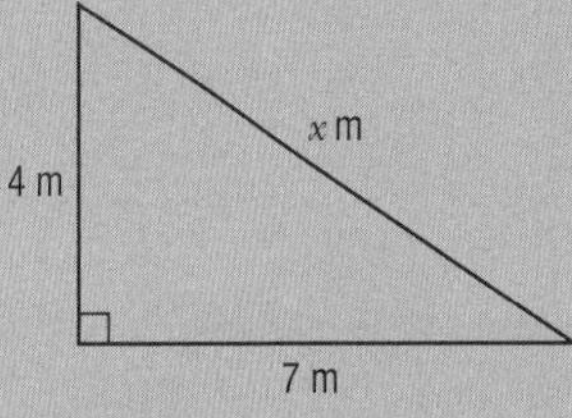

Example 2

Find the length of the diagonal of a square that has sides of length 25 centimetres.

Answer

Draw the square and the relevant right-angled triangle.
Let the length of the diagonal be d cm.

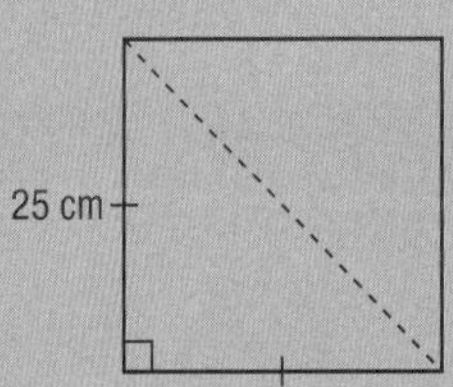

Pythagoras's rule for this triangle gives:

$d^2 = 25^2 + 25^2$

$d^2 = 625 + 625$

$d^2 = 1250$

Hence, $d = \sqrt{1250}$

i.e., $d = 35.36$ rounded to two decimal places

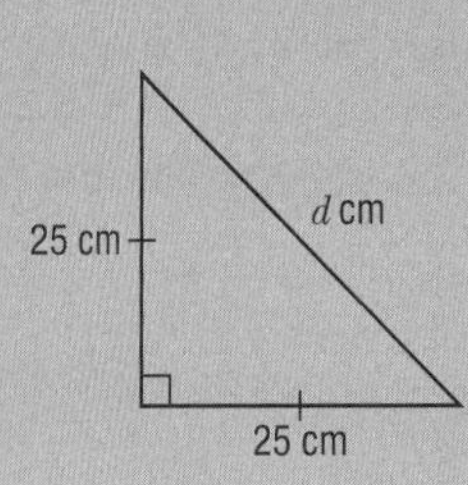

EXERCISE

1 Calculate the length of the hypotenuse for each of the following triangles, giving your answer as a surd value and also as a decimal rounded to two decimal places.

a 7 m, 4 m, *a* m

b

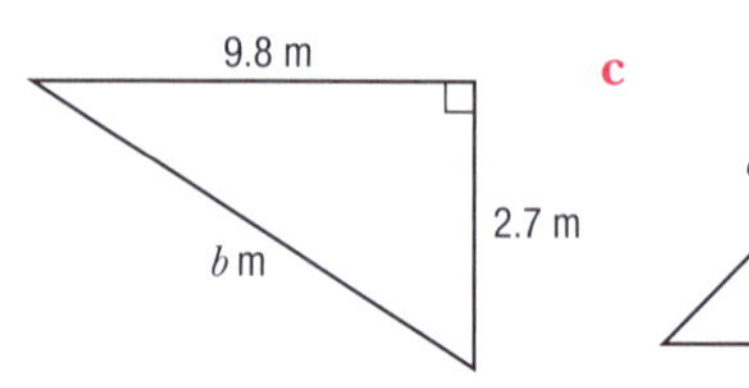

c *c* cm, 6 cm, 5 cm

d

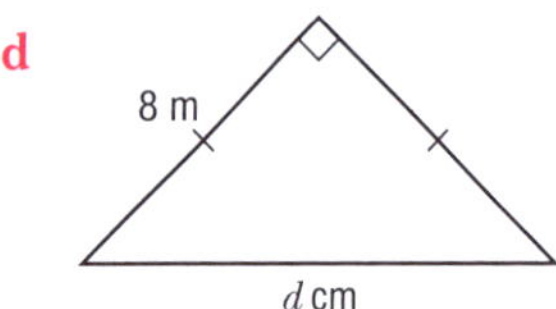

e

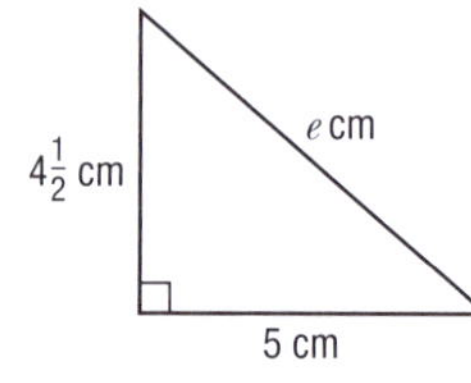

f

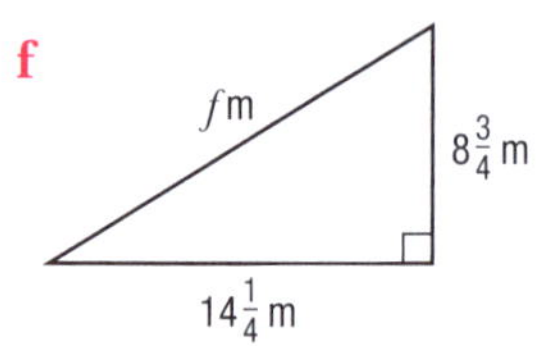

> **Remember**
> Change fractions to decimals.

2 Find the lengths of the diagonals for each of the following. Give surd and decimal answers rounded to two decimal places.

a 3 cm

b

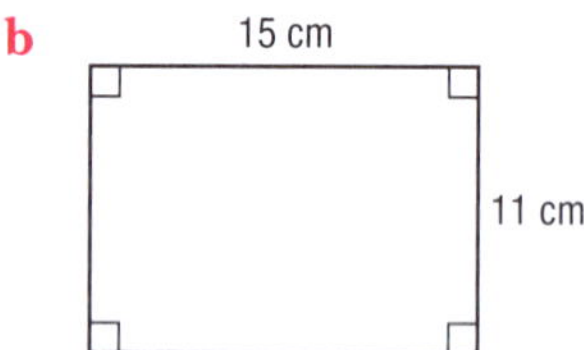

c A square of side length 5.6 metres.

d A rectangle with dimensions 15 centimetres × 20 centimetres.

3 By finding a relevant right-angled triangle, find the length marked x in each of the following. Round answers to two decimal places.

a x cm, 17 cm, 20 cm

b

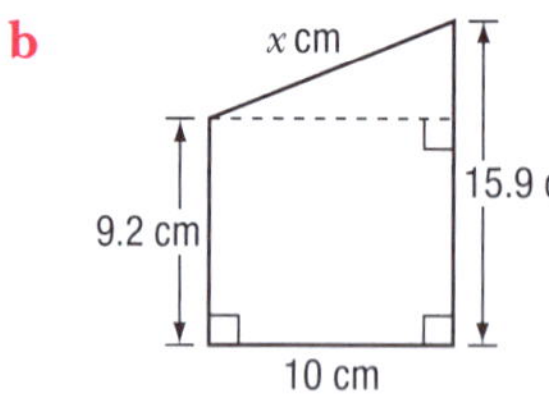

c

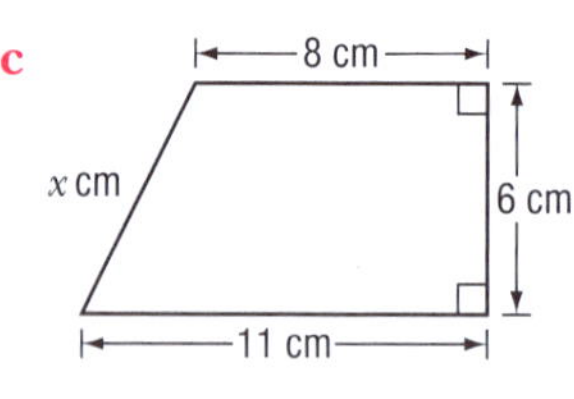

4 Find the perimeter of these figures giving your answers rounded to two decimal places.

a 8 cm, 15 cm

b

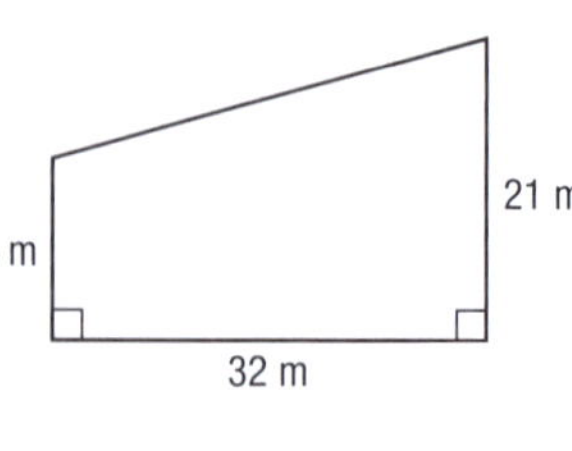

c

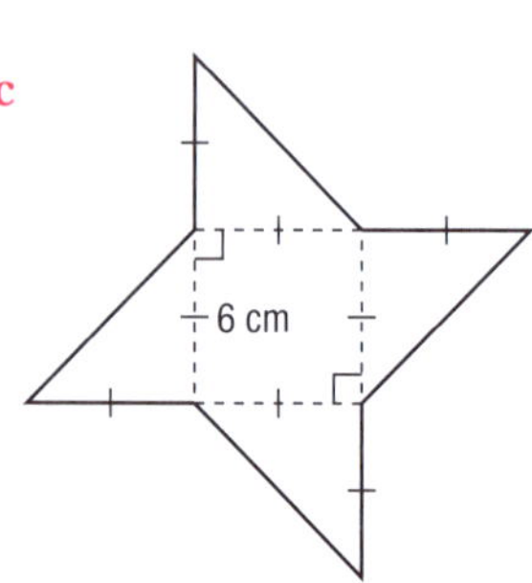

5 The kite in the diagram is to be strengthened with fine wire along all four outer edges. Find the length of wire required.

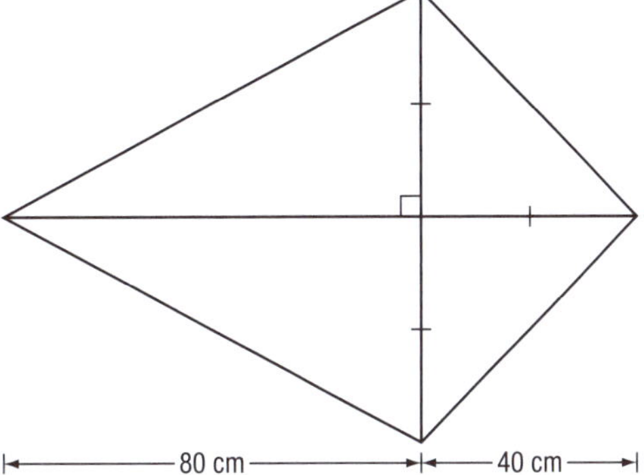

6 Find the lengths a, b and c in the diagram given. Give all the answers as exact values (i.e. as surds).

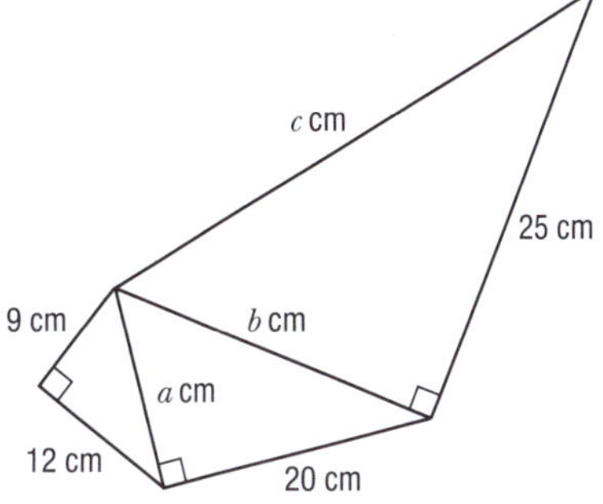

ACTIVITY The Pythagorean spiral: locating irrational numbers on the number line

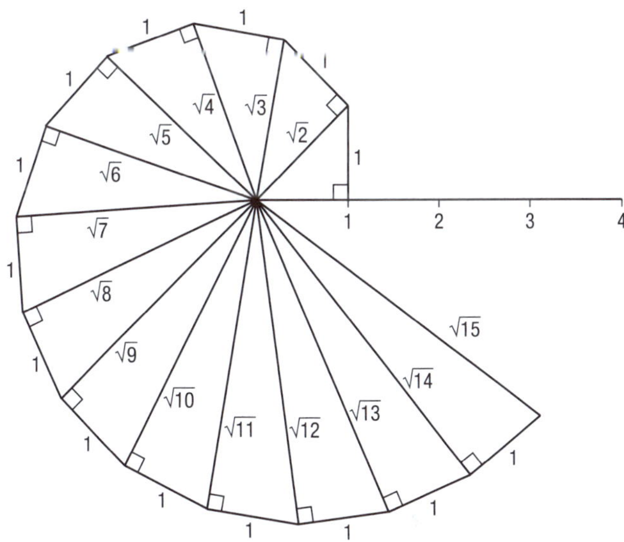

Equipment needed: ruler, protractor, compass, pencil, a 20 × 20 cm piece of paper (about half an A4 page).

Step 1: Starting in the middle of the area, rule a number line 4 units long using 1 unit = 2 cm. Mark the number line from 0 to 4.

Step 2: The first right-angled triangle is an isosceles right-angled triangle with the shorter sides being 1 unit in length. Using a protractor to measure the angle and a ruler to measure the length, construct the first right-angled triangle as shown in the diagram.

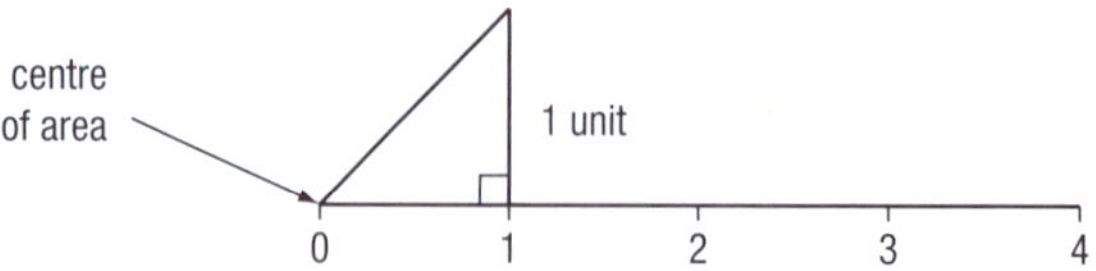

What is the length of the hypotenuse (x units) of this triangle?

$$x^2 = 1^2 + 1^2$$
$$x^2 = 1 + 1$$
$$\therefore\ x^2 = 2$$

Hence, $x = \sqrt{2}$

Step 3: To locate $\sqrt{2}$ on the number line, place your compass point on 0 and the pencil end on the other end of the hypotenuse and draw an arc of length $\sqrt{2}$ so that it reaches the number line. Mark this point on the number line.

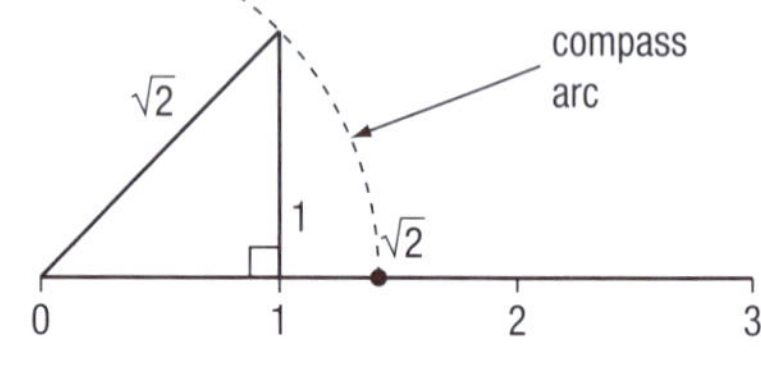

Step 4: Construct (using protractor and ruler) another right-angled triangle on the first one, as shown, adding a side of length 1 unit and then completing the hypotenuse.

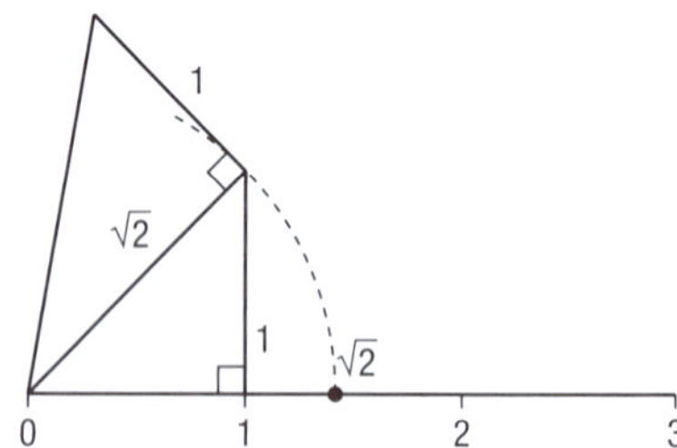

What is the length of the hypotenuse (x units) of this new triangle?

$$x^2 = (\sqrt{2})^2 + 1^2$$
$$x^2 = 2 + 1$$
$$\therefore\ x^2 = 3$$

Hence, $x = \sqrt{3}$

Use the compass to mark this length on the number line and label.

Step 5: Continue on in this manner constructing a right-angled triangle on the previous one, always adding a side of length one unit until you have a hypotenuse of length $\sqrt{16}$ = 4 units.

If you have accurately drawn your triangles then this length should correspond with 4 units on your number line. An earlier check of your accuracy would be $\sqrt{4}$ = 2 units.

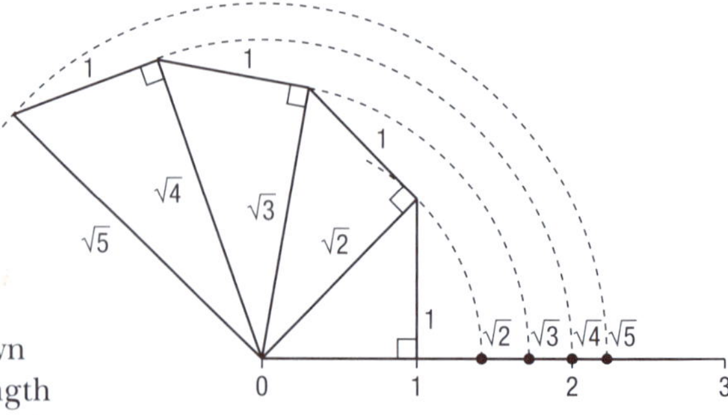

Using Pythagoras's rule to calculate the length of one of the shorter sides in a right-angled triangle

Example 1

Calculate the length of the side marked a in the given triangle.

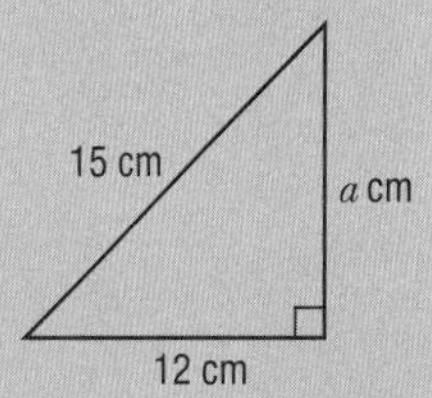

Answer

Always write down Pythagoras's rule for the triangle: $15^2 = a^2 + 12^2$.

We need to make a^2 the subject in this equation.

So as $15^2 = a^2 + 12^2$

then $15^2 - \mathbf{12^2} = a^2 + 12^2 - \mathbf{12^2}$ subract 12^2 from both sides

$15^2 - 12^2 = a^2$ simplifying

$225 - 144 = a^2$

$81 = a^2$

Hence, $a = \sqrt{81} = 9$

Note: The exact answer for a in this triangle is 9 as 81 is a perfect square. The answer $a = {}^{-}\sqrt{81}$ is discarded because a is a length.

Example 2

Calculate the perpendicular height of this isosceles triangle.

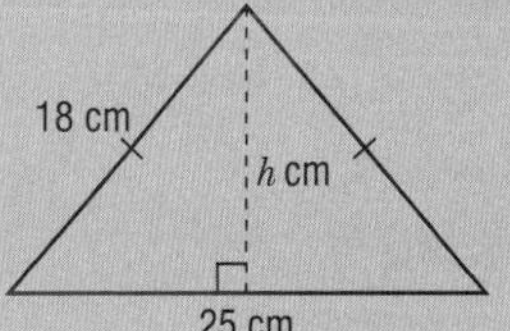

Answer

Drawing in the 'perpendicular height' line for an isosceles triangle produces two identical right-angled triangles with base length $\frac{25}{2} = 12.5$ cm. One of these triangles is shown.

Letting h represent the perpendicular height:

Pythagoras's rule gives

$18^2 = h^2 + 12.5^2$ subtract 12.52 from both sides of the equation

$18^2 - 12.5^2 = h^2$

$324 - 156.25 = h^2$

$167.75 = h^2$

Hence, $h = \sqrt{167.75} = 12.95$ correct to two decimal places

EXERCISE

Give your answers as decimal approximations, rounded to two decimal places.

1 Calculate the value of the pronumeral in each of the following.

a
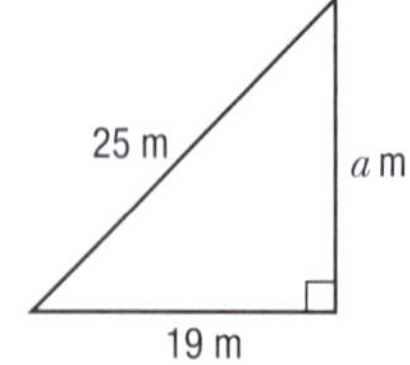

b
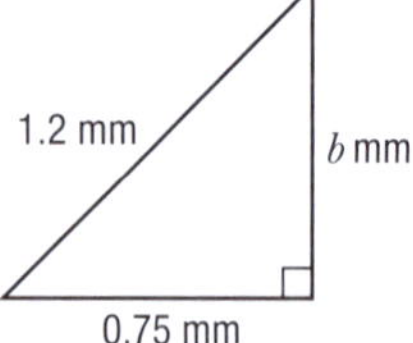

c
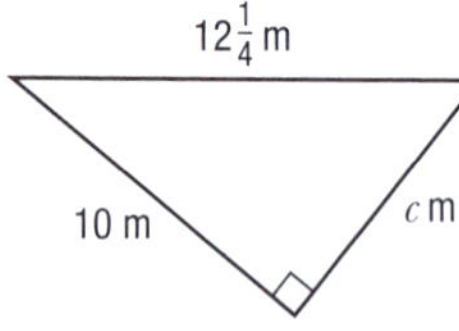

d
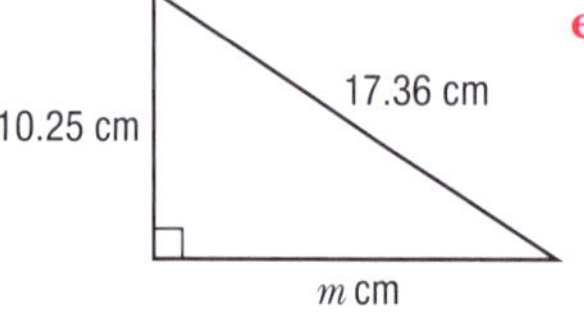

e
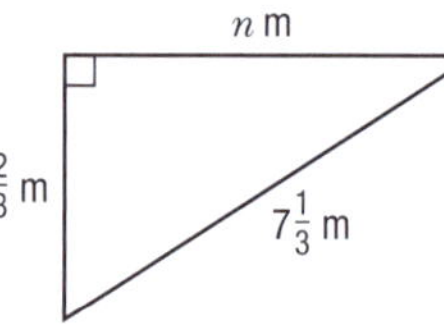

f
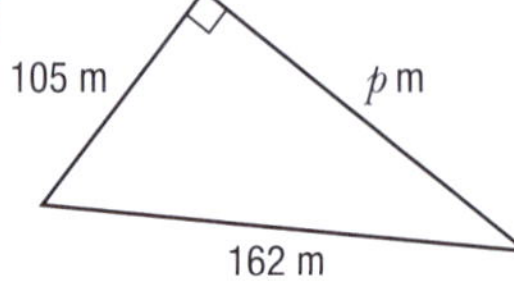

2 Calculate the value of the pronumeral in each of the following.

a
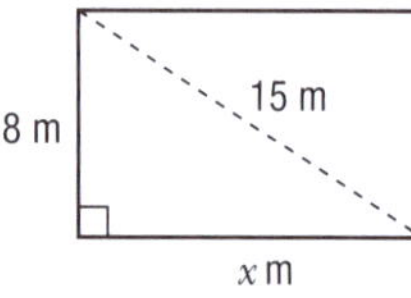

b
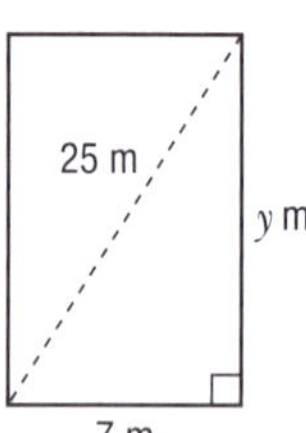

c
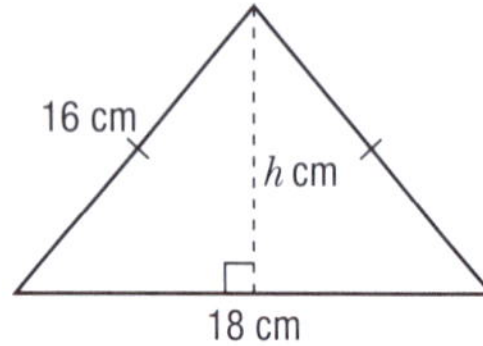

d
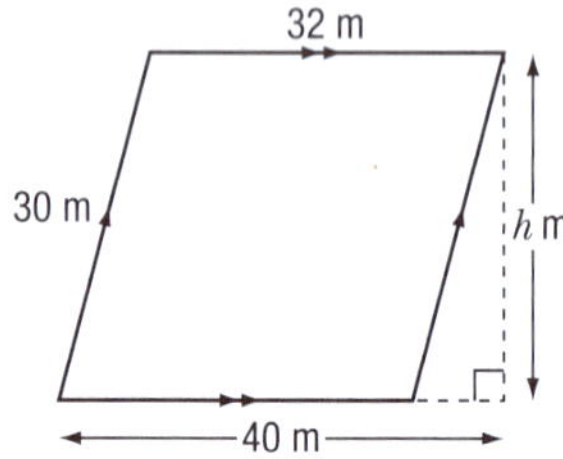

e
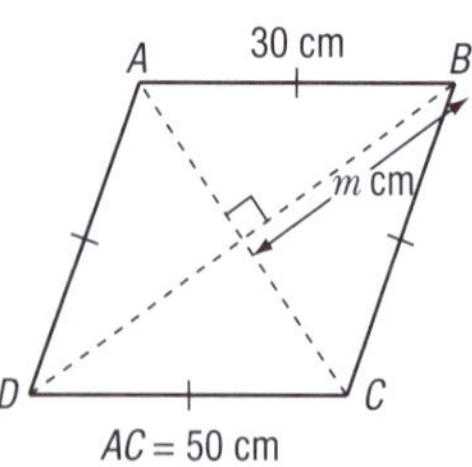

f
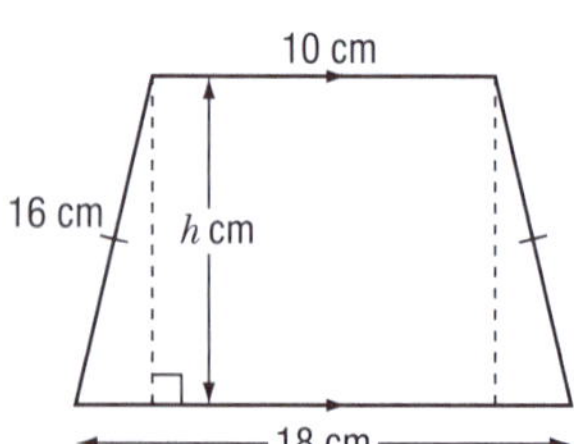

g
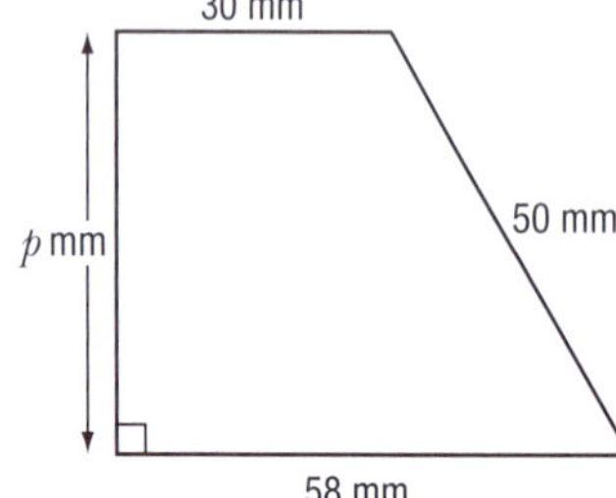

Applications of Pythagoras's rule

Example 1

The diagonal of a square has length 20 cm.
What is the length of the sides of the square?

Answer

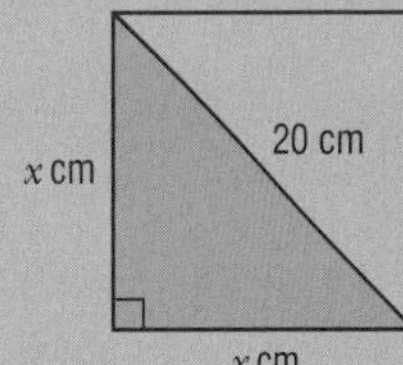

Pythagoras's rule for the triangle shaded gives:

$20^2 = x^2 + x^2$

$400 = 2x^2$

$200 = x^2$ dividing both sides by 2

Hence, $x = \sqrt{200}$

$x = 14.14$ rounded to two decimal places

Example 2

Find the value of the pronumeral, x, in the given triangle.

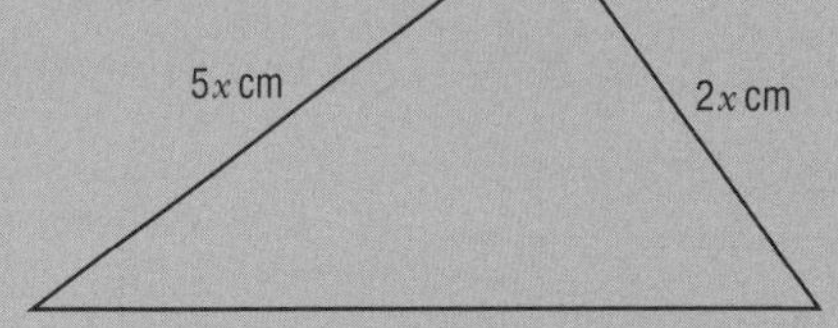

Answer

Pythagoras's rule for this triangle gives:

$34^2 = (5x)^2 + (2x)^2$

$1156 = 25x^2 + 4x^2$ square all values in the brackets

$1156 = 29x^2$

$\frac{1156}{29} = x^2$ dividing both sides by 29

$x^2 = 39.862\ldots$

$x = \sqrt{39.862}$

$x = 6.31$ rounded to two decimal places

Example 3

A ramp is to be built to reach a platform 1.5 m high from a point 9 m from the base of the platform. How long is the ramp? Give your answer in millimetres.

Answer

The first step is to draw a diagram and mark in the known lengths.

The length of the ramp, r m, is the hypotenuse in this right-angled triangle.

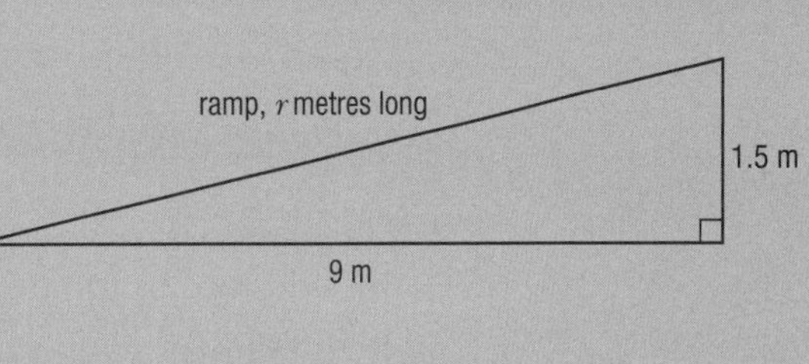

So, $r^2 = 9^2 + 1.5^2$

$r^2 = 81 + 2.25$

$r^2 = 83.25$

$r = \sqrt{83.25}$

$r = 9.1241\ldots$

The ramp is 9.124 m which is 9124 mm long.

Example 4

For the shape given:

a find the value of a, and then

b find the value of b, rounded to two decimal places.

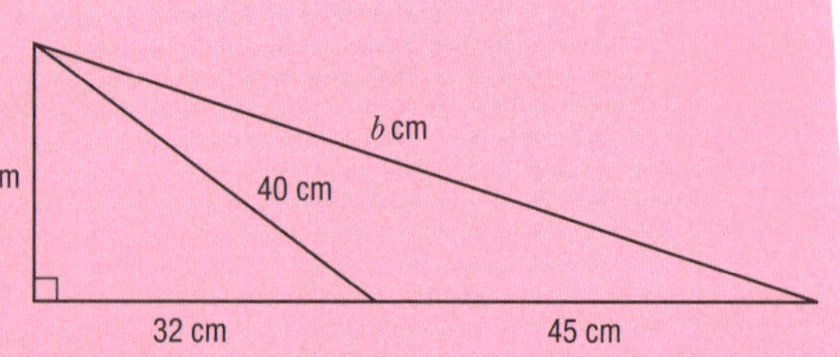

Answer

a To find a we will use Pythagoras's rule in the smaller triangle:

$$40^2 = a^2 + 32^2$$
$$40^2 - 32^2 = a^2$$
$$1600 - 1024 = a^2$$
$$a^2 = 576$$
$$a = \sqrt{576}$$
$$a = 24$$

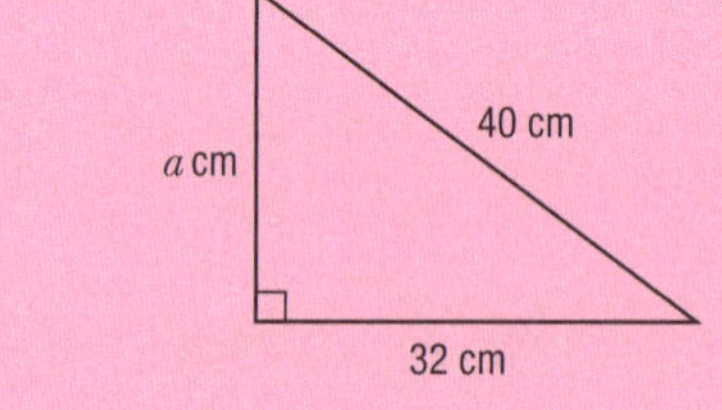

b Using the larger triangle to find b:

$$b^2 = 24^2 + 77^2$$
$$b^2 = 576 + 5929$$
$$b^2 = 6505$$
$$b = \sqrt{6505}$$
$$b = 80.65 \quad \text{rounded to two decimal places}$$

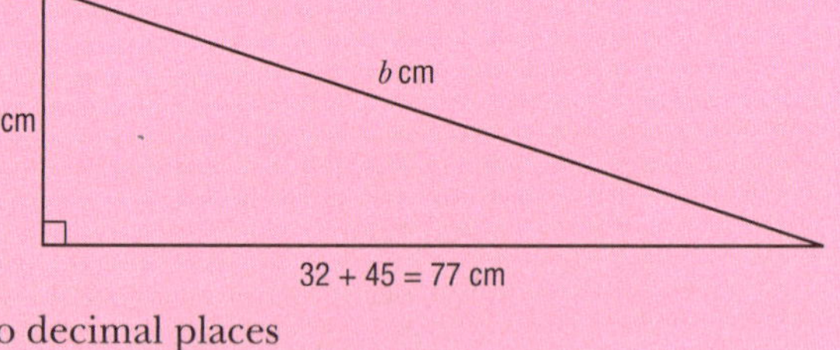

EXERCISE

Give your answers rounded to two decimal places unless stated otherwise.

1 Find the value of the pronumeral in each of the following triangles.

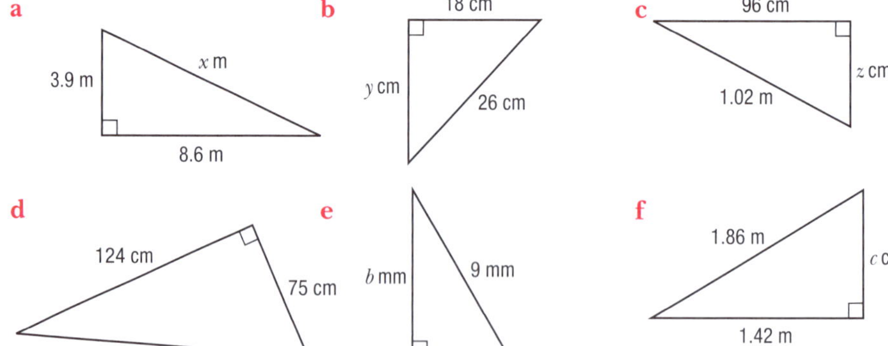

2 The diagonal of a square is 5 m long. Find the length of the sides of the square, giving your answer to the nearest millimetre.

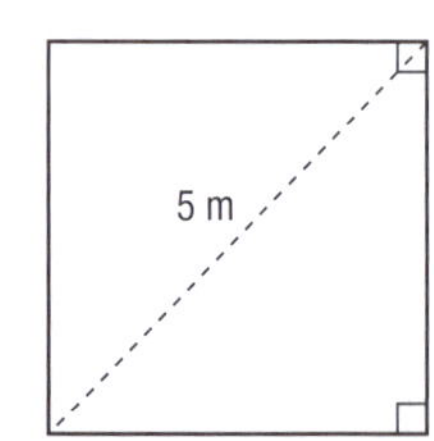

3 A rectangle is twice as long as it is wide. The diagonal is 30 cm long.

a If the width is x cm, what is the length?

b Find the value of x and hence the width and length of this rectangle.

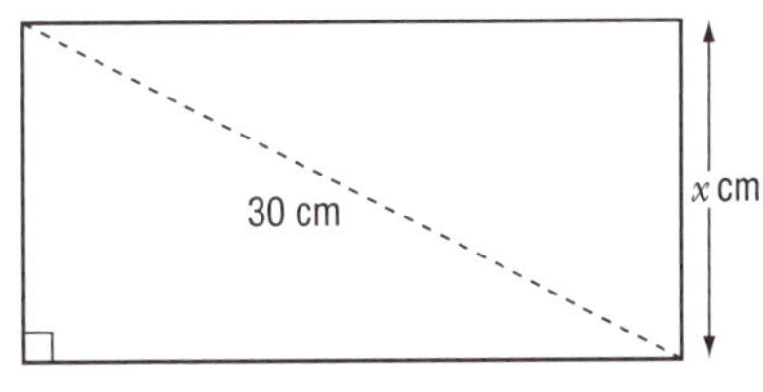

4 The diagonal of a rectangle measures 35 centimetres and one of the sides is 18 centimetres long. Find the length of the other side. (Draw the rectangle and a relevant triangle.)

5 Kenai has a fenced garden with dimensions 130 × 75 metres. He needs to get from one corner of the garden to the diagonally opposite corner and has the choice of walking along the fence line or diagonally across the garden. How much further is it to walk along the fence line?

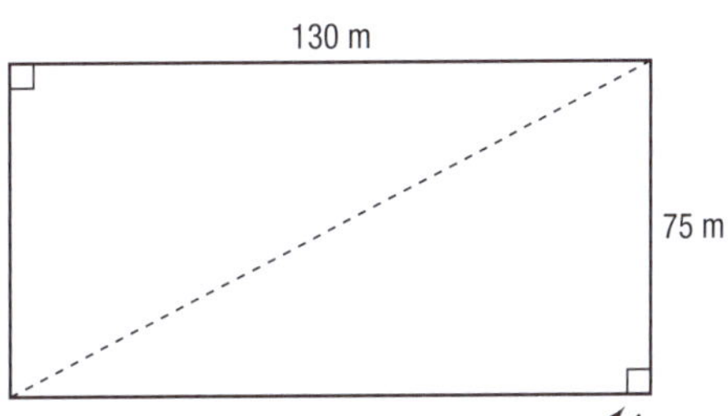

6 A 5-metre long ladder is to be used to reach the roof of a building that is 3.8 metres high. How far from the building should the bottom of the ladder be placed so that it overhangs the roof by 0.5 metres?

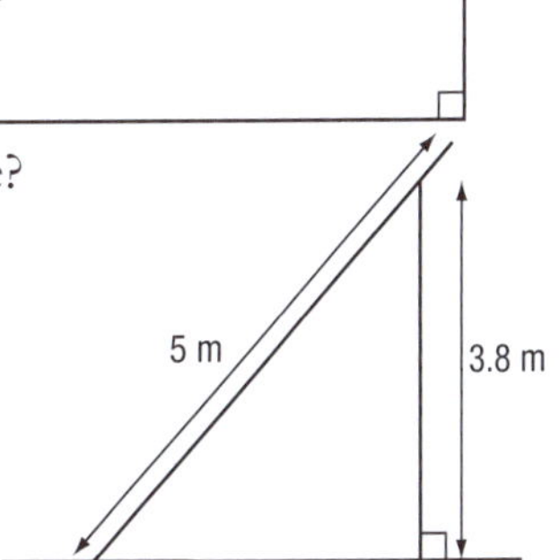

7 A yacht sails due east for 500 m then due north for 2 km. How far is the yacht from its starting point?

8 Jon travels at an average speed of 80 km/h. He can travel from Town A to Town C either by travelling along the road that goes directly between the towns or by going through Town B.

a How much shorter is the journey along the direct road?

b How much time will he save going by the direct road?

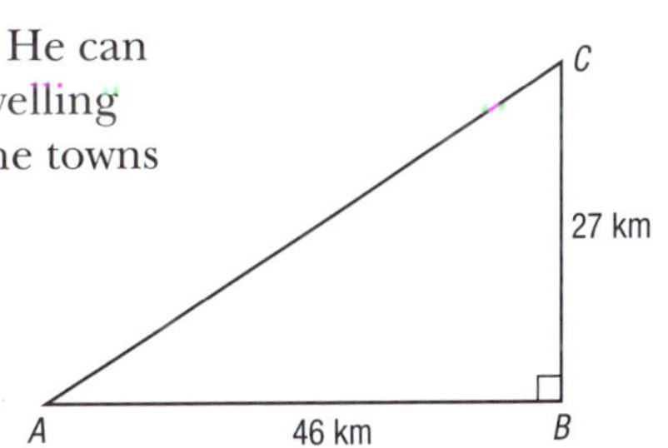

Remember

$$\text{Time} = \frac{\text{distance}}{\text{speed}}$$

9 A boat has drifted 14 m from where the anchor was dropped in a harbour. If the anchor chain is 20 m long, how deep is the harbour?

10 For each of the following:

i find the value of x, rounded to three decimal places if necessary

ii find the value of y, rounded to two decimal places. Assume all measurements are in metres.

a

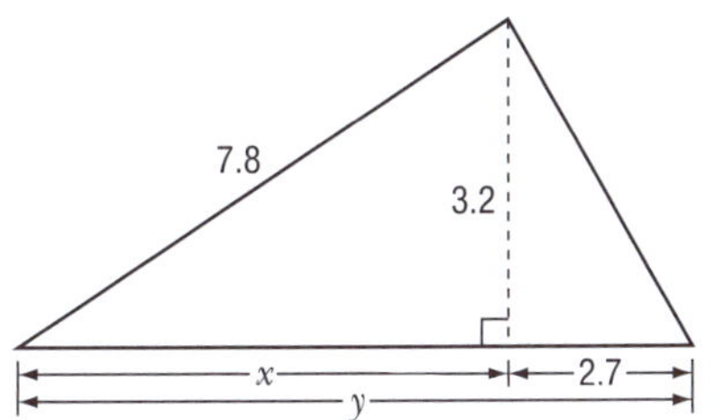

b

16
x
y
11
8

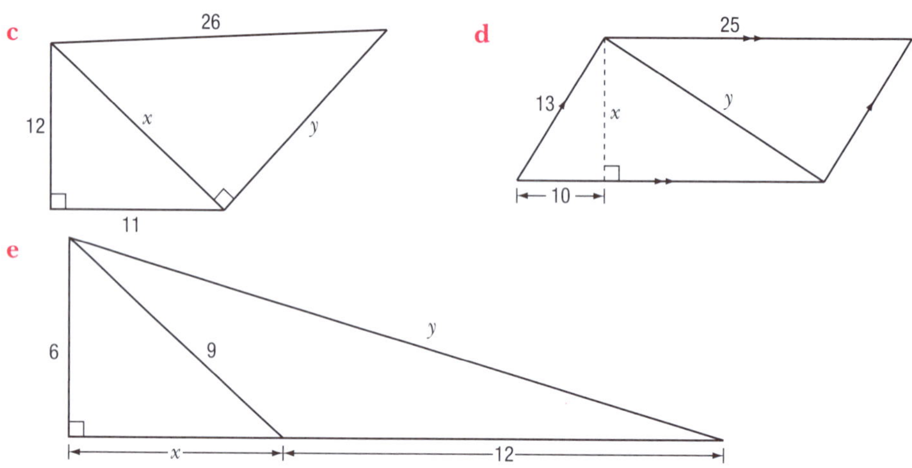

Pythagoras's rule in three dimensions

Three-dimensional diagrams are drawn in such a way that they appear to have depth as well as length and width and so can be used to represent solid figures. There are many opportunities within solid figures to apply Pythagoras's rule.

When calculating lengths using Pythagoras's rule in three-dimensional diagrams, it is important to identify and draw separately the relevant right-angled triangle.

Example

A cube has square faces, so all the edges are the same length. In the diagram the eight vertices of the cube are marked *ABCDEFGH*. Find the length of the diagonal *AG* if the edges all measure 10 cm.

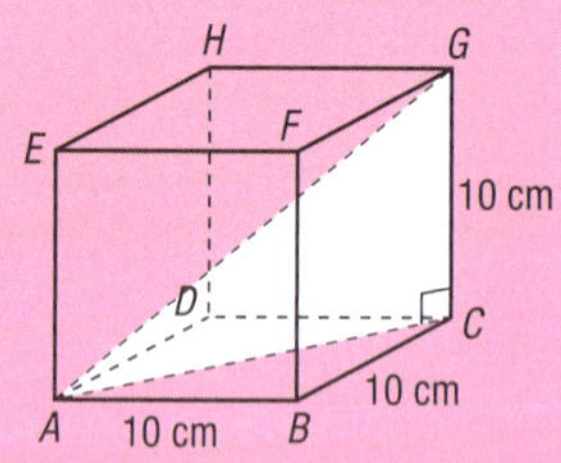

Answer

The diagonal *AG* is in the right-angled triangle *ACG*. However in this triangle we only know the length of the side *CG*. *CG* = 10 cm.

The length of side *AC* can be found using Pythagoras's rule in the right-angled triangle *ABC*.

The sides *AB* and *BC* in this triangle both measure 10 cm.

$AC^2 = 10^2 + 10^2$

$AC^2 = 100 + 100$

$AC^2 = 200$

Hence, $AC = \sqrt{200}$ cm

Now, going back to triangle *AGC* we can calculate the diagonal *AG*.

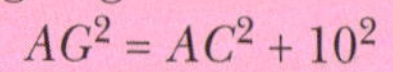

$AG^2 = AC^2 + 10^2$

$AG^2 = (\sqrt{200})^2 + 100$

$AG^2 = 200 + 100$

$AG^2 = 300$

Hence, $AG = \sqrt{300}$

$AG \approx 17.3$ cm

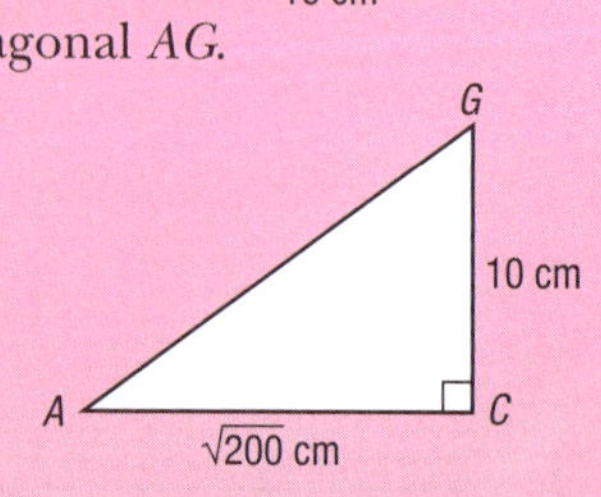

EXERCISE

Where necessary give answers rounded to two decimal places.

1 Practise drawing some common three-dimensional figures in your workbook following the steps given.

a a cuboid—all the faces are rectangles or squares

i Draw two identical rectangles, the second to the right and higher than the other.

ii Join the corners as shown.

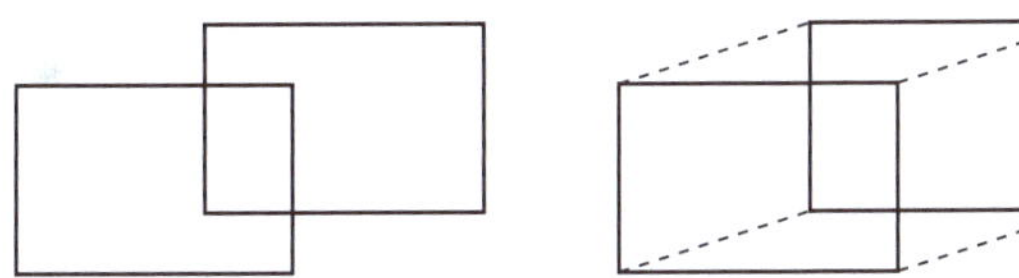

b a right rectangular pyramid—the base is a rectangle or square and the vertex is directly above the 'centre' of the base

i Draw a parallelogram to represent the base.

ii Find the centre of the base.

iii Draw a vertical line to give the pyramid height.

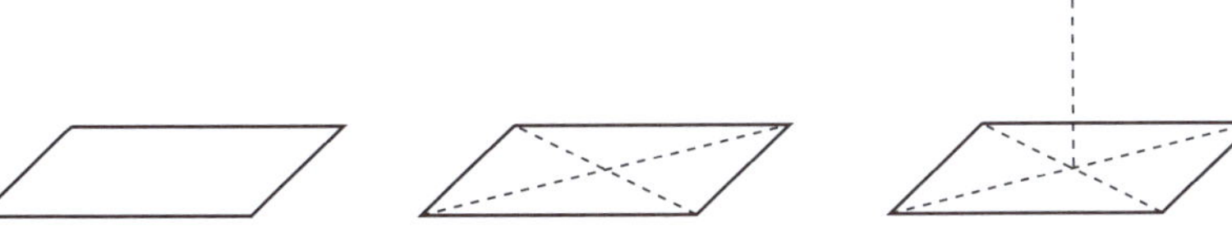

iv The highest point is called the **vertex**.
Join the vertex to the corners of the base.

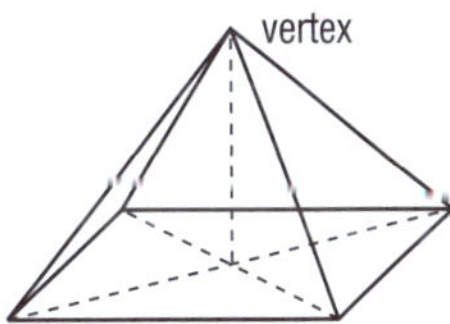

c an inclined plane—one side of an inclined plane is raised

i Draw a parallelogram for the base.

ii Draw two vertical lines to give height at the back.

iii Draw in the other three sides of the inclined plane.

2 A flagpole is to be supported by four wires as shown in the diagram. If the wires are to be attached to the ground 6 m from the base of the flagpole and reach a height of 9 m up the flagpole, find the total length of wire required.

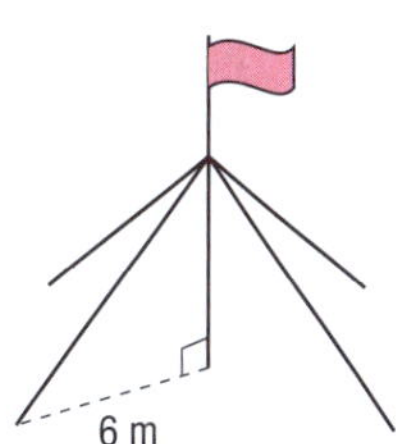

3 Find the length of the longest thin rod that will fit inside a cylinder of radius 3 cm and height 15 cm.

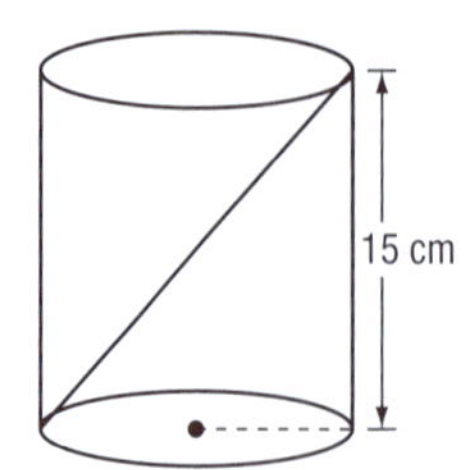

4 Draw a cuboid and label the vertices with letters and the sides with lengths, as shown.

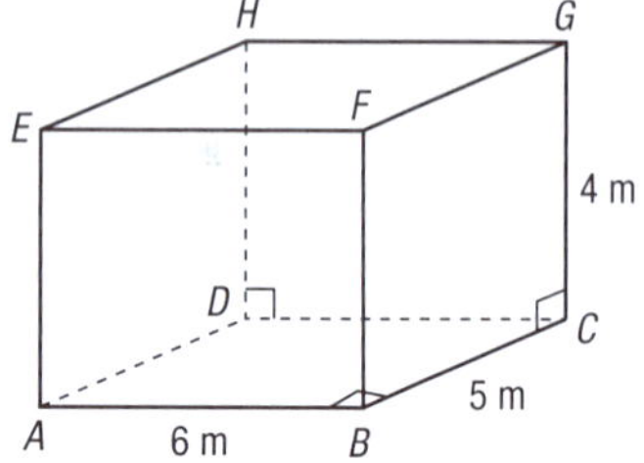

a Draw the triangle *ABC* and mark in the lengths of sides *AB* and *BC* and the right angle. Find the length of *AC*.

b Draw the triangle *ACG* and mark in the lengths *AC* and *CG* and the right angle. Find the length of the longest diagonal, *AG*.

c Find three other diagonals in this cuboid with the same length as *AG*.

5 A right pyramid has a rectangular base with dimensions 6 m × 8 m and height 8 m.

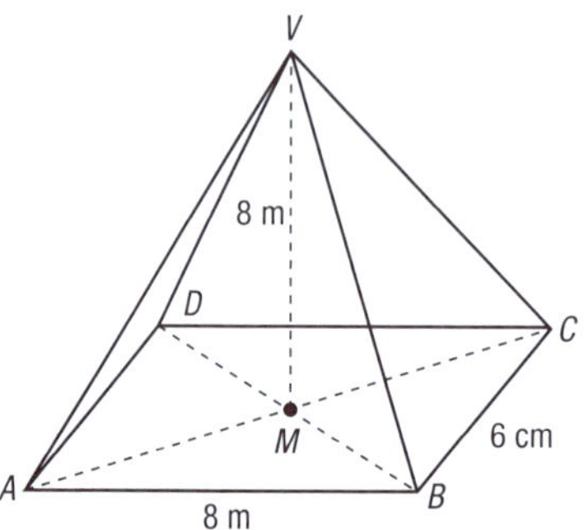

a Draw a pyramid, labelling the base *ABCD*, the centre of the base *M* and the vertex *V*. Mark in the lengths of the base and the height.

b Draw the triangle *ABC* and mark in the lengths of sides *AB* and *BC*. Find the length of *AC*.

c Find the length of *MC*.

d Draw the triangle *VMC*, mark in the lengths *VM* and *MC*. Hence find the length of the sloping edge of this pyramid.

6 **a** Draw an inclined plane, labelling the vertices and marking the lengths, as shown.

b Find the length of *BC*.

c Find the length of *DB*.

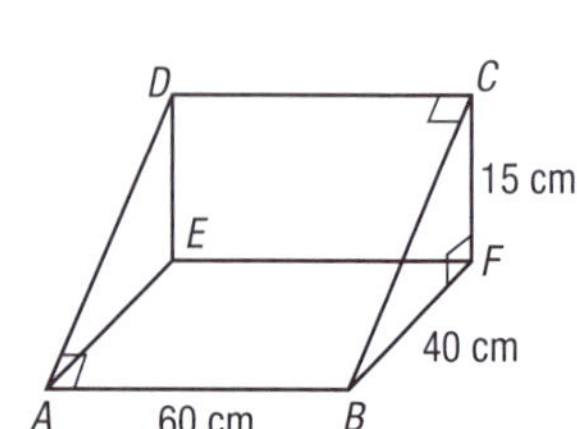

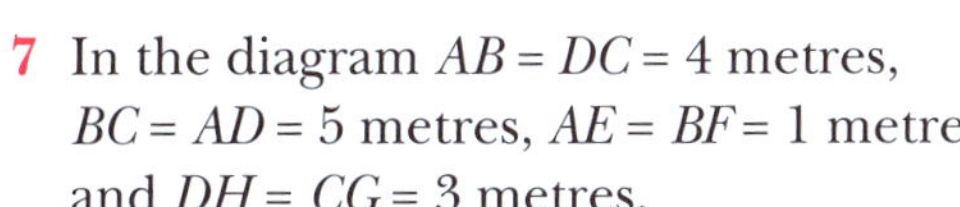

7 In the diagram $AB = DC = 4$ metres, $BC = AD = 5$ metres, $AE = BF = 1$ metre and $DH = CG = 3$ metres.

a Find the length of AC.

b Find the length of AG.

c Find the length of BG.

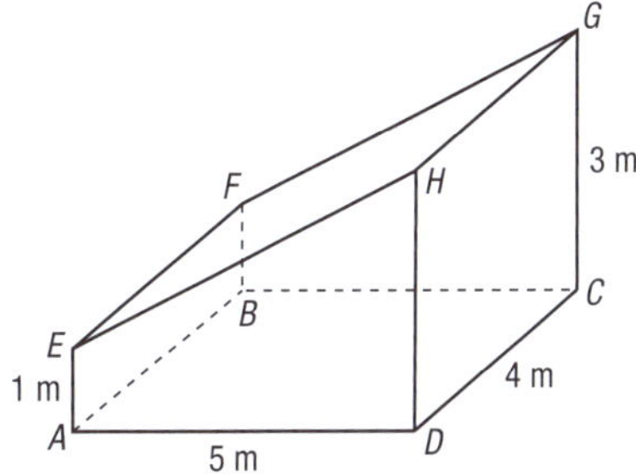

8 A right square pyramid has height 18 m and the length of the sloping edges is 30 m.

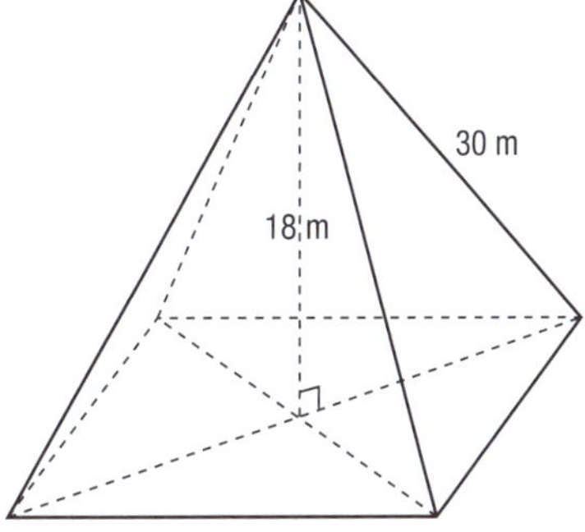

a Find the distance from the corner of the base to the centre of the base.

b Find the length of the diagonal of the base.

c Find the side length of the square base.

Challenge

9 The longest diagonal in a cube is 30 cm long. Find the length, x, of the edge of the cube.

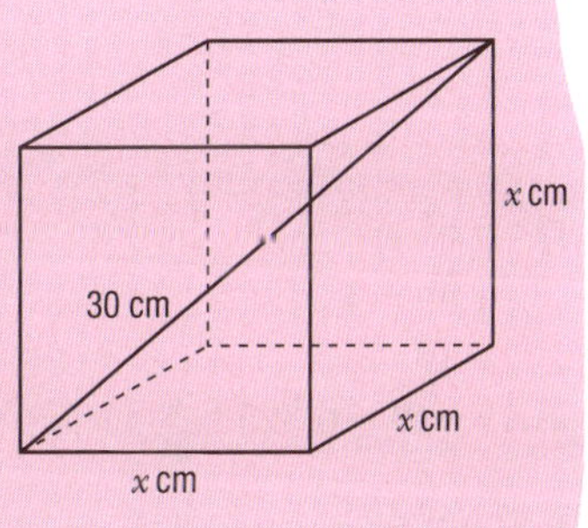

ACTIVITY: Plumbing

You are asked to run a copper pipe from the point A to the tap for the room illustrated.

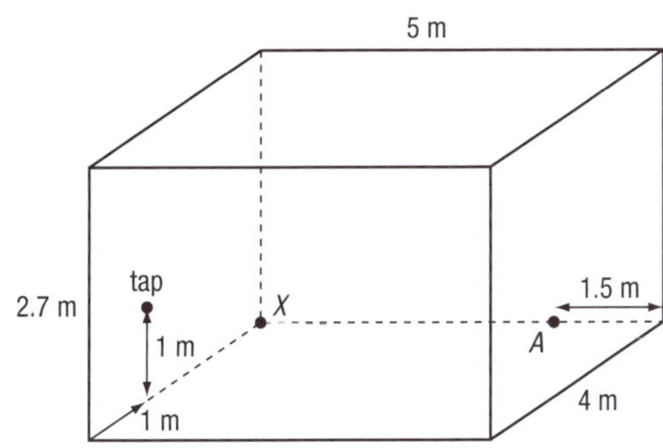

The room is in the shape of a cuboid with dimensions 5 × 4 × 2.7 metres. The tap is 1 metre from the floor and 1 metre from the nearest corner on the wall indicated in the diagram.

The copper pipe can run inside the walls or under the floor but not inside the room. You are to calculate the shortest length of copper pipe needed.

a Draw diagrams and calculate the length of copper pipe for at least three paths for the pipe.

b For the path that uses the shortest length of pipe, calculate the distance from the corner *X* to the point where the pipe crosses the line of intersection of the wall and the floor.

Lesson 13

The distance between two points on the Cartesian plane

Consider the two points *A*(1, 2) and *B*(7, 4) plotted on the Cartesian plane below. The straight-line distance between the two points can be found using **Pythagoras's rule**:

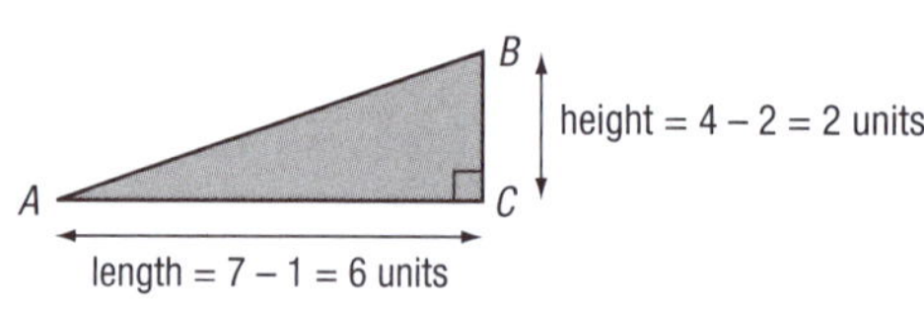

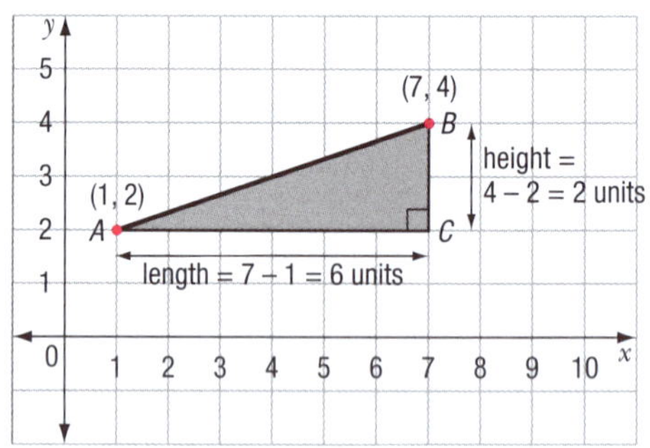

$$AB^2 = 6^2 + 2^2$$
$$= 36 + 4$$
$$= 40$$
$$AB = \sqrt{40} \approx 6.325$$

In general:

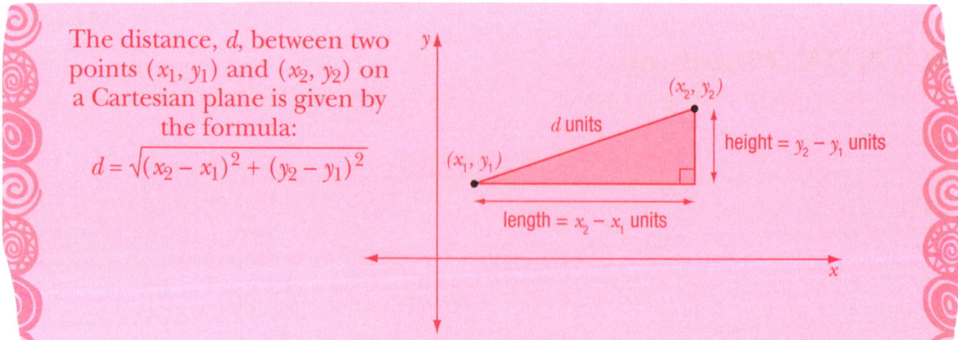
The distance, *d*, between two points (x_1, y_1) and (x_2, y_2) on a Cartesian plane is given by the formula:

$$d = \sqrt{(x_2 - x_1)^2 + (y_2 - y_1)^2}$$

Example

Use the formula to find the distance between the points (2, 5) and (⁻3, ⁻6).

Answer

Using (2, 5) as (x_1, y_1) and (⁻3, ⁻6) as (x_2, y_2) and substituting in the distance formula $d = \sqrt{(x_2 - x_1)^2 + (y_2 - y_1)^2}$ gives

$d = \sqrt{(^-3 - 2)^2 + (^-6 - 5)^2}$

$= \sqrt{(^-5)^2 + (^-11)^2}$

$= \sqrt{25 + 121}$

$= \sqrt{146}$ Use the square root table or your calculator to find the square root.

= 12.08, correct to two decimal places

EXERCISE

1 Use the distance formula to find the distance between the following pairs of points. Give your answer as an exact value (surd) and as a decimal correct to two decimal places.

  a $A(3, 1)$ and $B(4, 2)$
  b $P(^-1, 2)$ and $Q(5, 2)$
  c $M(0, ^-2)$ and $N(0, 3)$
  d $C(3, ^-2)$ and $D(^-1, ^-4)$

2 For each of the following, use the distance formula to classify triangle ABC as equilateral, isosceles or scalene.

  a $A(5, 3)$, $B(1, 8)$, $C(^-6, 1)$
  b $A(1, 0)$, $B(3, 1)$, $C(7, 3)$
  c $A(^-2, ^-1)$, $B(0, 3)$, $C(4, 1)$

3 Find the lengths of the sides of the triangle ABC and then use Pythagoras's rule to show that the points $A(1, 2)$, $B(3, 4)$ and $C(6, 1)$ are the vertices of a right-angled triangle.

4 Show that the points $A(2, 1)$, $B(2, 9)$, $C(1, 5)$ and $D(3, 5)$ are the vertices of a rhombus. Find the lengths of the diagonals of this rhombus.

5 Find two values of b so that the points $P(3, b)$ and $Q(^-1, 2)$ are $\sqrt{20}$ units apart.

6 The point (x, y) is a distance of 5 units away from the point (1, 2).

  a Find two possible points that could be the point (x, y).
  b If all the points that could be the point (x, y) were graphed, what would be the shape of the graph?
  c Substitute in the distance formula using (1, 2) as (x_1, y_1), (x, y) as (x_2, y_2) and 5 as d. Square both sides of this equation to give the formula of the shape from part **b**.

INTRODUCTION TO THE TRIGONOMETRIC RATIOS

Lesson 14

Trigonometry of the right-angled triangle: the sine ratio

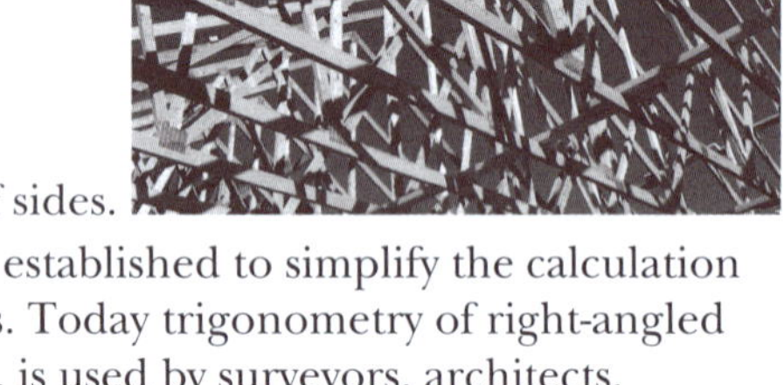

Similar triangles have corresponding angles that are equal in size, and their corresponding sides are in the same ratio; so the triangles 'look' the same but differ in size.

Right-angled triangles have been the basis of building construction from very early times. It was noted then that for **similar right-angled triangles** there was a set ratio between pairs of sides.

Trigonometry of the right-angled triangle was established to simplify the calculation of lengths and angles in right-angled triangles. Today trigonometry of right-angled triangles, as well as non-right-angled triangles, is used by surveyors, architects, engineers, geologists, navigators, cartographers, oceanographers, meteorologists and air-traffic controllers.

ACTIVITY

Both of these two right-angled triangles have an angle of 30° as one of their interior angles. The **hypotenuse** and the **side opposite the 30° angle** are labelled.

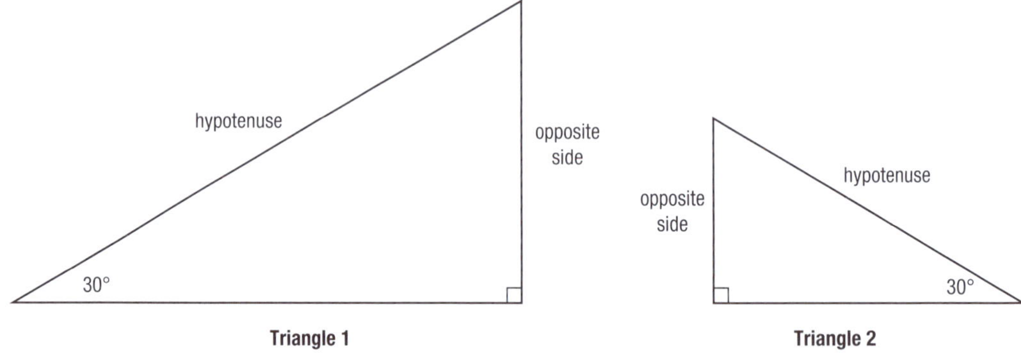

a Construct another right-angled triangle (Triangle 3) in your workbook, **similar** to the triangles above. Use a protractor to construct the right angle and the 30° angle. Label the hypotenuse and the side opposite the 30° angle.

b Use a ruler to measure, to the nearest millimetre, the length of the hypotenuse and the opposite side on all three triangles. Copy and complete the table below.

Triangle	Hypotenuse	Opposite side	$\frac{\text{Opposite side}}{\text{Hypotenuse}}$
1			
2			
3			

c What do you notice about the figure $\frac{\text{Opposite side}}{\text{Hypotenuse}}$ for the three triangles?

The ratio $\frac{\text{Opposite side}}{\text{Hypotenuse}} = \frac{1}{2}$ is called the **sine** of the angle 30°.

This is written as **sin 30°** = $\frac{1}{2}$. (*Note*: **sin** is an abbreviation of the word **sine**.)

In general for an angle θ in a right-angled triangle:

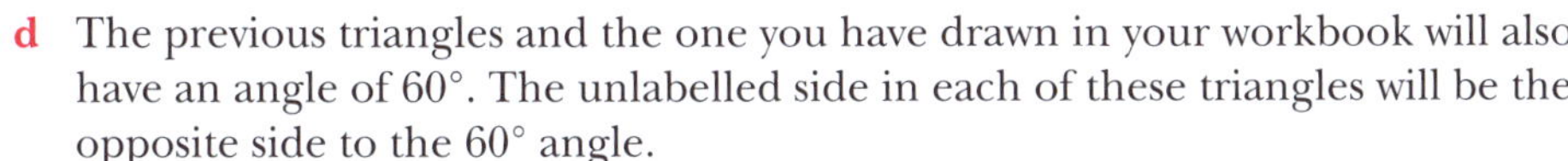

$$\sin\theta = \frac{\text{Length of the opposite side to angle }\theta}{\text{Length of the hypotenuse}}$$

d The previous triangles and the one you have drawn in your workbook will also have an angle of 60°. The unlabelled side in each of these triangles will be the opposite side to the 60° angle.

Use a ruler to measure, to the nearest millimetre, the length of the hypotenuse and the opposite side to the 60° angle, on all three triangles. Copy and complete the table below.

Triangle	Hypotenuse	Opposite side to 60° angle	$\frac{\text{Opposite side}}{\text{Hypotenuse}}$
1			
2			
3			

e Use your results to find an approximation for sin 60°.

A table of sine values for all angles from 0° to 90° is shown below. It is also possible to find the sine of an angle on your calculator.

You will see that sin 30° = 0.5000 and sin 60° = 0.8660; all the answers are given correct to four decimal places.

Table of sine values for angle 0° to 90°

Angle	Sine	Angle	Sine	Angle	Sine	Angle	Sine	Angle	Sine	Angle	Sine
0	0.0000	15	0.2588	30	0.5000	45	0.7071	60	0.8660	75	0.9659
1	0.0175	16	0.2756	31	0.5150	46	0.7193	61	0.8746	76	0.9703
2	0.0349	17	0.2924	32	0.5299	47	0.7314	62	0.8829	77	0.9744
3	0.0523	18	0.3090	33	0.5446	48	0.7431	63	0.8910	78	0.9781
4	0.0698	19	0.3256	34	0.5592	49	0.7547	64	0.8988	79	0.9816
5	0.0872	20	0.3420	35	0.5736	50	0.7660	65	0.9063	80	0.9848
6	0.1045	21	0.3584	36	0.5878	51	0.7771	66	0.9135	81	0.9877
7	0.1219	22	0.3746	37	0.6018	52	0.7880	67	0.9205	82	0.9903
8	0.1392	23	0.3907	38	0.6157	53	0.7986	68	0.9272	83	0.9925
9	0.1564	24	0.4067	39	0.6293	54	0.8090	69	0.9336	84	0.9945
10	0.1736	25	0.4226	40	0.6428	55	0.8192	70	0.9397	85	0.9962
11	0.1908	26	0.4384	41	0.6561	56	0.8290	71	0.9455	86	0.9976
12	0.2079	27	0.4540	42	0.6691	57	0.8387	72	0.9511	87	0.9986
13	0.2250	28	0.4695	43	0.6820	58	0.8480	73	0.9563	88	0.9994
14	0.2419	29	0.4848	44	0.6947	59	0.8572	74	0.9613	89	0.9998
										90	1.0000

EXERCISE

1 Use the table to find the sine of each of the following angles. Write your answers as shown in part **a**.

a 45° (For the answer write sin 45° = 0.7071.)

b 80° **c** 8° **d** 23° **e** 72°

f 67° **g** 90° **h** 0°

2 What do you notice about the size of the sine values in the table? What is the smallest value? What is the largest value?

For a right-angled triangle, we have discussed the **hypotenuse** and the **opposite side** to a particular angle; the third side is referred to as the **adjacent side**.

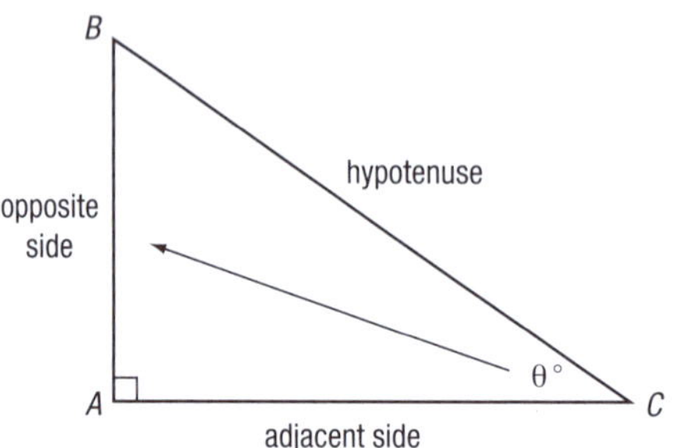

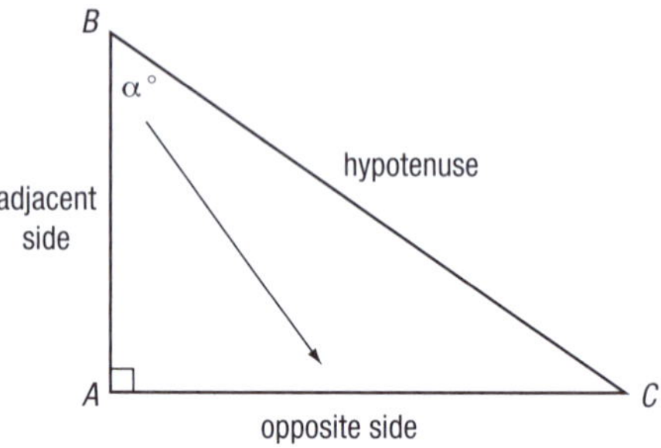

In the triangles above the sides are named differently because the name of each side relates to a particular angle:

$$\sin \theta = \frac{AB}{BC} \text{ but } \sin \alpha = \frac{AC}{BC}$$

3 Choose the correct title (hypotenuse, opposite side or adjacent side) for the side indicated in the diagram in relation to the angle marked *.

a

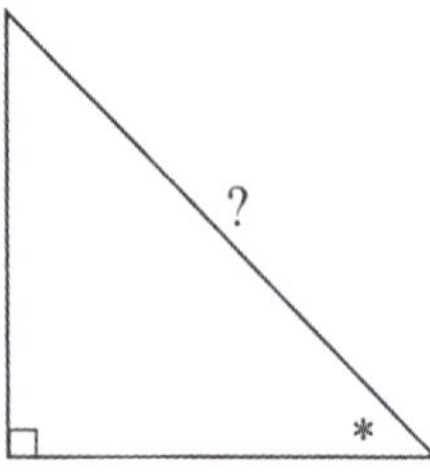

b

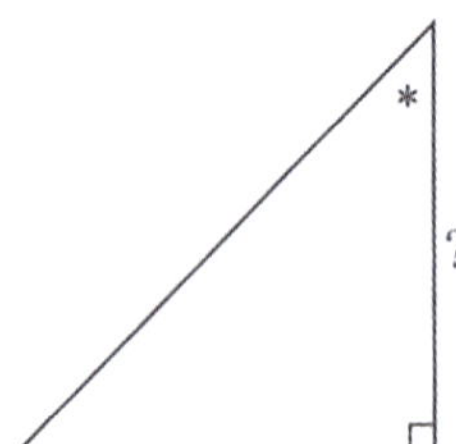

c

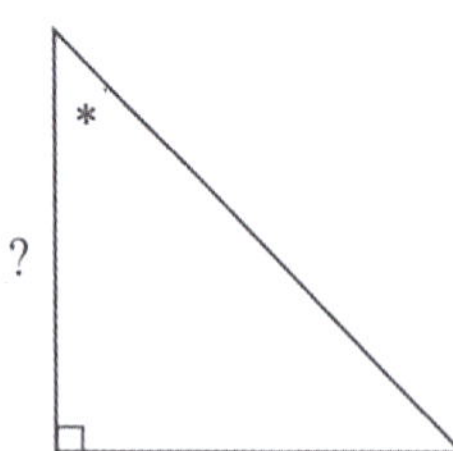

d

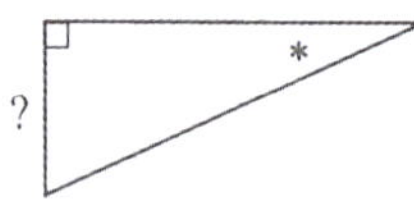

e

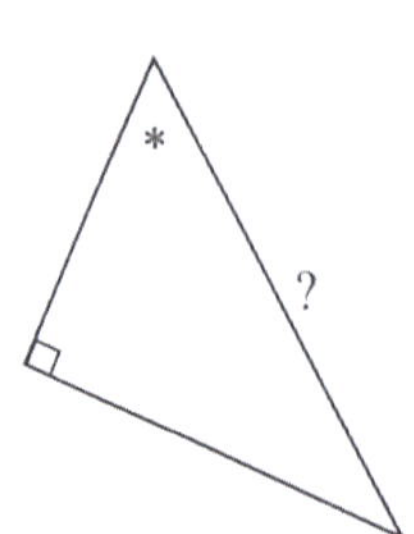

f

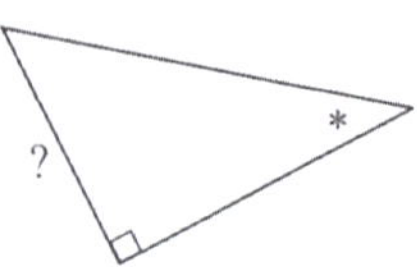

Using the sine ratio to find the length of sides in a right-angled triangle

Example 1

Find the value of x, correct to two decimal places, in the triangle.

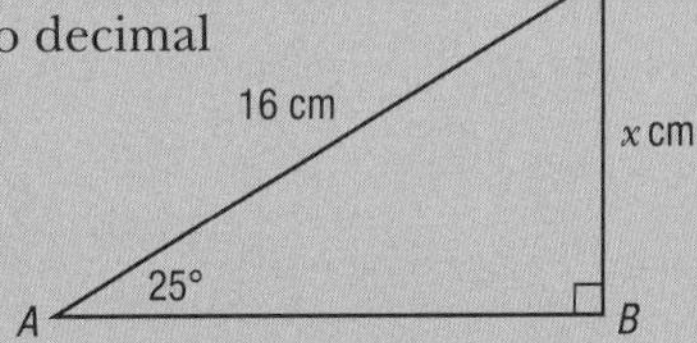

Answer

On the diagram the side marked x cm is the opposite side to the angle 25° and the hypotenuse has the length 16 cm.

Substituting these values in:

$$\sin\theta = \frac{\text{Length of the opposite side}}{\text{Length of the hypotenuse}}$$

$\sin 25° = \frac{x}{16}$ multiply both sides of the equation by 16

$16 \times \sin 25° = x$ substitute the value of sin 25° from the table

$16 \times 0.4226 = x$

$x = 6.7616$

$= 6.76$ correct to two decimal places

Example 2

Find the value of x, correct to one decimal place, in this triangle.

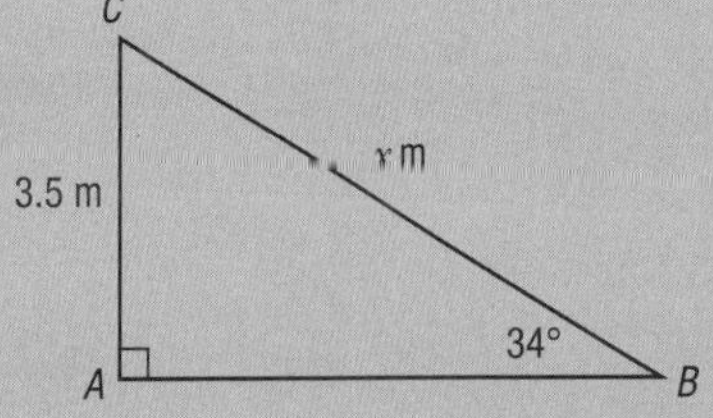

Answer

On the diagram the side marked x cm is the hypotenuse and the opposite side to the angle 34° has the length 3.5 m.

Substituting these values in:

$$\sin\theta = \frac{\text{Length of the opposite side}}{\text{Length of the hypotenuse}}$$

$\sin 34° = \frac{3.5}{x}$ multiply both sides of the equation by x

$x \times \sin 34° = 3.5$ divide both sides of the equation by sin 34°

$x = \frac{3.5}{\sin 34°}$ substitute the value of sin 34° from the table

$x = \frac{3.5}{0.5592}$

$= 6.259$

$= 6.3$ correct to one decimal place

EXERCISE

1 Find the value of x, correct to one decimal place, in the following triangles.

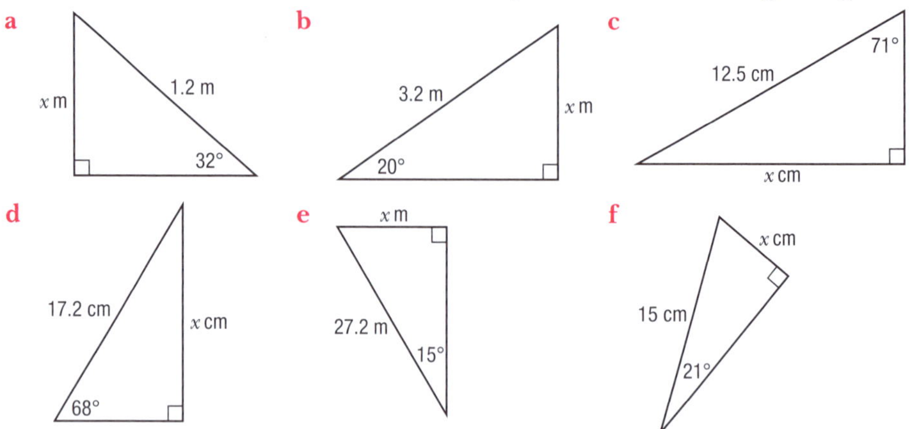

2 Find the value of x, correct to one decimal place, in each of the following diagrams.

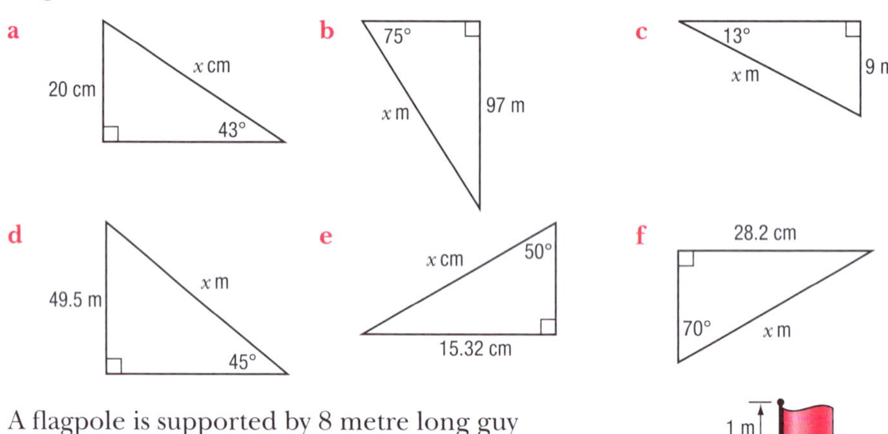

3 A flagpole is supported by 8 metre long guy ropes that are attached to the ground and to the flagpole at a point 1 metre from the top, as shown in the diagram. The guy ropes make an angle of 75° with the ground. Find the height of the flagpole in metres, correct to two decimal places.

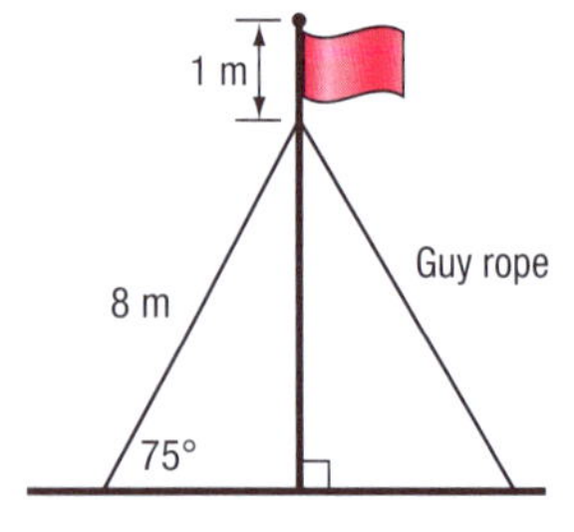

4 A ramp is built to the entrance to a store that is 92 cm above ground level. If the ramp makes an angle of 7° with the ground, find the length of the ramp in metres, correct to two decimal places.

x m
92 cm
7°

5 A hiker walks 7.4 km in the direction N60°E. Find the distance in kilometres, correct to two decimal places, that he is now north of his starting point.

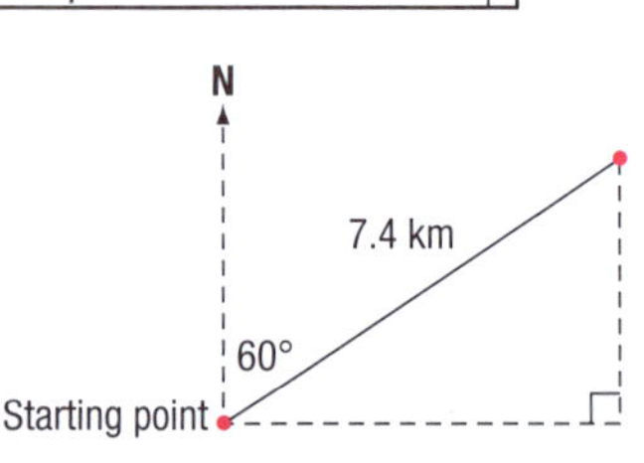

Finding the angle when the opposite side and the hypotenuse are known in a right-angled triangle

Lesson 16

Did you know?

Measurement of angles

There are 360 degrees in a revolution and each degree can then be divided into **minutes** where

1 degree = 60 minutes.

We can say that an angle of 36.5° is equivalent to 36 degrees and 30 minutes.

This can be written as 36°30′.

Each minute can be further divided into seconds where

1 minute = 60 seconds.

Although decimal degrees are now used in many situations, degrees, minutes and seconds are still used in navigation.

Example 1

Write 56.4° in degrees and minutes.

Answer

The degree part of 56.4° is 56°.

The minute part is 0.4 of 1° = 0.4 of 60 minutes = $0.4 \times 60 = 24$.

56.4° is equivalent to 56°24′.

Example 2

For the triangle at right find:

a the sine of angle θ

b the value of angle θ in degrees and minutes.

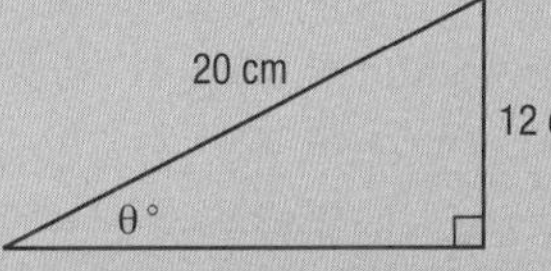

Answer

a The length of the opposite side (12 cm) to the angle θ and of the hypotenuse (20 cm) are known in this triangle.

We know that $\sin\theta = \dfrac{\text{Length of the opposite side to angle } \theta}{\text{Length of the hypotenuse}}$

Substituting gives $\sin\theta° = \dfrac{12}{20} = 0.6$

b Looking in the sine tables for the angle that gives a sine value closest to 0.6000:

0.5990 + 0.0009 = 0.5999 the value closest to 0.6000

NATURAL SINES

	0′	6′	12′	18′	24′	30′	36′	42′	48′	54′	1′	2′	3′	4′	5′
35	0.5736	5750	5764	5779	5793	5807	5821	5835	5850	5864	2	5	7	9	12
36	0.5878	5892	5906	5920	5934	5948	5962	5976	5990	6004	2	5	7	9	12
37	0.6018	6032	6046	6060	6074	6088	6101	6115	6129	6143	2	5	7	9	12
38	0.6157	6170	6184	6198	6211	6225	6239	6252	6266	6280	2	5	7	9	11
39	0.6293	6307	6320	6334	6347	6361	6374	6388	6401	6414	2	4	7	9	11

The angle is 36° + 48′ + 4′ = 36°52′.

Using a calculator: If $\sin \theta° = 0.6$ then $\sin^{-1} 0.6 = \theta$ and you can use the inverse sine ($\sin^{-1}$) key on your calculator to find the size of angle θ in decimal degrees, then change this to degrees, minutes and seconds using the DMS key.
Note: Calculators will give an answer in degrees, minutes and seconds. To round to the nearest minute, if the 'second' part of the answer is 30 or more then the minute part is increased by one.

EXERCISE

1 Convert each of the following angles in decimal degrees to degrees and minutes.

a 42.6° b 79.1° c 18.75° d 55.236°

2 For each of the following use the table of sine values on pages 337–9 to find the angle, in degrees and minutes, that has the sine value:

a 0.2672 b 0.6630 c 0.7260 d 1.0000
e 0.9575 f 0.1233 g 0.0123

3 For each of the following:

i find the sine of the angle θ
ii find the size of the angle θ in degrees and minutes.

a

b
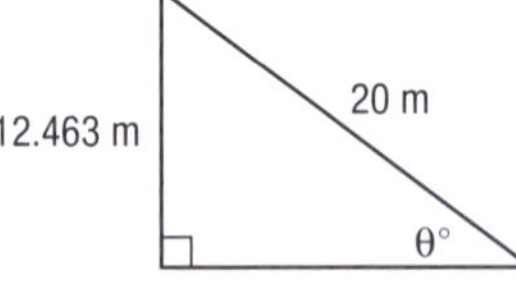

c
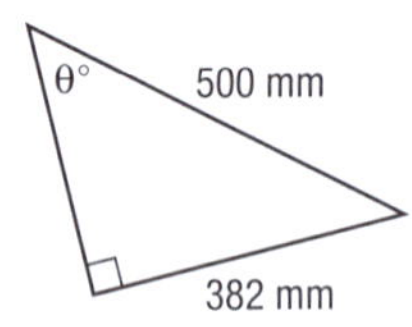

d
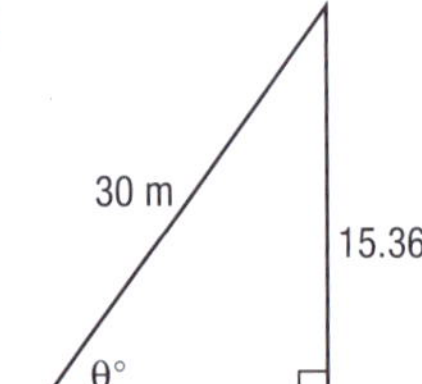

e
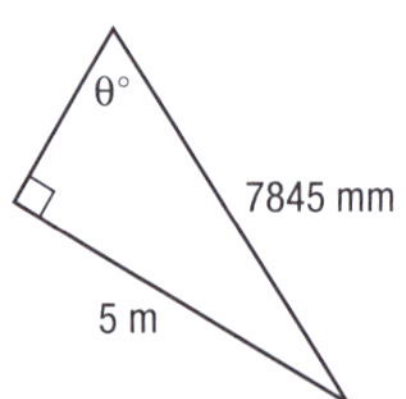

f
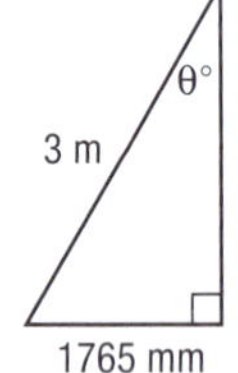

4 Find the value of x in the following rectangle.

5 Find the value of θ in this triangle.

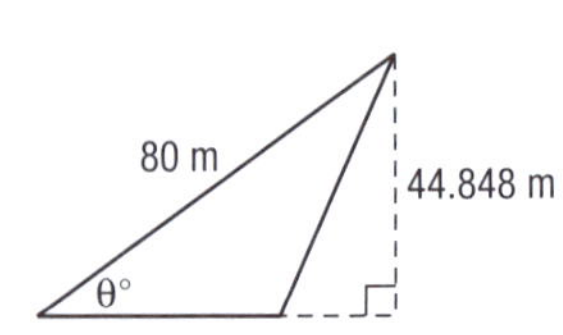

Trigonometry of the right-angled triangle: the cosine ratio

The cosine ratio is used when the adjacent side to the angle and the hypotenuse is known.

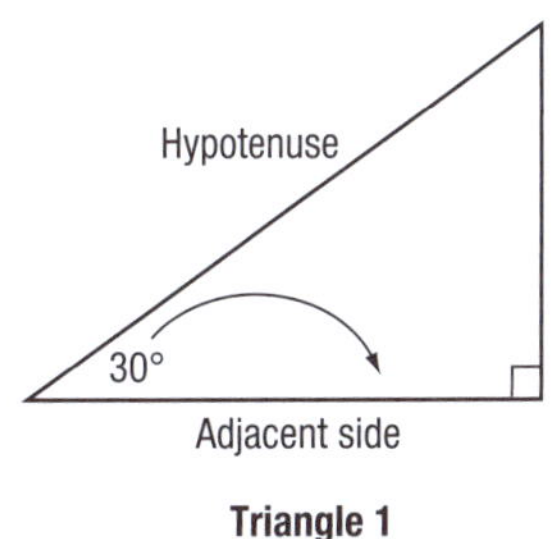

Triangle 1

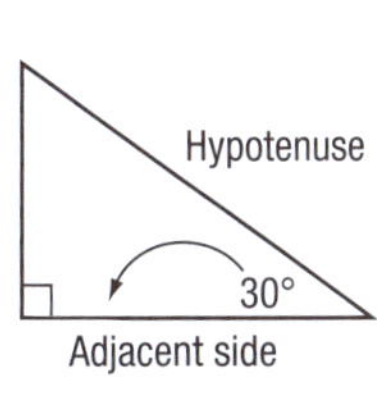

Triangle 2

If you calculate the ratio $\frac{\text{Length of adjacent side}}{\text{Length of the hypotenuse}}$ for each of the similar triangles above you will get a value of approximately 0.87.

This ratio is called the **cosine** of the angle 30°. We can write **cos 30° = 0.87** and looking at the table for cosines on pages 339–41 you will find that cos 30° = 0.8660, correct to four decimal places.

In general, for an angle θ in a right-angled triangle:

$$\cos \theta = \frac{\text{Length of the adjacent side to angle } \theta}{\text{Length of the hypotenuse}}$$

Using the cosine ratio to find the length of the sides in a right-angled triangle

Example 1

Find, correct to two decimal places, the value of x in the triangle.

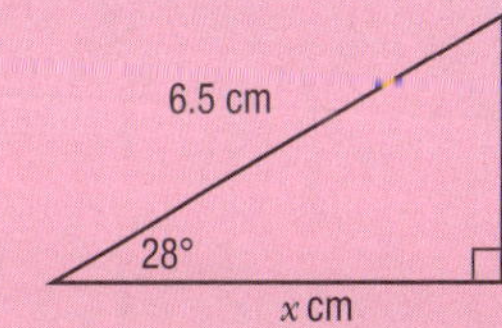

Answer

In the triangle, the side marked x cm is the adjacent side to the angle 28° and the length of the hypotenuse is 6.5 cm, so we can use the cosine ratio to find x.

Substituting in $\cos \theta = \frac{\text{Length of the adjacent side to angle } \theta}{\text{Length of the hypotenuse}}$

$\cos 28° = \frac{x}{6.5}$ multiply both sides of the equation by 6.5

$6.5 \times \cos 28° = x$ reading the value of cos 28° from the table or using the calculator

$6.5 \times 0.8829 = x$

$x = 5.7388\ldots$

$= 5.74$ correct to two decimal places

Example 2

Find, correct to two decimal places, the length of the hypotenuse in this triangle.

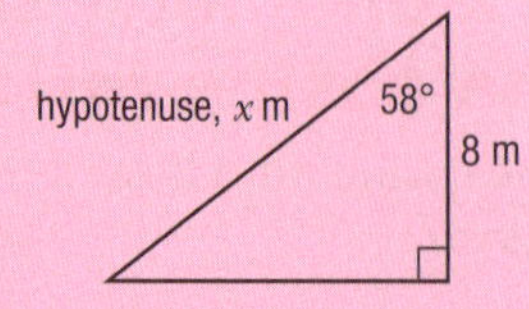

Answer

In the triangle, the side marked 8 m is the adjacent side to the angle 58° and the length of the hypotenuse is x m, so we can use the cosine ratio to find x.

Substituting in $\cos\theta = \dfrac{\text{Length of the adjacent side to angle } \theta}{\text{Length of the hypotenuse}}$

$\cos 58° = \dfrac{8}{x}$ multiply both sides of the equation by x

$x \cos 58° = 8$ divide both sides of the equation by cos 58°

$x = \dfrac{8}{\cos 58°}$ substitute the value of cos 58° or do the calculation on the calculator

$x = \dfrac{8}{0.5299}$

$= 15.097\ldots$

$= 15.10$ correct to two decimal places

EXERCISE

1 Use the table of cosines on pages 339–41 to find the value, correct to four decimal places, of each of the following:

a cos 36° **b** cos 24° **c** cos 78°36′ **d** cos 0° **e** cos 7°45′
f cos 90° **g** cos 60° **h** cos 45° **i** cos 85°40′ **j** cos 17°

2 What do you notice about the size of the cosine values in the table? What is the smallest value? What is the largest value?

3 For each of the following:

i write an equation using the formula
$\cos\theta = \dfrac{\text{Length of the adjacent side to angle } \theta}{\text{Length of the hypotenuse}}$

ii find the value of x, correct to one decimal place.

a

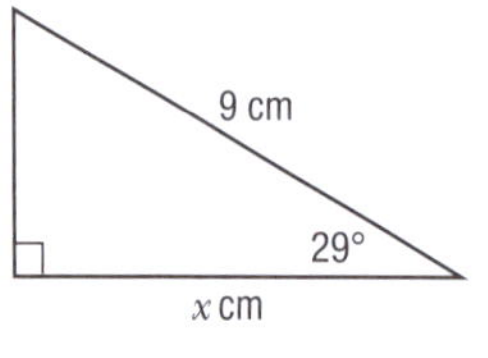

b

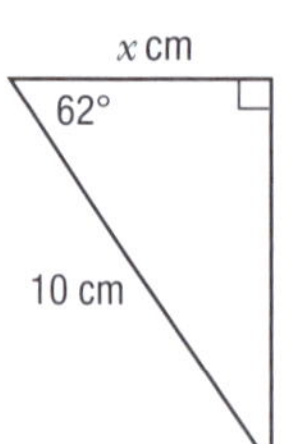

c

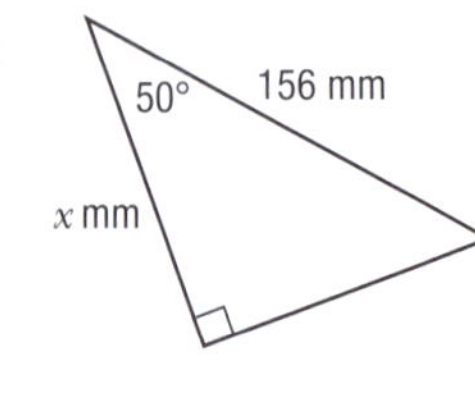

d

23 cm 56° x cm

e

40 m 65° x m

f

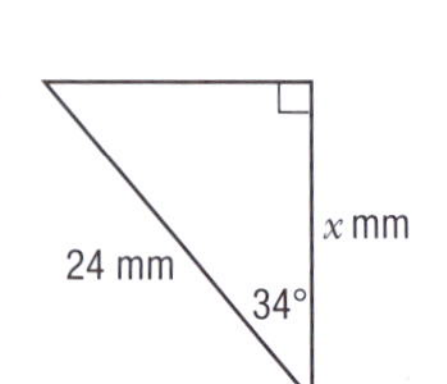

4 For each of the following:

i write an equation using the formula

$$\cos\theta = \frac{\text{Length of the adjacent side to angle } \theta}{\text{Length of the hypotenuse}}$$

ii find the value of x, correct to one decimal place.

a

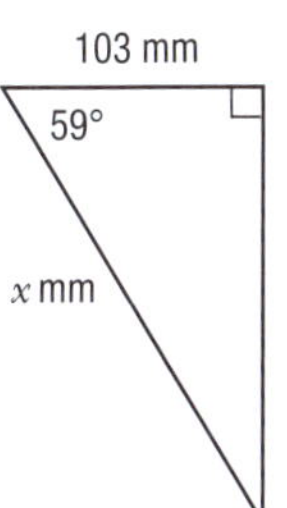

b

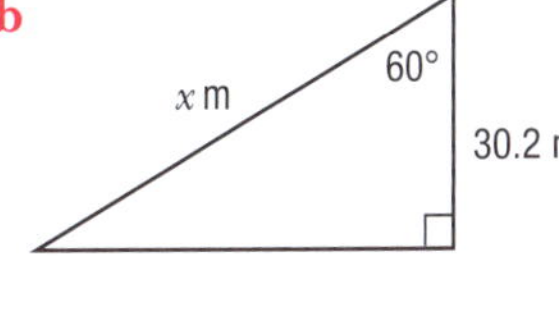

c

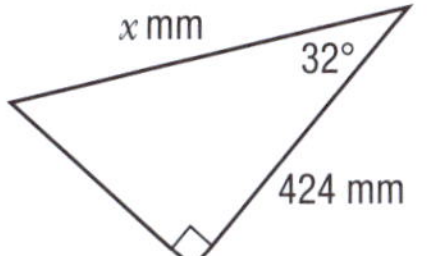

d

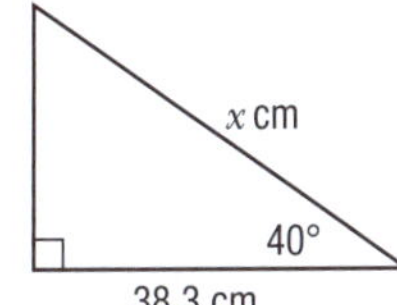

e

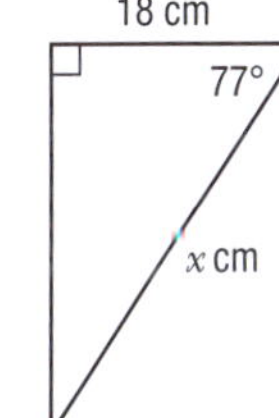

f

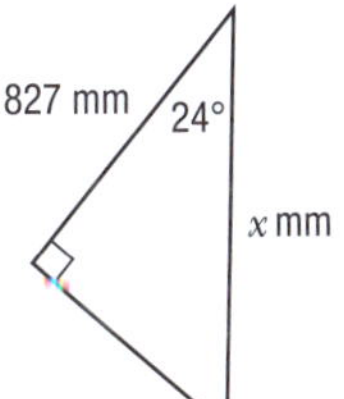

5 Find, correct to one decimal place, the value of x in this triangle.

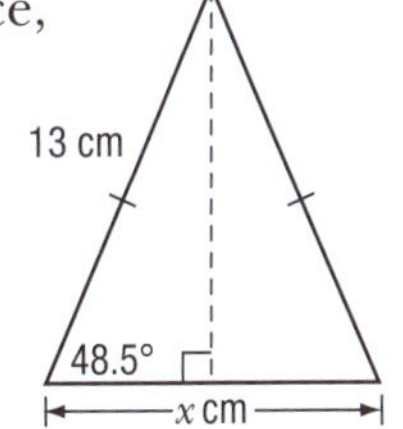

6 The sloping section of a children's slide makes an angle of 58° with the vertical and starts at a height of 3 m and finishes at a height of 35 cm. How long is the sloping section of the slide?

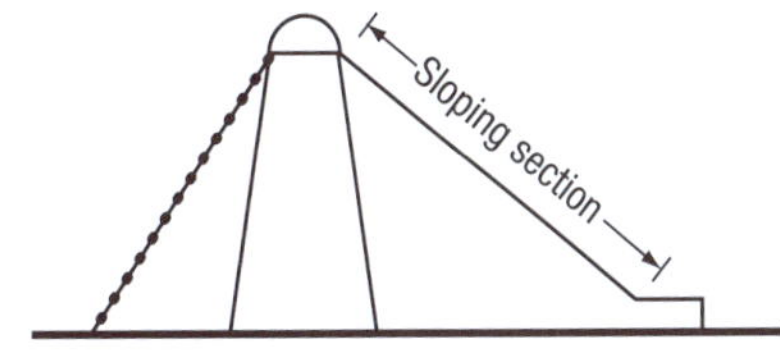

Finding the angle when the adjacent side and the hypotenuse are known in a right-angled triangle

Example

The length of the adjacent side to the angle, marked θ°, and of the hypotenuse are known in this triangle. Find the value of θ.

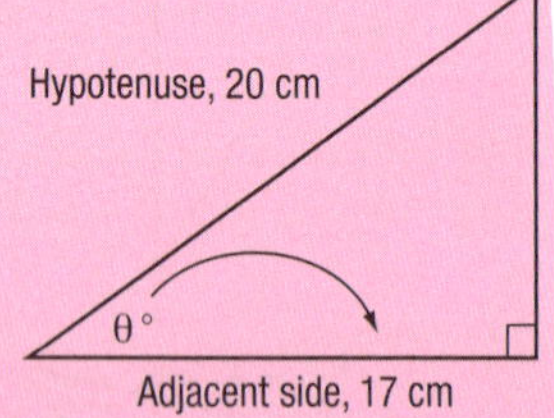

Answer

We know that $\cos\theta = \frac{\text{Length of the adjacent side to angle } \theta}{\text{Length of the hypotenuse}}$

Substituting the known values into this equation:

$\cos\theta° = \frac{17}{20} = 0.85$

Look in the cosine tables to find the angle with a cosine value closest to 0.8500.

For cosine we need to go to a value higher than 0.8500 and subtract the difference.

$0.8508 - 0.0008 = 0.8500$

NATURAL COSINES

	0′	6′	12′	18′	24′	30′	36′	42′	48′	54′	Subtract Differences 1′	2′	3′	4′	5′
30	0.8660	8652	8643	8634	8625	8616	8607	8599	8590	8581	1	3	4	6	7
31	0.8572	8563	8554	8545	8536	8526	8517	8508	8499	8490	2	3	5	6	8
32	0.8480	8471	8462	8453	8443	8434	8425	8415	8406	8396	2	3	5	6	8
33	0.8387	8377	8368	8358	8348	8339	8329	8320	8310	8300	2	3	5	6	8
34	0.8290	8281	8271	8261	8251	8241	8231	8221	8211	8202	2	3	5	7	8

The angle is $31° + 42′ + 5′ = 31°47′$

Using a calculator: If $\cos\theta° = 0.85$ then $\cos^{-1} 0.85 = \theta$ and you can use the inverse cosine ($\cos^{-1}$) button on your calculator: $\cos^{-1} 0.85 = 31°47′17″$ which rounds to $31°47′$.

EXERCISE

1 For each of the following find the size of the angle, in degrees and minutes, with the cosine value:

a 0.9945 b 0.9012 c 1 d 0.3843

e 0.3652 f 0 g 0.1140 h 0.5000

2 For each of the following find:

 i the cosine of the angle θ

 ii the size of the angle θ in degrees and minutes.

a

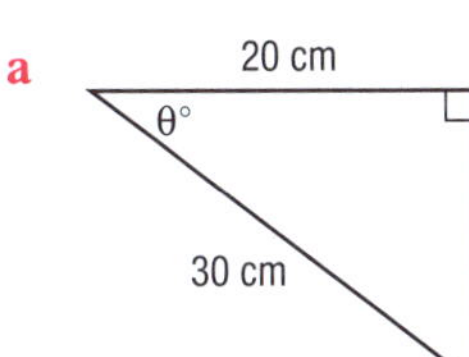

b

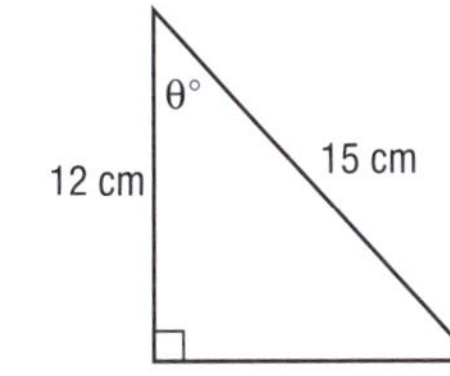

c

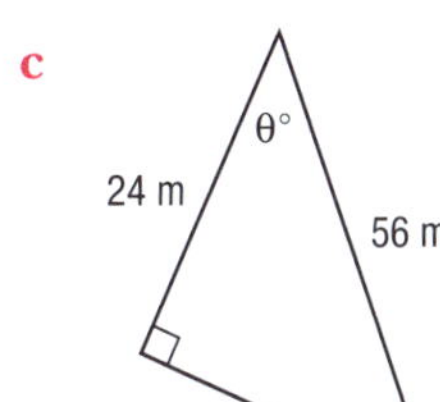

d

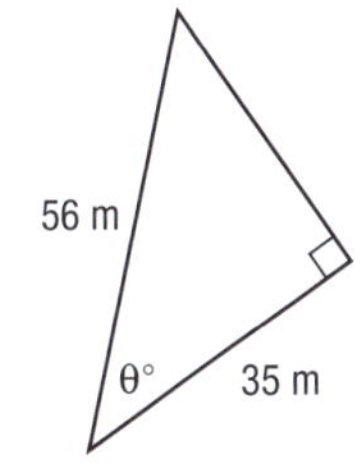

e

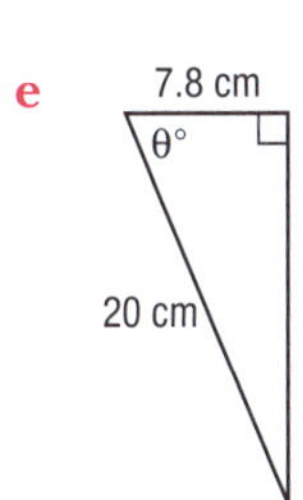

f

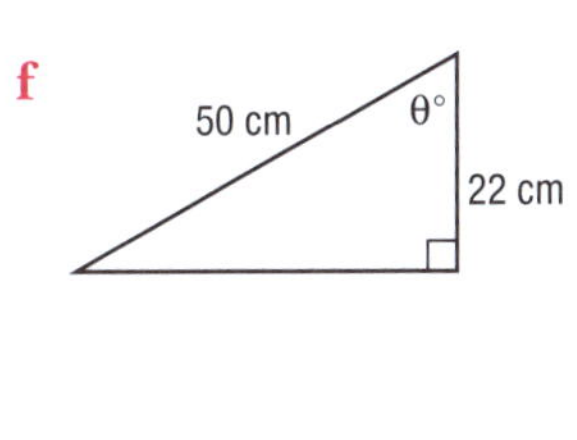

3 Find the value of θ, correct to the nearest minute, in this isosceles triangle.

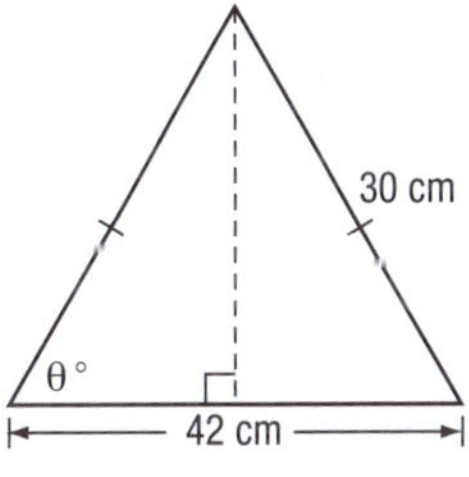

4 The diagonal of a rectangular gate is 2 m long and the width of the gate is 1238 mm. What angle, to the nearest degree, does the diagonal make with the horizontal?

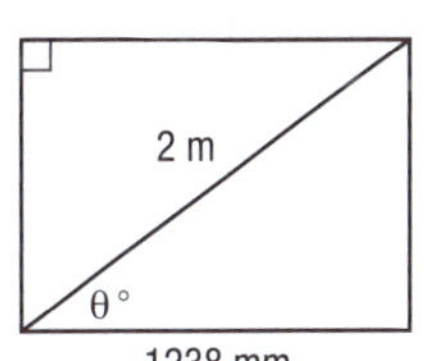

5 Find the inclination of a road, θ°, where travelling 5 km along the road results in a horizontal distance travelled of 4.6 km. Give your answer correct to the nearest degree.

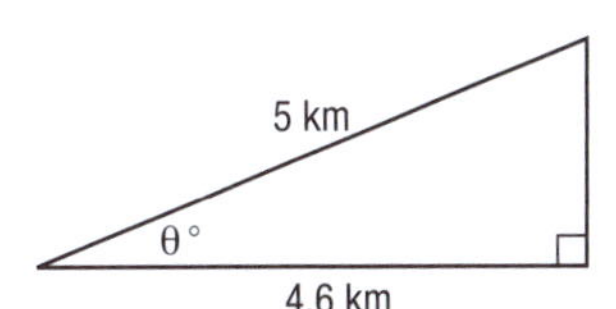

6 In the diagram at right find:

 i the value of x in metres, correct to two decimal places

 ii the value of θ in degrees and minutes.

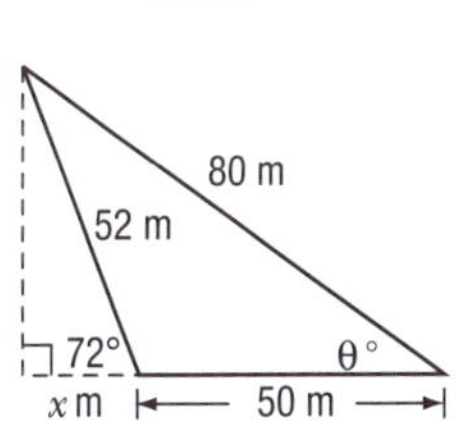

Lesson 19 Trigonometry of the right-angled triangle: the tangent ratio

The tangent ratio is used when the opposite and the adjacent side to the angle are known.

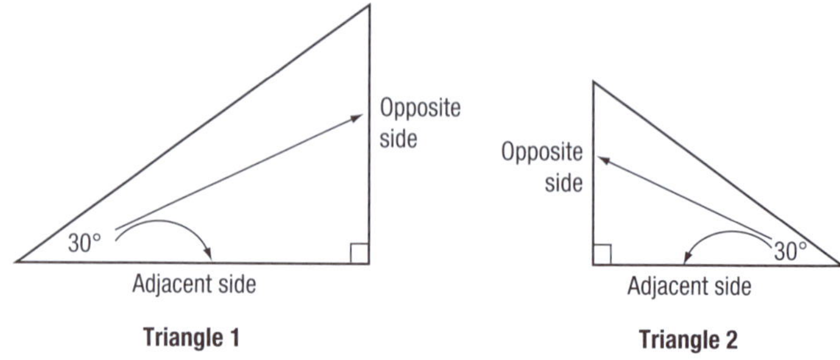

If you calculate the ratio $\frac{\text{Length of the opposite side}}{\text{Length of the adjacent side}}$ for each of the similar triangles above, you will get a value of approximately 0.58.

This ratio is called the **tangent** of the angle 30°. We can write **tan 30° = 0.58** and, looking at the table for tangents on pages 340–2, you will find that tan 30° = 0.5774, correct to four decimal places.

In general, for an angle θ in a right-angled triangle:

$$\tan\theta = \frac{\text{Length of the opposite side to angle }\theta}{\text{Length of the adjacent side to angle }\theta}$$

Using the tangent ratio to find the length of the sides in a right-angled triangle

Example 1

Find, correct to two decimal places, the value of x in the triangle.

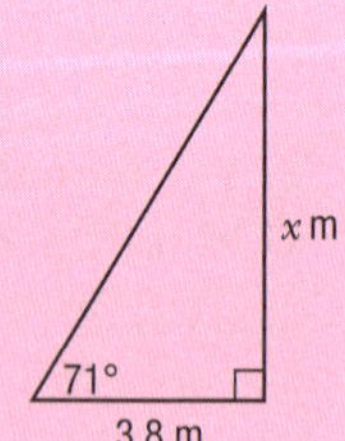

Answer

In the triangle, the side marked x cm is the **opposite side** to the angle 71° and the length of the adjacent side is 3.8 m, so we can use the tangent ratio to find x.

Substituting in $\tan\theta = \frac{\text{Length of the opposite side to angle }\theta}{\text{Length of the adjacent side to angle }\theta}$

$\tan 71^\circ = \frac{x}{3.8}$ multiply both sides of the equation by 3.8

$3.8 \times \tan 71^\circ = x$ substitute the value of tan 71° from the table

$3.8 \times 2.904 = x$

$x = 11.0352$

$= 11.04$ correct to two decimal places

Example 2

Find, correct to two decimal places, the value of x in this triangle.

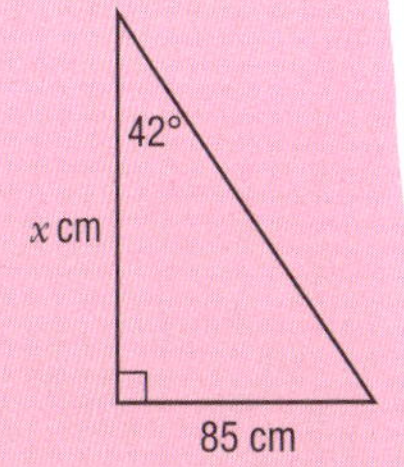

Answer

In the triangle, the side marked x cm is the **adjacent side** to the angle 42° and the length of the opposite side is 85 cm, so we can use the tangent ratio to find x.

Substituting in $\tan\theta = \dfrac{\text{Length of the opposite side to angle }\theta}{\text{Length of the adjacent side to angle }\theta}$

$\tan 42° = \dfrac{85}{x}$ multiply both sides of the equation by x

$x \tan 42° = 85$ divide both sides of the equation by tan 42°

$x = \dfrac{85}{\tan 42°}$ substitute the value of tan 42° from the table

$x = \dfrac{85}{0.9004}$

$= 94.4025\ldots$

$= 94.40$ correct to two decimal places

EXERCISE

1 Use the table on pages 340–2, or your calculator, to find the tangent of each of the following angles:

a 23° b 85° c 45° d 6°25′

e 55° f 30° g 90° h 73°45′

2 What do you notice about the size of the tangent values found in Question 1? How do these values differ from sine values and cosine values?

3 For each of the following:

i write an equation using $\tan\theta = \dfrac{\text{Length of the opposite side to angle }\theta}{\text{Length of the adjacent side to angle }\theta}$

ii solve the equation to find the value of x, correct to one decimal place.

a

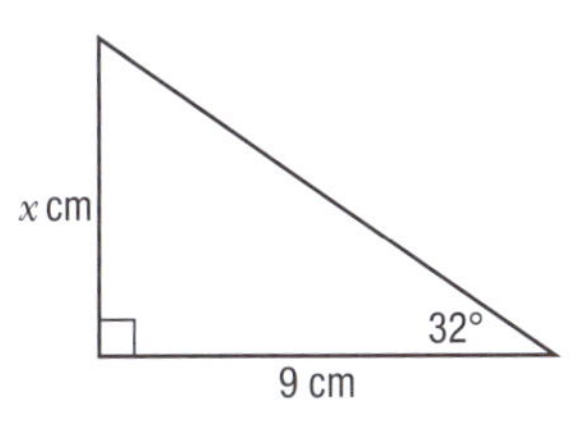

b

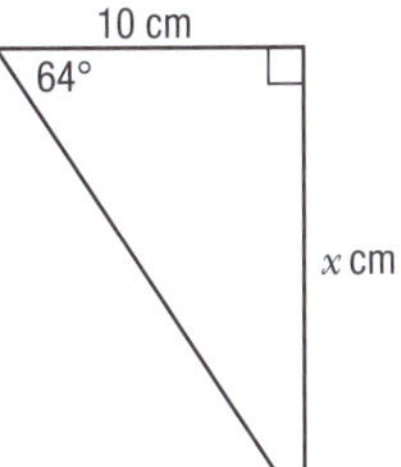

c

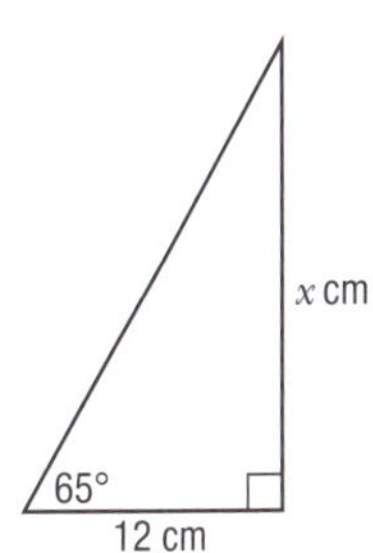

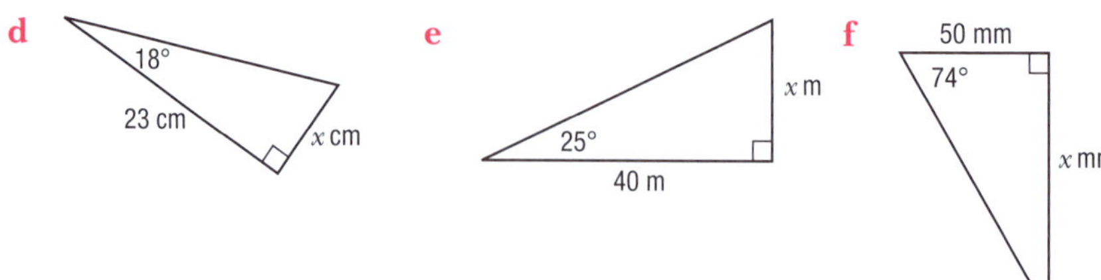

4 For each of the following:

i write an equation using the formula
$\tan \theta = \frac{\text{Length of the opposite side to angle } \theta}{\text{Length of the adjacent side to angle } \theta}$

ii find the value of x, correct to one decimal place.

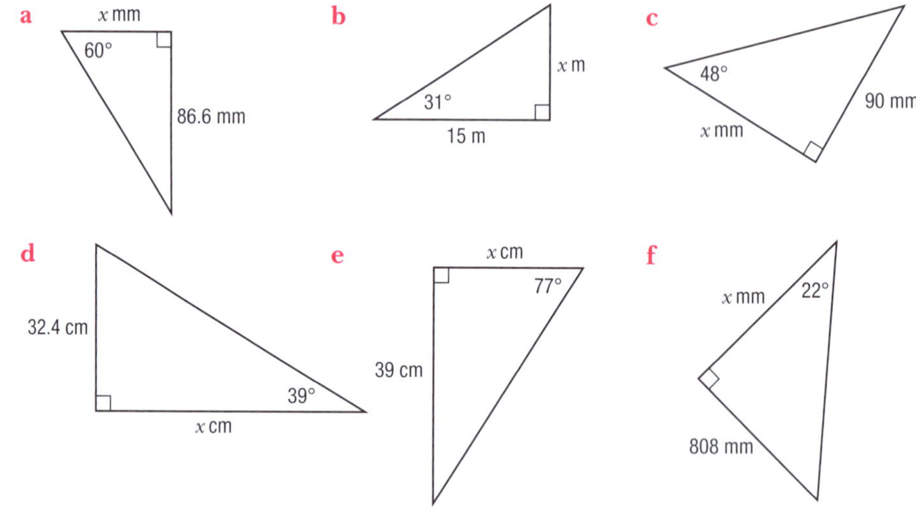

5 For the rectangle find the value of x, correct to one decimal place.

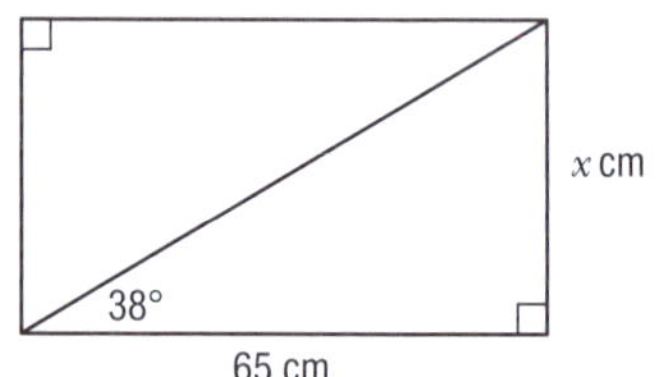

6 Find the height of this isosceles triangle. Give your answer in centimetres correct to one decimal place.

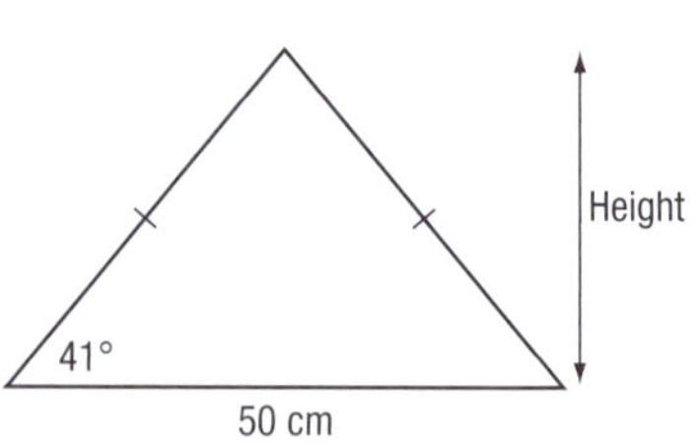

7 Find the value of x in this trapezium.

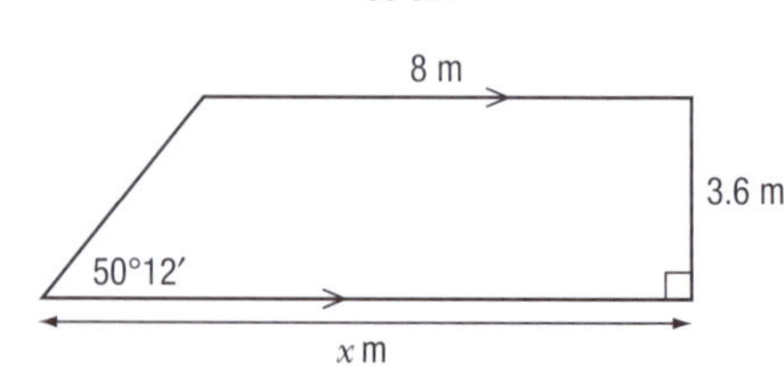

Finding the angle when the opposite side and the adjacent side are known in a right-angled triangle

Example

Find the value of θ in the given triangle.

5.4 m

θ°

9 m

Answer

In relation to the angle marked θ°, the opposite side (length 5.4 m) and the adjacent side (length 9 m) are known and we can use the tangent ratio to find θ.

Substituting in $\tan\theta = \dfrac{\text{Length of the opposite side to angle } \theta}{\text{Length of the adjacent side to angle } \theta}$:

$\tan\theta = \dfrac{5.4}{9} = 0.6$

Looking in the tables for the angle whose tangent value is 0.6 is 30°58′ to the nearest minute.

The closest figure to 0.6 is 0.6001.

0.6001 = 0.5985 + 0.0016

NATURAL TANGENTS

	0′	6′	12′	18′	24′	30′	36′	42′	48′	54′	1′	2′	3′	4′	5′
28	0.5317	5340	5362	5384	5407	5430	5452	5475	5498	5520	4	8	11	15	19
29	0.5543	5566	5589	5612	5635	5658	5681	5704	5727	5750	4	8	12	15	19
30	0.5774	5797	5820	5844	5867	5890	5914	5938	5961	5985	4	8	12	16	20
31	0.6009	6032	6056	6080	6104	6128	6152	6176	6200	6224	4	8	12	16	20
32	0.6249	6273	6297	6322	6346	6371	6395	6420	6445	6469	4	8	12	16	20
33	0.6494	6519	6544	6569	6594	6619	6644	6669	6694	6720	4	8	13	17	21
34	0.6745	6771	6796	6822	6847	6873	6899	6924	6950	6976	4	9	13	17	21

The angle is 30° + 54′ + 4′ = 30°58′

Using a calculator: To find the answer in degrees, minutes and seconds, the answer is 30°57′49″ which rounds to 30°58′ (there are 60″ in one minute so 30 or more seconds rounds up to the next minute).

EXERCISE

1 Use the tables or a calculator to find, to the nearest minute, the angle with a tangent of:

a 0.3269 **b** 0.6847 **c** 1.3713

d 2.103 **e** 0.3610 **f** 0

g 0.8352 **h** 10.78 **i** 2.630

2 For each of the following:

- i calculate the tangent of the angle θ
- ii find the size of the angle θ, in degrees and minutes.

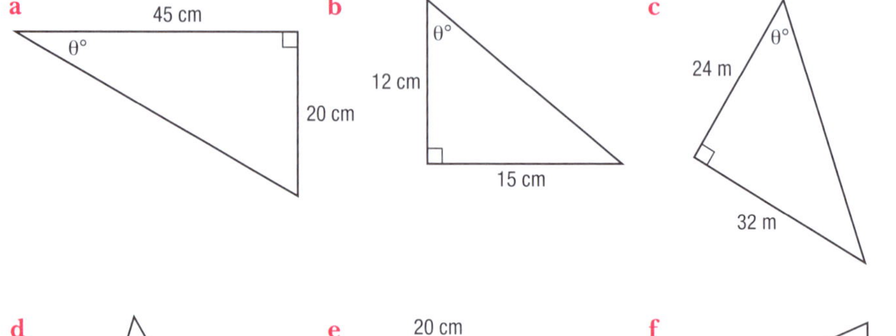

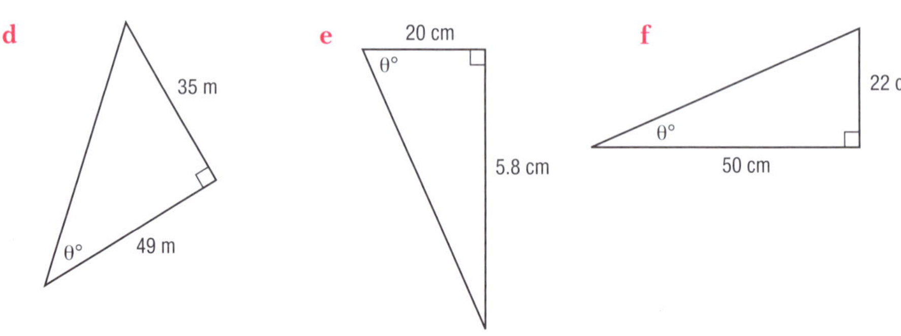

3 Find the value of θ, in degrees and minutes, in each of the following.

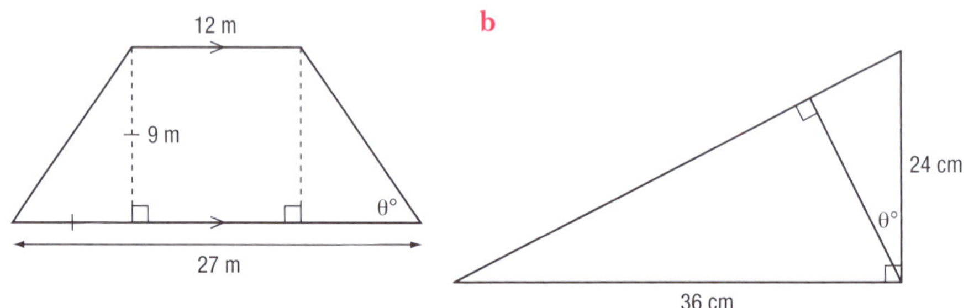

4 Find, to the nearest degree, the angle the roof of this house makes with the horizontal.

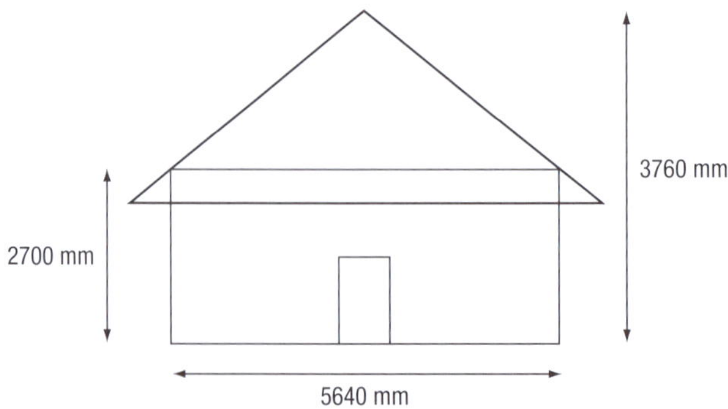

SOLVING RIGHT TRIANGLES USING TRIGONOMETRY

Using all three trigonometric ratios

Lesson 21

For this exercise you will need to decide which one the three trigonometric ratios will provide a solution to the problem. This will depend on the sides given in relation to an angle.

To help you remember which sides are associated with which ratio you can use SOH CAH TOA (said as sock-a toe-a) as explained below.

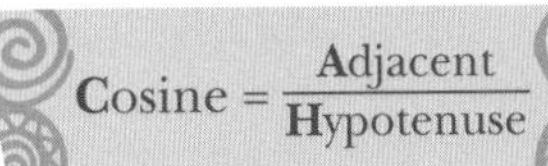

SOH	CAH	TOA
$\text{Sine} = \dfrac{\text{Opposite}}{\text{Hypotenuse}}$	$\text{Cosine} = \dfrac{\text{Adjacent}}{\text{Hypotenuse}}$	$\text{Tangent} = \dfrac{\text{Opposite}}{\text{Adjacent}}$

Example 1

Find the value of x, correct to two decimal places, in the triangle at right.

Answer

In relation to the angle of 18°, the side labelled x cm is the opposite side and the side of length 30 cm is the adjacent side. We need to use the tangent ratio to solve this problem.

Substituting in $\tan\theta = \dfrac{\text{Length of the opposite side to angle } \theta}{\text{Length of the adjacent side to angle } \theta}$:

$$\tan 18^\circ = \frac{x}{30} \quad \text{multiply both sides of the equation by 30}$$

$$30 \times \tan 18^\circ = x$$

$$30 \times 0.3249 = x$$

$$9.747 = x$$

$$x = 9.75 \quad \text{correct to two decimal places}$$

Example 2

Find the value of θ, correct to the nearest minute, in this triangle.

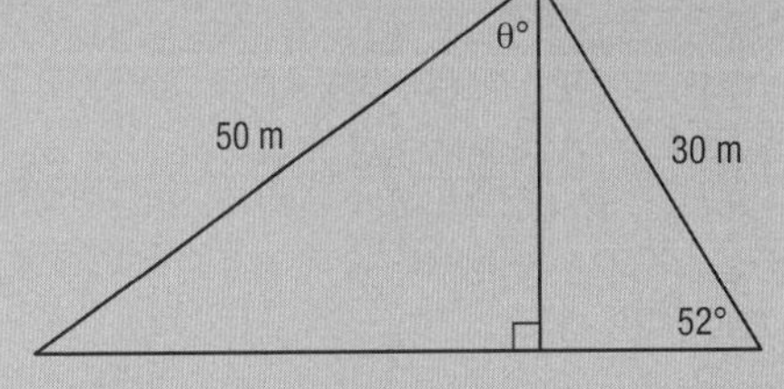

Answer

We will need to do two calculations to find the value of θ.

First, in triangle BDC we will need to find the value of x and then use this value in triangle ABD to find the value of θ.

In triangle BDC, the side labelled x m is the opposite side to the angle 52° and the side of length 30 m is the hypotenuse, so we will use the sine ratio.

Substituting in $\sin\theta = \dfrac{\text{Length of the opposite side}}{\text{Length of the hypotenuse}}$:

$\sin 52° = \dfrac{x}{30}$ multiply both sides of the equation by 30

$30 \times \sin 52° = x$

$x = 30 \times 0.7880 = 23.64$

In triangle ABD the side labelled x m is the adjacent side to angle θ, and the side of length 50 m is the hypotenuse so we will use the cosine ratio to find θ:

Substituting in $\cos\theta = \dfrac{\text{Length of the adjacent side}}{\text{Length of the hypotenuse}}$

$\cos\theta = \dfrac{23.64}{50}$

$\cos\theta = 0.4728$

Using the tables or a calculator this gives a value for θ of 61°47′.

EXERCISE

Give your answer correct to one decimal place or in degrees and minutes.

1 Find the value of the pronumeral in each of the following right-angled triangles.

a 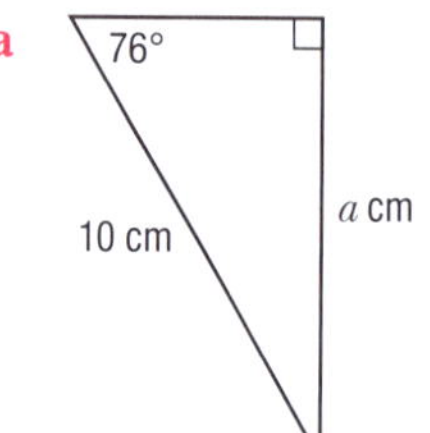

b 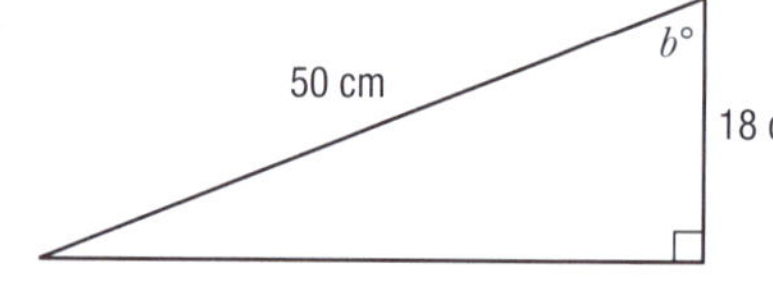

c 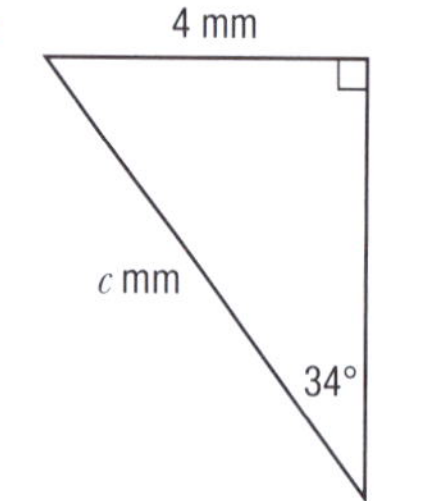

d 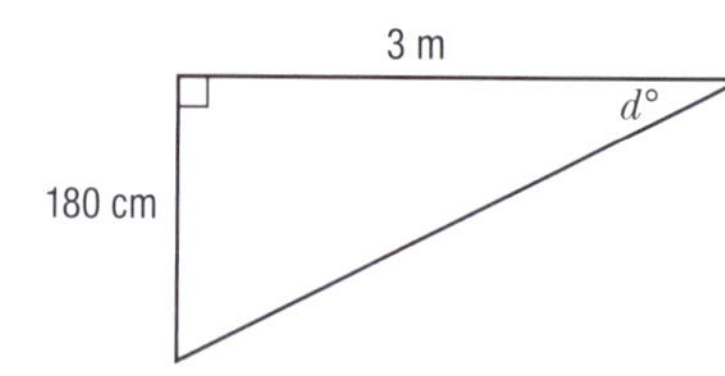

e 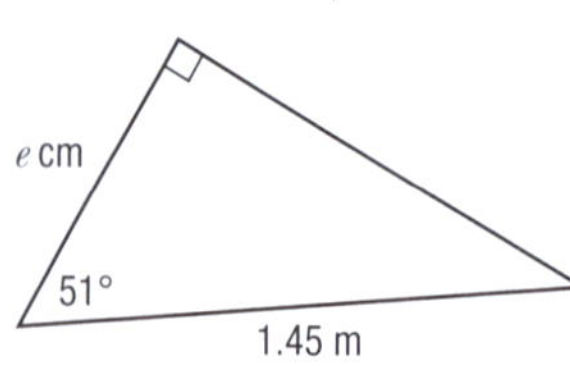

f 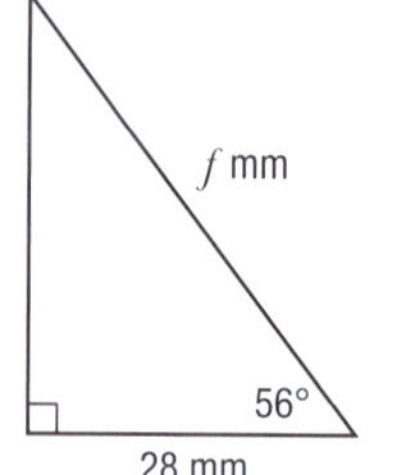

2 Find the height of an equilateral triangle of side length 20 cm.

Height

3 The length of the diagonals of a kite are given on the diagram. Find the value of θ and α if the shorter diagonal cuts the longer diagonal in the ratio 1 : 3.

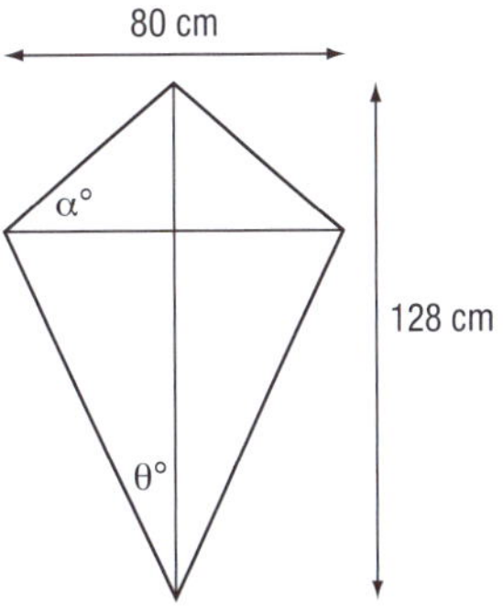

4 Find the height of a regular pentagon of side length 20 cm.

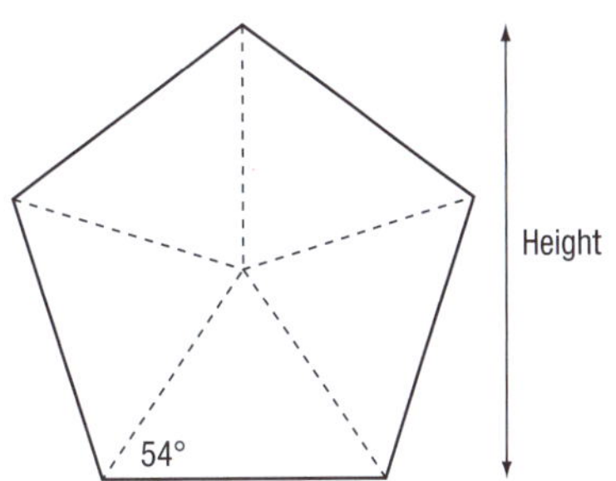

5 The pitch of a roof is 35° and the height of the roof is 1360 mm, as shown in the diagram. Find the length that the builder needs to make the rafters if they need to overhang the wall by 300 mm. Give your answer to the nearest millimetre.

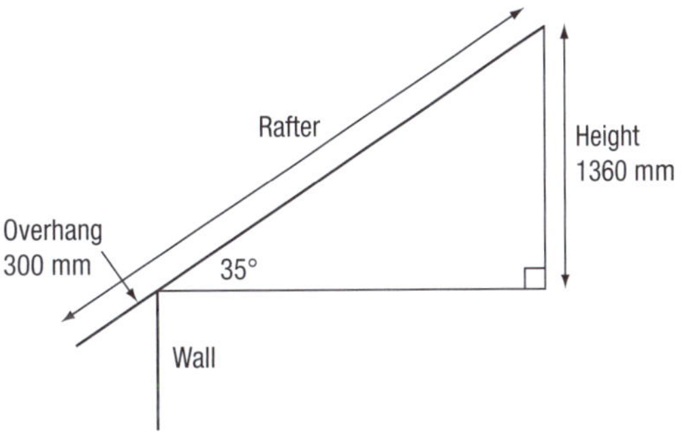

6 Find the width of the base, w metres, of the isosceles triangle at right.

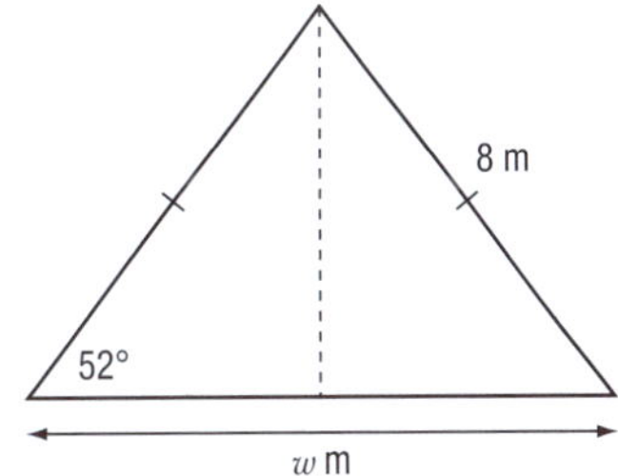

7 A straight access road to a beach is to have a gradient of no more than 9°. If the access road is to start at a point 14.2 metres above the beach, what is the minimum length of the access road?

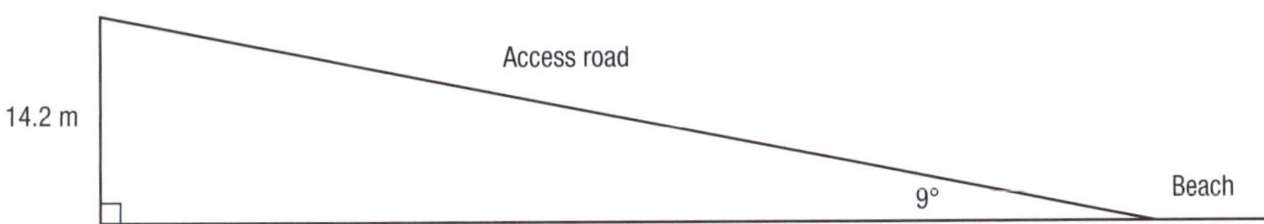

8 In the diagram at right, find the value of:

a θ b x c y

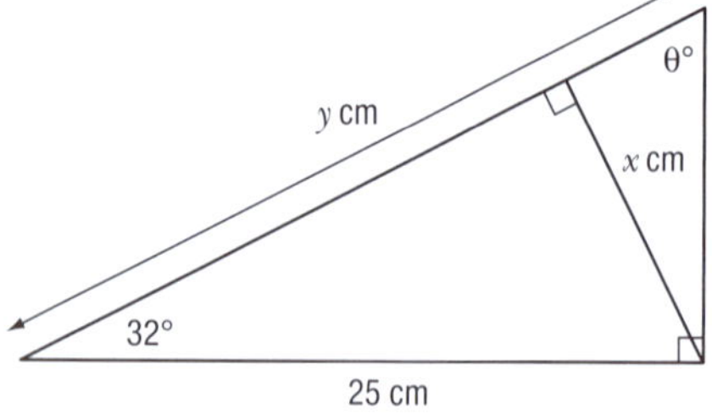

9 Sam has to walk from a point A to a point B. There is a path that goes from A to B through point C or he can go across rough ground directly from point A to point B.

a Find the distance Sam will walk if he goes in a straight line from A to B. Give your answer in kilometres correct to one decimal place.

b How much further will Sam walk if he takes the path from A to C to B?

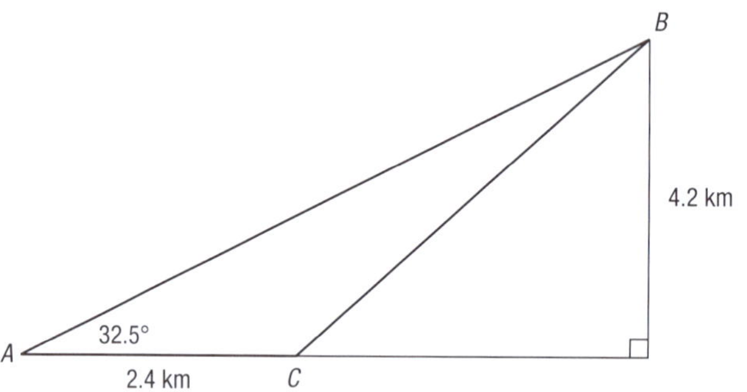

10 A roof truss has measurements and angles as shown on the diagram.

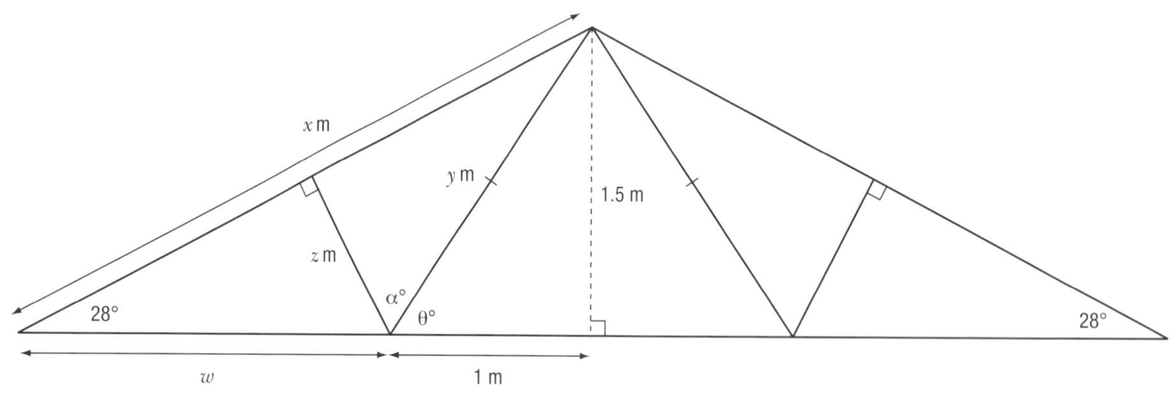

a Find the value of:

i θ ii α iii x

iv y v z vi w

b Calculate the total length of wood needed to construct this roof truss.

Applications of Pythagoras's rule and the trigonometric ratios in three dimensions

Example

A rectangular field, *ABCD*, with dimensions 100 × 65 metres, is inclined at an angle of 18° to the horizontal.

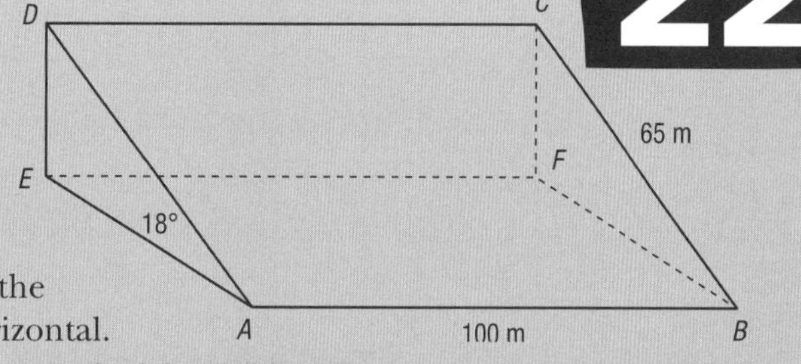

a Find the height (*ED*) of the corner, *D*, above the horizontal.

b Find the length of the diagonal of the field (*BD*).

c Find the inclination of the diagonal *BD* to the horizontal ($\angle DBE$).

Answer

a *ED* is the opposite side to the 18° angle in the triangle *DAE*. The hypotenuse is the length 65 m. We will use the sine ratio to find *DE*.

Substituting in $\sin\theta = \dfrac{\text{Length of the opposite side}}{\text{Length of the hypotenuse}}$

$$\sin 18° = \frac{DE}{65}$$

$$DE = 65 \times \sin 18° = 20.09 \text{ m}$$

b *BD* is the diagonal of a rectangle of dimensions 100 × 65 m.

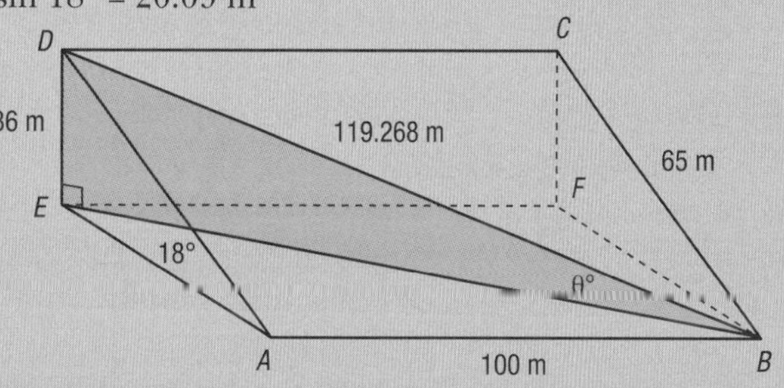

Using Pythagoras's rule:

$BD^2 = 100^2 + 65^2$

$BD^2 = 14225$

$BD = \sqrt{14225} \approx 119.2$ m

c In the right-angled triangle *EDB*, the angle *DBE* is the inclination of the diagonal to the horizontal.

Substituting in $\sin\theta = \dfrac{\text{Length of the opposite side}}{\text{Length of the hypotenuse}}$

$$\sin\theta = \frac{20.086}{119.268} = 0.1684$$

Using the tables or your calculator:

$$\theta = \sin^{-1}(0.1684) = 9.695° = 9°42'$$

EXERCISE

Give your length answers correct to two decimal places and angles to the nearest minute.

1 A conical-shaped hill rises 28 m to the summit for a horizontal distance of 63 m.

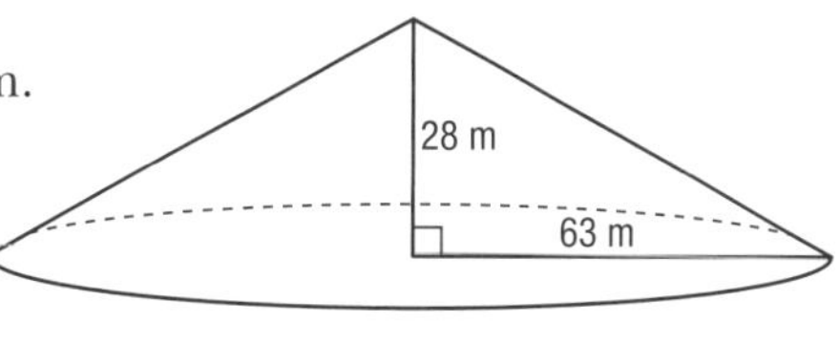

a If you walk the shortest distance from the base of the hill to the summit how far, to the nearest metre, will you walk?

b Find, to the nearest degree, the inclination of the hill to the horizontal.

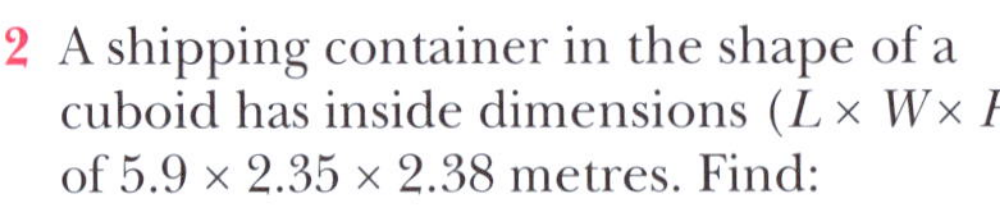

2 A shipping container in the shape of a cuboid has inside dimensions ($L \times W \times H$) of $5.9 \times 2.35 \times 2.38$ metres. Find:

- **a** the length of the longest beam that will fit across the bottom of the container
- **b** the length of the longest beam that will fit inside this container
- **c** the angle that the longest beam makes with the horizontal.

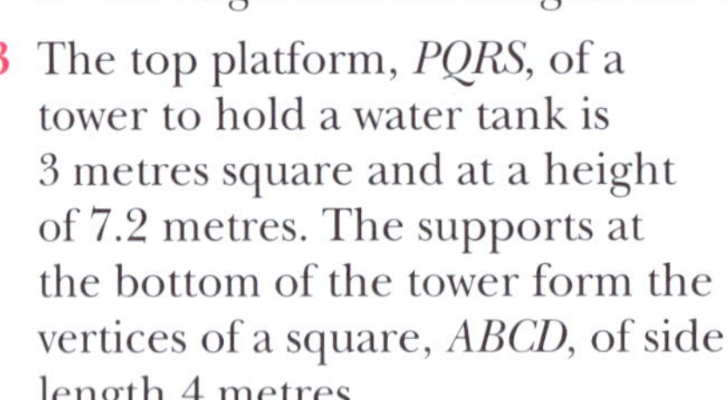

3 The top platform, *PQRS*, of a tower to hold a water tank is 3 metres square and at a height of 7.2 metres. The supports at the bottom of the tower form the vertices of a square, *ABCD*, of side length 4 metres.

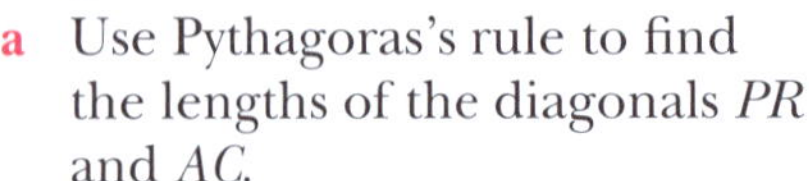

- **a** Use Pythagoras's rule to find the lengths of the diagonals *PR* and *AC*.
- **b** Calculate the length *AM*.
- **c** Calculate the angle that a sloping support makes with the horizontal ground.
- **d** Find the length of one of the sloping supports.

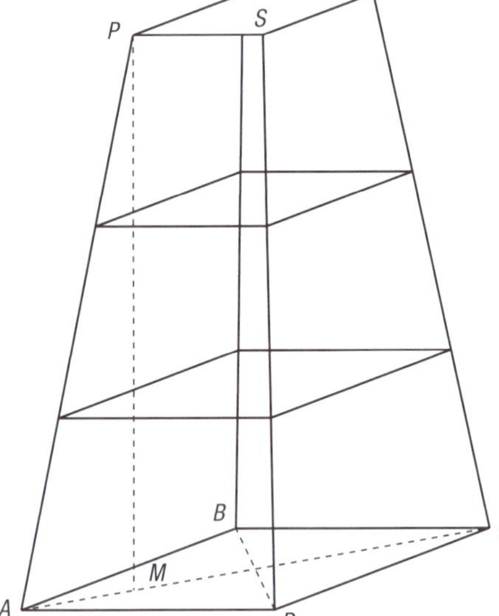

4 A pyramid has a square base, *ABCD*, of side length 6 metres and of height (*VO*) 4.5 metres.

- **a** Find the length of the diagonal of the base (*DB*).
- **b** Find the length of a slant edge (*VB*).
- **c** Find the inclination of a slant edge to the horizontal ($\angle VBO$).
- **d** Find the inclination of a sloping face to the horizontal ($\angle VMO$).

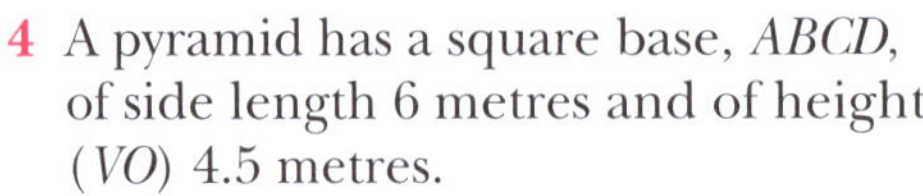

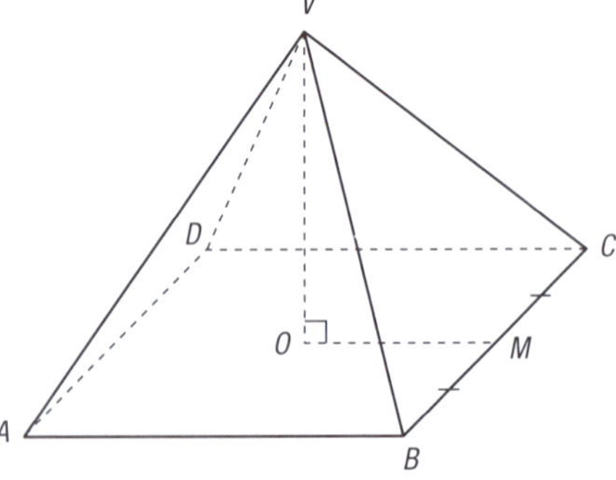

5 The measurements for the roof of a building are given at right.

- **a** Find the angle that a sloping face of the roof makes with the horizontal ($\angle PQR$).
- **b** Find the angle that a sloping edge of the roof makes with the horizontal ($\angle ABC$).

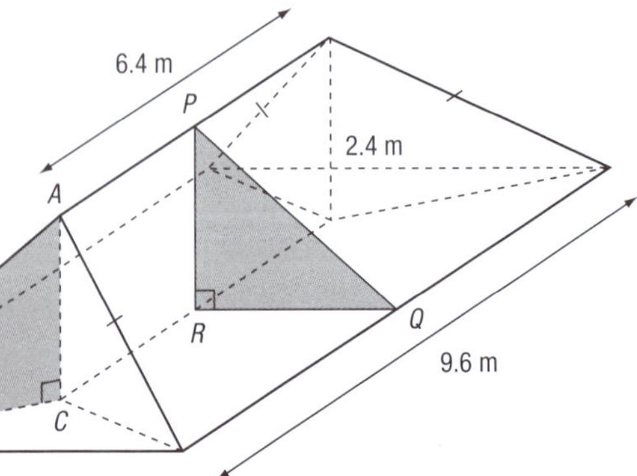

REVISION AND ASSESSMENT

Similar triangles and Pythagoras's rule

Multiple-choice questions

1 Which two of the following triangles are similar?

I
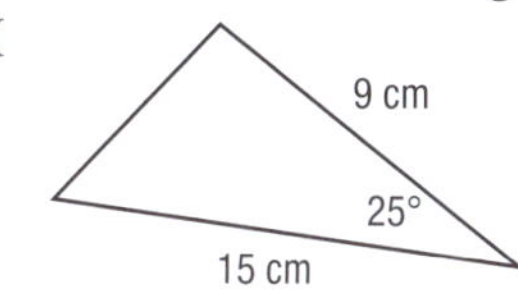

II
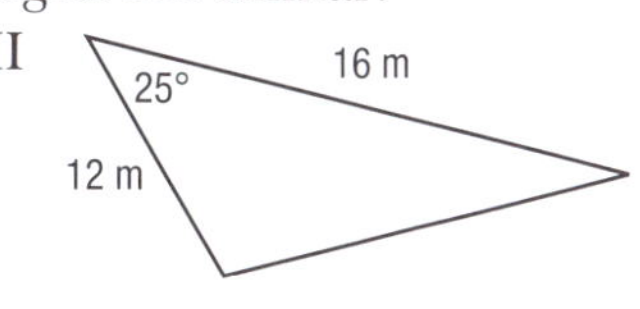

III
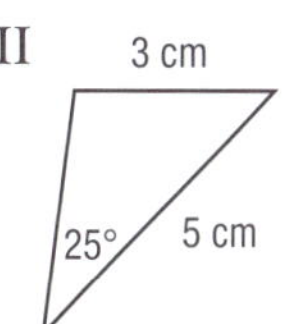

IV
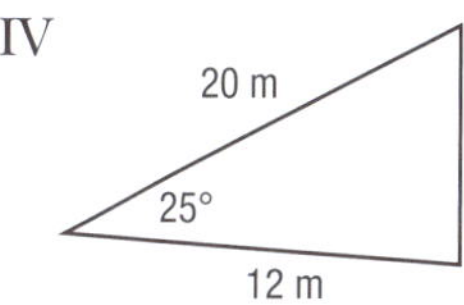

A I and II
B II and III
C III and IV
D I and IV

2 On the diagram, the angle that is the same as $\angle ACD$ is:

A $\angle CDB$
B $\angle BAD$
C $\angle ADB$
D $\angle CBE$

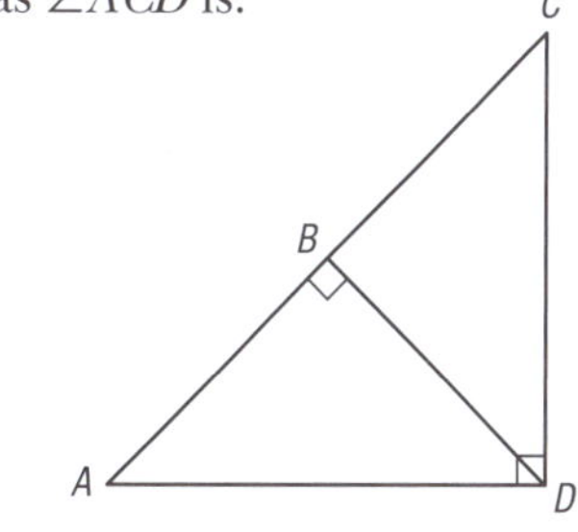

The following information refers to Questions 3 and 4.

The two triangles illustrated are similar.

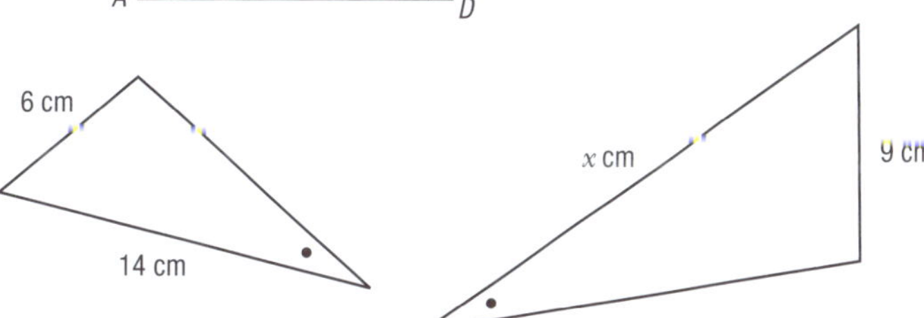

3 The scale factor for these two triangles is:

A 1.5 B $\frac{2}{3}$ C 6 D 3

4 The value of x is:

A $9\frac{1}{3}$ B 18 C 21 D $23\frac{1}{3}$

5 The value of x in this triangle is:

A 13
B 5
C 10
D 12

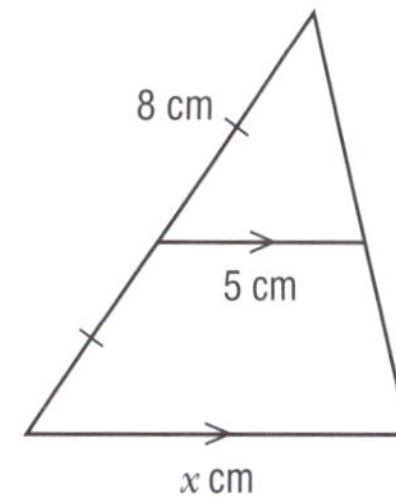

6 The value of x in the given triangle is:

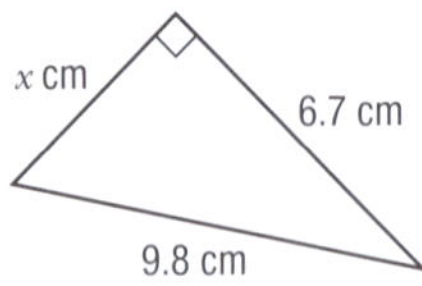

A $\sqrt{51.15}$ cm

B 3.1 cm

C 11 871 cm rounded to three decimal places

D 4.062 cm rounded to three decimal places

7 In the cube, which of the following angles is *not* a right angle?

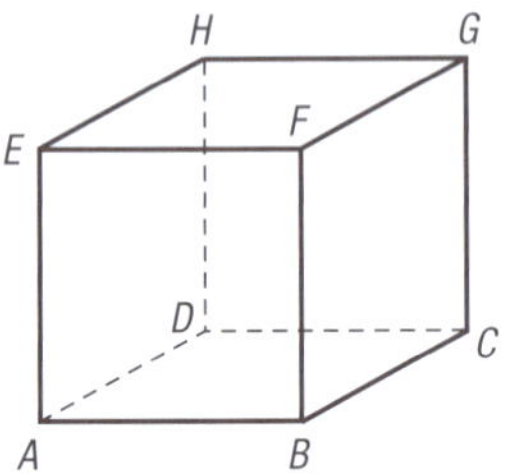

A $\angle EGC$

B $\angle ABC$

C $\angle HCF$

D $\angle EHC$

8 Which of the following is *not* true?

A A surd is an irrational number.

B Recurring decimals are surds.

C Surds cannot be written as ratios of whole numbers.

D $\sqrt{25}$ is not a surd because 25 is a perfect square.

9 The length of the diagonal of an 8×10 metre rectangle is:

A $\sqrt{18}$ metres

B 6 metres

C 12.806 metres rounded to three decimal places

D $\sqrt{80}$ metres

10 The length of the hypotenuse of this triangle is:

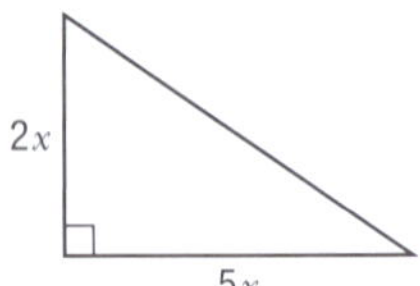

A $7x$

B $29x$

C $\sqrt{29x}$

D $\sqrt{29}x$

11 The distance between the two points on the Cartesian plane is given by:

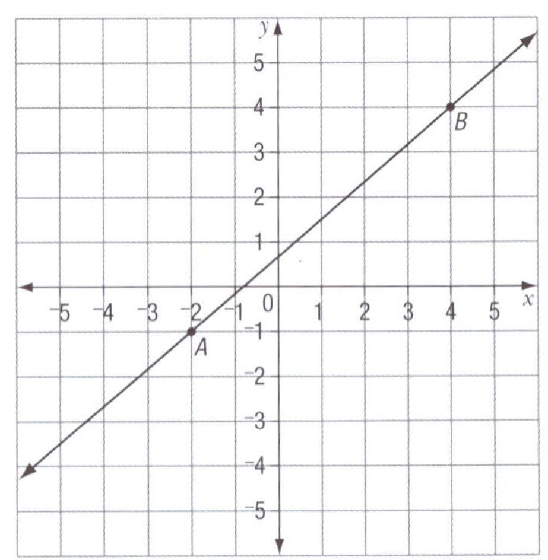

A $\sqrt{(4-2)^2+(4+1)^2}$

B $\sqrt{(4+2)^2+(4+1)^2}$

C $\sqrt{(4+2)^2+(4-1)^2}$

D $\sqrt{(4-2)^2+(4-1)^2}$

Short answer questions

1 The two triangles illustrated are similar. By measuring the length of a pair of corresponding sides, find the scale factor.

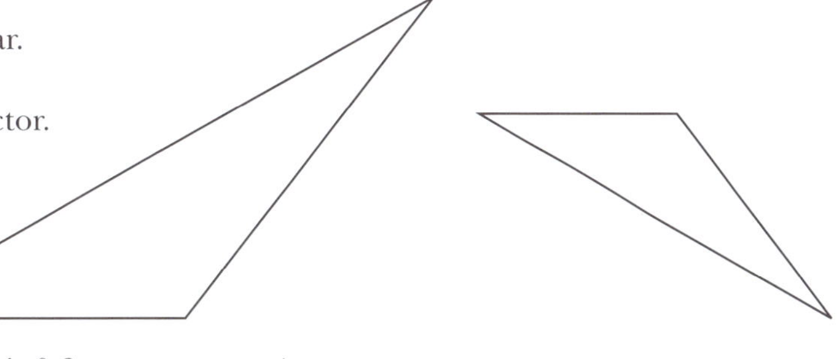

2 The scale factor of these two triangles is 0.6. Find the value of:

a x **b** y

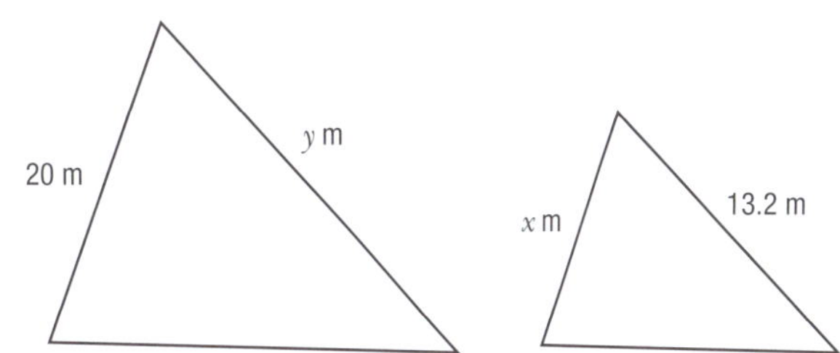

3 **a** For the diagram at right draw the two similar triangles separately. Include all measurements.

b Find the value of:

i x **ii** y

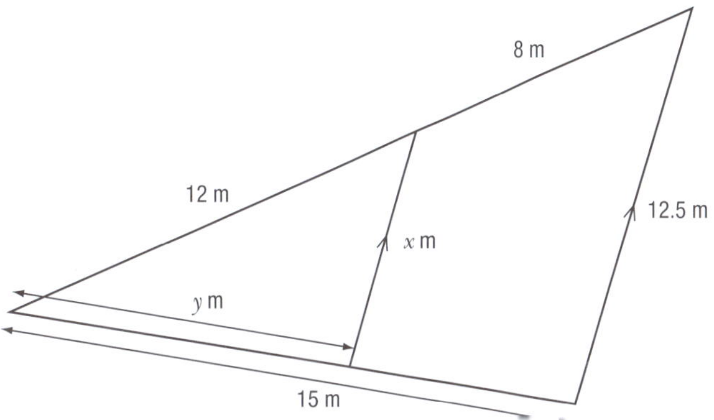

4 Find the value of x in each of the following triangles giving your answers as both a surd value and a decimal approximation, correct to two decimal places.

a

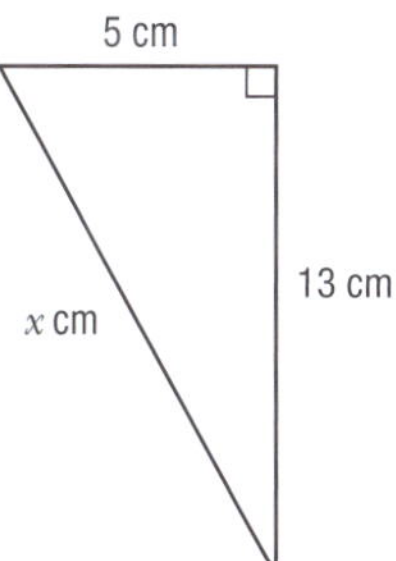

b

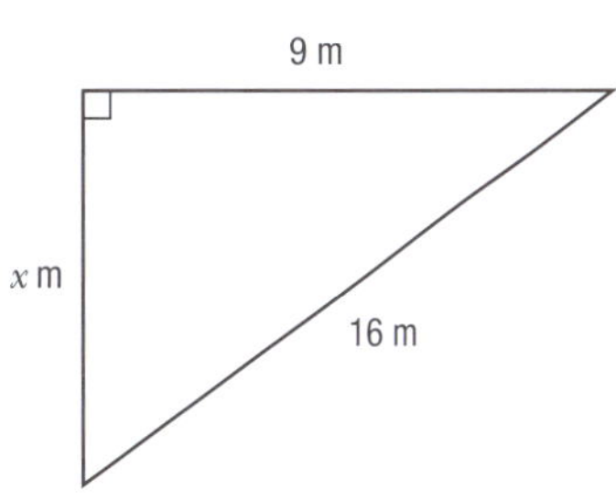

5 The diagonal of a rectangle has length 8.5 cm and one of the sides is 3.6 cm long. What is the length of the other side?

6 A kite on a 50-metre string is flying directly over a landmark that is 20 metres from the person holding the string. If the string is held 2 metres above the ground and is fully extended, how high is the kite?

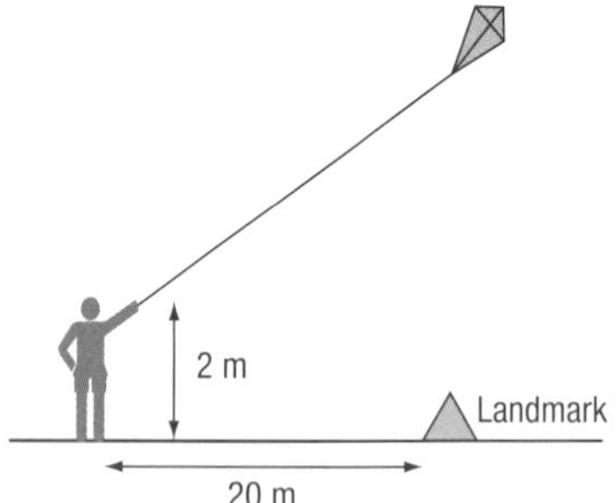

7 Find the length of x and then the length of y in the diagram.

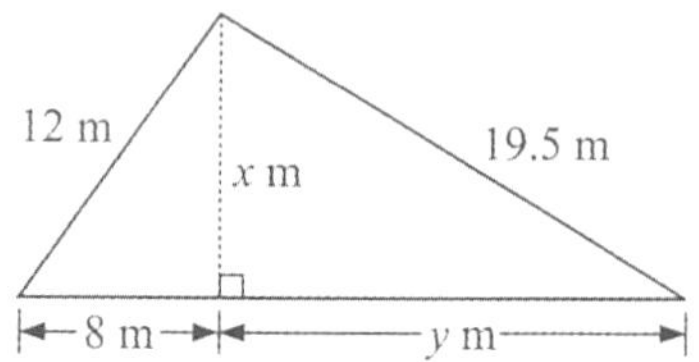

8 An equilateral triangle has sides of length 15 cm. Find the perpendicular height of this triangle.

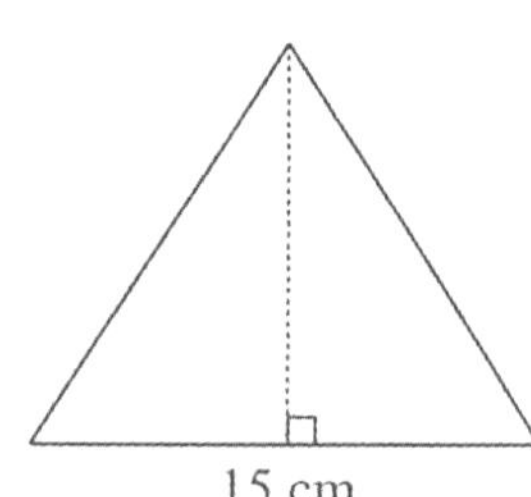

9 A boat has drifted with the tide until it is stationary at a point 16 metres from where the anchor was dropped. The anchor chain is 25 metres long. Calculate the depth of the water, correct to two decimal places.

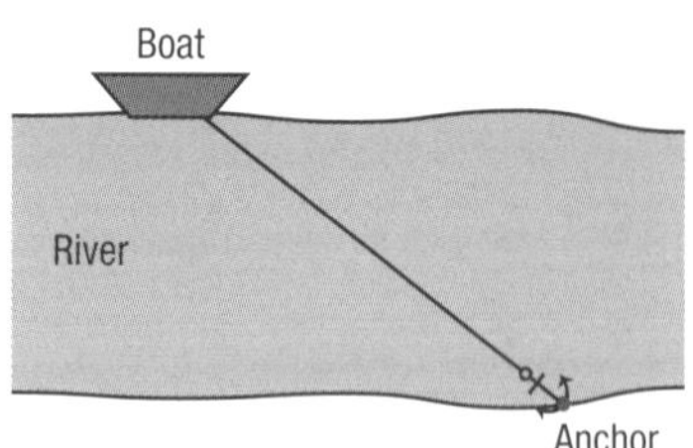

10 The three points $A(^-1, 2)$, $B(3, 1)$ and $C(4, 5)$ plotted on Cartesian axes form the vertices of a triangle.

a Find the lengths of the sides of the triangle ABC and show that it is isosceles.

b Use Pythagoras's rule to show that the triangle is a right-angled triangle.

11 A cuboid has the dimensions given on the diagram. M is the midpoint of the edge HG. Find the length of MB, giving your answer correct to two decimal places.

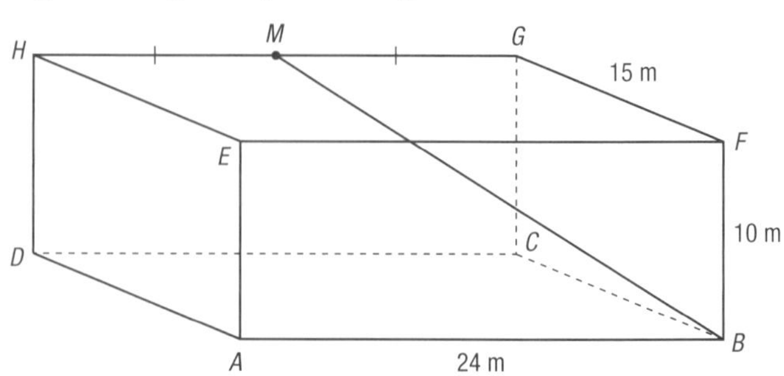

Trigonometric ratios

Multiple-choice questions

1 In relation to the angle θ in the right-angled triangles illustrated, which triangle has the sides correctly labelled? (*A* is adjacent, *O* is opposite and *H* is the hypotenuse.)

A 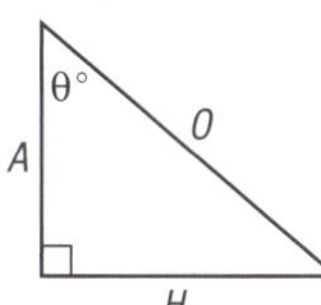

B 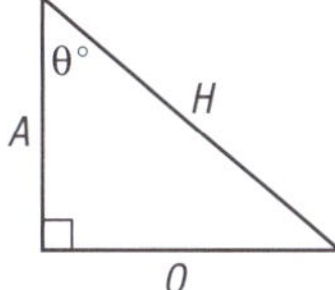

C 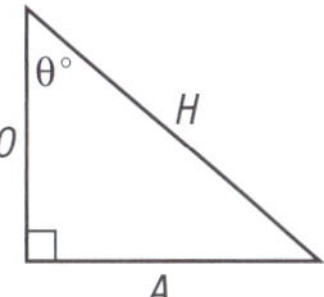

D 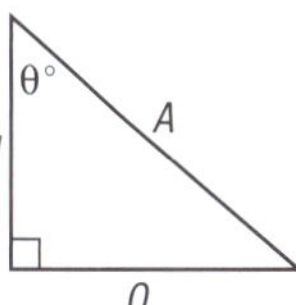

2 The value of sin 32°16′, correct to four decimal places, is:

A 0.5920 B 0.8468 C 0.6313 D 0.5339

3 The correct trigonometric ratio to use to find *x* in this right-angled triangle is:

A sine
B cosine
C tangent
D Pythagoras's rule

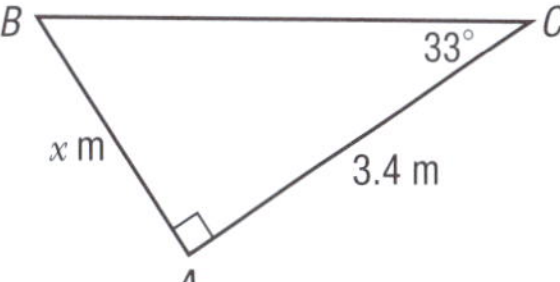

4 56.4° in degrees and minutes is:

A 56°40′ B 56°48′ C 56°24′ D 56°4′

5 The angle, in degrees and minutes, with a cosine of 0.69 is:

A 46°18′ B 46°22′ C 46°26′ D 53°4′

6 The value of sin θ in this triangle is:

A $\frac{5}{12}$
B $\frac{5}{13}$
C $\frac{12}{13}$
D $\frac{13}{12}$

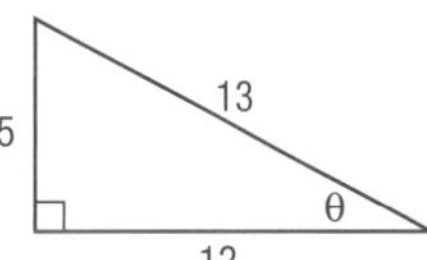

7 The value of *x* in this triangle is:

A $\frac{17}{\cos 38°}$
B $\frac{17}{\sin 38°}$
C 17 sin 38°
D 17 cos 38°

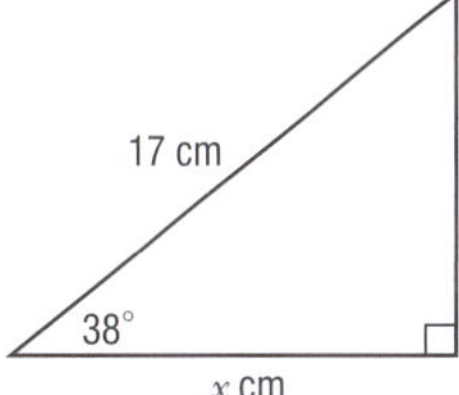

8 A navigator calculates an angle as 57°36′41″. Which one of the following would be a correct rounding of this angle?

A 57° B 57°36′ C 58°37′ D 58°

9 Which one of the following equations could be used to find an angle in the triangle PQR?

A $\sin(\angle QPR) = \frac{5}{13}$

B $\sin(\angle PQR) = \frac{12}{13}$

C $\cos(\angle QPR) = \frac{12}{13}$

D $\tan(\angle PQR) = \frac{5}{12}$

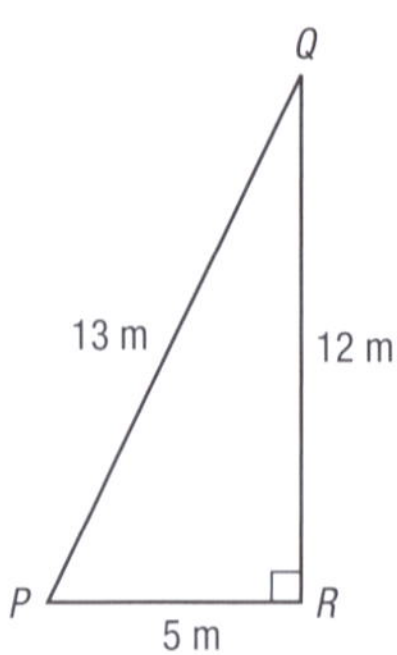

10 PQR is an isosceles triangle. The length of the side PQ can be found using:

A $\frac{18}{\sin 66°}$

B $\frac{36}{\cos 66°}$

C $\frac{18}{\cos 66°}$

D $36 \tan 66°$

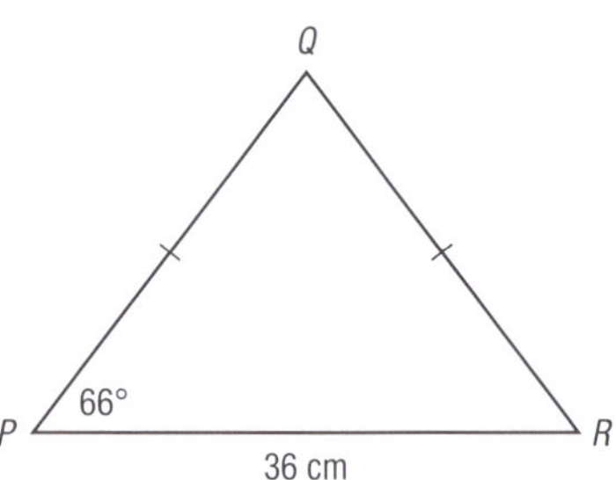

11 Which one of the following equations does not have a solution?

A $\sin \theta = 1.235$ B $\cos \theta = 0.4672$ C $\tan \theta = 0.8821$ D $\tan \theta = 2.4379$

12 $VABCD$ is a right, square pyramid and the point O is the centre of the base. Which one of the following angles is not a right angle?

A $\angle DOA$

B $\angle VOB$

C $\angle VDA$

D $\angle DAB$

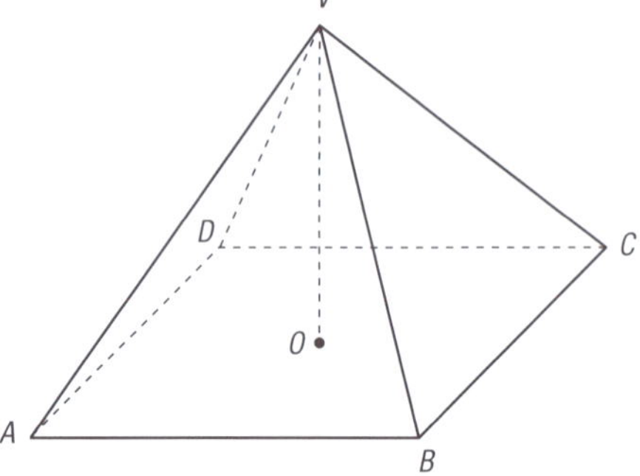

Short answer questions

1 Triangle ABC has a 90° angle at B.
Copy this diagram into your workbook.

a Mark in the 90° angle.

Using $\angle ACB$, mark in:

b the hypotenuse (H)

c the side adjacent (A) to the angle

d the side opposite (O) the angle.

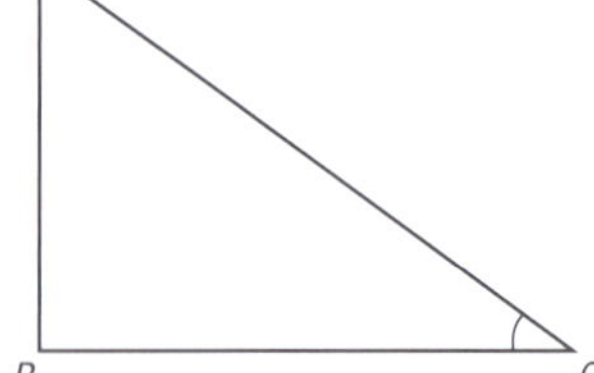

Again using $\angle ACB$, state which trigonometric ratio is represented by:

e $\frac{AB}{AC}$ f $\frac{AB}{BC}$ g $\frac{BC}{AC}$

2 Write the following, correct to three decimal places.

a $\cos 34°$ b $\sin 78°$ c $\tan 65°$

3 Find, in degrees and minutes, the angle whose:

a cosine is 0.5 b tangent is 0.622 c sine is 0.9860

4 A train track has a gradient of 1 in 25. That means that the track rises 1 metre for every 25 metres travelled horizontally. What angle does the track make with the horizontal?

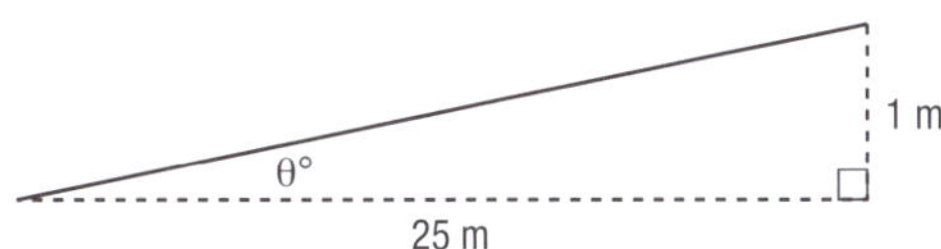

5 Find the value of θ, in degrees and minutes, in the triangle at right.

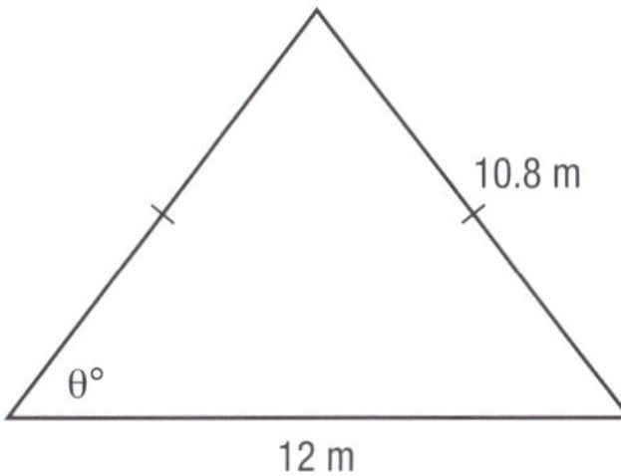

6 The diagonals of a rhombus bisect each other at right angles. If the diagonals of this rhombus are 20 cm and 30 cm long, find the value of θ, in degrees and minutes.

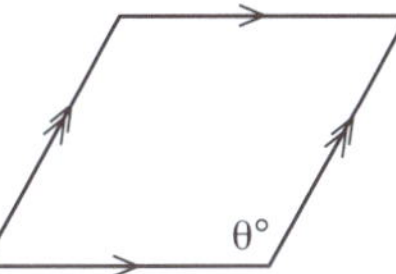

7 *ABCDEF* is an inclined plane with lengths as shown on the diagram. *M* is the midpoint of the side *DC*. Find, to the nearest minute, the size of:

a $\angle DAE$

b $\angle DMA$

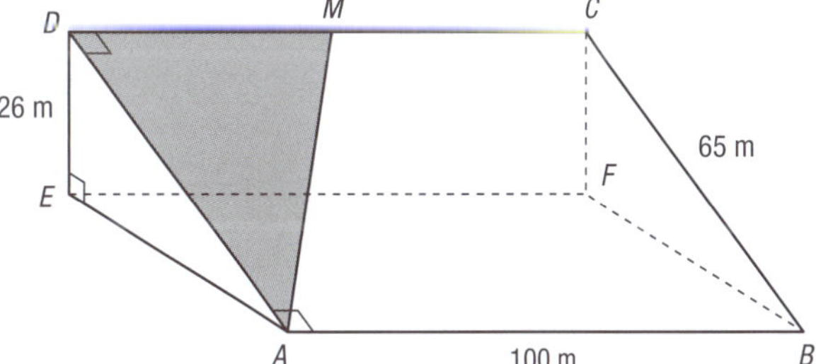

8 In the diagram at right, find the value of:

a θ

b *x*

c *y*

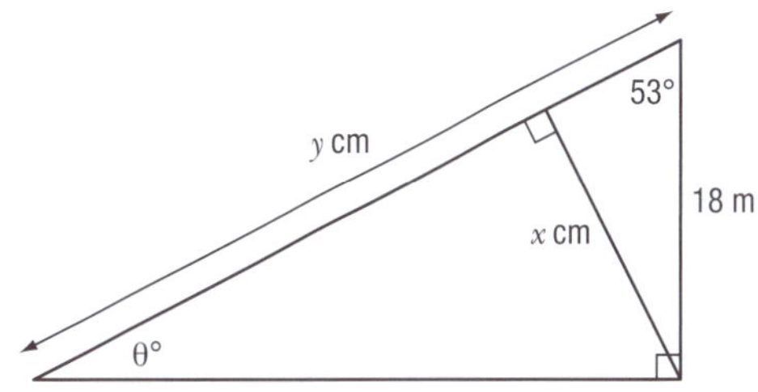

9 A hiker travels for 9.8 km in the direction N 72° E. How far east of her starting point is she now?

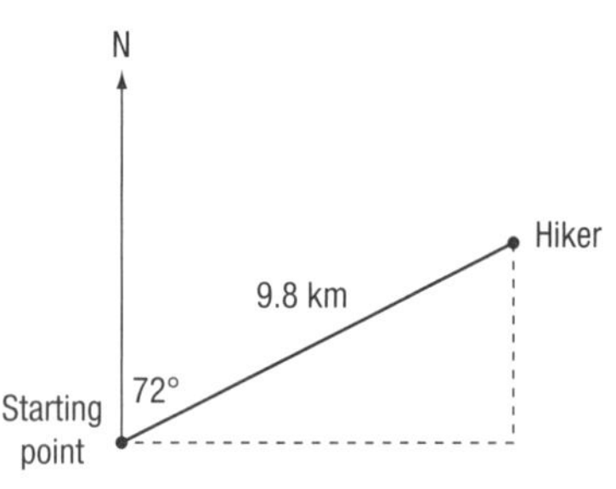

10 A garden in the shape shown needs to be fenced on all sides. Calculate the amount of fencing required, giving your answer correct to the nearest metre.

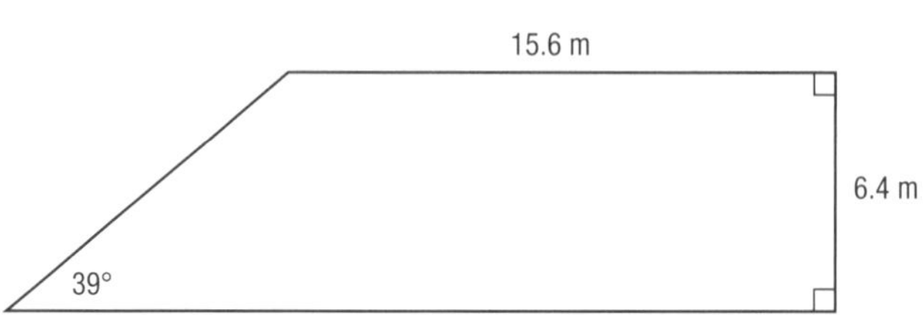

OPTION A: SURVEYING

The measurement of land distances and areas, and the production of maps, has increasingly become important to the way of life in Papua New Guinea. Most accurate surveying today is done with sophisticated instruments, however good estimates of distances and areas can be found using some basic, easily obtainable instruments and a knowledge of geometry and trigonometry.

This option introduces the trigonometry of non-right-angled triangles, map drawing and some of the methods of surveying that are the foundations of modern surveying techniques.

Angles of elevation and depression

An **angle of elevation** is measured from the horizontal 'upwards'.

B
Angle of elevation of point *B* from point *A*
A
Horizontal

An **angle of depression** is measured from the horizontal 'downwards'.

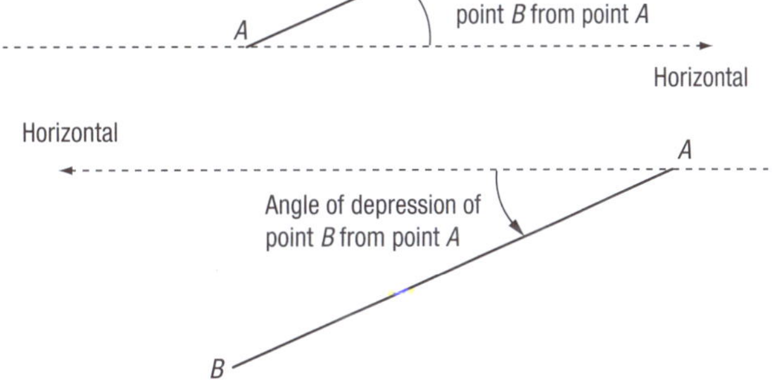

Example 1

The angle of elevation from a point 30 metres from the base of a building to the top of the building is 56°. How high is the building?

Answer

If the height of the building in the diagram is h metres, then h can be found using:

$$\tan 56° = \frac{h}{30}$$

$$h = 30 \tan 56°$$

$$= 44.48 \quad \text{correct to two decimal places}$$

Building height, h
Angle of elevation
56°
30 m

Example 2

From a point on top of a cliff two boats are visible in the same line of sight. If the cliff is 25 metres high and the angle of depression to the closer boat is 36° and to the other boat 21°, how far apart are the boats? (Answer to the nearest metre.)

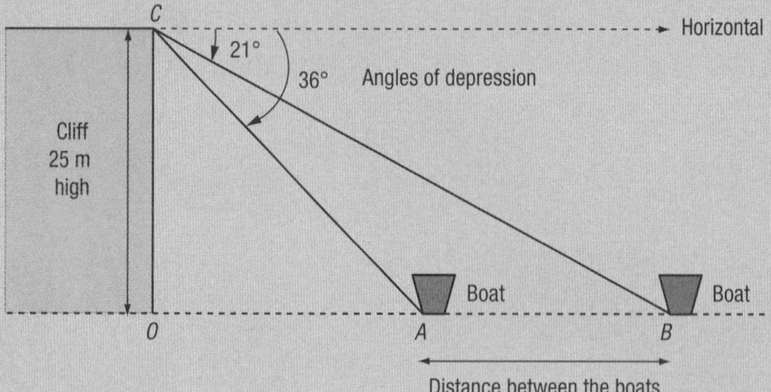

Answer

From the diagram the distance between the boats is $OB - OA$.
Drawing the triangles COB and COA separately to calculate OB and OA:

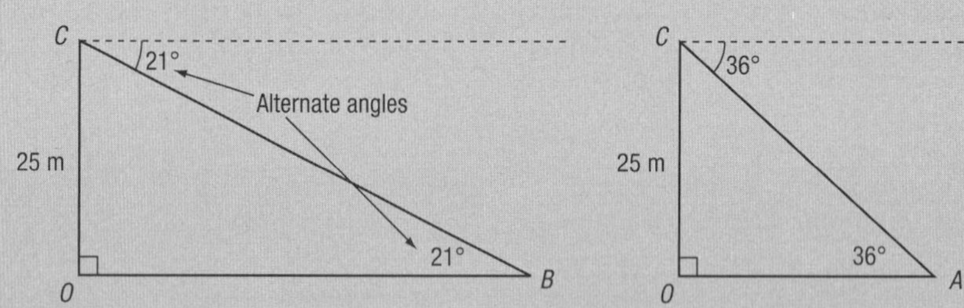

$$\tan 21° = \frac{25}{OB} \qquad \tan 36° = \frac{25}{OA}$$

$$OB = \frac{25}{\tan 21°} \qquad OA = \frac{25}{\tan 36°}$$

$$= 65.127 \text{ metres} \qquad = 34.410 \text{ metres}$$

Distance between the boats $= OB - OA$

$= 65.127 - 34.410$

$= 30.717$

The distance between the boats is 31 metres to the nearest metre.

EXERCISE

1 From a point 20 m from the base of its trunk, the angle of elevation to the top of a tree is 18°. How tall is the tree?

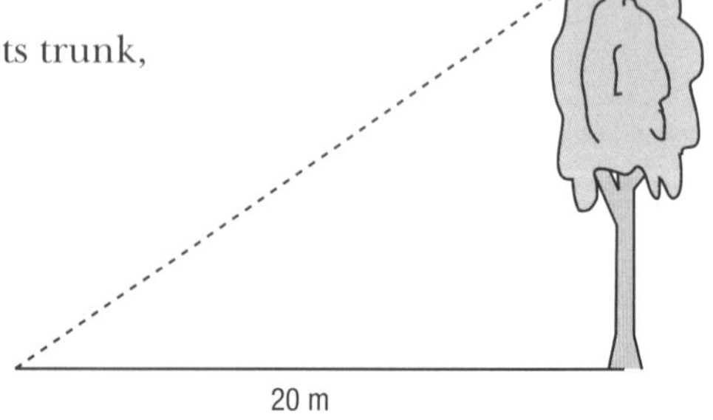

2 The angle of depression from the top of a building 30.2 m high to a landmark at ground level is 24.6°. Find the distance from the base of the building to the landmark.

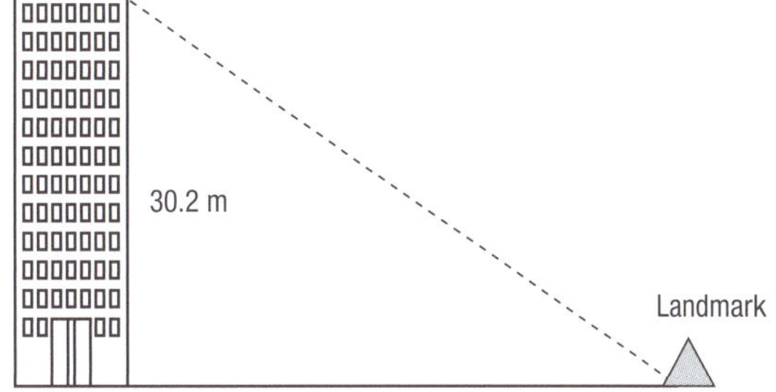

3 From the top of Building A the angle of elevation to the top of Building B is 34°, and the angle of depression to the base of Building B is 47°. If Building A and Building B are 52 m apart, find the height of Building B.

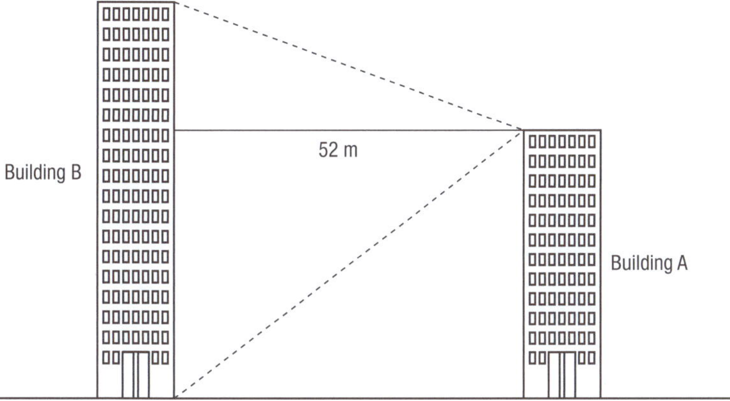

4 From the observation deck of a lighthouse the angle of depression of a ship travelling in a straight line towards the lighthouse is 12.6°. An hour later the angle of depression is 34.8°. If the observation deck of the lighthouse is 54 metres above sea level, how far has the ship travelled in the hour between observations?

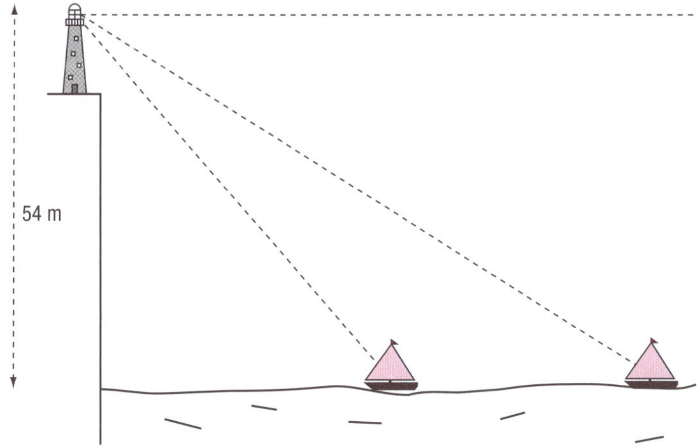

5 An aeroplane is observed directly overhead flying at 6000 m and, from the same position 1 minute later, the angle of elevation of the plane is observed to be 25°. How far has the plane flown in the minute it was observed?

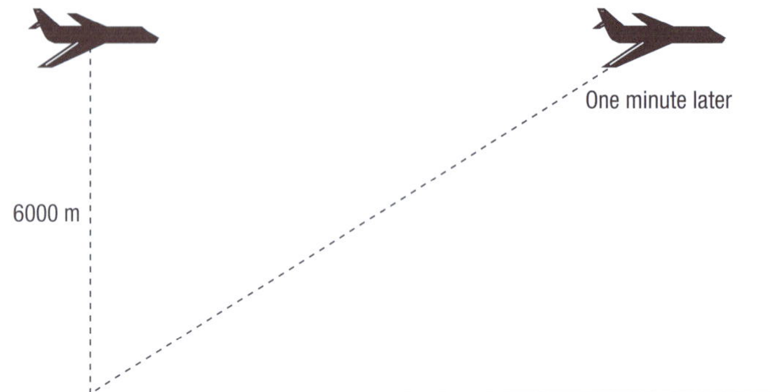

Lesson 2

The sine rule

For right-angled triangles, the lengths of sides and the magnitude of angles can be found using Pythagoras's rule and the trigonometric ratios (SOH CAH TOA). The **sine rule** is used to find the lengths of sides and the magnitude of angles in triangles that **are not right-angled triangles**.

Conventional labelling of triangles

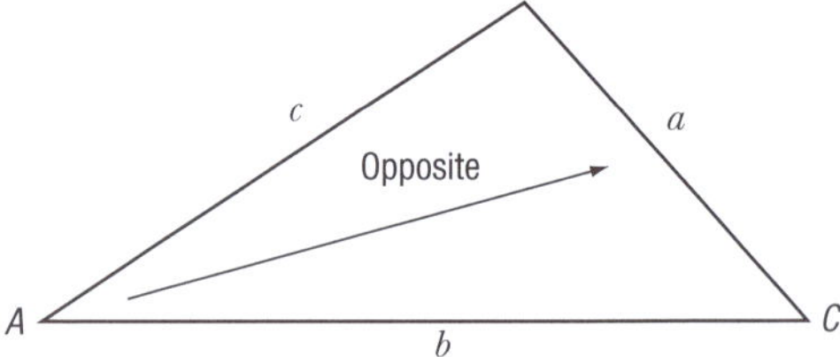

In the triangle *ABC*:

- the size of the angles is represented by the capital letters *A*, *B* and *C*
- the lengths of the sides of the triangle are represented by the lower case letters *a*, *b* and *c*
- the side of length *a* is 'opposite' the angle of size *A*, the side of length *b* is 'opposite' the angle of size *B*, and the side of length *c* is 'opposite' the angle of size *C*.

The sine rule

The sine rule tells us that in any triangle, the ratio of the length of a side and the sine of the angle opposite that side is the same. For any conventionally labelled triangle the sine rule can be written in two ways:

$$\frac{a}{\sin A} = \frac{b}{\sin B} = \frac{c}{\sin C} \quad \text{OR} \quad \frac{\sin A}{a} = \frac{\sin B}{b} = \frac{\sin C}{c}$$

The sine rule is used to find unknown side lengths and angles in a triangle where:

- two side lengths and a non-included angle are known, OR

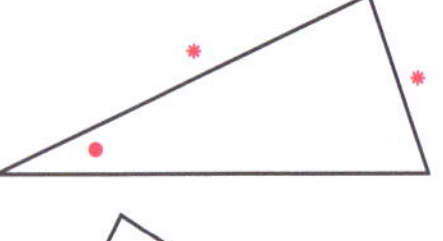

- two angles (and hence three angles) and one side are known.

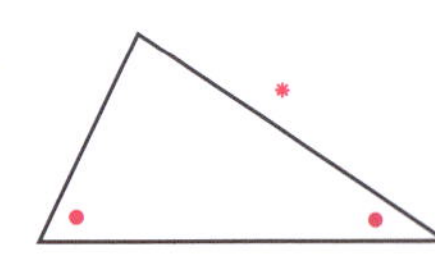

Note: Only two components of this formula are used at one time.

We use the formula $\frac{a}{\sin A} = \frac{b}{\sin B} = \frac{c}{\sin C}$ if we want to find the length of a side, and the formula $\frac{\sin A}{a} = \frac{\sin B}{b} = \frac{\sin C}{c}$ if we want to find the size of an angle.

Example 1

For the triangle given, find *B*, and hence *A*.

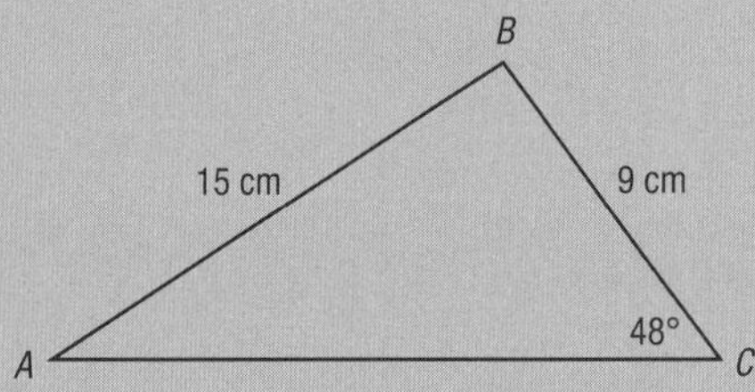

Answer

B and *A* are both angles in this triangle.

For this triangle we are given the length of two of the sides and the magnitude of a non-included angle: $b = 9$ cm, $c = 15$ cm and $C = 48°$.

To use the formula we need to know the magnitude of one pair of 'angle and its opposite side', which we have with *C* and *c*.

Since we are also given the side length *b*, we can find the magnitude of the opposite angle *B*.

The components of the sine rule that we will use are:

$$\frac{\sin B}{b} = \frac{\sin C}{c}$$

Substituting the known values:

$$\frac{\sin B}{9} = \frac{\sin 48°}{15}$$

Making sin *B* the subject:

$$\sin B = \frac{9 \sin 48°}{15}$$

$$= 0.4459$$

Using the tables or a calculator:

$B = \sin^{-1}(0.4459)$

$= 26.48°$ correct to two decimal places

$= 26°29'$ correct to the nearest minute

To find angle *A*, we use the fact that the three angles of a triangle sum to 180°.

$A = 180° - B - C$

$= 180 - 26°29' - 48°$

$= 105°31'$ to the nearest minute

$= 105.52°$ correct to two decimal places

Example 2

Find x in the given triangle.

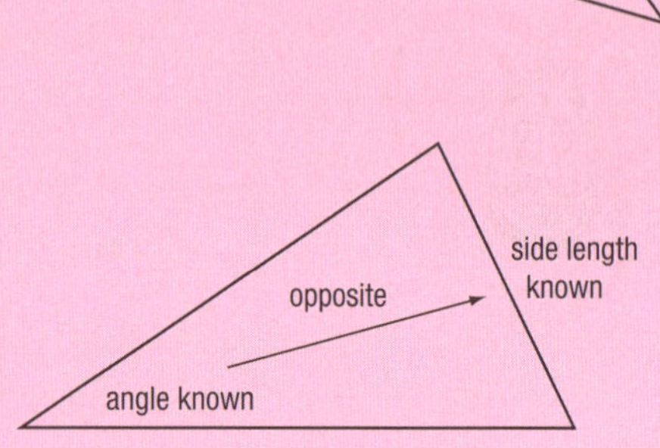

Answer

Recognising that each of the components of the formula is the ratio of a pair of *side length* and *the sine of the angle opposite*, and that to use the formula we need to know one of these pairs and another angle or side length, it is possible to use the sine rule without labelling the triangle in the conventional way.

Remember

One pair must be known to be able to use the sine rule.

In this triangle we have one pair of known 'angle and opposite side': 42° and 12 m.

The angle opposite the side marked x is not given but can easily be found using 'the three angles of a triangle sum to 180°'.

The angle opposite x is 180 – 91 – 42 = 47°.

We now know one 'pair' (42° and 12 m) and the angle opposite the side of length x.

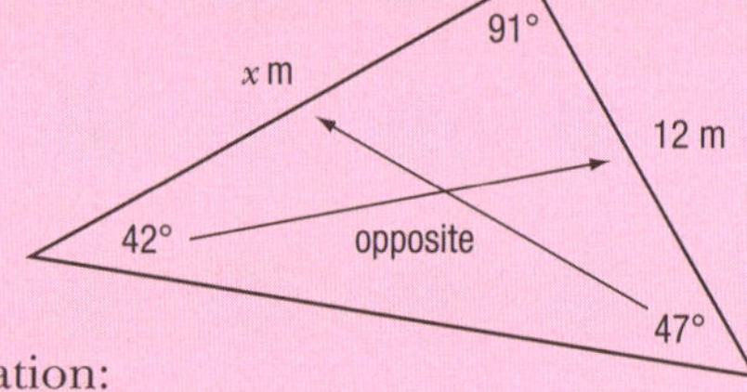

Using the sine rule to calculate x gives the equation:

$$\frac{x}{\sin 47^\circ} = \frac{12}{\sin 42^\circ}$$

$$x = \frac{12 \sin 47^\circ}{\sin 42^\circ}$$

$$= 13.115\ldots$$

The side marked x is 13.12 metres long, correct to two decimal places.

Using tables to find the sine and cosine of angles that are greater than 90° but less than 180°

The sine and cosine tables on pages 337–41 are designed to be used for finding the lengths of sides and angles in **right-angled triangles**. Right-angled triangles have one angle that is 90° and the other two angles are less than 90°, hence the tables only give values for angles in the range 0° to 90°.

However there are triangles that contain angles that are greater than 90° but less than 180° and it is possible to find sine and cosine values for these angles.

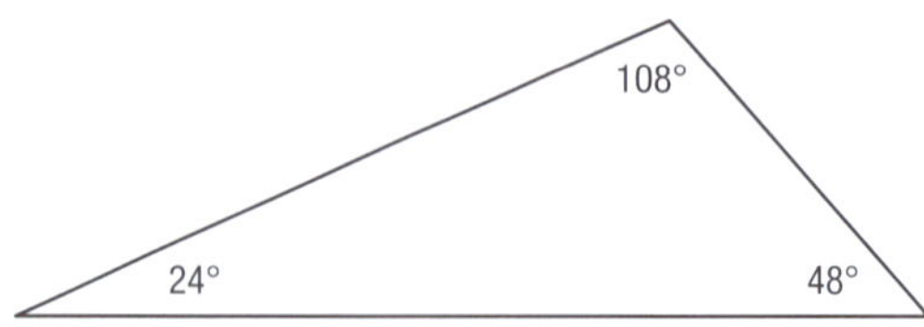

For example, to find a side length (using the sine rule or the cosine rule) in this triangle it may be necessary to find sin 108° or cos 108°. The mathematics that is the foundation for finding these values will be studied in senior years.

Sine and cosine values for angles that are greater than 90° but less than 180° can be found from the tables using either of the following rules.

If θ is greater than 90° but less than 180° then:

$\sin\theta = \sin(180° - \theta)$

$\cos\theta = {}^{-}\cos(180° - \theta)$

Example

Use the tables to find:

a sin 108°

b cos 145°.

Answer

a $\sin 108° = \sin(180° - 108°)$
$= \sin 72°$
$= 0.9511$ reading the value of sin 72° from the table

b $\cos 145° = {}^{-}\cos(180° - 145°)$
$= {}^{-}\cos 35°$
$= {}^{-}0.8192$ reading the value of cos 35° from the table. Note that the cosine value is negative.

Using the cosine tables to find the angle when the cosine value is negative

If you have a negative cosine value for an angle in a triangle then the angle is going to be between 90° and 180°.

To find this angle:

i Ignore the negative sign and use the table to find the angle between 0° and 90° associated with the positive value.

ii Subtract this angle from 180° to give the required angle in the range 90° to 180°.

Example

Find the angle in a triangle whose cosine is ⁻0.4561.

Answer

Ignoring the negative sign and finding the angle whose cosine is 0.4561:

$\cos^{-1}(0.4561) = 62°52'$

The required angle will be $180° - 62°52' = 117°8'$.

EXERCISE

Give all answers correct to two decimal places.

1 Copy and complete the following calculations using the sine rule.

a Find b (a side length).

We know the side length $c = 20$ m and its opposite angle $C = 31°$.

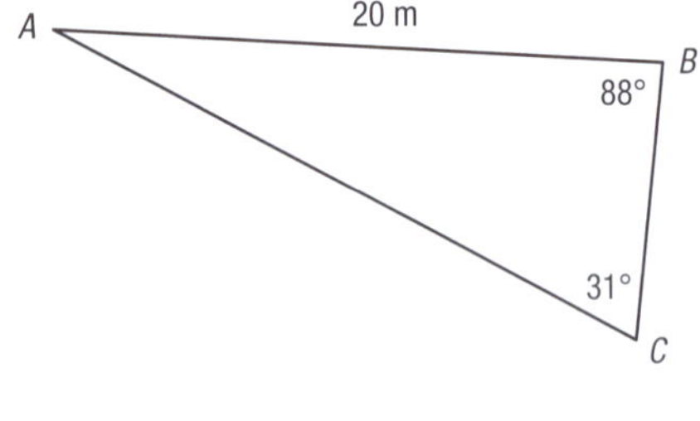

$$\frac{b}{\sin B} = \frac{c}{\sin C}$$

$$\frac{b}{\sin 88°} = \frac{\quad}{\sin \quad}$$

$$b = \frac{\quad}{\sin \quad} \times \sin 88°$$

$$b = \ldots\ldots$$

b Find A (an angle).

We know the side length $b = 40$ cm and its opposite angle $B = 60°$.

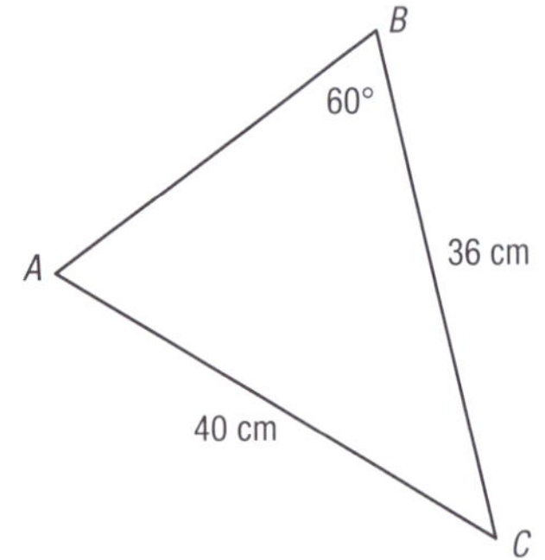

$$\frac{\sin A}{a} = \frac{\sin B}{b}$$

$$\frac{\sin A}{36} = \frac{\sin \quad}{\quad}$$

$$\sin A = \ldots \times 36$$

$$\sin A = \ldots\ldots$$

$$A = \ldots\ldots$$

2 Use the sine rule in the form $\frac{a}{\sin A} = \frac{b}{\sin B} = \frac{c}{\sin C}$ to find the specified lengths in the following triangles.

a side a

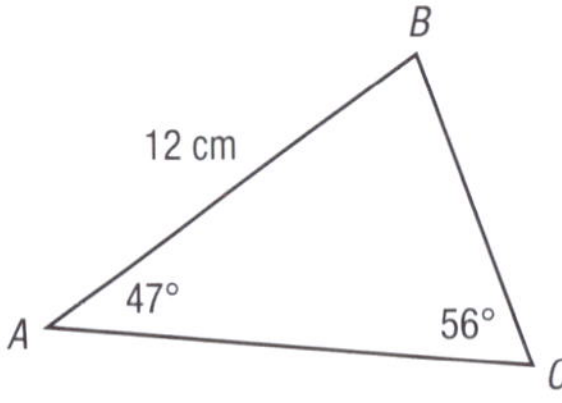

b side a

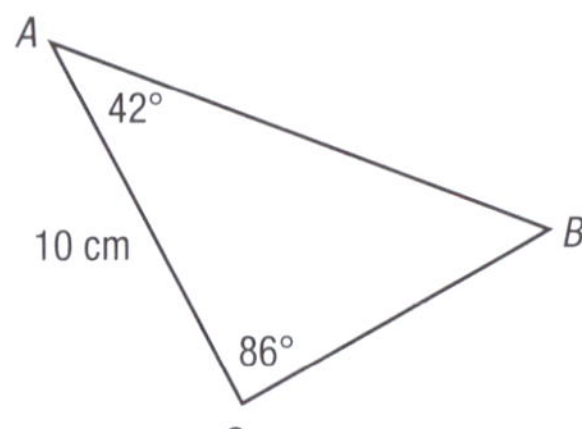

c side c

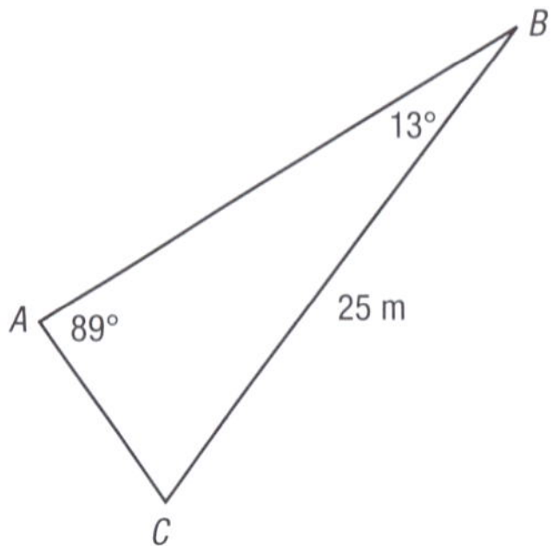

d side c

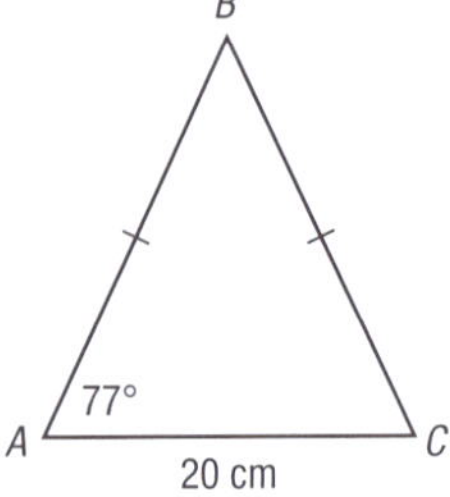

3 Use the sine rule in the form $\frac{\sin A}{a} = \frac{\sin B}{b} = \frac{\sin C}{c}$ to find the unknown angles in the following triangles.

Remember

To use the sine tables for angles between 90° and 180°: if 90° < θ < 180°, use sin θ = sin (180° − θ).

a **b** **c**

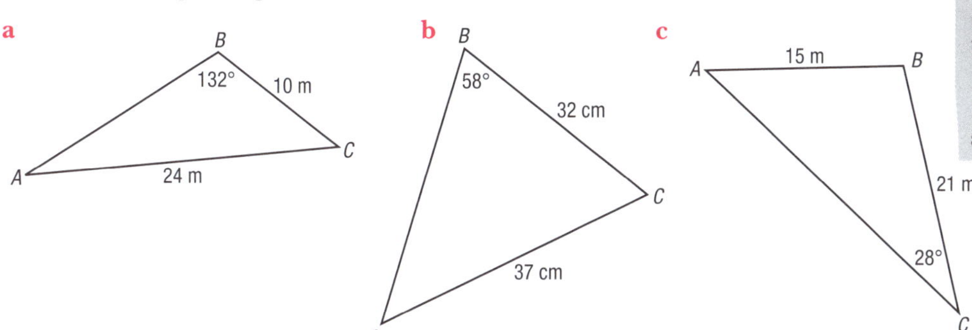

4 In a conventionally marked triangle ABC, $a = 33$ cm, $b = 25$ cm and $A = 110°$. Draw the triangle ABC and find B, C and c.

5 Use the sine rule to find the value of x in each of the following triangles.

a

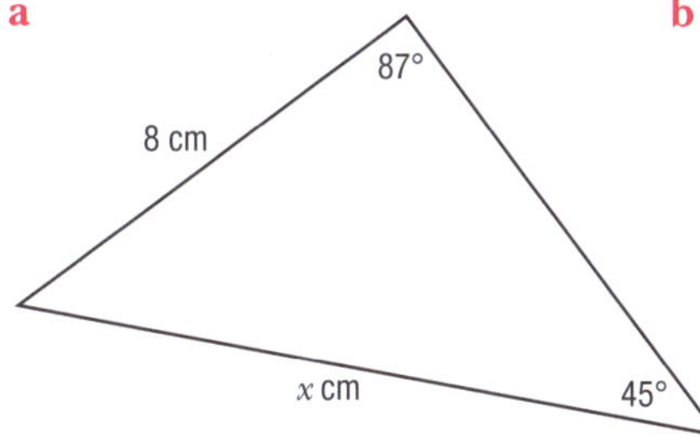

b

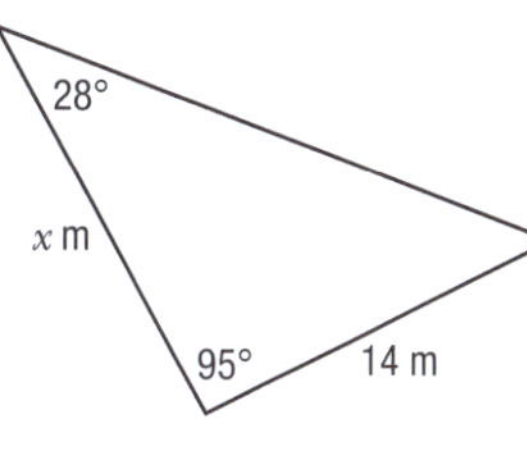

c

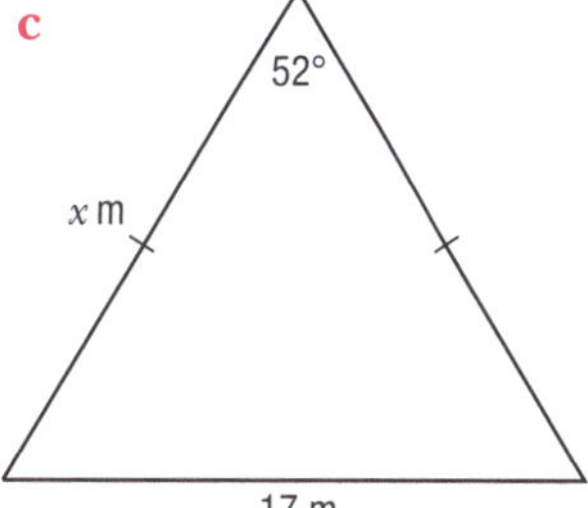

6 Use the sine rule to find the value of θ in each of the following triangles.

a

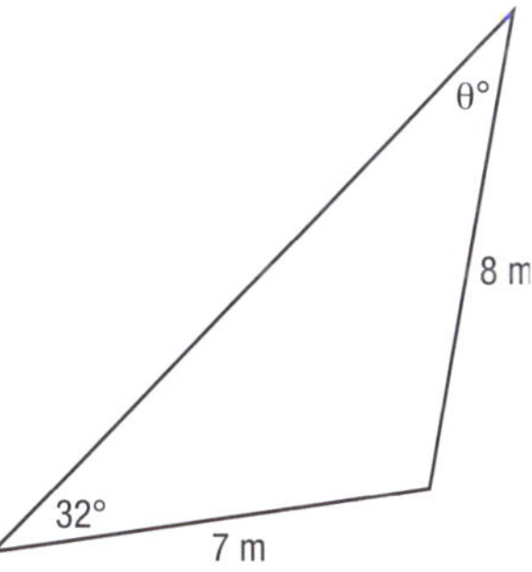

b

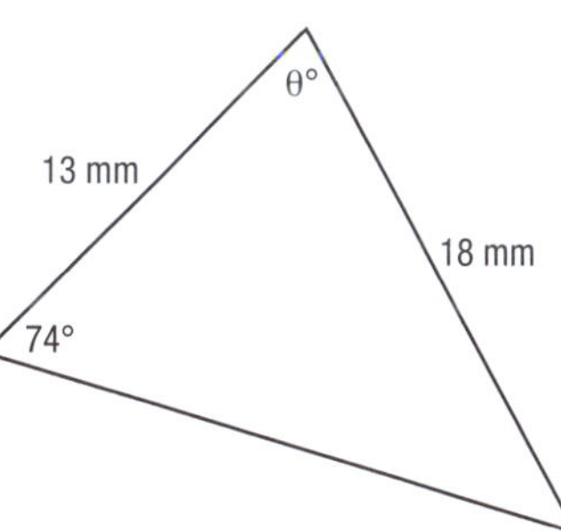

c

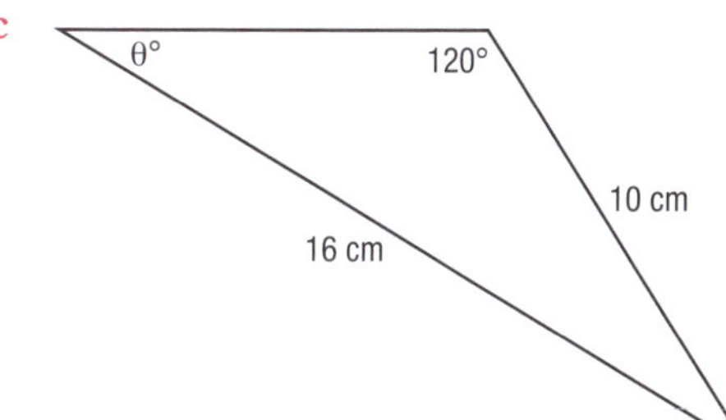

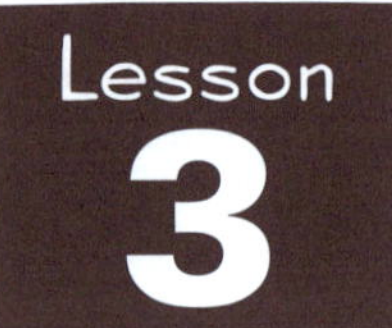

Lesson 3

The cosine rule

The **cosine rule** is used to find the length of sides and the magnitude of angles in triangles that **are not right-angled triangles**.

The cosine rule is used:

1 to find the length of the third side when the length of two sides is known and the magnitude of the included angle is known. (The included angle is opposite the unknown side.)

2 to find any of the angles in a triangle when the length of all three sides is known.

To find the length of the third side

For a conventionally labelled triangle with b, c and A known, the cosine rule to find a is:

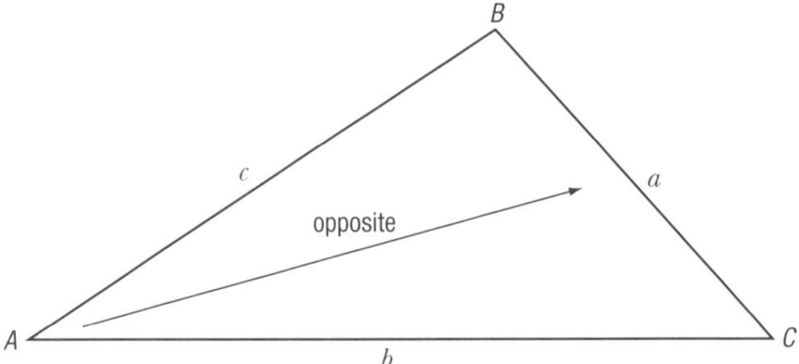

$$a^2 = b^2 + c^2 - 2bc \cos A$$

To find b, given values for a, c and B the cosine rule would be $b^2 = a^2 + c^2 - 2ac \cos B$

To find c, given values for a, b and C the cosine rule would be $c^2 = a^2 + b^2 - 2ab \cos C$

Note: It is important to see the 'pattern' in this formula and not rely on labelling all triangles in the conventional way.

To find an angle

If the lengths of the three sides of the triangle are known, the rule can be re-arranged to make cos A the subject:

$$\cos A = \frac{b^2 + c^2 - a^2}{2bc}$$

The rule in this form would be used to find angle A.

To find B, given values for a, b and c: $\cos B = \dfrac{a^2 + c^2 - b^2}{2ac}$

To find C, given values for a, b and c: $\cos C = \dfrac{a^2 + b^2 - c^2}{2ab}$

Once again it is important to see the pattern in the rule and not rely on labelling all triangles in the conventional way.

Example 1

Find the value of a for the given triangle.

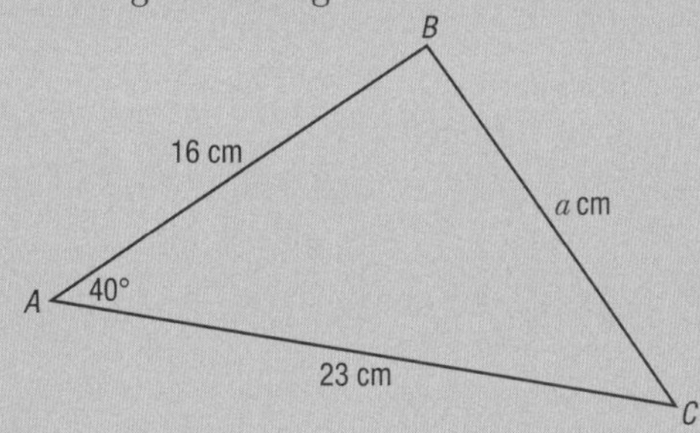

Answer

Two side lengths, $b = 23$ cm and $c = 16$ cm, and the included angle, $A = 40°$, are known so we have sufficient information to use the cosine rule to find a.

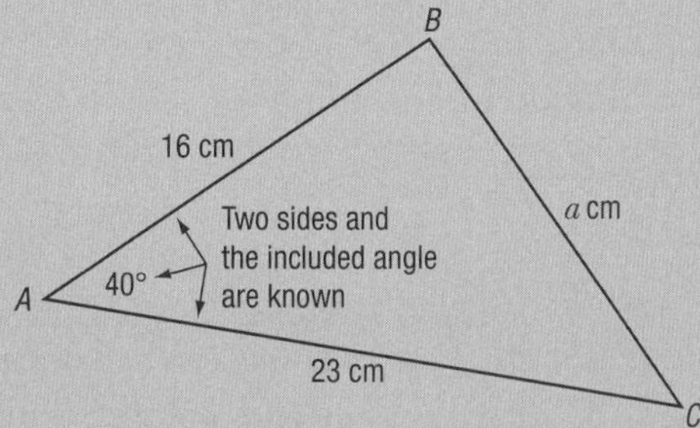

Substituting in $a^2 = b^2 + c^2 - 2ab \cos A$ gives:

$a^2 = 23^2 + 16^2 - 2 \times 23 \times 16 \times \cos 40°$

$= 221.1912...$

$a = \sqrt{221.1912}$

$= 14.873\ldots$

Hence a is 14.87 cm, correct to three decimal places.

Example 2

Find x in the given triangle.

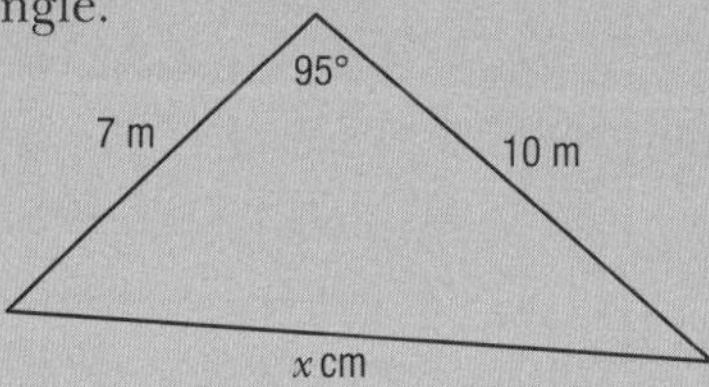

Answer

This is a situation where the cosine rule can be used to find the length of x because it is opposite the given angle and the lengths of the other two sides are known.

The cosine rule:

$x^2 = 7^2 + 10^2 - 2 \times 7 \times 10 \times \cos 95°$

$= 161.2018$

$x = \sqrt{161.2018}$

$= 12.696\ldots$

The side marked x is 12.70 metres, correct to two decimal places.

Example 3

Find the magnitude of the angles in this triangle.

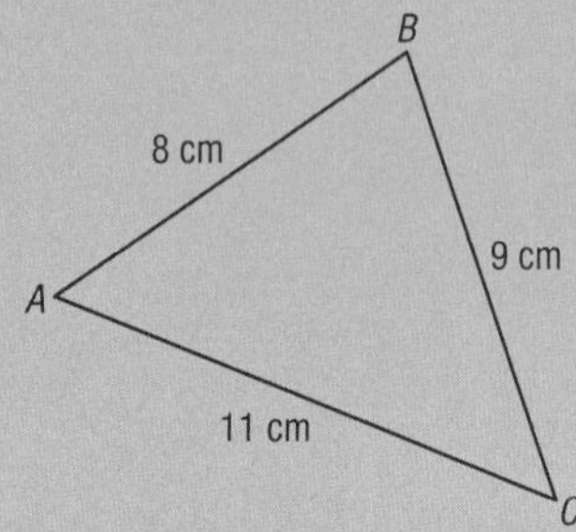

Answer

The lengths of all three sides are known, $a = 9$ cm, $b = 11$ cm and $c = 8$ cm, so we can use the cosine rule to find angles.

To find C we can substitute the known values in $\cos C = \dfrac{a^2 + b^2 - c^2}{2ab}$.

$$\cos C = \frac{9^2 + 11^2 - 8^2}{2 \times 9 \times 11}$$
$$= 0.6969\ldots$$
$$C = \cos^{-1}(0.6969\ldots)$$
$$= 45.82° \quad \text{correct to two decimal places}$$
$$= 45°49' \quad \text{to the nearest minute}$$

A similar process is used to find B:

$$\cos B = \frac{a^2 + c^2 - b^2}{2ac}$$
$$\cos B = \frac{9^2 + 8^2 - 11^2}{2 \times 9 \times 8}$$
$$= 0.1666\ldots$$
$$B = \cos^{-1}(0.1666\ldots)$$
$$= 80.41° \quad \text{correct to two decimal places}$$
$$= 80°24' \quad \text{to the nearest minute}$$

To find A we can use the fact that the three angles of a triangle sum to 180°.

$$A = 180° - 45.816° - 80.406°$$
$$= 53.778°$$
$$= 53.78° \quad \text{correct to two decimal places}$$
$$= 53°47'$$

Note: In any triangle the longest side is opposite the largest angle, and the shortest side is opposite the smallest angle. This provides a good check for your answers. In this triangle, $b = 11$ cm is the longest side and it is opposite the largest angle, $B = 80.41°$. Similarly, the smallest angle $C = 45.82°$ is opposite the smallest side, $c = 8$ cm.

EXERCISE

Give all answers correct to two decimal places.

1 Use the cosine rule to find the length of the unknown side in these triangles.

a

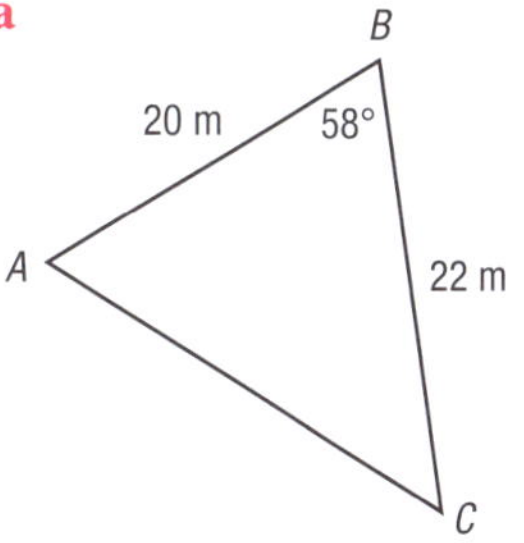

b

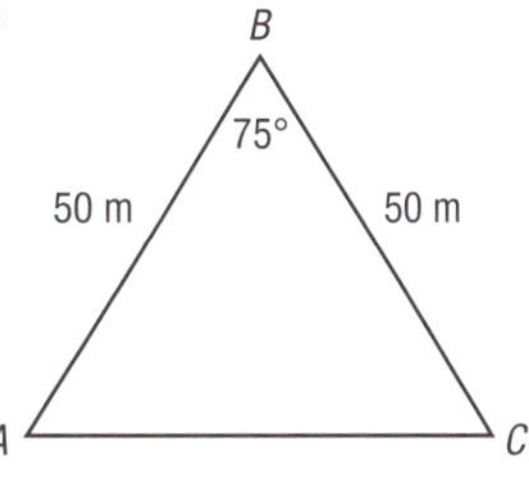

c

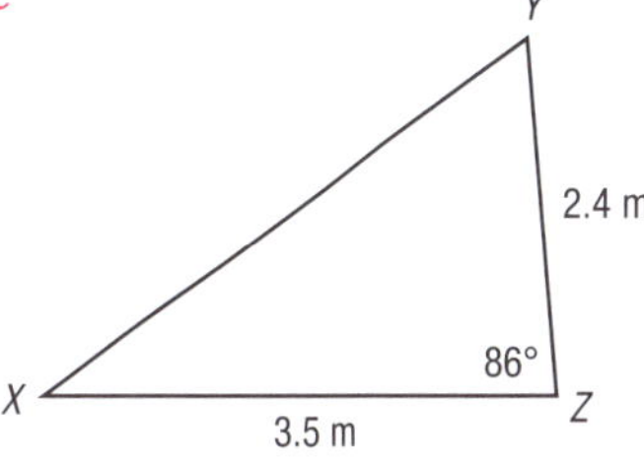

2 If $a = 41$ cm, $b = 28$ cm and $C = 73°$ in a conventionally labelled triangle, draw the triangle and find c.

3 Find the size of all the angles in the following triangles.

a

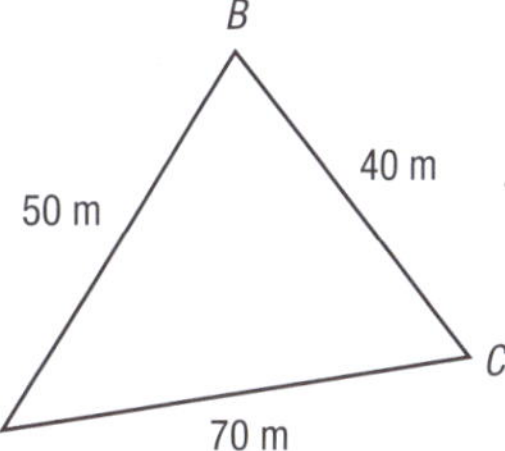

b

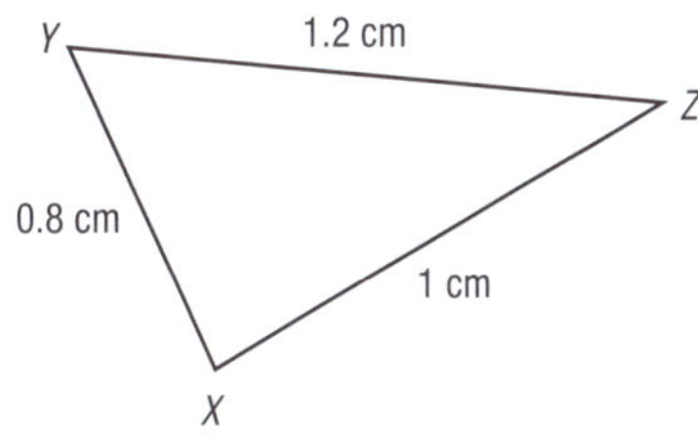

4 A conventionally labelled triangle has sides $a = 41$ m, $b = 37$ m and $c = 64$ m. Draw the triangle and find A.

5 Find the length of the shorter diagonal of this parallelogram.

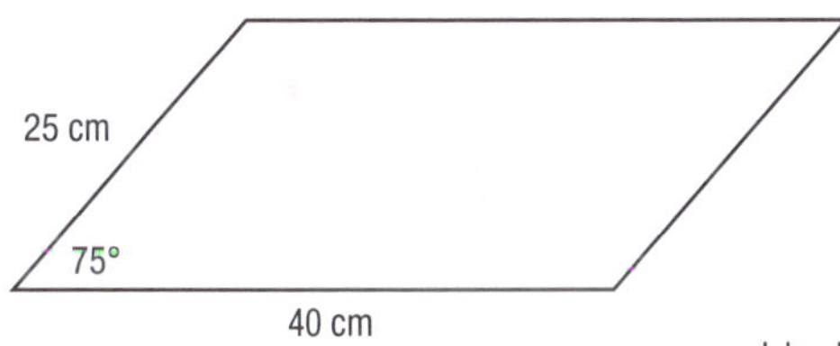

6 A boat travels for 5 km due east from its home port and then changes course and travels 7 km to reach an island. The boat sets a straight course from the island for the 10 km to the home port. Find the value of θ and hence the compass bearing of the direction the boat travels on the trip back to home port.

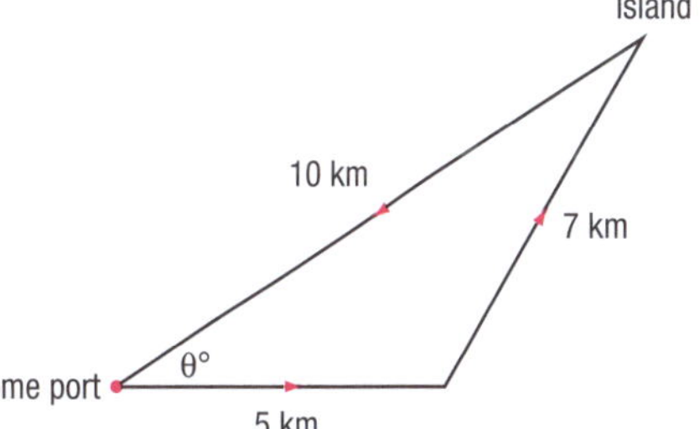

7 In this triangle, find the magnitude of the smallest angle.

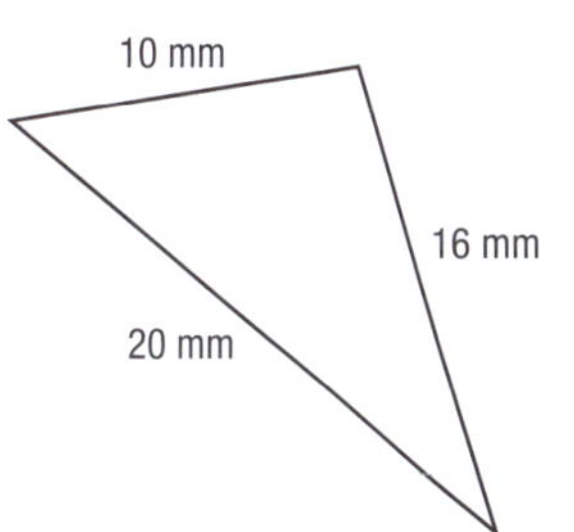

Remember

In any triangle the longest side is opposite the largest angle and the shortest side is opposite the smallest angle.

*Using the sine and cosine rules in more complex problems

For problems that are not straightforward, it is important to develop a strategy for solving the problem before starting. More than one calculation is usually necessary to solve each of the problems in the exercise, and you will need to decide whether to use the sine or the cosine rule. A calculator is recommended.

Example 1

Find AD in the given triangle.

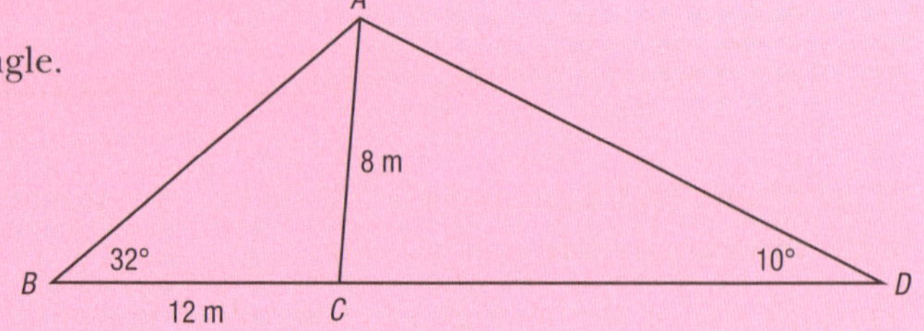

Answer

In the triangle ACD containing the side AD we need the angle opposite AD, i.e. $\angle ACD$.

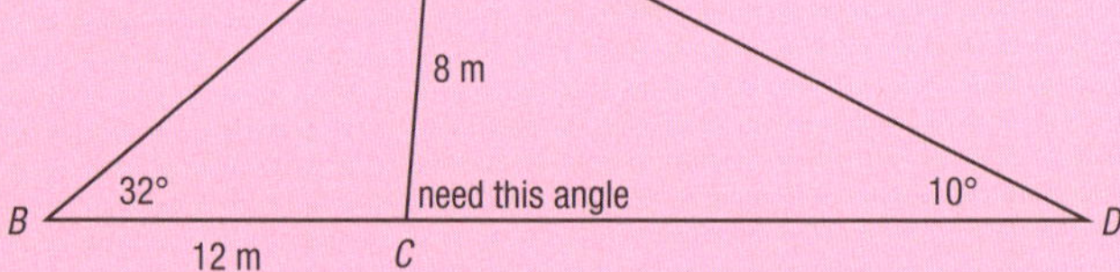

To find $\angle ACD$ we can use the sine rule in triangle ABC to find θ, then $\angle ACB$, then $\angle ACD$.

$$\frac{\sin\theta}{12} = \frac{\sin 32^\circ}{8}$$

$$\sin\theta = \frac{12 \times \sin 32^\circ}{8}$$

$$\theta = 52.6438\ldots$$

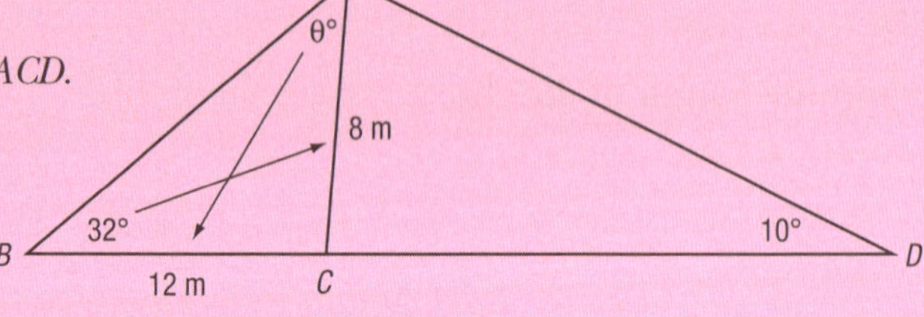

Hence $\angle ACB = 180^\circ - 32^\circ - 52.644^\circ = 95.356^\circ$
and $\angle ACD = 180^\circ - 95.356^\circ = 84.644^\circ$

Using the sine rule to find AD:

$$\frac{AD}{\sin 84.644^\circ} = \frac{8}{\sin 10^\circ}$$

$$AD = \frac{8 \sin 84.644^\circ}{\sin 10^\circ} = 45.869\ldots$$

The side AD is 45.87 m long, correct to two decimal places.

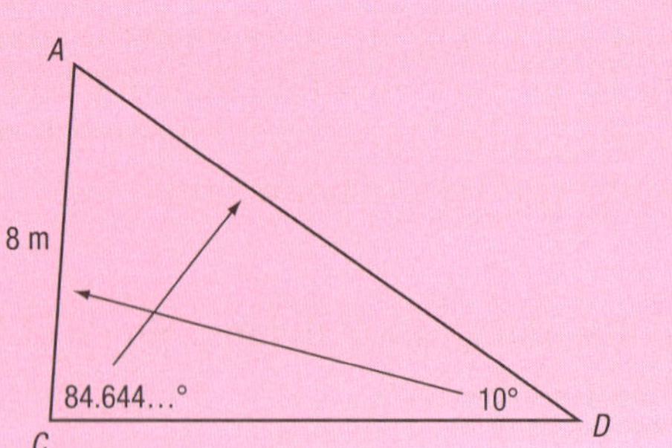

Example 2

A parallelogram has adjacent sides of length 15 cm and 21 cm, and the included angle is 65°. Find the length of the diagonals of this parallelogram.

Answer

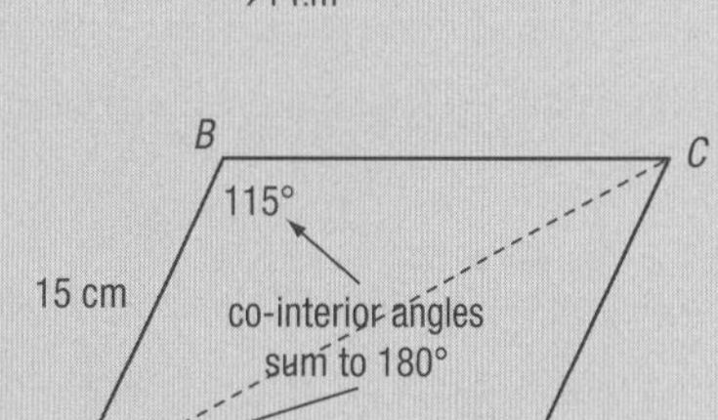

The shortest of the diagonals is the unknown side in the triangle *BAD*. We know two sides and the included angle in this triangle, so we can use the cosine rule to find the other side.

$BD^2 = 15^2 + 21^2 - 2 \times 15 \times 21 \times \cos 65°$

$= 399.750$

$BD = \sqrt{399.750}$

$= 19.934\ldots$

The shortest diagonal is 19.934 cm, correct to three decimal places.

The other diagonal is *AC* and we will need to know either $\angle ABC$ or $\angle ADC$ to find this length.

In this parallelogram $\angle BAD$ (= 65°) and $\angle ABC$ are co-interior angles and so sum to 180°.

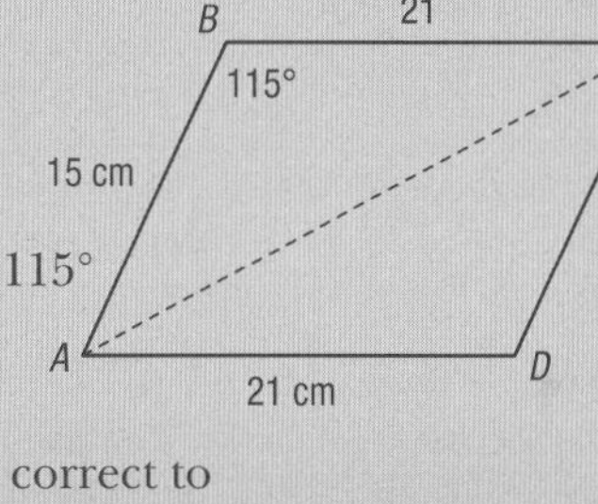

$\angle ABC = 180 - \angle BAD$

$= 180° - 65°$

$= 115°$

Using the cosine rule:

$AC^2 = 15^2 + 21^2 - 2 \times 15 \times 21 \times \cos 115°$

$= 932.2495\ldots$

$AC = 30.5327\ldots$

The longest diagonal is 30.533 cm, correct to three decimal places.

B 21 C
115°
15 cm
A 21 cm D

Remember

Using the cosine table for angles between 90° and 180°:

$\cos\theta$
$= {}^{-}\cos(180 - \theta)$

$\cos 115°$
$= {}^{-}\cos 65°$
$= {}^{-}0.4226$

EXERCISE

1 The adjacent sides of a parallelogram are 18 and 27 cm long, and the angle between these sides is 58°. Draw the diagram and calculate the length of the longest diagonal of this parallelogram.

2 Find the size of the angle *ABC* and hence the length *BD* in the given figure.

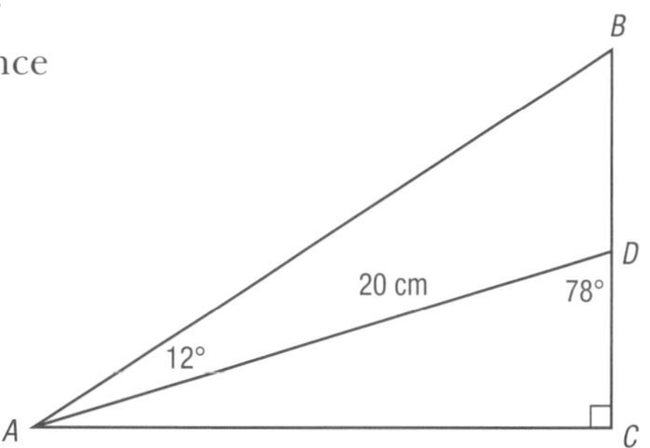

3 Find the value of x in the given figure.

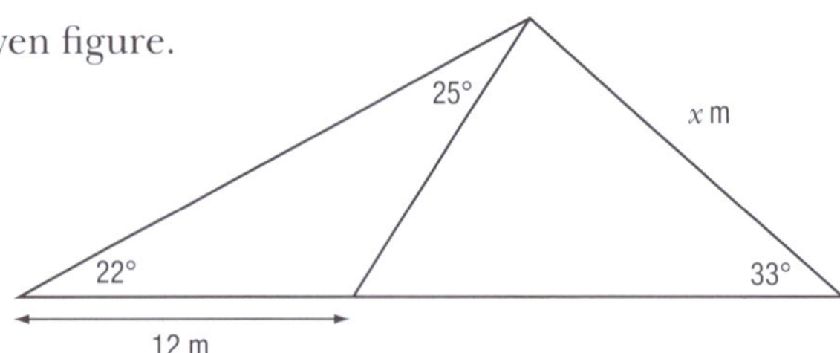

4 *ABCDE* is a regular pentagon of side length 10 cm.

a Find the size of the interior angle of the pentagon.

b Use the cosine rule to find the length *AC*.

c Calculate the size of $\angle ACE$.

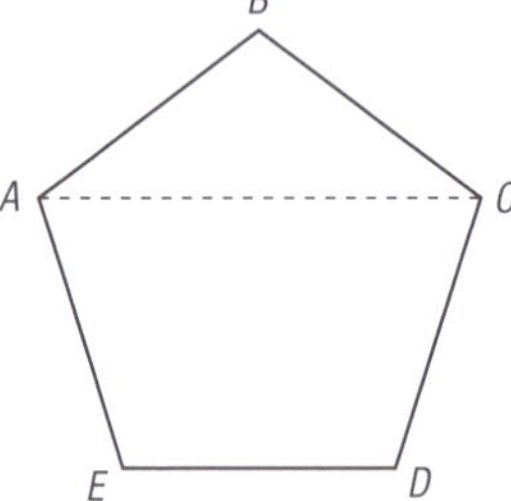

5 Find:

a the length of *BD*

b the size of $\angle ABC$ given that $AC = 18$ m.

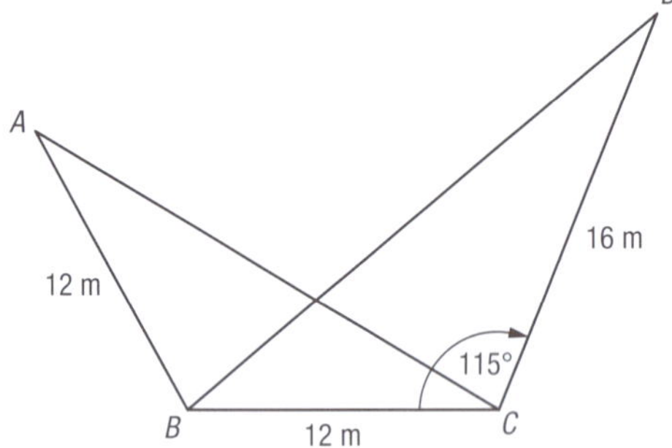

6 Find the values of x and y in the given figure.

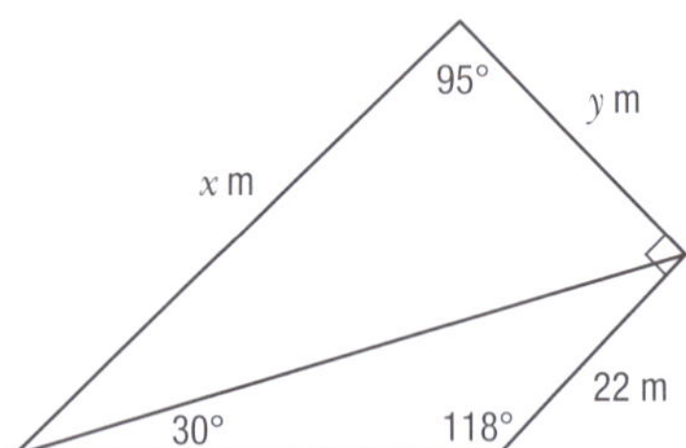

7 Find the value of x.

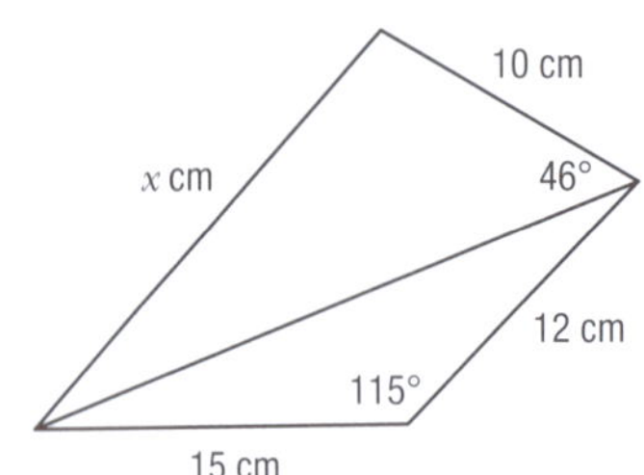

8 Find the length of the side:

a AD b AC.

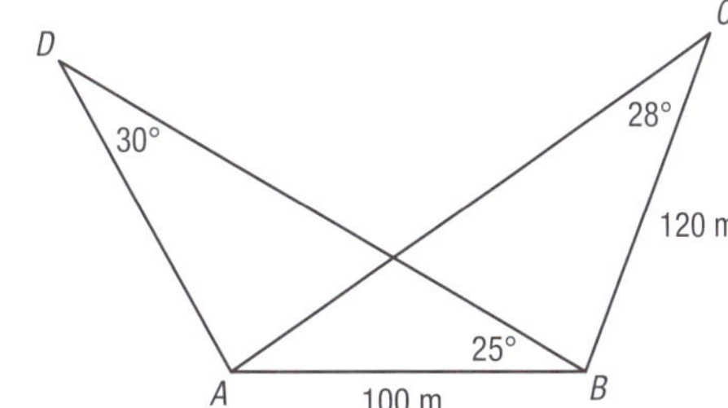

9 Find the value of x in the given figure.

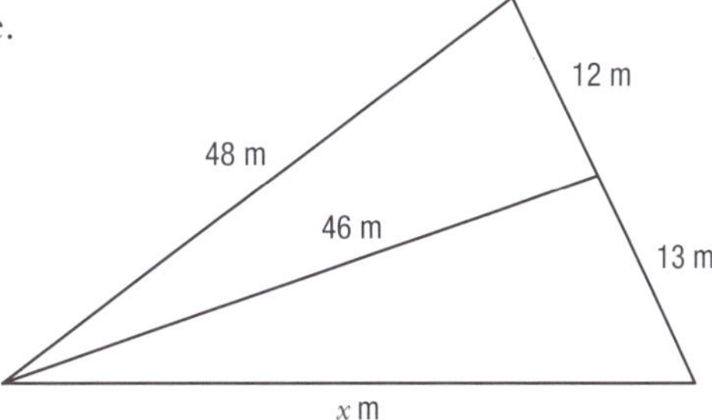

Area of a triangle

You should be familiar with the formula below for finding the area of a triangle.

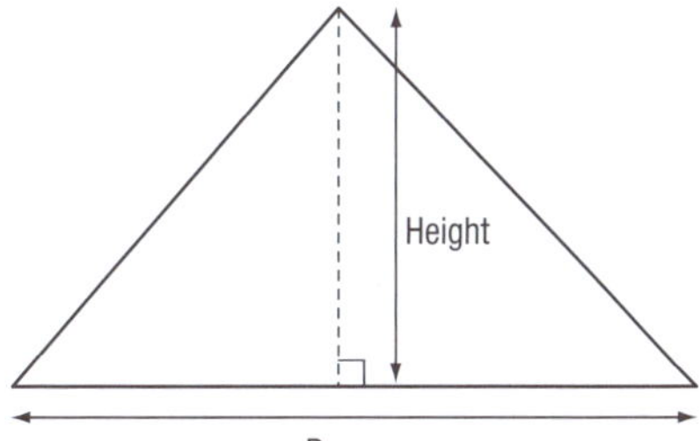

$$\text{Area of triangle} = \frac{1}{2} \times \text{base} \times \text{height}$$

For many triangles the perpendicular height is not known, but the length of some of the sides and the magnitude of one or more of the angles is known. We can find a formula for the area of a triangle in these cases.

Consider a conventionally labelled triangle with the perpendicular height, h.

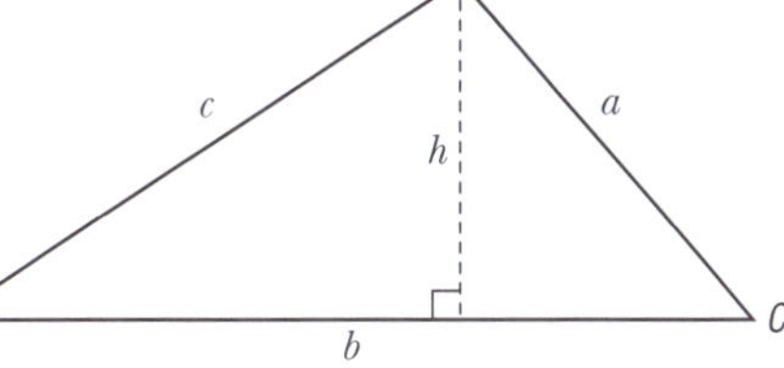

Area of triangle $= \frac{1}{2}b \times h$ ....(equation i)

but $\sin A = \frac{h}{c}$, so $h = c \sin A$ ....(equation ii)

Substituting (ii) into (i) gives:

$$\text{Area of triangle} = \frac{1}{2}bc \sin A$$

This formula could be used to find the area of the triangle when the length of two of the sides, b and c, and the magnitude of the included angle, A, is known.

Similarly, if a, b and C are known then the formula is:

Area of triangle $= \frac{1}{2}ab \sin C$

and if a, c and B are known then the formula is:

Area of triangle $= \frac{1}{2}ac \sin B$

In all cases it is necessary to know the length of two of the sides and the magnitude of the included angle to use this formula to find the area.

Example 1

Find the area of the following triangles.

a

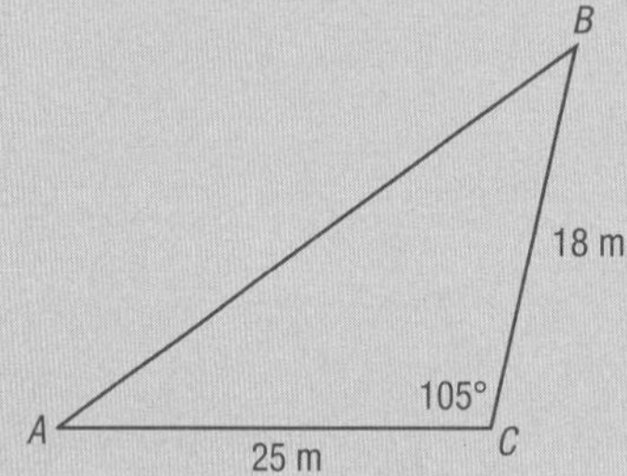

b

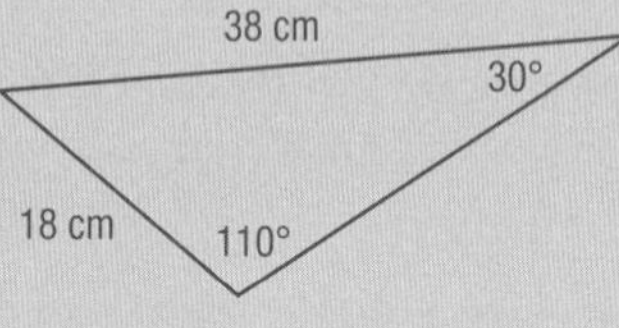

Answer

a We know the length of two sides, $a = 18$ cm and $b = 25$ cm, and the magnitude of the included angle, $C = 105°$, so we can use the formula:

$$\text{Area of triangle} = \frac{1}{2}ab \sin C$$
$$= \frac{1}{2} \times 18 \times 25 \times \sin 105°$$
$$= 217.33 \text{ cm}^2 \text{ correct to two decimal places}$$

b We are given two side lengths for this triangle but not the included angle. The included angle can be found using 'the angles of a triangle sum to 180°'.

$$\text{Included angle} = 180° - 30° - 110°$$
$$= 40°$$
$$\text{Area of triangle} = \frac{1}{2} \times 38 \times 18 \times \sin 40°$$
$$= 219.83 \text{ cm}^2 \text{ correct to two decimal places}$$

Example 2

Use the sine rule to find the other angles in this triangle and hence the area of the triangle.

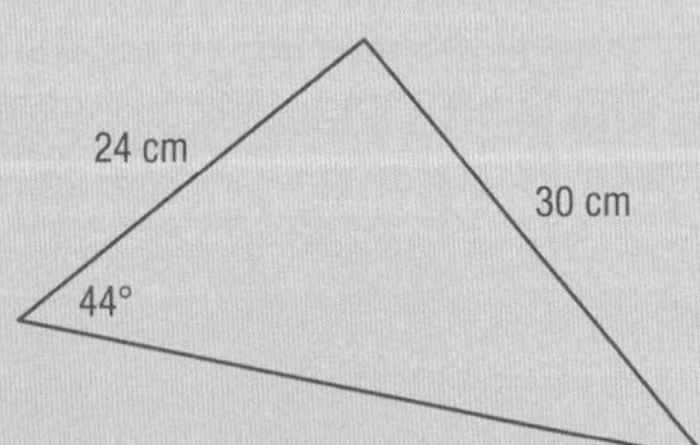

Answer

The sine rule can be used to find θ and hence the angle that is the included angle for the two given sides.

$$\frac{24}{\sin \theta} = \frac{30}{\sin 44°}$$
$$\sin \theta = \frac{24 \sin 44°}{30}$$
$$\sin \theta = 0.5557$$
$$\theta = 33.76$$

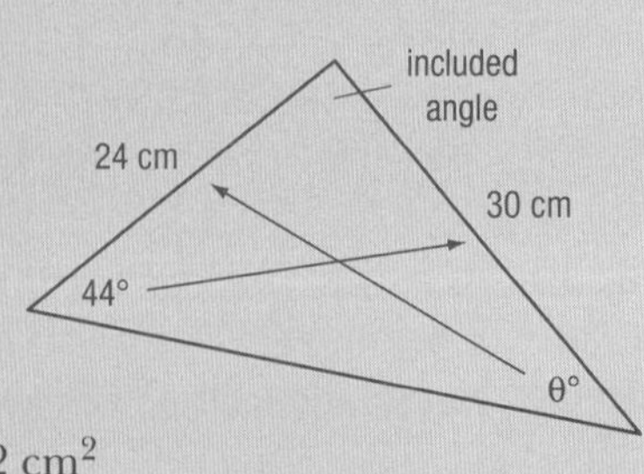

Included angle = 180° – 44° – 33.76° = 102.24°

Area of triangle $= \frac{1}{2} \times 24 \times 30 \times \sin 102.24° = 351.82 \text{ cm}^2$

Example 3

Find the area of this triangle.

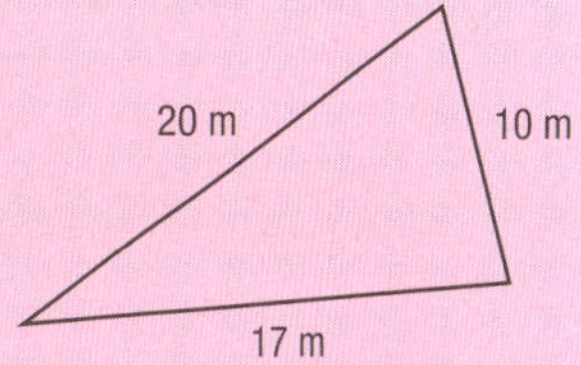

Answer

The lengths of the three sides of this triangle are known. To find the area we need one of the angles—it can be any one of the angles—so that we have 'two sides and the included angle'.

The angle θ can be found using the cosine rule:

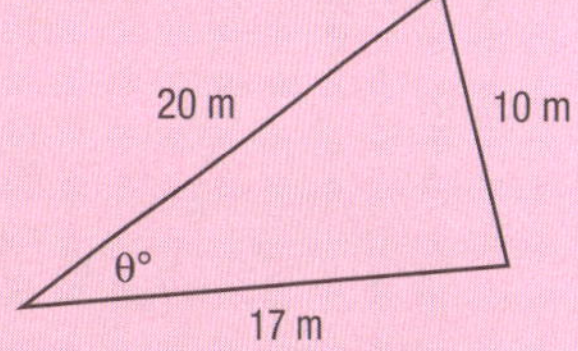

$$\cos\theta = \frac{20^2 + 17^2 - 10^2}{2 \times 20 \times 17}$$
$$= 0.8662$$
$$\theta = \cos^{-1}(0.8662)$$
$$= 29.98\ldots° \ (29°59')$$

We now have the length of two sides and the magnitude of the included angle, $\theta = 29.98$

$$\text{Area of triangle} = \frac{1}{2} \times 20 \times 17 \times \sin 29.98\ldots°$$
$$= 84.96 \text{ m}^2 \text{ correct to two decimal places}$$

EXERCISE

Give all answers correct to two decimal places.

1 Find the area of each of the following triangles.

a

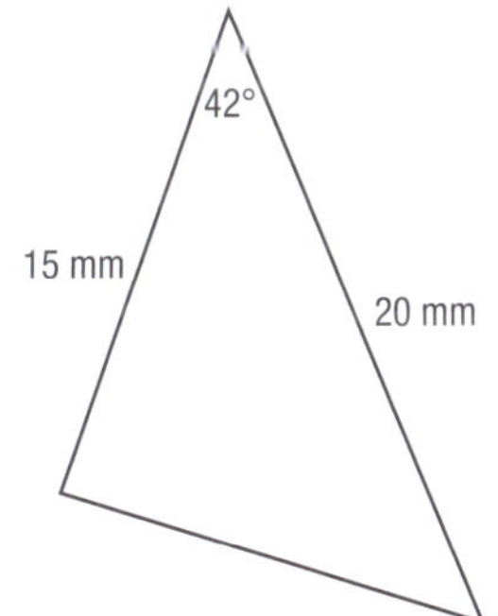

b

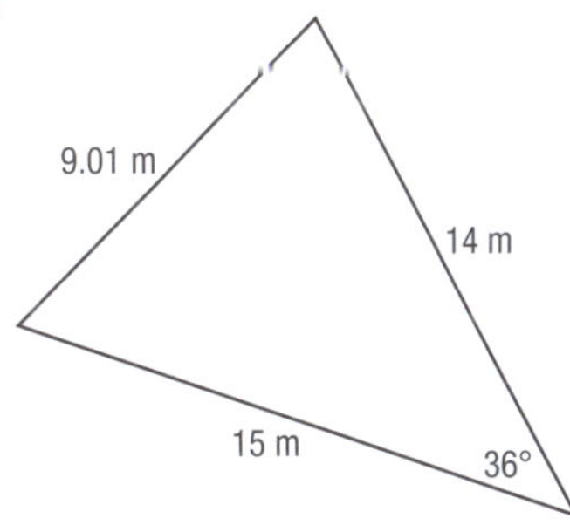

c

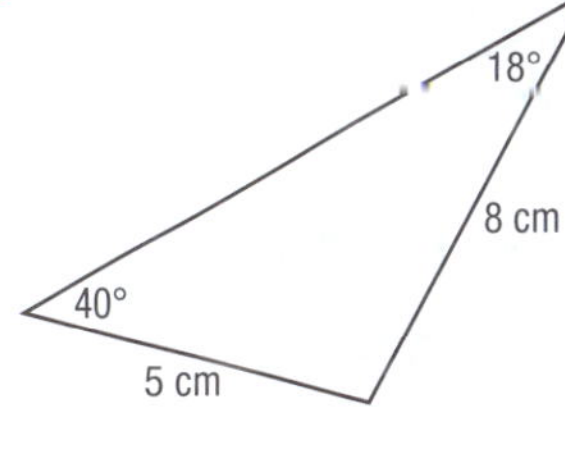

d

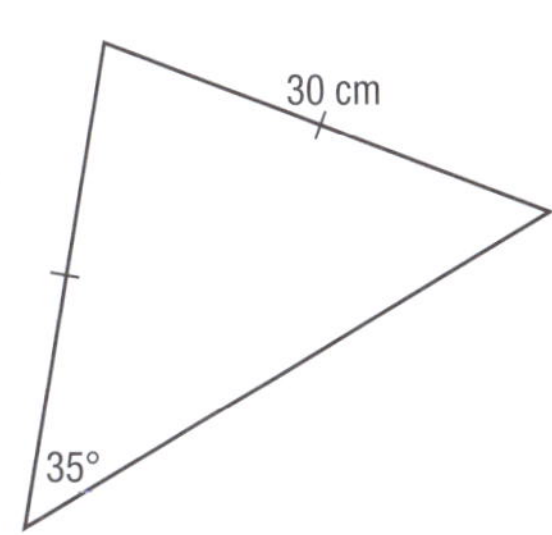

e

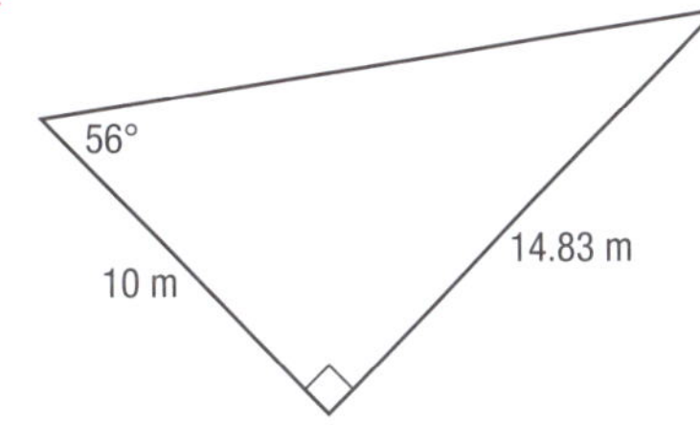

f

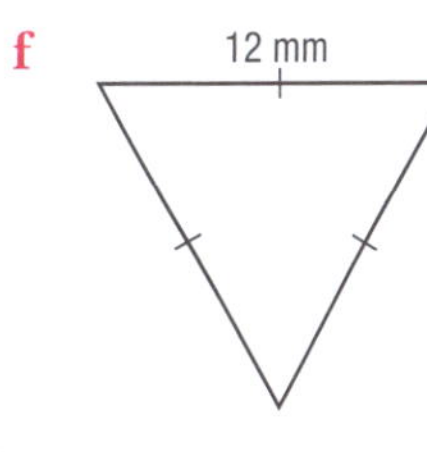

2 Find the area of each of the following triangles. You may need to find an appropriate side length or angle using a trigonometric ratio, or the sine or cosine rule before you can calculate the area.

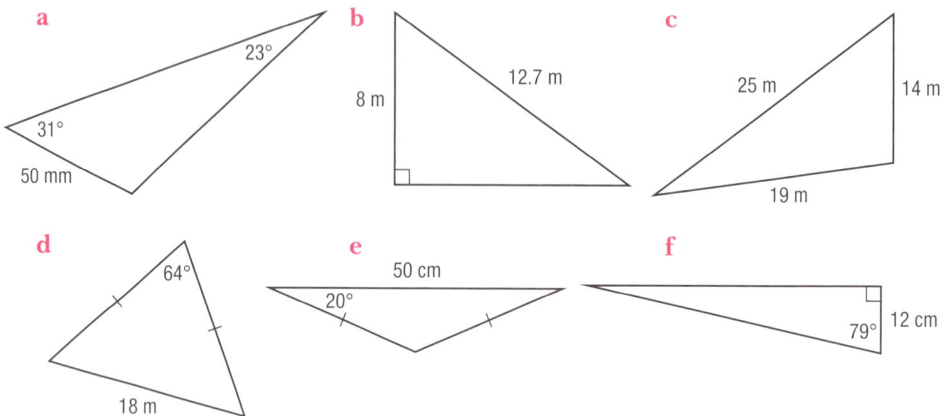

3 Find the area of each of the following figures.

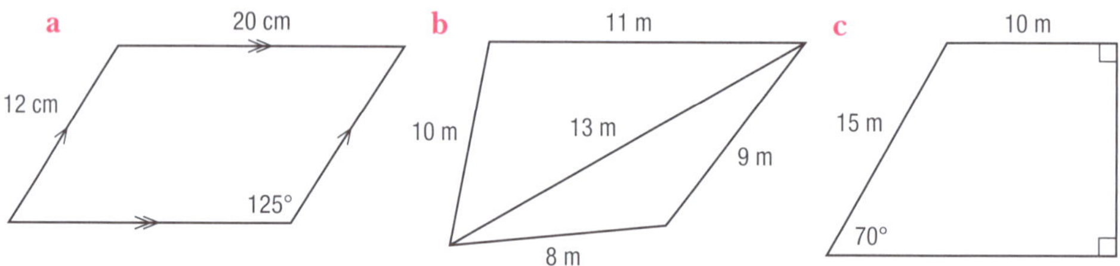

Lesson 6

Maps and scale drawings

Scales on maps are usually given as either a **ratio scale** or a **simple scale**.

For example, the ratio scale 1 : 1 000 000 means that 1 unit on the map represents 1 000 000 units in actual distance.

The ratio scale 1 : 1 000 000 can be expressed in a more practical way:

1 unit on the map ≡ 1 000 000 units in actual distance

1 centimetre on the map ≡ 1 000 000 cm in actual distance

= 1000 m in actual distance

= 10 km in actual distance

So each centimetre on the map represents 10 kilometres in actual distance.

Did you know?

The symbol ≡ means 'is equivalent to'.

A **simple scale** for the same map would look like this.

Each of the partitions on the simple scale is 1 centimetre long, and each represents 10 kilometres.

EXERCISE

1 a The ratio scale for the map of Bougainville Island is 1 : 2 500 000. Convert this so that you can say 'One centimetre on the map represents .... kilometres'.

b Construct a simple scale for this map.

c Measure the straight line distance on the map, to the nearest millimetre, between:

i Buka Passage and Arawa

ii Panguna and Buin.

d Convert the map distances from part c to actual distances in kilometres.

2 Draw a simple scale for each of the following ratio scales:

a 1 : 1000 b 1 : 150 000 c 1 : 25 000

3 Convert the following simple scales to ratio scales.

a 

b 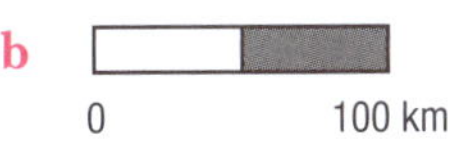

c 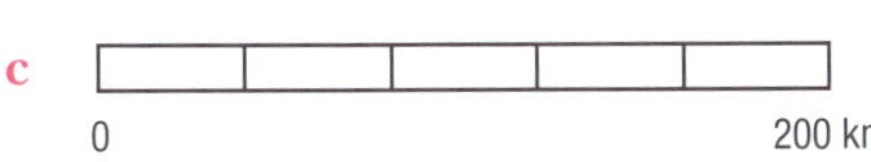

4 Two locations that are known to be 360 m apart are drawn on a map with a ratio scale of 1 : 5000. What is the distance between the locations on the map?

5 The straight-line distance between two cities is known to be 610 km. If the simple scale on the map is shown below, how far apart are the two cities on the map?

6 On a map two cities are 8.4 cm apart, and the actual distance between the two cities is known to be 294 km. What is the ratio scale of the map?

7 A triangular block of land has two adjacent boundaries of length 18.2 m and 45.8 m. The angle between these two boundaries is 80°.

a Draw a scale drawing of this triangular block of land using a scale of 1 : 500. Use a protractor to construct the angle.

b Measure the length of the third side of the triangle and convert this to an actual measurement in metres.

c Calculate the length of the third side of the triangle using the cosine rule. Compare this value with your answer from part **b**.

8 This scale drawing of a block of land has a ratio scale of 1 : 1500.

a Find the actual length of the perimeter of the block.

b Use a protractor to measure two of the angles, then use the formula Area of triangle $= \frac{1}{2}bc \sin A$ to find an estimate of the area of the block.

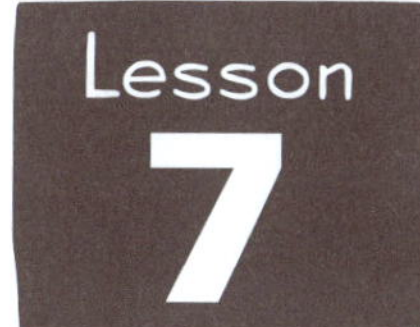

The clinometer

A **clinometer** is an instrument used to measure 'vertical' angles, that is the **angle of elevation** or the **angle of depression**.

A simple clinometer (or inclinometer) can be made using a wooden ruler, a protractor or cut-out of a protractor, string and a weight (a sinker or washer) that can be tied to the end of the string.

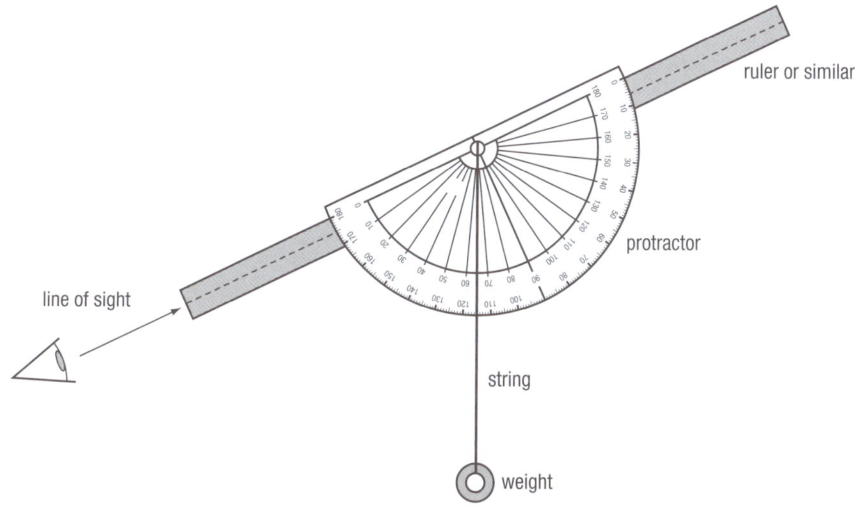

Using your clinometer to measure the height of a tree

Example

1 Mark your position on the ground from the base of the object whose height you are trying to find.

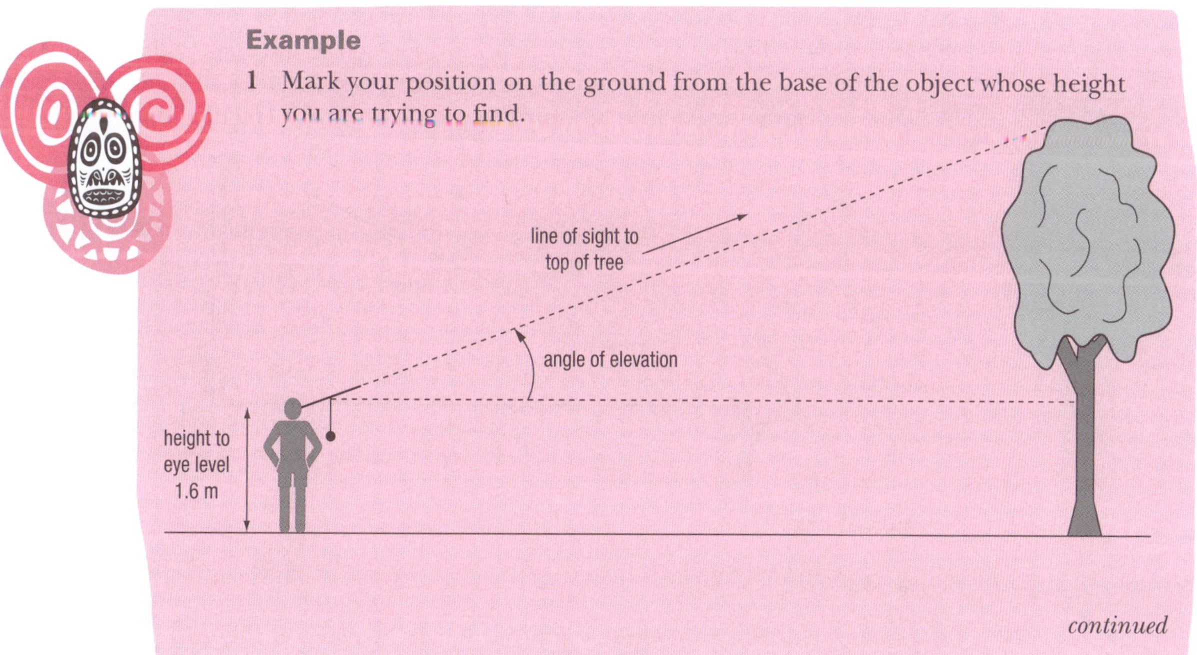

continued

Example *(continued)*

2 Line-up the top of the object with the clinometer and measure the angle of elevation.

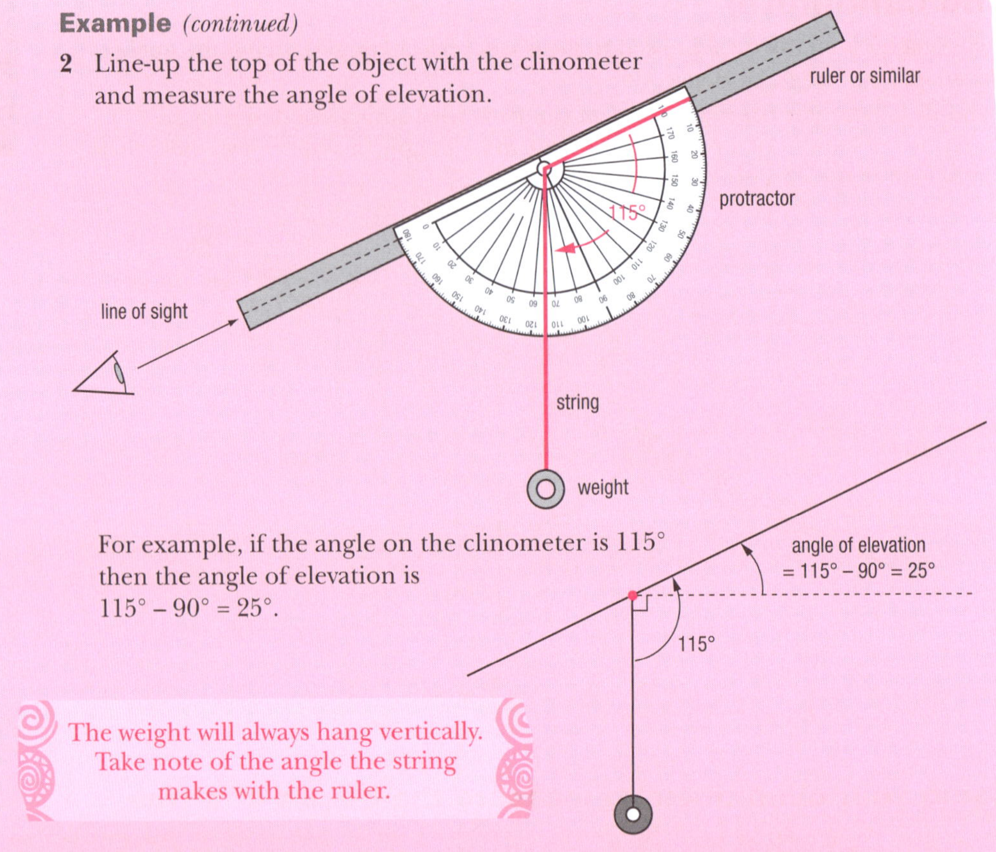

For example, if the angle on the clinometer is 115° then the angle of elevation is 115° – 90° = 25°.

The weight will always hang vertically. Take note of the angle the string makes with the ruler.

3 Measure the horizontal distance from your position to the object (20 m in this example). You can do this with a tape measure or by pacing.

4 Draw a diagram that shows all your measurements and calculate the value of x:

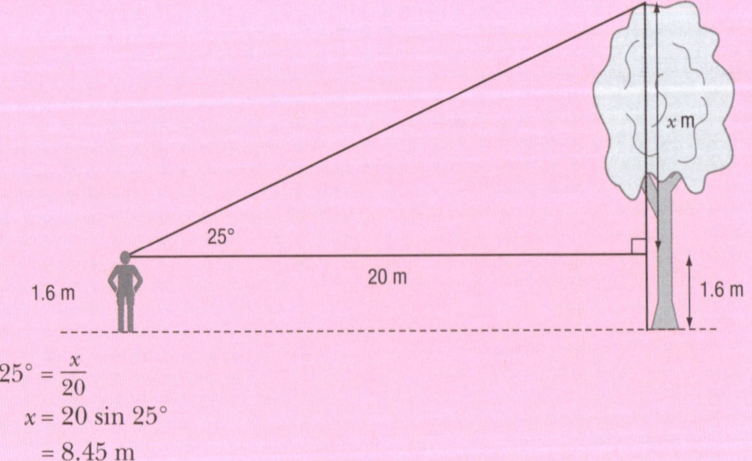

$$\sin 25° = \frac{x}{20}$$

$$x = 20 \sin 25°$$

$$= 8.45 \text{ m}$$

To the value of x you need to add your height at eye level (here 1.6 m).

The height of the tree = 8.45 + 1.6

= 10.05 m

EXERCISE

Constructing a clinometer and testing its accuracy

1 In pairs, construct a clinometer. You will need a wooden ruler (or similar), a protractor*, a length of string (about 30 cm) and a weight that can be tied to the string (a washer or sinker).

*It is a good idea to use a cut-out of a protractor pasted on cardboard because the string needs to be attached at the centre of the protractor.

The more care that you take constructing your clinometer, the more accurate your answers will be. Take particular care to attach the string at the 'centre' of your 'protractor'.

2 Choose some tall objects in your school grounds, and a particular place to measure the angles from. Find the angle of elevation to the top of these objects.

3 Compare your results with other students and discuss the reasons why your answers differ. Compare clinometers and see where you can make modifications so that they are more accurate.

Using a plane table for surveying

A **plane table** is a simple apparatus used to survey (find lengths and areas) a reasonably large plot of land.

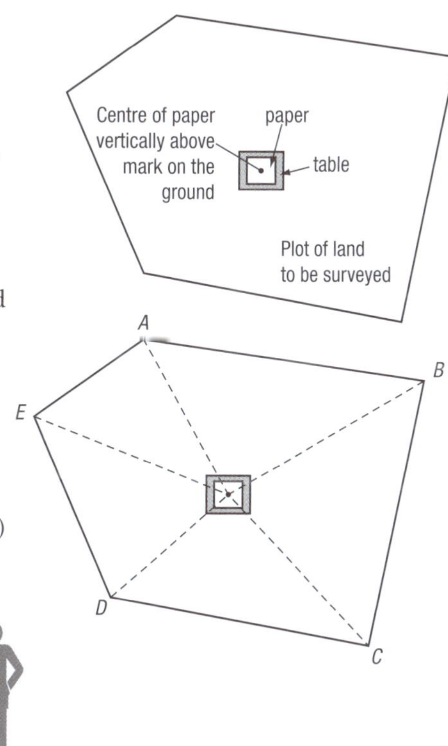

A plane table consists of a horizontal 'table' onto which a large sheet of paper is attached (to be removed later). Place the table in a convenient spot somewhere near the centre of the plot of land being surveyed.

Locate a point near the centre of the sheet of paper (point *O*) and place it directly over a mark on the ground.

Draw lines on the paper from this centre (*O*) along the sight lines to each of the corners of the plot of land (*A*, *B*, *C*, *D* and *E*).

It is helpful to have someone stand on each corner holding a straight stick so that the position of the corner is visible.

A simple instrument, called an **alidade**, can also be used to 'line up' the corners. You could make this from a wooden ruler with two pins placed on the edge.

Position the alidade on the plane table so that the edge is going through the point *O* and the two pins and the stick on the corner are lined up.

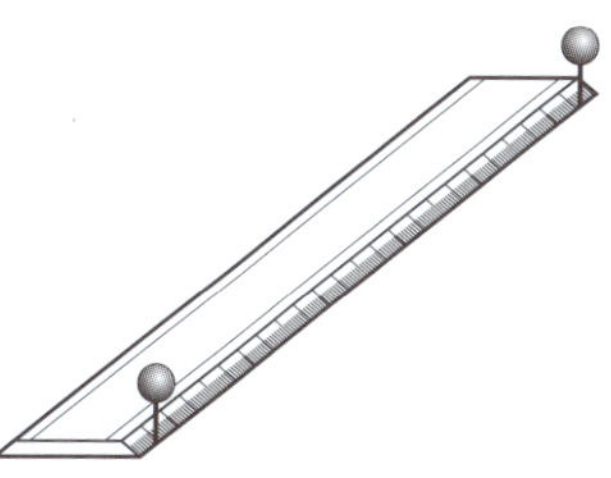

On the ground from the point directly below point O, measure the distance to each of the corners. You can do this using a long tape measure or by pacing. Record the measurements on the sheet of paper.

Sheet of paper from the plane table

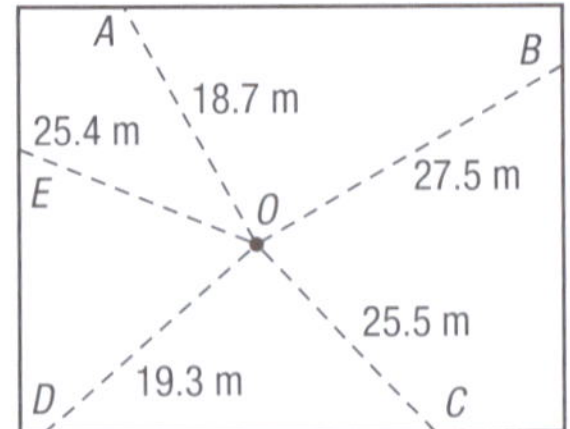

Measure the angles between the lines using a protractor. The angles should sum to 360°.

You can now make a scale drawing of the plot of land.

A scale of 1 : 500 (1 centimetre on the map represents 5 metres in actual distance) has been used for this map.

The line OB is represented by a line of $\frac{27.5}{5} = 5.5$ cm.

Sheet of paper from the plane table

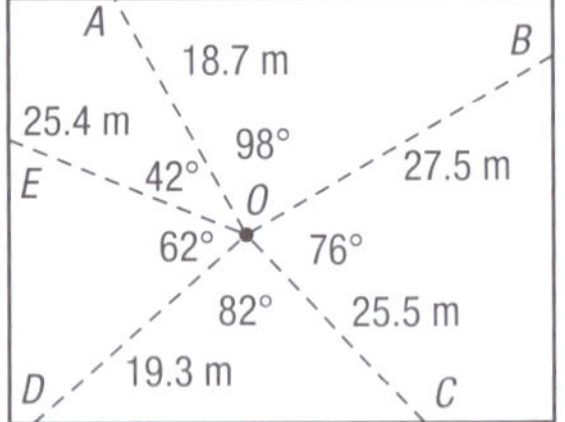

The boundaries of the plot of land can now be measured on the scale diagram and converted to actual measurements. For example, the boundary AB measures 7.1 cm on the scale diagram. This converts to $7.1 \times 5 = 35.5$ metres in actual distance.

Alternatively, the length of the boundaries can be calculated using the cosine rule:

$c = \sqrt{a^2 + b^2 - 2ab \cos C}$

A B E 42° 98° O 76° 62° 82° D C

The length of $AB = \sqrt{18.7^2 + 27.5^2 - 2 \times 18.7 \times 27.5 \times \cos 98°} = 35.3$ metres.

The area of the plot of land can be found by finding the area of the five triangles.

Using the formula: Area $= \frac{1}{2}ab \sin C$

Area of triangle: $AOB = \frac{1}{2} \times 18.7 \times 27.5 \times \sin 98°$

$= 254.6 \text{ m}^2$

EXERCISE

Constructing a plane table and testing its accuracy

1 In small groups of two or three, collect the equipment necessary to construct a plane table and alidade.

2 Using the instructions in the example above, conduct a survey of a plot of land in your school ground that has distinct corners. This could be a rectangular area such as a basketball court or an irregular area that is easily accessible. All groups are to survey the same plot. Remember that your plane tables do not have to be placed in the same position within the plot.

  a When you have recorded the angles and distances for the plot of land, construct a scale drawing making sure that you include the scale. Calculate the length of the boundaries from your scale diagram.

  b Calculate the length of the boundaries using the cosine rule. How do these calculations compare with the calculations from the scale diagram? Comment on any differences found and investigate the reasons for this.

  c Calculate the area of the plot of land using the formula: Area $= \frac{1}{2}ab \sin C$ (even if your plot is rectangular).

  d Compare your answers with other groups. If your answers are very different from others then you will need to find the reason for this and perhaps modify your plane-table set-up.

Triangulation*

Lesson 9

Triangulation is a method used by surveyors to find lengths and areas where part of the land they are surveying is inaccessible.

Surveying using triangulation involves using a baseline of known length and measuring the angles of points of interest (which may be inaccessible) from both ends of the baseline. This could be done using a plane table.

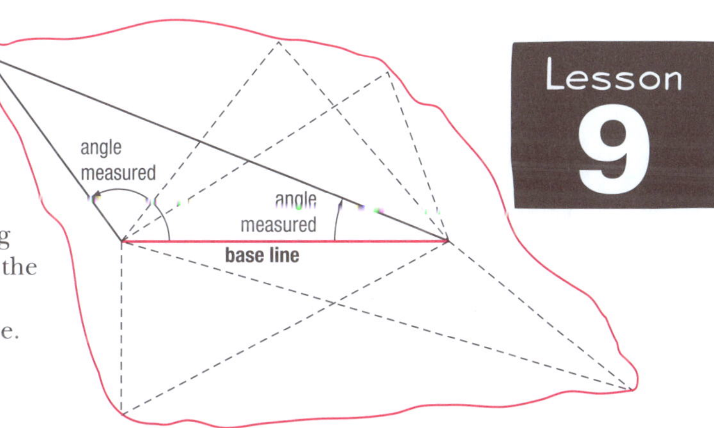

Example 1

An irregularly shaped block of land, $ABCD$, has been surveyed using triangulation methods, taking AB as the baseline. AB = 50 metres and the angles to points C and D have been measured from both ends of AB, as shown in the diagram. Find:

a the length of the boundary AD

b the length of the boundary BC

c the length of AC and hence the length of boundary DC

d the area of the block of land $ABCD$.

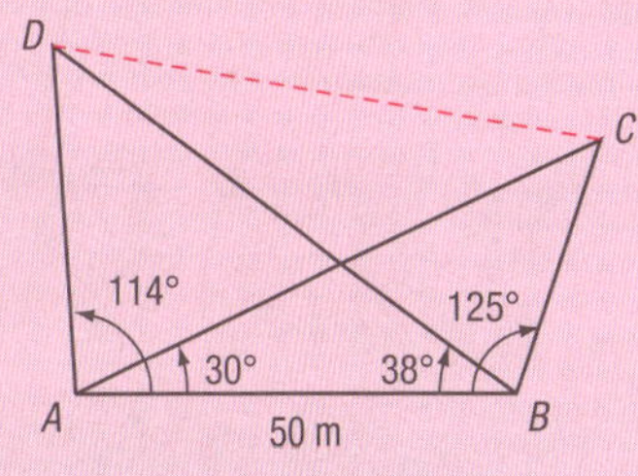

Answer

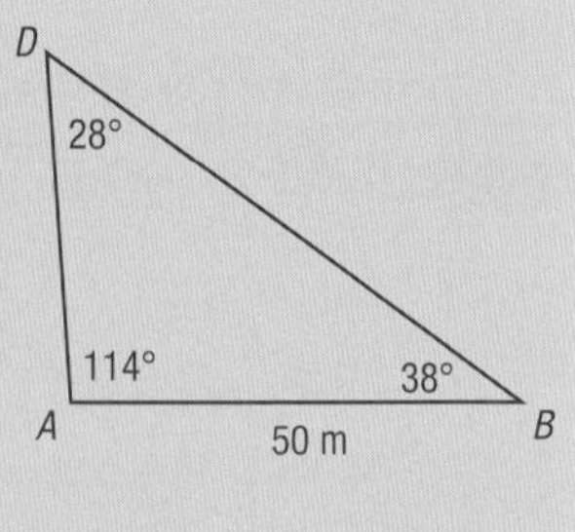

a $\angle ADB = 180° - 114° - 38° = 28°$

The sine rule can be used to find AD:

$$\frac{AD}{\sin 38°} = \frac{50}{\sin 28°}$$

$$AD = \frac{50 \sin 38°}{\sin 28°}$$

$$\approx 65.570 \text{ metres}$$

b $\angle ACB = 180° - 125° - 30° = 25°$

Using the sine rule to find BC:

$$\frac{BC}{\sin 30°} = \frac{50}{\sin 25°}$$

$$BC = \frac{50 \sin 30°}{\sin 25°}$$

$$\approx 59.155 \text{ metres}$$

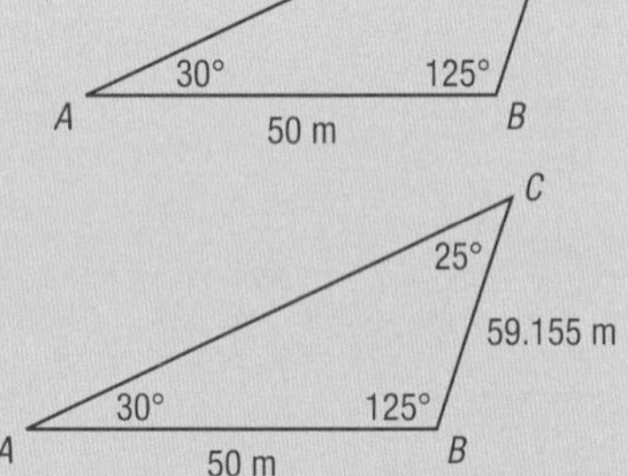

c Using the cosine rule to find AC:

$$AC^2 = 50^2 + BC^2 - 2 \times 50 \times BC \times \cos 125°$$

$$= 9392.312\ldots$$

$$AC \approx 96.914 \text{ metres}$$

In triangle ADC, $\angle DAC = 114° - 30° = 84°$ and AD and AC are known so the cosine rule can be used to find the length of the boundary DC.

$$DC^2 = 65.570^2 + 96.914^2 - 2 \times 65.570 \times 96.914 \times \cos 84°$$

$$DC^2 = 12\,356.615$$

$$DC \approx 111.16 \text{ m, correct to two decimal places}$$

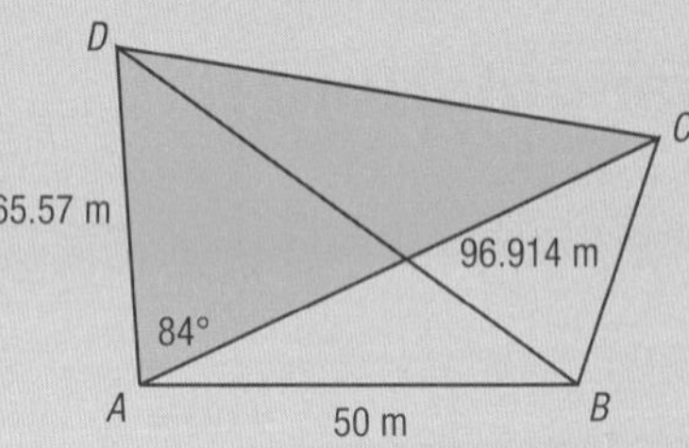

d By summing the areas of the triangles ADC and ABC the area of the block of land can be found.

Area of triangle: $ADC = \frac{1}{2} \times AC \times AD \times \sin 84°$

$$= \frac{1}{2} \times 96.914 \times 65.570 \times \sin 84°$$

$$= 3159.920 \text{ m}^2$$

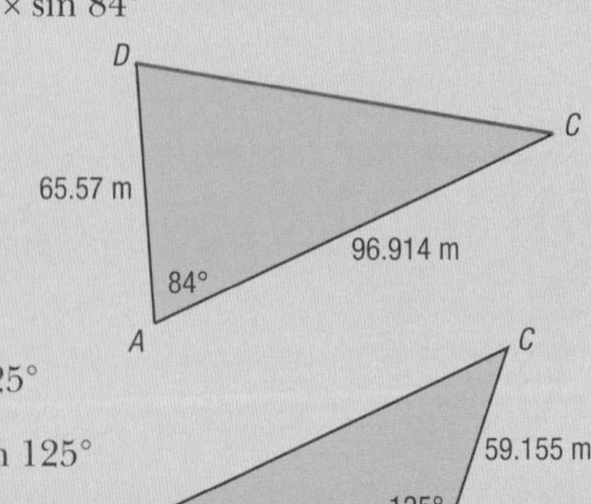

Area of triangle: $ABC = \frac{1}{2} \times AB \times BC \times \sin 125°$

$$= \frac{1}{2} \times 50 \times 59.155 \times \sin 125°$$

$$= 1211.423 \text{ m}^2$$

Hence the area of the block of land $ABCD$ is $3159.920 + 1211.423 = 4371.34 \text{ m}^2$.

EXERCISE

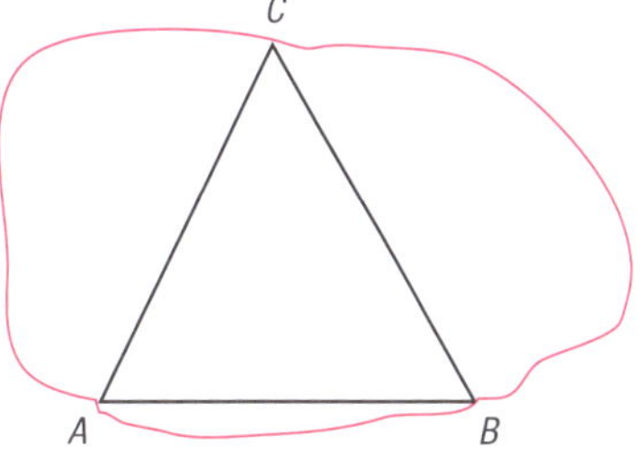

1 A surveyor has measured the angles to an inaccessible point, C, from both sides of a baseline AB. $\angle CAB = 68°$ and $\angle ABC = 82°$. If $AB = 112$ m, find:
   - **a** the length of AC
   - **b** the length of BC
   - **c** the area of triangle ABC.

2 A baseline, AB, is measured on one side of a river and the angle to a landmark, C, on the other side of the river, is noted from each end of the baseline. $AB = 20$ m, $\angle CAB = 61°$ and $\angle CBA = 72°$.
   - **a** Find the length AC.
   - **b** Find the shortest distance across the river.

3 Triangulation is used to find an estimate of the dimensions and area of the block of land shown. Find: 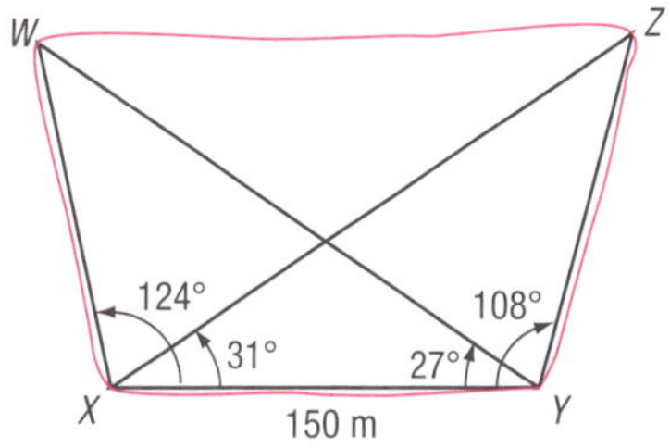
   - **a** the length of YZ
   - **b** the length of XW
   - **c** the length of WY
   - **d** the length of WZ
   - **e** the area of triangles WYZ and XWY and hence an estimate of the area of the block of land.

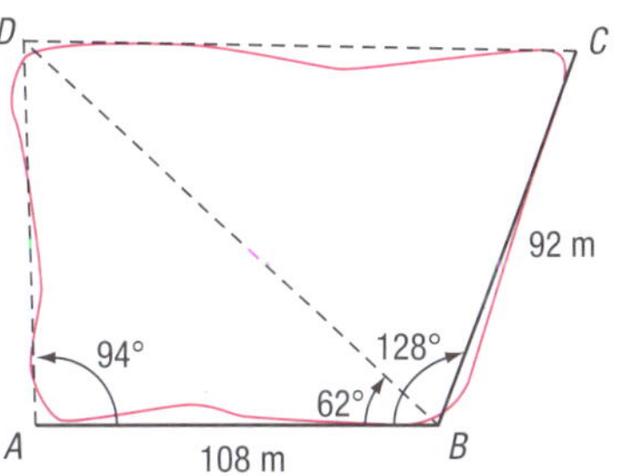

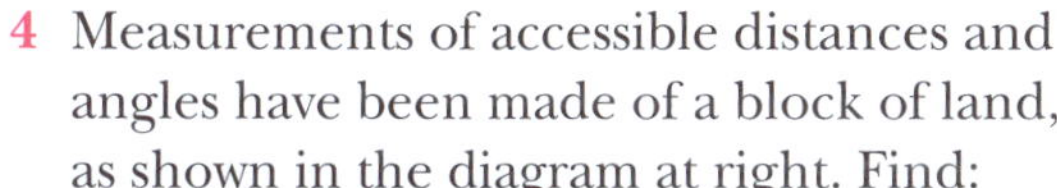

4 Measurements of accessible distances and angles have been made of a block of land, as shown in the diagram at right. Find:
   - **a** the length of AD
   - **b** the length DB
   - **c** the area of triangle ADB
   - **d** the area of triangle DBC
   - **e** the length of DC.

Traverse surveys

Lesson 10

Traverse surveys are also sometimes called **off-set surveys**. They are used by surveyors to measure irregular blocks of land. The measurements recorded are used to make estimates of the perimeter or area of the land.

To carry out a traverse survey of an irregular block of land:

1 The length of the straight line joining the two points furthest apart is measured and recorded. This line, AB on the diagram, is called the **traverse** or **baseline**. Its direction is also recorded.

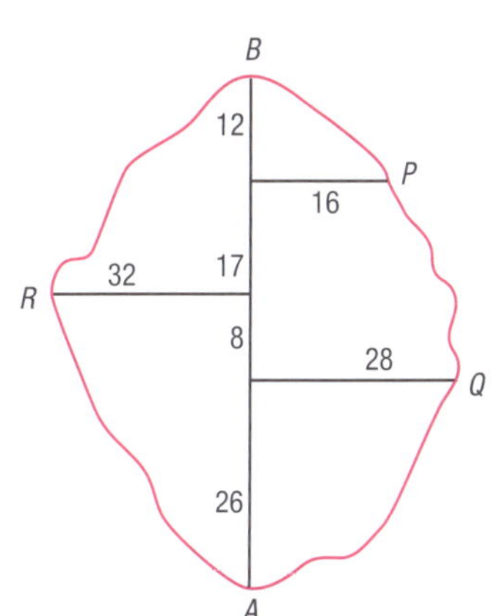

2 Measurements, along AB and **perpendicular** to AB, are then recorded for various features (P, Q and R) of the block of land.

The measurements perpendicular to the baseline are called **offset distances**. In this example they are measured in metres.

A surveyor would record these measurements as a field sketch in one of the following ways. The measurements along the baseline are from point A.

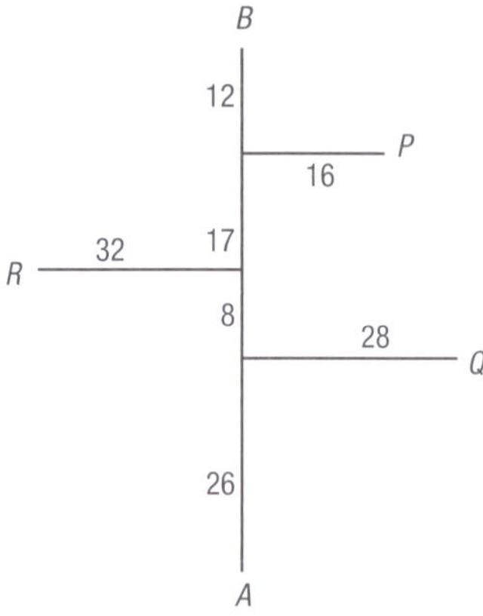

	B	
	63	
	51	16
32	34	
	26	28
	A	

The measurements along the baseline are from point A

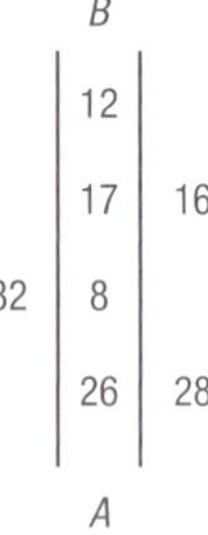

Example

A traverse survey has been carried out on a block of land. The shape of the block and the measurements, in metres, are shown. Use the measurements to find an estimate of:

- **a** the perimeter
- **b** the area
- **c** the distance from R to P for this block of land.

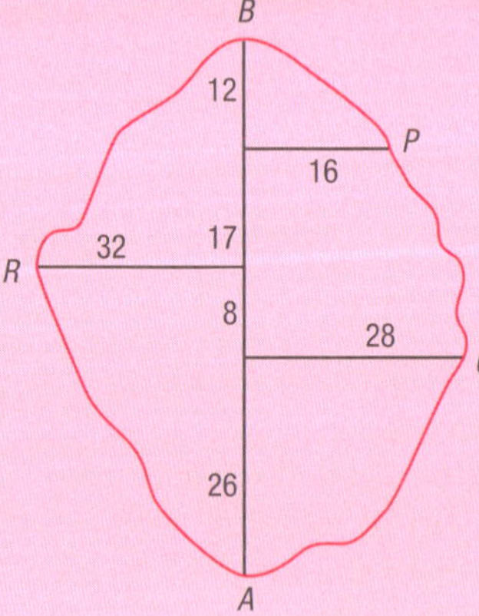

Answer

a Drawing straight lines between adjacent points divides the block of land into five areas that are either right-angled triangles or trapeziums.

Using Pythagoras's rule to find the length BP:

$BP^2 = 12^2 + 16^2$

$= 400$

$BP = \sqrt{400} = 20$ m

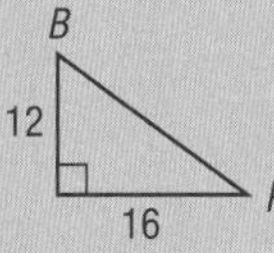

Similarly,

$QA = \sqrt{26^2 + 28^2} = 38.21$ m

$AR = \sqrt{32^2 + 34^2} = 46.69$ m

$RB = \sqrt{32^2 + 29^2} = 43.19$ m

PQ is the side of a trapezium but these lengths can also be found using Pythagoras's rule.

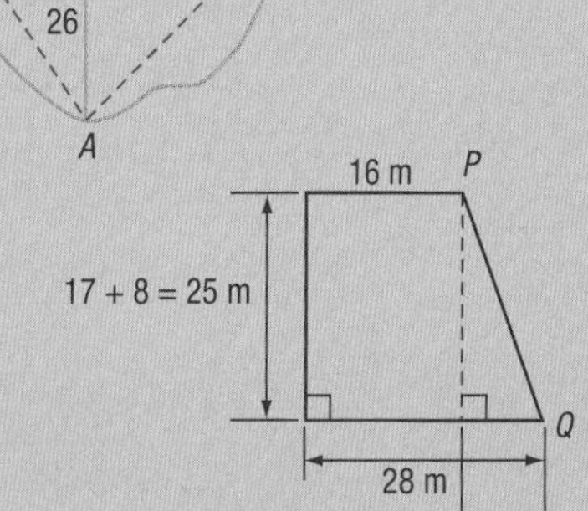

$PQ^2 = 25^2 + 12^2$

$PQ = \sqrt{769} \approx 27.73$ m

Hence an estimate of the perimeter of the block of land is:

$20 + 27.73 + 38.21 + 46.69 + 43.19 = 175.82$ m

b The area of the block can be estimated by finding the area of the four right-angled triangles and the trapezium.

The area of the triangle containing side BP

$= \frac{1}{2} \times 16 \times 12 = 96 \text{ m}^2$

Similarly,

the area of the triangle containing the side $QA = \frac{1}{2} \times 26 \times 28 = 364 \text{ m}^2$

the area of the triangle containing the side $AR = \frac{1}{2} \times 34 \times 32 = 144 \text{ m}^2$

the area of the triangle containing the side $RB = \frac{1}{2} \times 32 \times 29 = 464 \text{ m}^2$

The area of a trapezium is found using Area $= \frac{1}{2}h(a + b)$ where a and b are the lengths of the parallel sides and h is the perpendicular height.

The area of the trapezium containing the side PQ

$= \frac{1}{2} \times 25 \times (16 + 28) = 550 \text{ m}^2$

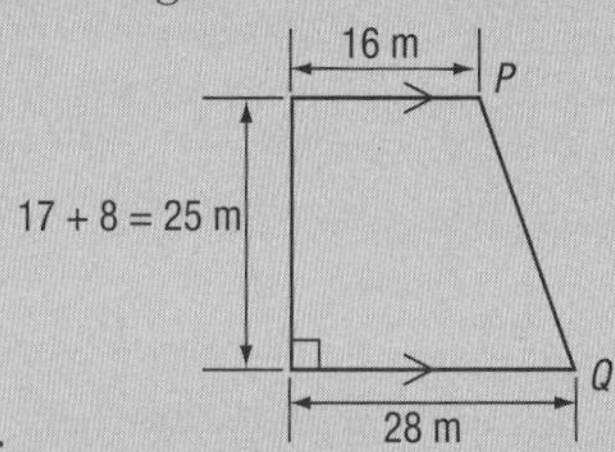

Hence an estimate of the area of the block of land is:

$96 + 364 + 144 + 464 + 550 = 1618 \text{ m}^2$

c To find RP Pythagoras's rule can be used for the triangle RPO:

$RO = 32 + 16 = 48$ m and $OP = 17$ m

$$RP^2 = RO^2 + OP^2$$
$$= 48^2 + 17^2$$
$$= 2593$$
$$RP = \sqrt{2593}$$
$= 50.92$ m correct to two decimal places

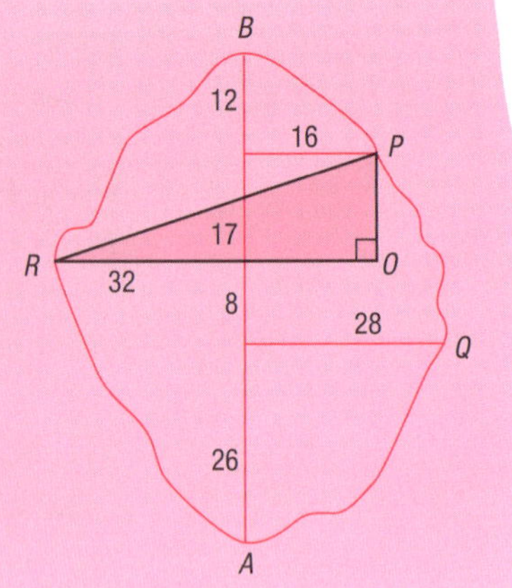

EXERCISE

1 Draw a scale diagram of the areas recorded by the following field sketches of traverse surveys. All measurements are in metres.

a

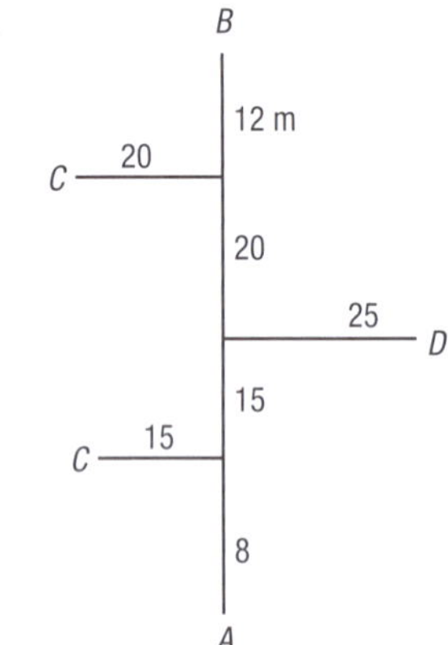

b

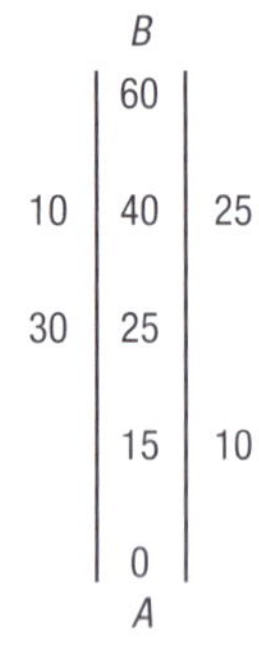

2 An irregular block of land has been surveyed using traverse-surveying techniques as shown in the diagram. All measurements are in metres.

a Find an estimate of the area $ABCDE$ using the measurements recorded in the traverse survey.

b Find the distance between the points:

i A and B

ii E and C

iii D and B.

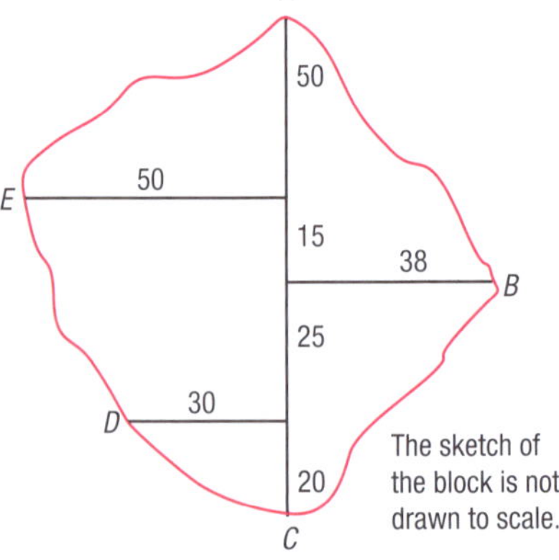

3 A field sketch for a traverse survey is shown. Measurements are in metres.

a Draw a scale drawing of the block of land from the field sketch.

b Find an estimate of the perimeter of the block of land from the scale drawing.

c Find an estimate of the perimeter of the block of land using trigonometry and the measurements recorded in the traverse survey.

d Find an estimate of the area that was surveyed.

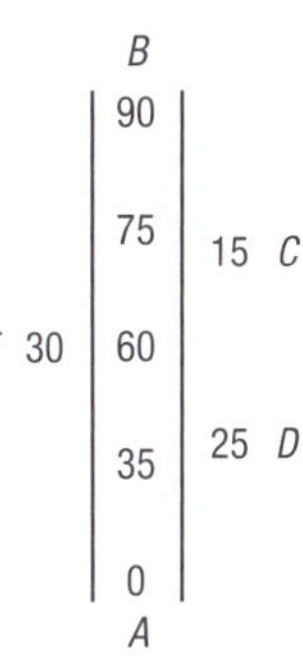

Surveying: practice for the project

Lesson 11

Assessment of this option is a group project where you are asked to survey a local, non-rectangular area using one or more of the techniques that you have learnt in the previous lessons. The results of your survey are to be presented in writing.

You will need to:

- demonstrate appropriate investigation skills
- choose and apply relevant mathematical techniques
- make an effective communication of the project results.

The exercise below tests some of the skills you will need to demonstrate in your project.

EXERCISE

1 The island in the diagram on the right appears on a map that has a ratio scale of 1 : 1000.

a Using *AD* as the traverse line draw in the offset lines to the points marked and, by carefully measuring the lines, construct a field sketch for your traverse survey.

b Estimate the area and perimeter of the island using the measurements on your traverse survey.

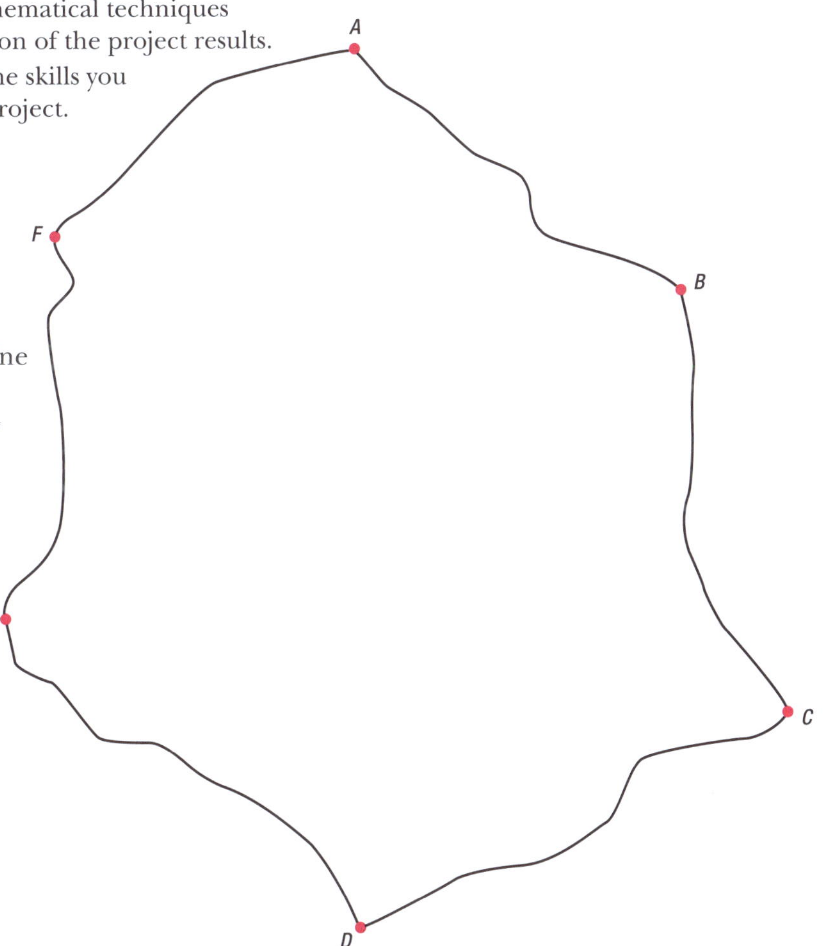

2 The following sheet of notes was taken from a plane table survey of a plot of land.

a Choose a suitable scale and draw a scale diagram of the plot of land. Make sure you measure the angles accurately.

b Measure the perimeter of the plot of land on the map. Convert the map distance to an estimate of the actual perimeter, in metres correct to one decimal place.

c Calculate the perimeter of the plot of land using the cosine rule. Compare this with the estimate from part **b**.

d Calculate an estimate of the area of the plot of land using the original measurements, angles and the area formula: Area of triangle $= \frac{1}{2}bc \sin A$.

3 The angles to the bottom of an inaccessible landmark, L, from a baseline AB are measured and shown on the diagram.

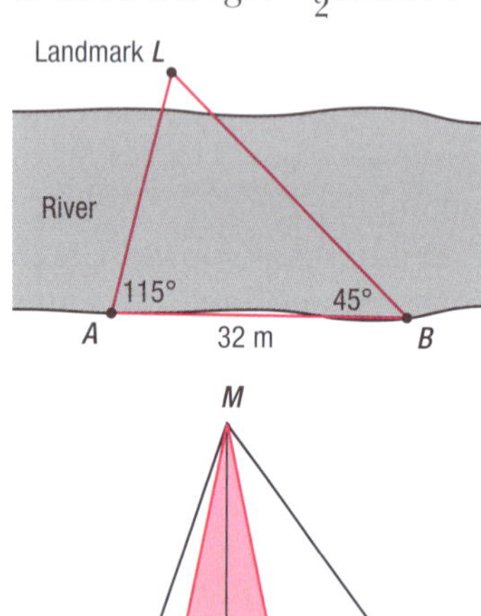

a Use the measurements to calculate the distances AL and BL.

b Calculate the perpendicular distance from AB to the landmark.

c To calculate the height of the landmark, LM, the angle of elevation from A has been measured as 43°, using a clinometer. Calculate the height of the landmark.

d Calculate the angle of elevation of point M from the point B.

4 *The vertices of a roughly triangular plot of land are inaccessible, so triangulation from a plane table is used to survey the plot. The points A and B are 30 m apart.

a Find the size of $\angle AQB$, $\angle APB$ and $\angle ARB$.

b i Use the sine rule to calculate the length of BR.

ii In triangle QAB use the sine rule to calculate the length of QB.

iii In triangle QBR find the angle QBR.

iv Use the cosine rule to find the length of QR.

c Using the same steps as in part **b**, find the length of PQ.

d Using the same steps as in part **b**, find the length of PR.

e Calculate the area of the triangular plot of land, PQR.

OPTION B: NAVIGATION

Lessons 1–6 for Option B are the same as those for Option A.

Bearings

Lesson 7

Relative position

To specify the position of a point *B* relative to a point *A* both the **distance** and the **direction** of point *B* from point *A* must be given.

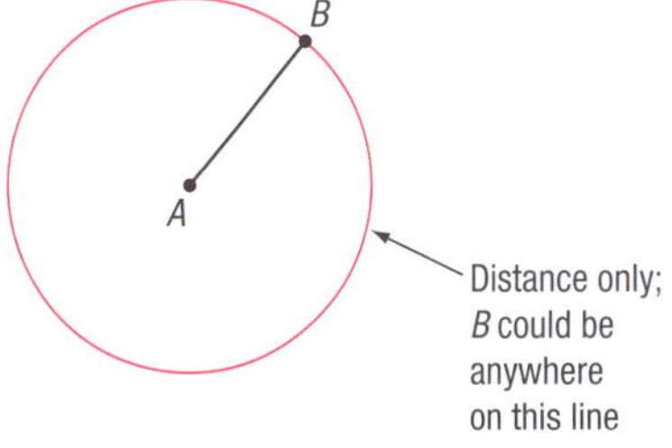

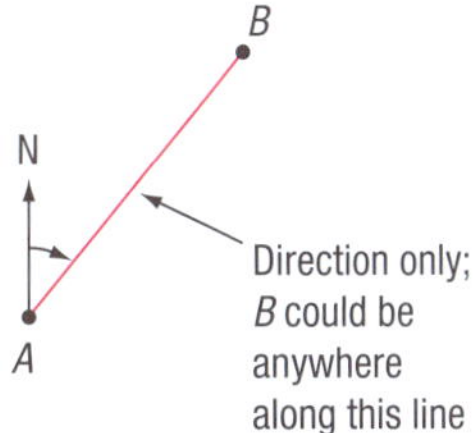

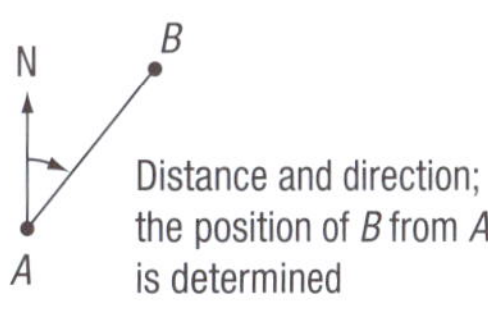

Direction and bearings

Compass bearings

Direction is often quoted as an angle east or west of north or south.

Directions quoted in this manner are called **compass bearings**.

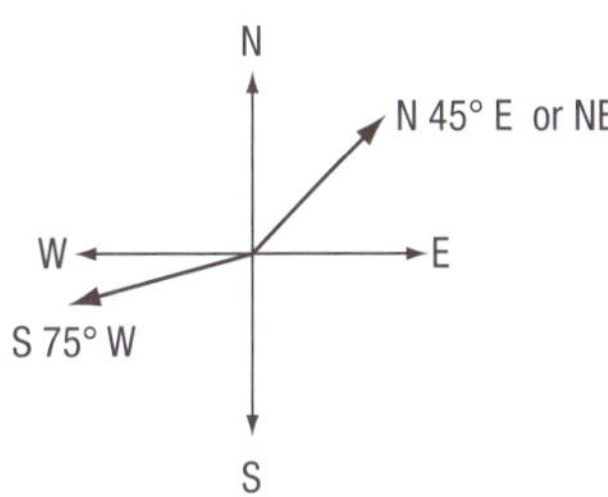

For example, if the direction of Town B from Town A is N 48° E (we say: 'north, 48 degrees, east') then the diagram depicts this direction.

We are considering the direction of Town B **from Town A** so the compass points are drawn at Town A. Starting at north rotate towards east through an angle of 48°, and draw the direction line from Town A to Town B.

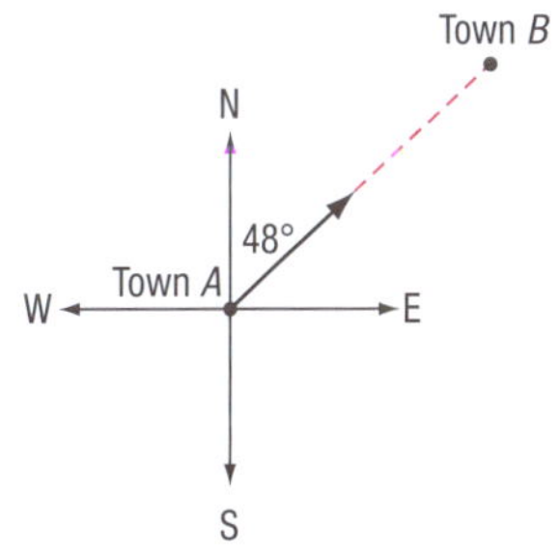

The direction of Town A **from Town B** would be S 48° W ('south, 48 degrees, west').

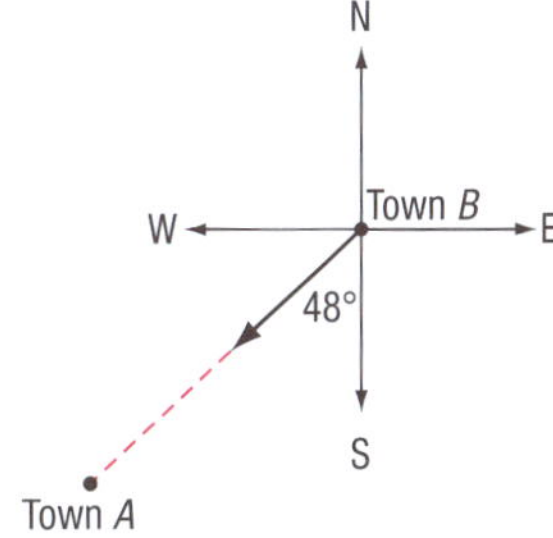

True bearings (sometimes called **three-figure bearings**)

A true bearing gives the direction relative to north only, the angle being measured in a clockwise direction.

For example, if the direction of Town B from Town A is N 48° E then the **true bearing of Town B from Town A** is given as 048° (sometimes written as 048°T; the 'T' standing for true bearings).

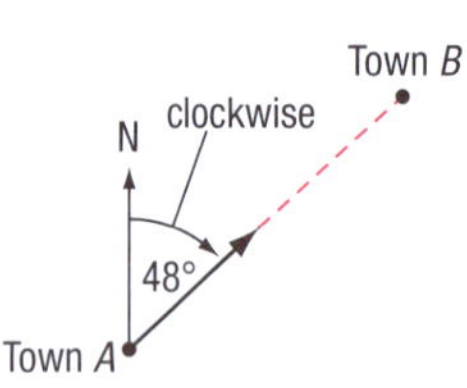

The bearing of Town A from Town B would be 228°.

Note: Bearings can be angles from 000° (north) up to 360° (which would also be north). It is customary to give all bearings as three digits so that there is no confusion that a digit has been accidentally omitted.

The word 'true' is often omitted when quoting this type of bearing.

Example 1

For the diagram given, find the bearing of:

a *B* from *A*

b *C* from *B*

c *A* from *C*

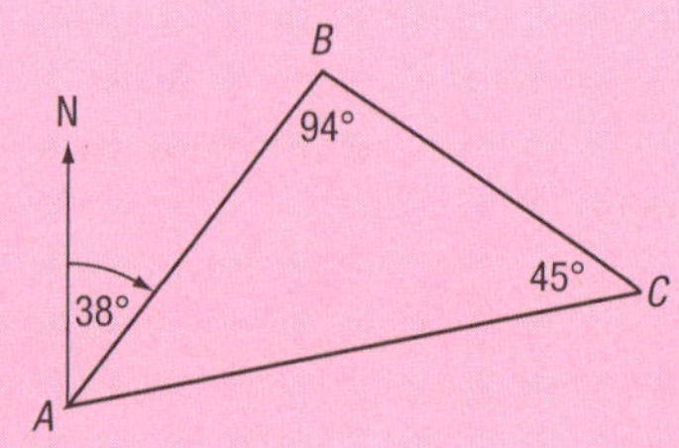

Answer

a In the diagram a line showing the direction of north at *A* is given. The bearing of *B* from *A* will be the angle between this north line at point *A* and the line joining *A* to *B*, measured in a clockwise direction. This is the angle shown as 38° on the diagram, hence the **bearing of *B* from *A* is 038°**.

b A line showing the direction of north has been drawn at *B*. The bearing of *C* from *B* will be the angle between this line and the line joining *B* to *C*.

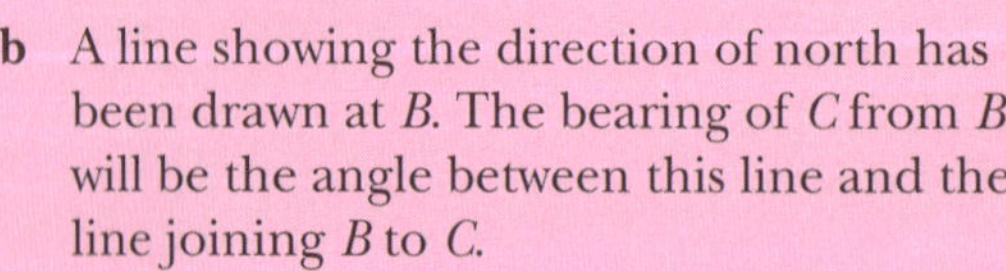

The bearing of *C* from *B*

= 360° – 94°– 142°

= 124°

c A north line is drawn at *C* and the angle measured from this line to the line joining *C* to *A*, in the clockwise direction.

The bearing of *A* from *C*

= 360° – 45° – 56°

= 259°

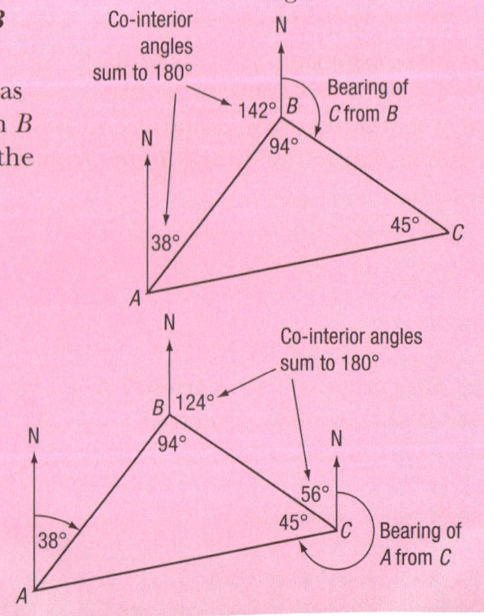

Example 2

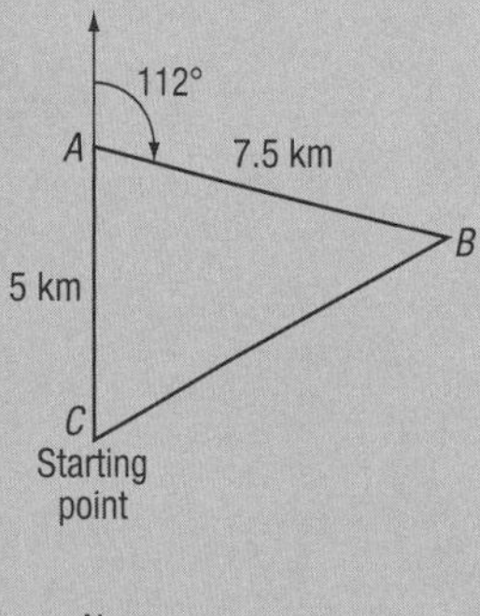

A hiker has mapped out a triangular course, and for the first leg of the hike he walks 5 km due north to point *A*. He then changes direction and walks 7.5 km to point *B* on a bearing of 112°. On the third leg of his hike he returns to his starting point, *C*.

a Find the length of the third leg of the hike.

b Find the bearing of the starting point (*C*) from point *B*.

Answer

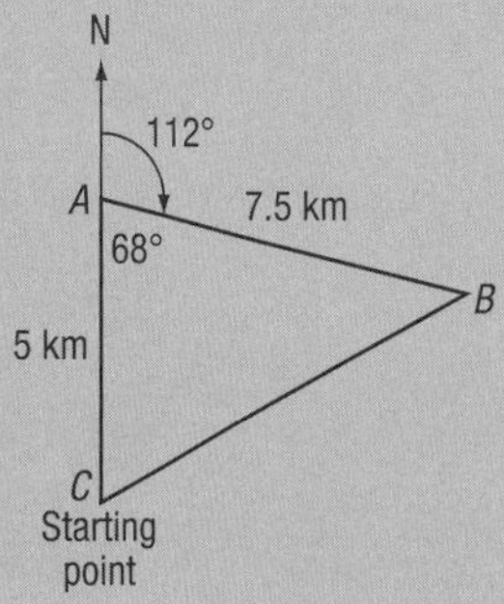

a In the triangle *CAB* we are given *CA* = 5 km and *AB* = 7.5 km. The included angle, $\angle CAB$, has magnitude 68° because it forms a straight angle with the bearing of *B* from *A*, 112°.

We can find the length of *CB* using the cosine rule:

$CB^2 = 5^2 + 7.5^2 - 2 \times 5 \times 7.5 \times \cos 68°$

$= 53.1545\ldots$

So *CB* = 7.291 kilometres, correct to three decimal places.

b To find the bearing of *C* from *B* (marked on the diagram) we need to find the magnitude of angle θ using the sine rule:

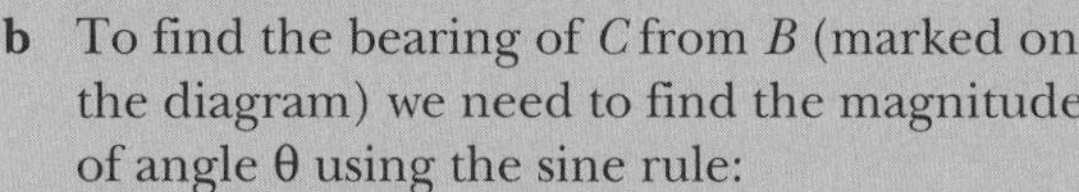

$\frac{5}{\sin\theta} = \frac{7.291}{\sin 68°}$

$\sin\theta = \frac{5 \sin 68°}{7.291}$

$= 0.6359$

$\theta = 39.48°$

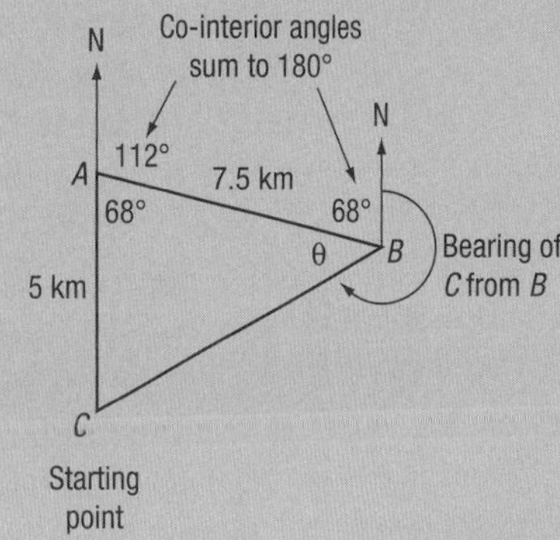

The bearing of *C* from *B*

$= 360° - 68° - 39.48°$

$= 252.52°$

$= 252°31'$ to the nearest minute.

Note: The angle θ can also be found using the cosine rule. The lengths of all three sides are known.

EXERCISE

Give answers correct to two decimal places where necessary.

1 Draw a diagram to show the direction then convert the following compass bearings to true bearings.

a S 50° E **b** N 14° W **c** S 33° W

2 Draw a diagram and state the direction in compass bearings for each of these true bearings.

a 097° **b** 196° **c** 302°

3 In the diagram given find the bearing of:
  a B from A
  b B from C
  c A from B
  d A from C

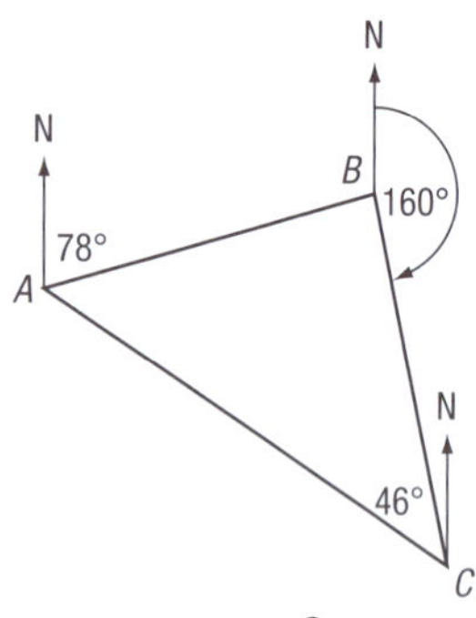

4 $ABCDEF$ is a regular hexagon with the point B being due north of A. Find the bearing of the point:
  a B from A
  b C from B
  c D from C
  d E from D
  e F from E
  f A from F
  g D from B
  h E from B

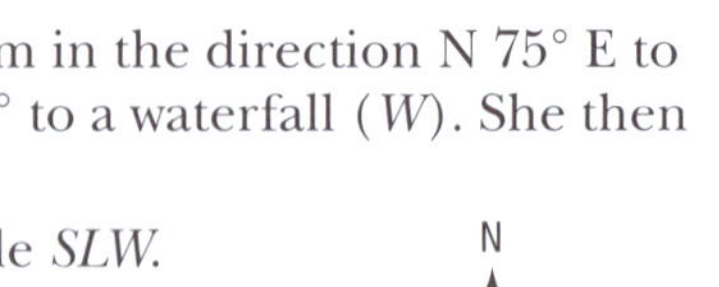

5 From her starting point (S), a hiker walks 12 km in the direction N 75° E to a landmark (L) then 8 km on a bearing of 132° to a waterfall (W). She then returns to her starting point.
  a Find the size of the angle SLW in the triangle SLW.
  b Use the cosine rule to find the distance from the waterfall to the starting point.
  c Use the sine rule to calculate the size of the angle LWS in triangle SLW.
  d Find the true bearing of the starting point from the waterfall.

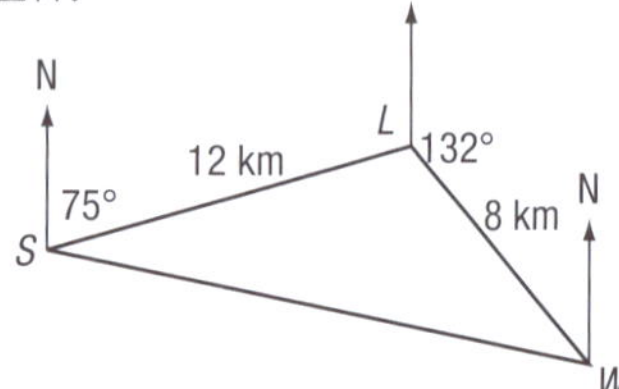

6 An orienteer has a triangular course to run. On the first leg of the course he has to travel 6.3 km due north, then for the second leg he changes direction and has to travel 5.4 km on a bearing of 096°. He is to travel back to his starting point for the third leg of course.
  a Draw a diagram showing the directions and distances of the course.
  b Find the distance travelled on the third leg of the course.
  c Find the bearing that he will need to travel on the third leg of the course.

7 A ship is 8 km on a bearing of 077° from an oil rig, and another ship is 12 km on a bearing of 094° from the oil rig. Draw a diagram and find the distance between the ships.

8 From point A the bearing of C is 081° and from point B, due east of A, the bearing of C is 308°. Draw a diagram and find the angle ACB.

9 A line PQ has length 40 m and the bearing from P to Q is 070°. From the point P the bearing of the point R is 020° and from point Q the bearing of point R is 300°.
  a Draw a diagram representing the triangle PQR. Include all the known angles and lengths.
  b Find the size of the interior angles of the triangle PQR.
  c Calculate the length of the side PR.

Position on the Earth's surface

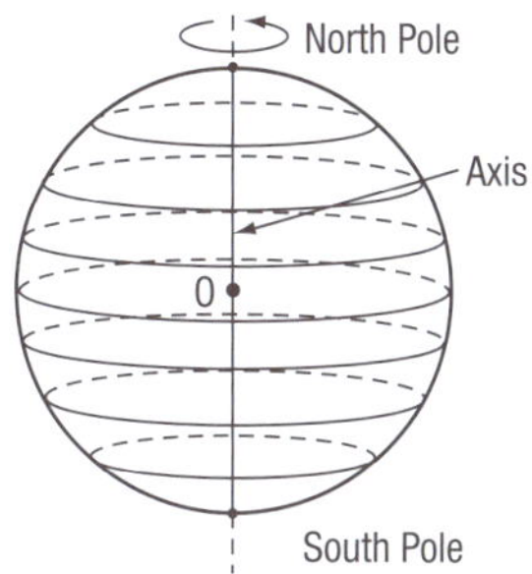

The shape of the Earth is very close to a sphere, with a radius of approximately 6400 kilometres.

The Earth rotates about an axis that passes through the North Pole and the South Pole.

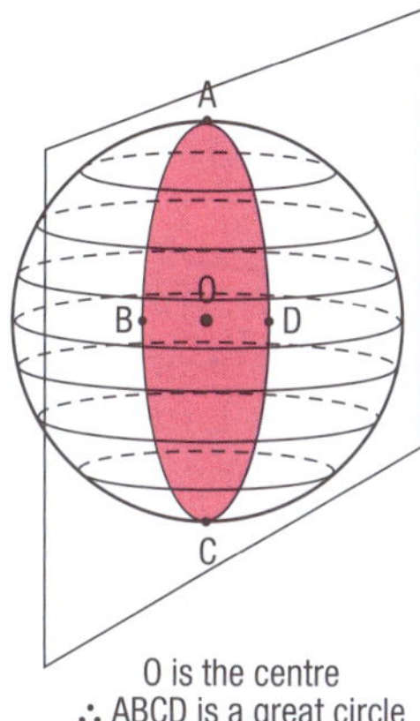

O is the centre
∴ ABCD is a great circle

If the Earth is cut by a **plane**, then the cross-section will be circular. Planes that cut the Earth and go through the **centre** of the Earth are called **great circles**.

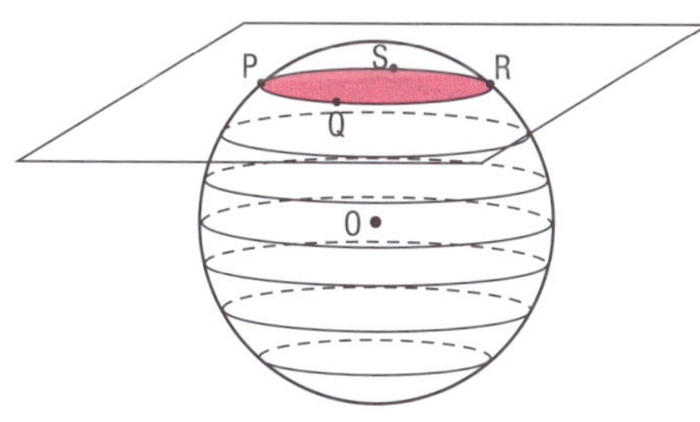

PQRS is a small circle

Planes that cut the Earth but do not go through the centre are called **small circles**.

Planes that are perpendicular to the Earth's axis are called **parallels of latitude**. All parallels of latitude are **small circles** except the parallel of latitude that goes through the **equator**, which is a great circle.

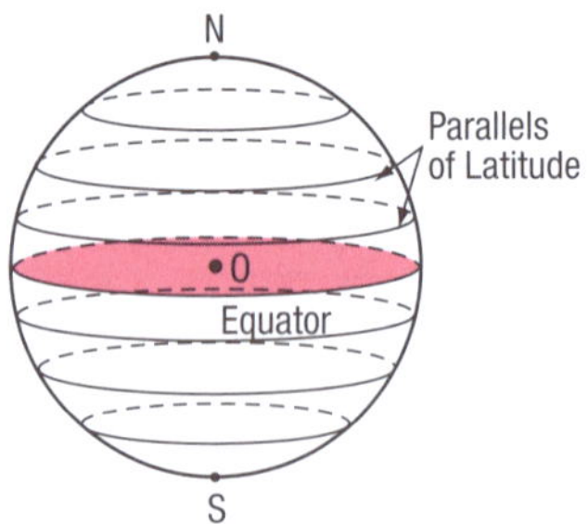

Great circles that pass through the North Pole and the South Pole are cut in half at the poles. These semicircles form the **meridians of longitude**.

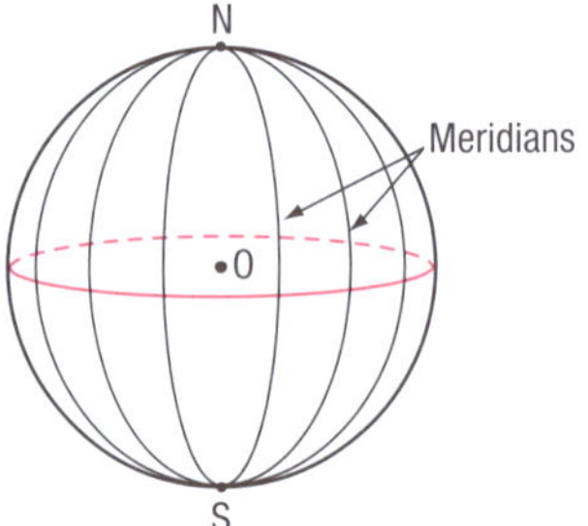

The parallels of latitude and the meridians of longitude form an imaginary grid on the Earth's surface. This means that the position of any point on the Earth's surface can be determined.

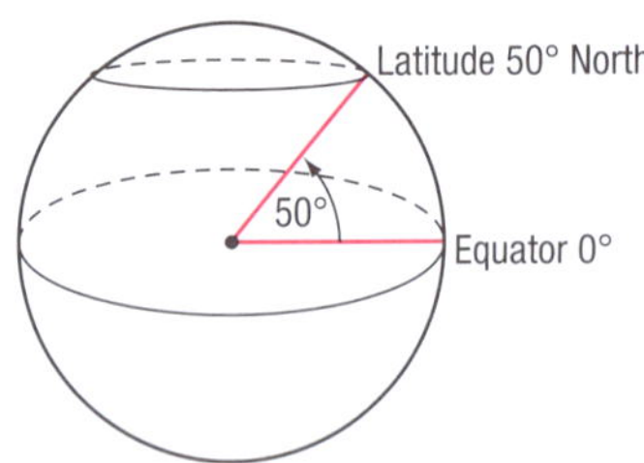

The latitude of every point on the equator is 0° and the latitudes of other points are given as degrees, up to 90°, north or south of the equator. All points on the same parallel of latitude have the same latitude.

The **Prime Meridian** is the meridian that goes through the observatory at Greenwich in England. The Prime Meridian is given the longitude 0° and all other points are given a longitude up to 180° west or 180° east. All points on the same meridian have the same longitude.

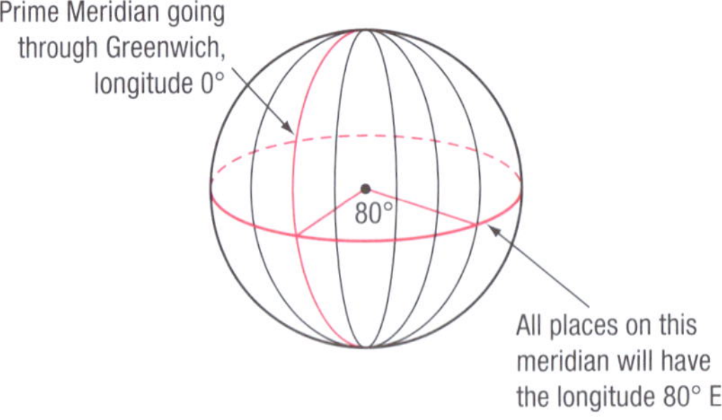

The position of a point on the Earth's surface is given as (latitude, longitude). Angles are measured in degrees, minutes and seconds (60 minutes in a degree, 60 seconds in a minute). It is convention now to use decimal minutes rather than seconds, for example, $32°15'12''$ would be written as $32°15.2'$.

Example

Find the coordinates of Arawa on the navigation chart.

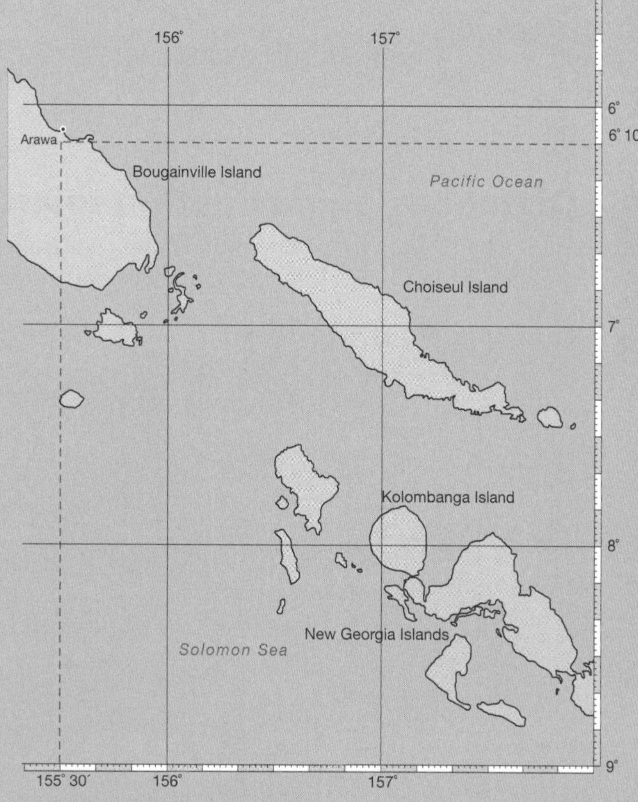

Answer

Drawing straight lines from Arawa to the latitude (vertical) and longitude (horizontal) scales gives the coordinates of Arawa as (6°10′ S, 155°30′ E).

Note: There are six markings between each degree (60 minutes) on the scales, so each of these represents 10 minutes.

EXERCISE

Use a globe of the Earth or an atlas for this exercise.

1 Find the position (latitude, longitude) to the nearest degree of each of the following places in Papua New Guinea.

  a Madang **b** Mount Hagen **c** Arawa
  d Wewak **e** Rabaul

2 Name another town in PNG that has the same latitude as:

  a Mount Hagen **b** Arawa **c** Lai.

3 Name another town or city in PNG that has the same longitude as:

  a Madang **b** Kavieng.

4 Name the city or town in PNG that is closest to:
   a (10°S, 150°E)   b (7°S, 146°E)   c (8°30″S, 151°E)

5 Name the city or town in Australia that is closest to:
   a (35°S, 149°E)   b (12°S, 131°E)   c (38°S, 147°E)

6 a Name two cities that have the same latitude as Paris, France.
   b Name two cities that have the same longitude as Berlin, Germany.

Lesson 9

Distances on navigation maps

Map makers have the problem of drawing or projecting the surface of a sphere onto a flat surface without distorting distances, directions and areas.

Mercator's projection

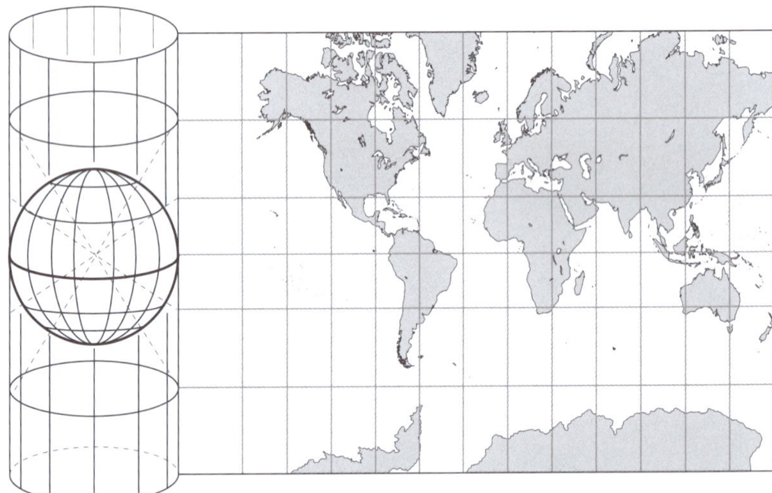

The most commonly used projection for sea navigation is known as Mercator's projection, named after the Belgian cartographer Gerhardt Mercator (1512–94).

The main advantage of the Mercator projection for navigation is that the lines of latitude and longitude are at right angles, so that direction is preserved. This means a compass bearing is represented on the map by a straight line.

Because the meridians are shown as parallel lines, when in fact they get closer together as they approach the poles, areas become more distorted the further they are from the equator. For example, on the map Greenland appears about as large as South America whereas, in fact, it has an area about 12% of the area of South America. For this reason Mercator's projection is used between the latitude 60°S and 60°N.

Distances on navigation maps are in **nautical miles** (abbreviation **n mile**). The nautical mile was originally defined as the length of a great circle arc which subtends an angle of 1 minute at the centre of the Earth.

Now the nautical mile is defined as the distance on the Earth's surface corresponding to one minute on the latitude scale.

1 nautical mile = 1852 metres = 1.852 kilometres

On a Mercator projection map the parallels of latitude are drawn at increasing intervals apart as we approach the poles, so the latitude scale will vary at different latitudes.

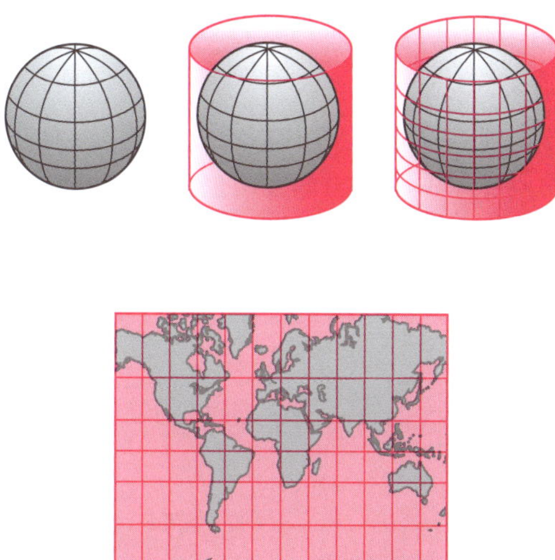

The **vertical scale (latitude scale)** depicted on the right demonstrates the distortion. The two little pink markers on the left of the ruler have precisely the same size, the upper one measures only 0.64 degrees (= 38.4 n mile) while the other measures 1.00 degrees (= 60 n mile). For this reason distances measured on the chart should be converted to nautical miles using the latitude scale (vertical) adjacent to the two points whose distance apart is being measured.

40°
35°
30°
25°

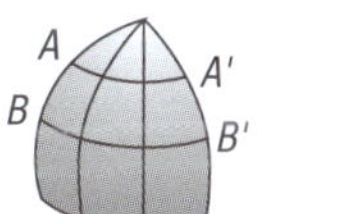

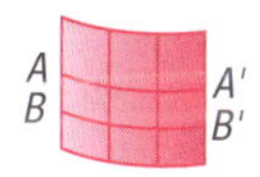

The horizontal scale is only valid for one particular latitude in the chart, and can therefore only be used for the coordinates (a point but not a line). If you divide the surface of the Earth into eight pieces, lift one out and project it, you end up with the figure above. The result is that both *A*–*A*′ and *B*–*B*′ are now as long as the bottom of the chart and are 'too long'.

When travelling on water, the unit for measuring speed is the knot, where:

1 knot = 1 nautical mile per hour = 1.852 kilometres per hour

Radio Mast
(Red Lights)
X
Tower
Water Tank
Big Sister Is.
Trees (conspic.)
35°
55′
Shark
Bay
Wambarra Lt.
Beacon
Y
Little Sister Is.
Peak
5′
Washpool Rocks
(always breaks)
Pine (conspic.)
Stewart Lt.
Z
TV Tower
Mt. Pleasant
50′
55′
Longitude 152° East from Greenwich

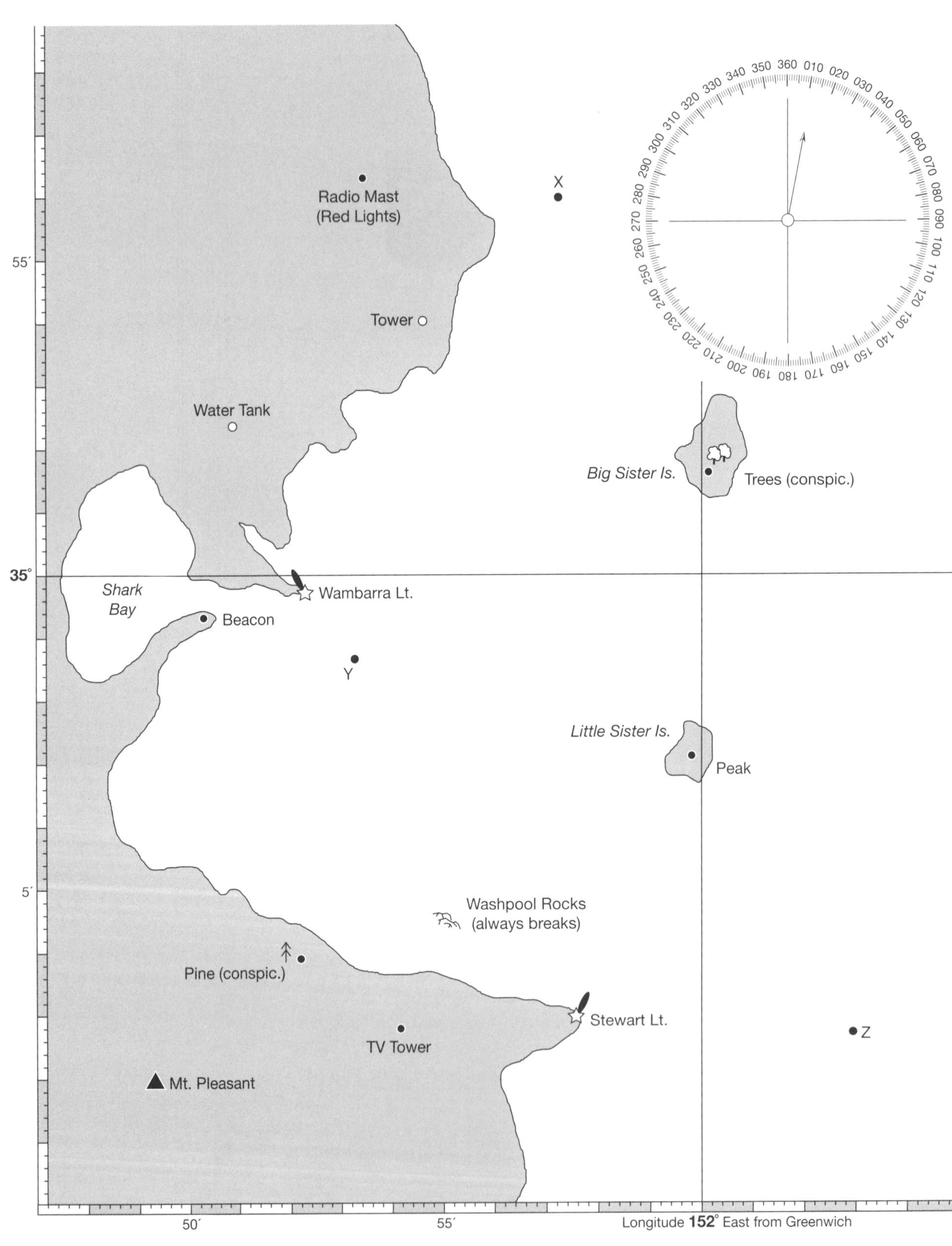

Chart 1

Example

Use Chart 1 to answer the following questions.

a Name the place with position coordinates (34°56′S, 151°54.6′E).

b Give the position coordinates of Stewart Lighthouse.

c Use the latitude scale to find the distance, in nautical miles and in kilometres, from the northern tip of Little Sister Island to the beacon on Shark Bay.

d If a boat took 20 min to travel from the northern tip of Little Sister Island to the southern-most tip of Big Sister Island, what was the average speed of the boat in km/h?

Answer

a The angles of the parallels of latitude are increasing as you go down the page. 35° goes through Shark Bay so 34°56.2′S is north of (above) this bay. Similarly 151°54.6′E is to the left of the 152° meridian. The coordinates are for the tower.

b Drawing horizontal and vertical lines from Stewart Lighthouse to the longitude and latitude scales gives the coordinates (35°S, 151°57.5′E).

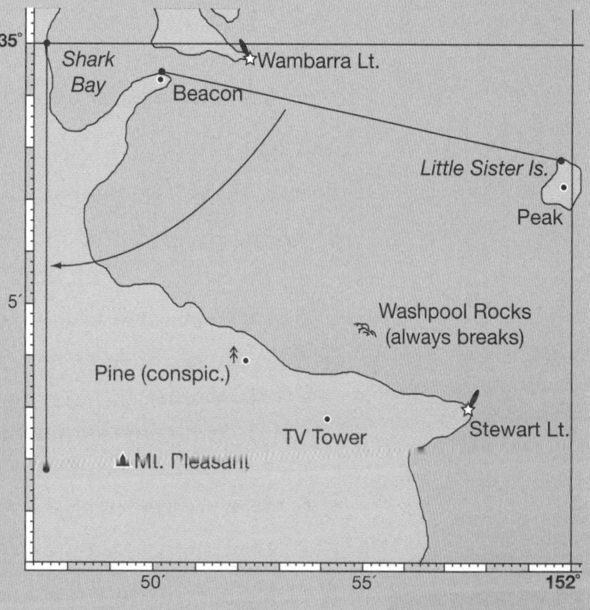

c Measuring the distance on the map and then transferring this distance to the vertical scale gives 8.2′ on the scale so the distance is 8 nautical miles (n mile).

To convert nautical miles to kilometres we multiply by 1.852.

Distance = 8.2 n mile
= 8.2 × 1.852 km
= 15.186 km

d Measuring the distance on the map and then transferring this distance to the vertical scale gives 3.5′ on the scale, so the distance is 3.5 n mile.

Distance = 3.5 n mile
= 3.5 × 1.852 km
= 6.482 km

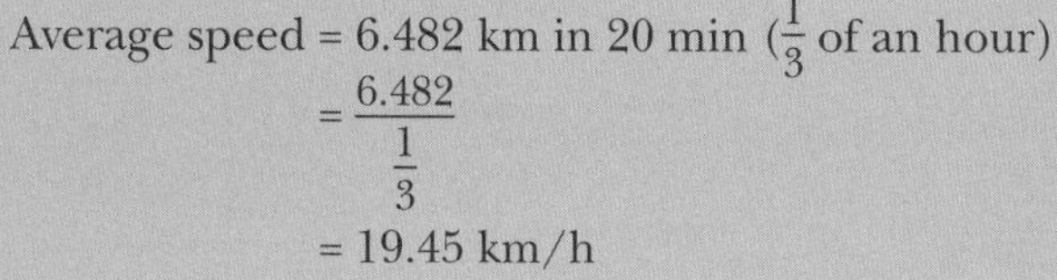

Average speed = 6.482 km in 20 min ($\frac{1}{3}$ of an hour)

$$= \frac{6.482}{\frac{1}{3}}$$

= 19.45 km/h

Remember

Transfer the distance to that part of the vertical scale between the two islands.

EXERCISE

1 Convert the following distances in nautical miles to kilometres:

a 2 n mile
b 25 n mile
c 220 n mile
d 0.4 n mile

2 Convert the following speeds, in knots, to kilometres per hour:

a 10 knots
b 12 knots
c 5 knots
d 6.5 knots

3 Two locations have the same latitude but have longitudes 150°6′E and 150°18.6′E. How far apart are these locations in:

a nautical miles?
b kilometres?

4 Two locations have the same longitude but have latitudes 10°5′N and 5°50′S. How far apart are these locations in:

a nautical miles?
b kilometres?

5 Calculate the speed in:

i knots
ii km/h for the journeys:

a 12 n mile in 1.5 hours
b 210 n mile in 14 hours
c 1.2 n mile in 45 minutes

Use Chart 1 to answer Questions 6 to 8.

6 How do you know that Chart 1 is of a location south of the equator?

7 Locate the position coordinates of:

a the peak on Little Sister Island
b Stewart Lighthouse
c the Beacon at the entrance to Shark Bay

8 Find the distance in:

i n mile
ii km from the peak on Little Sister Island to:

a the Beacon at the entrance to Shark Bay
b Stewart Lighthouse
c the Tower at the position 34°56.1′S, 151°54.6′E.

9 Calculate the time it would take a boat travelling at 12 knots to go from:

a point X to point Y
b point Y to point Z
c point Z to point X

Remember

Remember

If speed $= \dfrac{\text{distance}}{\text{time}}$

then

time $= \dfrac{\text{distance}}{\text{speed}}$

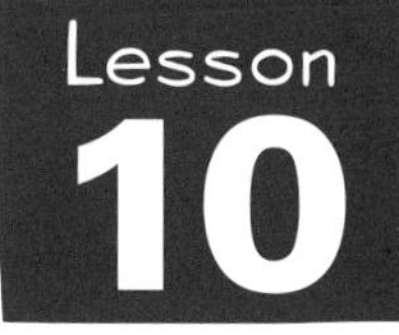

Bearings on nautical maps

A **navigation rose (compass rose)** similar to the one shown is found on navigation charts. The navigation rose has an angle scale from 0° to 360° with the line through 0° and 180° showing **due north**. It is used to measure and draw **true bearings**.

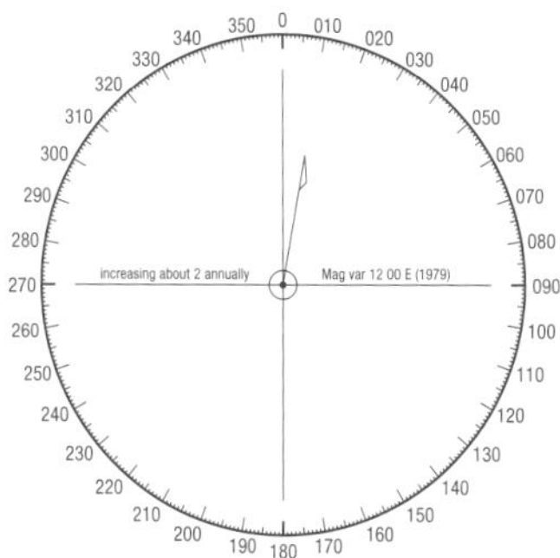

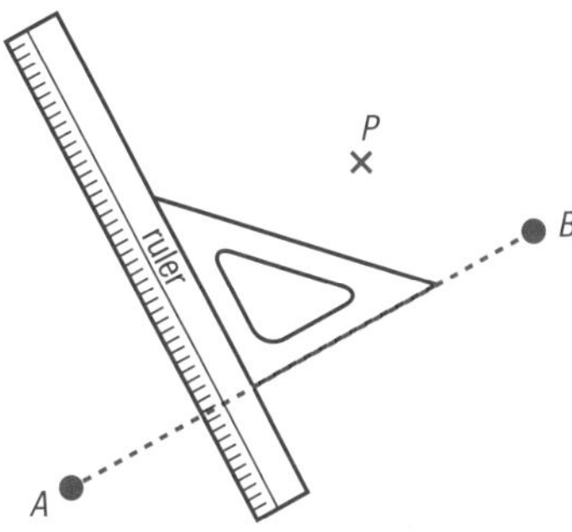

A ruler and set square are useful items to help you find bearings on navigation maps.

Example 1

Use the navigation rose to find the bearing of *P* from *B*.

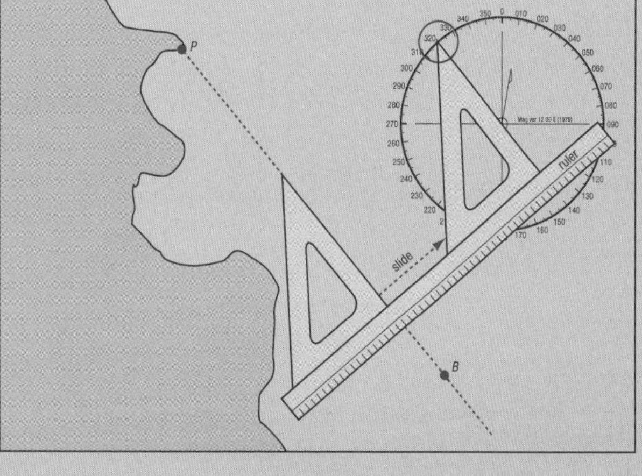

Answer

To find the bearing of *P* from *B*:

a Rule a straight line from *B* to *P*.

b Position the ruler and set square so that one edge of the set square is along the line *BP*, as shown in the diagram.

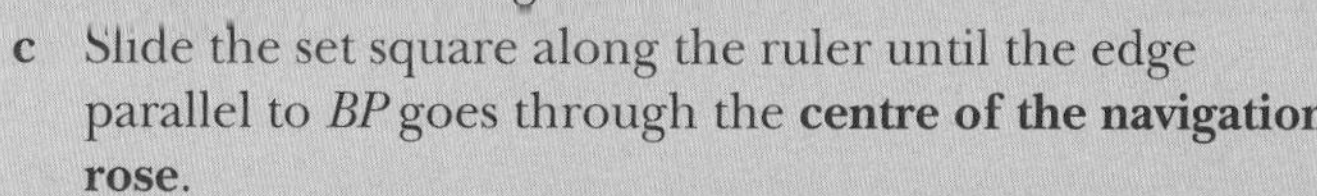

c Slide the set square along the ruler until the edge parallel to *BP* goes through the **centre of the navigation rose**.

d Read the bearing, 320° in this example, on the compass rose.

Note: If we were finding the bearing of *B* from *P* the set square would be placed on the other side of the ruler. Do this and confirm that the bearing of *B* from *P* is 140°.

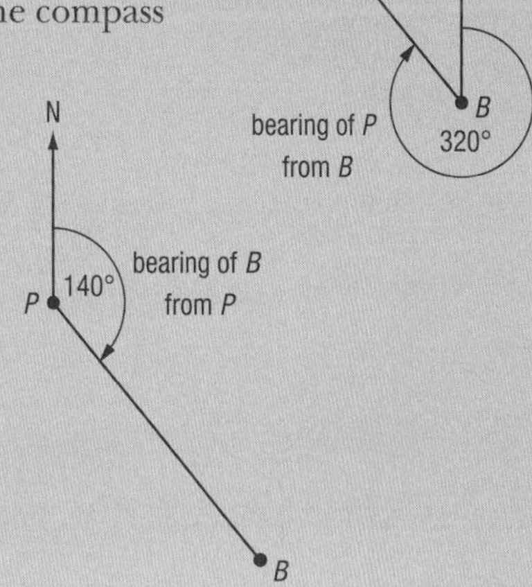

Example 2

From a ship the point Q is on a bearing of 210°. Find the **position** line of the ship.

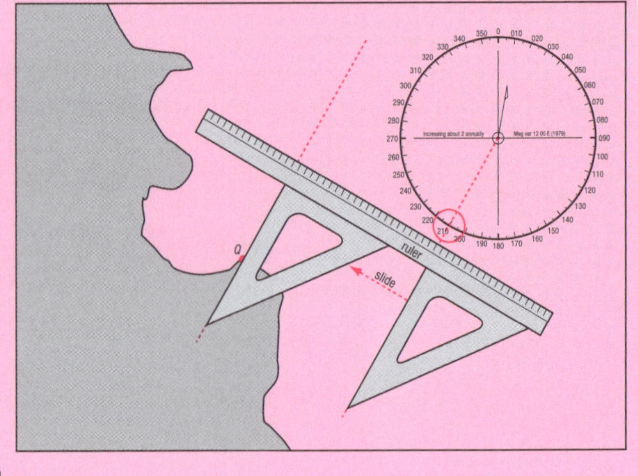

Answer

The **position line** of the ship means that the ship's position is somewhere along this line.

To find the position line of the ship:

a Rule a line from the centre of the navigation rose through the bearing 210°.

b Place the ruler and set square as shown with one edge of the set square on the ruled line.

c Slide the set square along the ruler until it goes through point Q.

d Rule the line through point Q (the dotted line on the diagram) and extend it to show the position line of the ship. The ship's position is somewhere along this line.

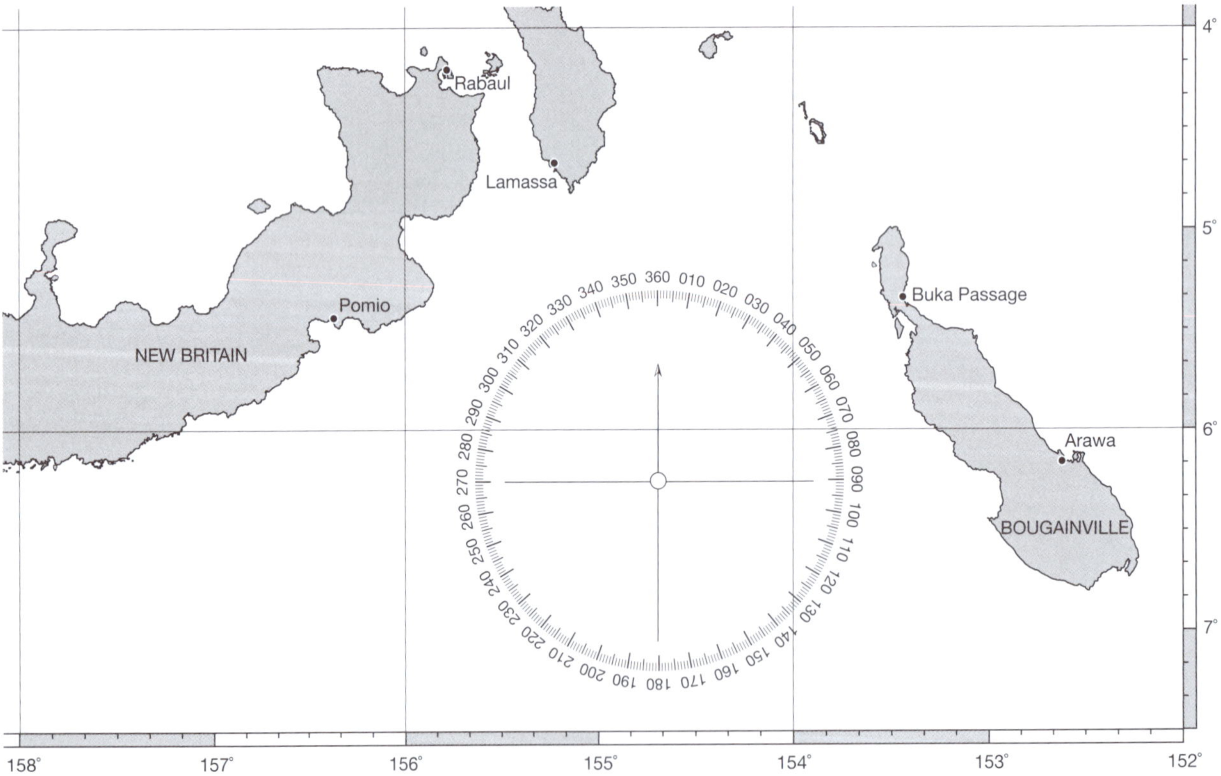

Chart 2

EXERCISE

Use Chart 2 for this exercise.

1 Use the navigation rose, a ruler and set square to find the bearing of Rabaul from:
   a Arawa
   b Pomio
   c Buka Passage
   d Lamassa

2 Use the navigation rose, a ruler and set square to find the bearing of:
   a Buka Passage from Lamassa
   b Lamassa from Pomio
   c Pomio from Rabaul
   d Pomio from Buka Passage

3 Using a copy of Chart 2 (or carefully drawing in pencil on the chart) find, and rule on the chart, the position lines of a ship that is on a bearing of:
   a 200° from Lamassa
   b 025° from Buka Passage
   c 130° from Pomio
   d 045° from Arawa

4 Use the latitude (vertical scale) to find and mark on Chart 2 the position of a ship that is:
   a 120 n mile on a bearing of 200° from Lamassa
   b 40 n mile on a bearing of 025° from Buka Passage
   c 100 n mile on a bearing of 130° from Pomio
   d 20 n mile on a bearing of 045° from Arawa

Remember

1 nautical mile (1 n mile) is measured as 1 minute on the latitude (vertical) scale.

Challenge

5 A ship is on a bearing of 110° from Pomio and on a bearing of 195° from Lamassa. Find the position of the ship on the map and give its coordinates.

6 At 12 noon a ship is 65 km on a bearing of 165° from Pomio. If the ship is travelling to Lamassa maintaining a speed of 15 knots, what time will it get to Lamassa?

Lesson 11

Plotting straight-line courses

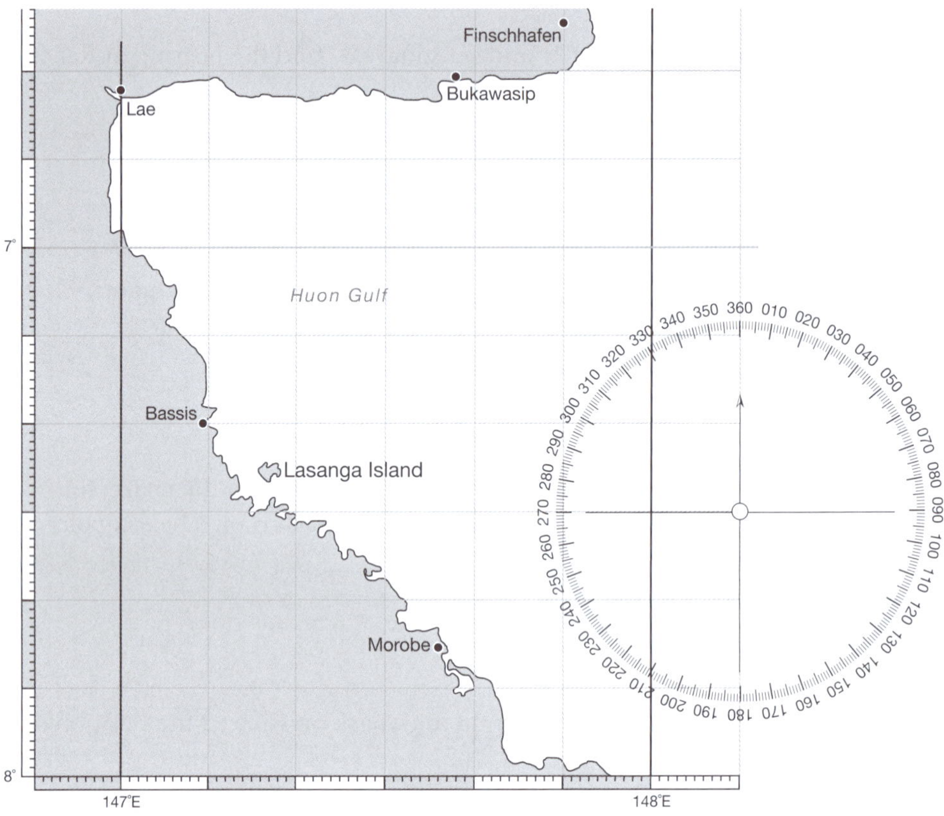

Chart 3

Example

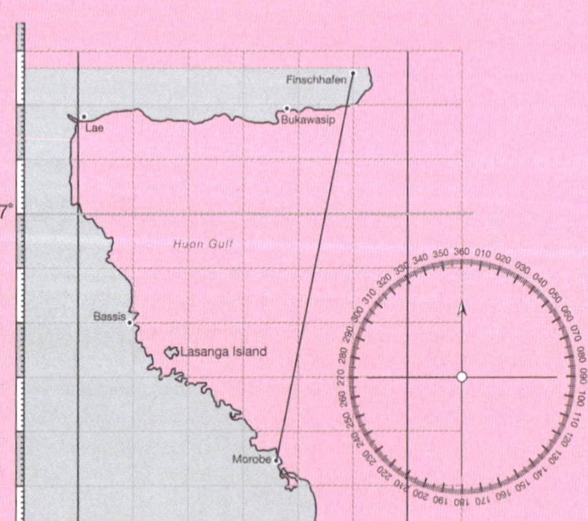

A light plane is to fly from Finschhafen to Morobe.

a Use Chart 3 to find the bearing of Morobe from Finschhafen.

b Use the longitude scale to calculate the straight-line distance from Finschhafen to Morobe.

c If the plane flies at an average speed of 200 km/h, find the time it will take to fly from Finschhafen to Morobe.

Answer

a Drawing a straight line from Finschhafen to Morobe and using the compass rose on the chart (or a 360° protractor), the direction can be read as 192°.

b Using the longitude scale: the distance measures 72′ on the longitude scale so the distance is 72 n mile.

$72 \times 1.852 = 133.3$ km

c If speed $= \frac{\text{distance}}{\text{time}}$ then time $= \frac{\text{distance}}{\text{speed}}$

so time $= \frac{133.3}{200} = 0.67$ h $= 0.67 \times 60$ min $= 40$ min

Remember

1 n mile = 1.852 km

EXERCISE

1 Use your atlas and a 360° protractor (you may need to make a 360° 'compass rose' using your 180° protractor).

a Find the bearing of Madang from Port Moresby taking care to put the 000°–180° line (due north) parallel to a longitude line and the centre of the protractor at Port Moresby.

b Calculate the straight-line distance in kilometres from Port Moresby to Madang using either the scale on the map or the longitude scale.

c If an aeroplane flies at an average speed of 500 km/h, find the time it will take to fly from Port Moresby to Madang.

2 Use your atlas and a 360° protractor.

a Find the bearing of Alotau from Lae taking care to put your 000°–180° line (due north) along the longitude line.

b Calculate the straight-line distance in kilometres from Lae to Alotau using either the scale on the map or the longitude scale.

c If an aeroplane flies at an average speed of 200 km/h, find the time it will take to fly from Lae to Alotau.

3 Use Chart 4.

a Find the bearing of Kerema from Madang.

b Calculate the straight-line distance in kilometres from Madang to Kerema using the longitude scale.

c If an aeroplane flies at an average speed of 240 km/h, find the time it will take to fly from Madang to Kerema.

4 Use Chart 4.

a Find the bearing of Madang from Mt. Hagen.

b Calculate the straight-line distance in kilometres from Mt Hagen to Madang using the longitude scale.

c If an aeroplane flies at an average speed of 300 km/h, find the time it will take to fly from Mt Hagen to Madang.

144°
146°
148°
4°
6°
8°
Wewak
Angoram
Awar
Bogia
EAST SEPIK
Kinim
Bunabun
Sarang
MADANG
ENGA
Madang
Wabag
WESTERN
HIGHLANDS
Mount Hagen
Saidor
Kundiawa
Goroka
Wasu
Mendi
Gumine
EASTERN
HIGHLANDS
Kaiapit
Kabwum
SOUTHERN
HIGHLANDS
SIMBU
Okapa
Finschhafen
MOROBE
Lae
Wonenara
GULF
Mumeng
Marawaka
Menyamya
Kikori
Aseki
Wau
Baimuru
Morobe
Ihu
Garaina
Kerema
Malalaua
Kukipi
NORTHERN
Tapini
Bereina
CENTRAL
Popondetta
Kairuku
Kokoda
Afore
NATIONAL
CAPITAL
DISTRICT
Sogeri
Port Moresby

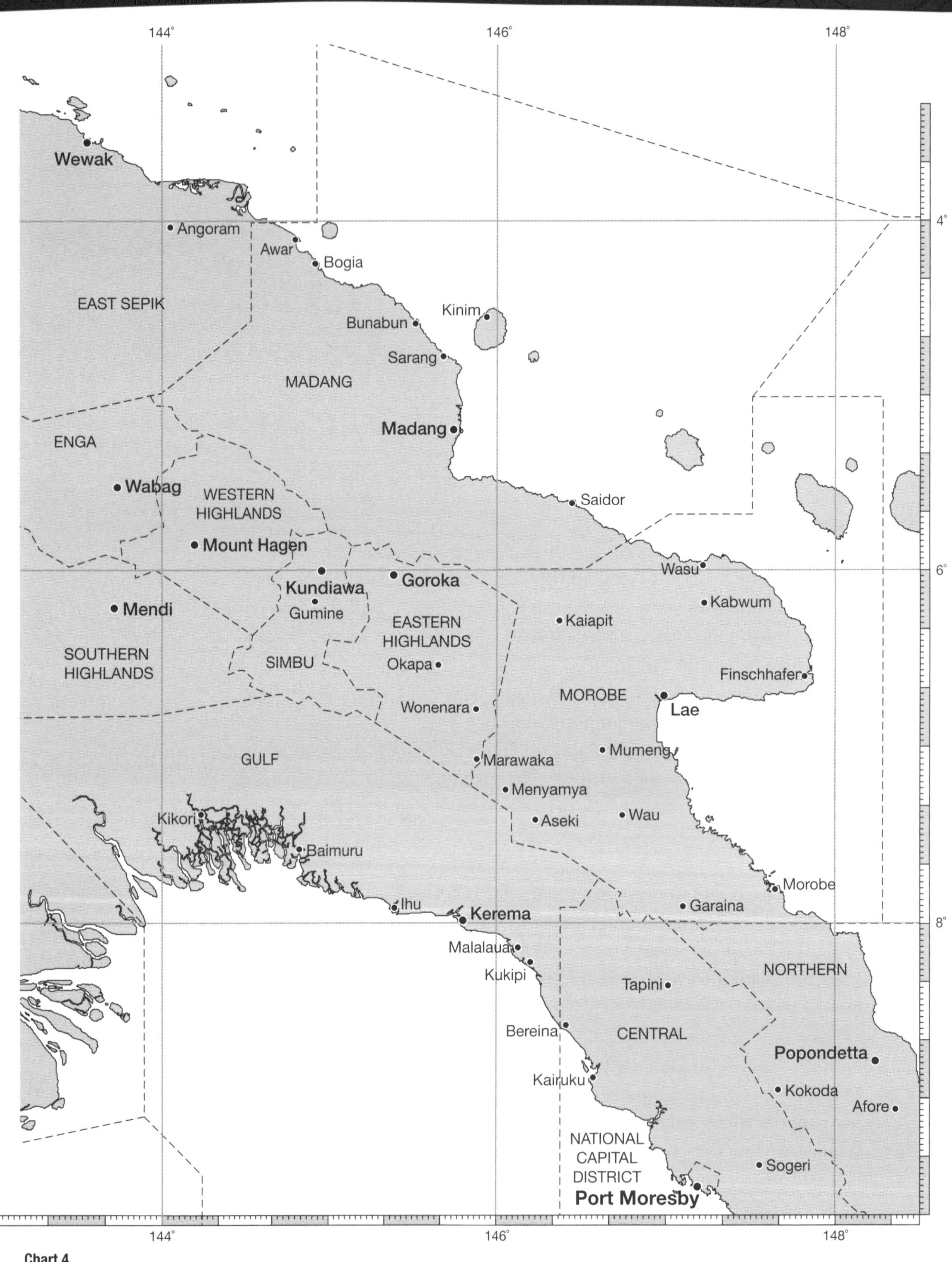

Chart 4

Navigation at sea

Lesson

12

Often it is not possible to plot a straight course from one location to another when travelling by sea. Peninsulas and islands are often 'in the way' of a straight-line course, so a journey may consist of several 'legs' that each have their own direction and distance.

Example

a Devise a route that you could use for a journey by sea from Tufi to Rabaraba (see map).

b Find the directions of each leg of the route. Calculate the distance travelled on each leg of the course.

c If the boat can travel at an average speed of 12 knots, find the time the journey will take.

d Find the point on the route where you will be nearest to land and the distance from land of this point.

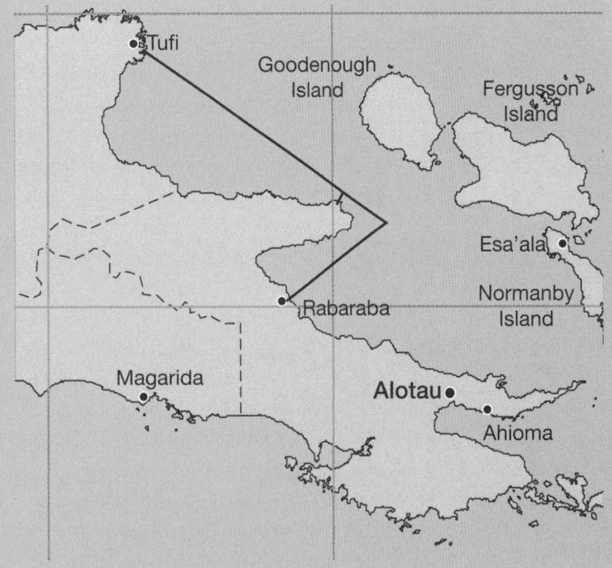

Answer

Final calculations will depend on the map.

a A route with two legs has been drawn on the map. Other routes are possible.

b Using a 360° protractor with the 000°–180° line parallel to one of the longitude lines and the centre going through Tufi, the direction of the first leg is read as 123°. With the centre of the protractor on the point where the direction changes and the 000°–180° line parallel to one of the longitude lines on the map, the direction of the second leg of the route is read as 233°.

The distance on the map for the two legs of the route are 43 mm and 18 mm respectively.

On the map the distance between the latitude lines is 39 mm, so we can say 39 mm on the map is equivalent to 60 n mile in actual distance.

If 39 mm ≡ 60 n mile then 1 mm ≡ $\frac{60}{39}$ n mile

The distance travelled on the first leg will be $43 \times \frac{60}{39} = 66.154$ n mile

(66.154 n mile ≡ 66.154 × 1.852 km = 122.5 km)

The distance travelled on the second leg will be $18 \times \frac{60}{39}$ n mile = 27.692 n mile

(27.692 n mile ≡ 27.692 × 1.852 km = 51.29 km)

continued

c The ship averages 12 knots = 12 n mile/h

Using time = $\frac{\text{distance}}{\text{speed}}$:

The time for the first leg of the course is $\frac{66.154}{12}$ hours = 5.513 hours = 5 h 31 min.

The time for the second leg of the course is $\frac{27.692}{12}$ h = 2.308 hours = 2 h 18 min.

d The point where the route is nearest to land is shown on the map as a line drawn perpendicular to the first leg of the course. The length of this line is measured on the map as 2.0 mm, so the distance is

$2.0 \times \frac{60}{39} = 3.077$ n mile $= 3.077 \times 1.852$ km $= 5.70$ km.

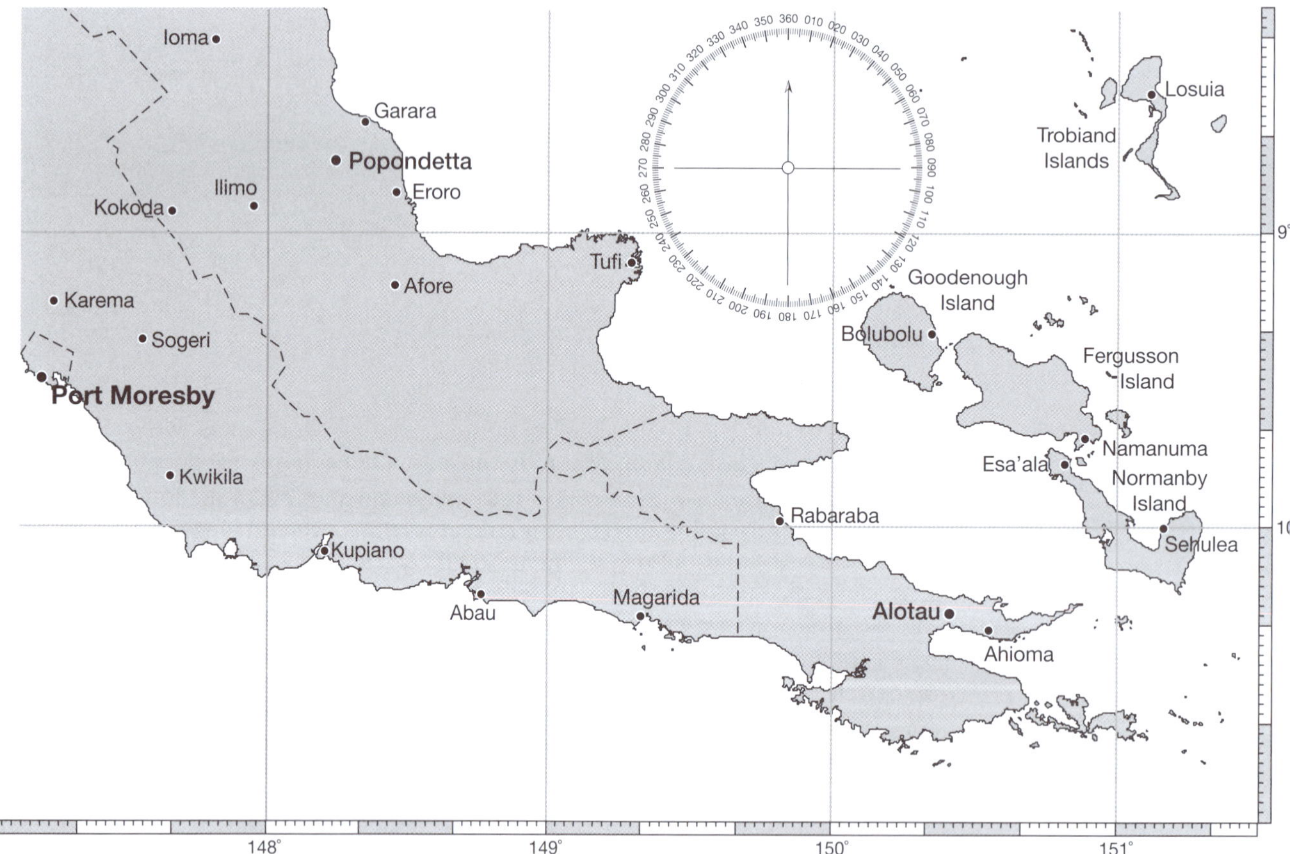

Chart 5

EXERCISE

Using Chart 5 for each of the following:

- **a** Devise a route that you could use for a journey by sea.
- **b** Find the directions of each leg of the course.
- **c** Calculate the distance travelled on each leg of the course.
- **d** If the boat can travel at an average speed of 12 knots, find the time the journey will take.
- **e** Find the point on the course where you will be nearest to land and the distance from land of this point.

1. Port Moresby to Kupiano
2. Tufi to Bolubolu
3. Rabaraba to Alotau
4. Alotau to Esa'ala
5. Tufi to Losuia to Esa'ala.

Navigation: practice for the project

Assessment of this option is a group project where you are asked to chart a navigation route using one or more of the techniques that you have encountered in the previous lessons. The results of your project are to be presented in writing.

You will need to:

- demonstrate appropriate investigation skills
- choose and apply relevant mathematical techniques
- make an effective communication of the project results.

The exercise below tests some of the skills you will need to demonstrate in your project.

EXERCISE

1. A yacht race has three legs and the yachts are required to locate and go around two buoys and then return to the starting point. The first buoy has a bearing of 273° and is 5 n mile from the starting point. The second buoy has a bearing of 168° and is 7 n mile from the first buoy.

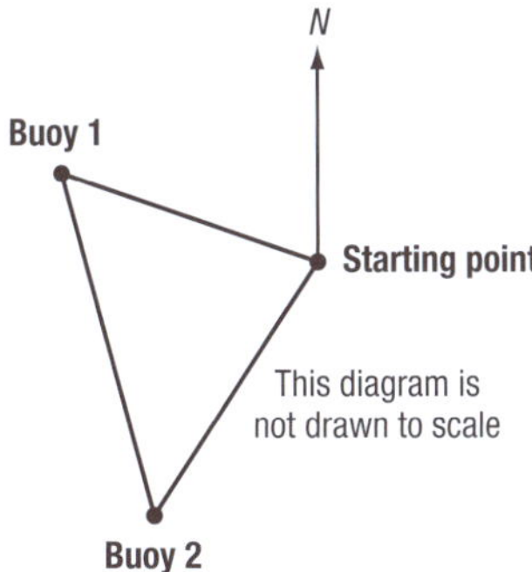

- **a** Draw a **scale** diagram showing the starting point and the two buoys.
- **b** Find the angle inside the triangular course at buoy 1.

c Use the cosine rule to calculate the distance, in nautical miles, travelled on the third leg of the race (from buoy 2 back to the starting point). Compare this answer with the distance found from the scale diagram.

d Measure the angle inside the triangular course at buoy 2 and then calculate this angle using the sine rule. Compare and comment on your answers.

e Find the bearing of the starting point from buoy 2.

f Calculate the distance travelled on this yacht race giving your answer in nautical miles and kilometres.

g A particular yacht can travel at an average speed of 15 knots. How long would it take for this yacht to finish the race? Give your answer in hours and minutes.

2 Use Chart 2 or your atlas.

a i Find the bearing of Arawa from Rabaul.

ii Calculate the straight line distance, in kilometres, from Rabaul to Arawa.

iii If an aeroplane flies at an average speed of 300 km/h, find the time it will take to fly from Rabaul to Arawa.

b i Devise a route that you could use for a journey by sea from Rabaul to Arawa.

ii Find the directions of each leg of the route.

iii Calculate the distance travelled, in nautical miles and kilometres, on each leg of the route.

iv If a boat can travel at an average speed of 15 knots, find the time the journey will take in hours and minutes.

v Find three points on the route where you will be nearest to land and the distance from land of these points.

Appendix 1

SQUARES

	0	1	2	3	4	5	6	7	8	9	1	2	3	4	5	6	7	8	9
1.0	1.000	1.020	1.040	1.061	1.082	1.102	1.124	1.145	1.166	1.188	2	4	6	8	11	13	15	17	19
1.1	1.210	1.232	1.254	1.277	1.300	1.322	1.346	1.369	1.392	1.416	2	5	7	9	12	14	16	18	21
1.2	1.440	1.464	1.488	1.513	1.538	1.562	1.588	1.613	1.638	1.664	3	5	8	10	13	15	18	20	23
1.3	1.690	1.716	1.742	1.769	1.796	1.822	1.850	1.877	1.904	1.932	3	5	8	11	14	16	19	22	24
1.4	1.960	1.988	2.016	2.045	2.074	2.102	2.132	2.161	2.190	2.220	3	6	9	12	15	17	20	23	26
1.5	2.250	2.280	2.310	2.341	2.372	2.402	2.434	2.465	2.496	2.528	3	6	9	12	16	19	22	25	28
1.6	2.560	2.592	2.624	2.657	2.690	2.722	2.756	2.789	2.822	2.856	3	7	10	13	17	20	23	26	30
1.7	2.890	2.924	2.958	2.993	3.028	3.062	3.098	3.133	3.168	3.204	4	7	11	14	18	21	25	28	32
1.8	3.240	3.276	3.312	3.349	3.386	3.422	3.460	3.497	3.534	3.572	4	7	11	15	19	22	26	30	33
1.9	3.610	3.648	3.686	3.725	3.764	3.802	3.842	3.881	3.920	3.960	4	8	12	16	20	23	27	31	35
2.0	4.000	4.040	4.080	4.121	4.162	4.202	4.244	4.285	4.326	4.368	4	8	12	16	21	25	29	33	37
2.1	4.410	4.452	4.494	4.537	4.580	4.622	4.666	4.709	4.752	4.796	4	9	13	17	22	26	30	34	39
2.2	4.840	4.884	4.928	4.973	5.018	5.062	5.108	5.153	5.198	5.244	5	9	14	18	23	27	32	36	41
2.3	5.290	5.336	5.382	5.429	5.476	5.522	5.570	5.617	5.664	5.712	5	9	14	19	24	28	33	38	42
2.4	5.760	5.808	5.856	5.905	5.954	6.002	6.052	6.101	6.150	6.200	5	10	15	20	25	29	34	39	44
2.5	6.250	6.300	6.350	6.401	6.452	6.502	6.554	6.605	6.656	6.708	5	10	15	20	26	31	36	41	46
2.6	6.760	6.812	6.864	6.917	6.970	7.022	7.076	7.129	7.182	7.236	5	11	16	21	27	32	37	42	48
2.7	7.290	7.344	7.398	7.453	7.508	7.562	7.618	7.673	7.728	7.784	6	11	17	22	28	33	39	44	50
2.8	7.840	7.896	7.952	8.009	8.066	8.122	8.180	8.237	8.294	8.352	6	11	17	23	29	34	40	46	51
2.9	8.410	8.468	8.526	8.585	8.644	8.702	8.762	8.821	8.880	8.940	6	12	18	24	30	35	41	47	53
3.0	9.00	9.06	9.12	9.18	9.24	9.30	9.36	9.42	9.49	9.55	1	1	2	2	3	4	4	5	5
3.1	9.61	9.67	9.73	9.80	9.86	9.92	9.99	10.05	10.11	10.18	1	1	2	3	3	4	4	5	6
3.2	10.24	10.30	10.37	10.43	10.50	10.56	10.63	10.69	10.76	10.82	1	1	2	3	3	4	5	5	6
3.3	10.89	10.96	11.02	11.09	11.16	11.22	11.29	11.36	11.42	11.49	1	1	2	3	3	4	5	5	6
3.4	11.56	11.63	11.70	11.76	11.83	11.90	11.97	12.04	12.11	12.18	1	1	2	3	3	4	5	6	6
3.5	12.25	12.32	12.39	12.46	12.53	12.60	12.67	12.74	12.82	12.89	1	1	2	3	4	4	5	6	6
3.6	12.96	13.03	13.10	13.18	13.25	13.32	13.40	13.47	13.54	13.62	1	1	2	3	4	4	5	6	7
3.7	13.69	13.76	13.84	13.91	13.99	14.06	14.14	14.21	14.29	14.36	1	2	2	3	4	5	5	6	7
3.8	14.44	14.52	14.59	14.67	14.75	14.82	14.90	14.98	15.05	15.13	1	2	2	3	4	5	5	6	7
3.9	15.21	15.29	15.37	15.44	15.52	15.60	15.68	15.76	15.84	15.92	1	2	2	3	4	5	6	6	7
4.0	16.00	16.08	16.16	16.24	16.32	16.40	16.48	16.56	16.65	16.73	1	2	2	3	4	5	6	6	7
4.1	16.81	16.89	16.97	17.06	17.14	17.22	17.31	17.39	17.47	17.56	1	2	2	3	4	5	6	7	7
4.2	17.64	17.72	17.81	17.89	17.98	18.06	18.15	18.23	18.32	18.40	1	2	3	3	4	5	6	7	8
4.3	18.49	18.58	18.66	18.75	18.84	18.92	19.01	19.10	19.18	19.27	1	2	3	3	4	5	6	7	8
4.4	19.36	19.45	19.54	19.62	19.71	19.80	19.89	19.98	20.07	20.16	1	2	3	4	4	5	6	7	8
4.5	20.25	20.34	20.43	20.52	20.61	20.70	20.79	20.88	20.98	21.07	1	2	3	4	5	5	6	7	8
4.6	21.16	21.25	21.34	21.44	21.53	21.62	21.72	21.81	21.90	22.00	1	2	3	4	5	6	7	7	8
4.7	22.09	22.18	22.28	22.37	22.47	22.56	22.66	22.75	22.85	22.94	1	2	3	4	5	6	7	8	9
4.8	23.04	23.14	23.23	23.33	23.43	23.52	23.62	23.72	23.81	23.91	1	2	3	4	5	6	7	8	9
4.9	24.01	24.11	24.21	24.30	24.40	24.50	24.60	24.70	24.80	24.90	1	2	3	4	5	6	7	8	9

SQUARES (continued)

	0	1	2	3	4	5	6	7	8	9	1	2	3	4	5	6	7	8	9
5.0	25.00	25.10	25.20	25.30	25.40	25.50	25.60	25.70	25.81	25.91	1	2	3	4	5	6	7	8	9
5.1	26.01	26.11	26.21	26.32	26.42	26.52	26.63	26.73	26.83	26.94	1	2	3	4	5	6	7	8	9
5.2	27.04	27.14	27.25	27.35	27.46	27.56	27.67	27.77	27.88	27.98	1	2	3	4	5	6	7	8	9
5.3	28.09	28.20	28.30	28.41	28.52	28.62	28.73	28.84	28.94	29.05	1	2	3	4	5	6	7	9	10
5.4	29.16	29.27	29.38	29.48	29.59	29.70	29.81	29.92	30.03	30.14	1	2	3	4	5	7	8	9	10
5.5	30.25	30.36	30.47	30.58	30.69	30.80	30.91	31.02	31.14	31.25	1	2	3	4	6	7	8	9	10
5.6	31.36	31.47	31.58	31.70	31.81	31.92	32.04	32.15	32.26	32.38	1	2	3	5	6	7	8	9	10
5.7	32.49	32.60	32.72	32.83	32.95	33.06	33.18	33.29	33.41	33.52	1	2	3	5	6	7	8	9	10
5.8	33.64	33.76	33.87	33.99	34.11	34.22	34.34	34.46	34.57	34.69	1	2	4	5	6	7	8	9	11
5.9	34.81	34.93	35.05	35.16	35.28	35.40	35.52	35.64	35.76	35.88	1	2	4	5	6	7	8	10	11
6.0	36.00	36.12	36.24	36.36	36.48	36.60	36.72	36.84	36.97	37.09	1	2	4	5	6	7	8	10	11
6.1	37.21	37.33	37.45	37.58	37.70	37.82	37.95	38.07	38.19	38.32	1	2	4	5	6	7	9	10	11
6.2	38.44	38.56	38.69	38.81	38.94	39.06	39.19	39.31	39.44	39.56	1	3	4	5	6	8	9	10	11
6.3	39.69	39.82	39.94	40.07	40.20	40.32	40.45	40.58	40.70	40.83	1	3	4	5	6	8	9	10	11
6.4	40.96	41.09	41.22	41.34	41.47	41.60	41.73	41.86	41.99	42.12	1	3	4	5	6	8	9	10	12
6.5	42.25	42.38	42.51	42.64	42.77	42.90	43.03	43.16	43.30	43.43	1	3	4	5	7	8	9	10	12
6.6	43.56	43.69	43.82	43.96	44.09	44.22	44.36	44.49	44.62	44.76	1	3	4	5	7	8	9	11	12
6.7	44.89	45.02	45.16	45.29	45.43	45.56	45.70	45.83	45.97	46.10	1	3	4	5	7	8	9	11	12
6.8	46.24	46.38	46.51	46.65	46.79	46.92	47.06	47.20	47.33	47.47	1	3	4	5	7	8	10	11	12
6.9	47.61	47.75	47.89	48.02	48.16	48.30	48.44	48.58	48.72	48.86	1	3	4	6	7	8	10	11	13
7.0	49.00	49.14	49.28	49.42	49.56	49.70	49.84	49.98	50.13	50.27	1	3	4	6	7	8	10	11	13
7.1	50.41	50.55	50.69	50.84	50.98	51.12	51.27	51.41	51.55	51.70	1	3	4	6	7	9	10	11	13
7.2	51.84	51.98	52.13	52.27	52.42	52.56	52.71	52.85	53.00	53.14	1	3	4	6	7	9	10	12	13
7.3	53.29	53.44	53.58	53.73	53.88	54.02	54.17	54.32	54.46	54.61	1	3	4	6	7	9	10	12	13
7.4	54.76	54.91	55.06	55.20	55.35	55.50	55.65	55.80	55.95	56.10	1	3	4	6	7	9	10	12	13
7.5	56.25	56.40	56.55	56.70	56.85	57.00	57.15	57.30	57.46	57.61	2	3	5	6	8	9	11	12	14
7.6	57.76	57.91	58.06	58.22	58.37	58.52	58.68	58.83	58.98	59.14	2	3	5	6	8	9	11	12	14
7.7	59.29	59.44	59.60	59.75	59.91	60.06	60.22	60.37	60.53	60.68	2	3	5	6	8	9	11	12	14
7.8	60.84	61.00	61.15	61.31	61.47	61.62	61.78	61.94	62.09	62.25	2	3	5	6	8	9	11	13	14
7.9	62.41	62.57	62.73	62.88	63.04	63.20	63.36	63.52	63.68	63.84	2	3	5	6	8	10	11	13	14
8.0	64.00	64.16	64.32	64.48	64.64	64.80	64.96	65.12	65.29	65.45	2	3	5	6	8	10	11	13	14
8.1	65.61	65.77	65.93	66.10	66.26	66.42	66.59	66.75	66.91	67.08	2	3	5	7	8	10	11	13	15
8.2	67.24	67.40	67.57	67.73	67.90	68.06	68.23	68.39	68.56	68.72	2	3	5	7	8	10	12	13	15
8.3	68.89	69.06	69.22	69.39	69.56	69.72	69.89	70.06	70.22	70.39	2	3	5	7	8	10	12	13	15
8.4	70.56	70.73	70.90	71.06	71.23	71.40	71.57	71.74	71.91	72.08	2	3	5	7	8	10	12	14	15
8.5	72.25	72.42	72.59	72.76	72.93	73.10	73.27	73.44	73.62	73.79	2	3	5	7	9	10	12	14	15
8.6	73.96	74.13	74.30	74.48	74.65	74.82	75.00	75.17	75.34	75.52	2	3	5	7	9	10	12	14	16
8.7	75.69	75.86	76.04	76.21	76.39	76.56	76.74	76.91	77.09	77.26	2	4	5	7	9	11	12	14	16
8.8	77.44	77.62	77.79	77.97	78.15	78.32	78.50	78.68	78.85	79.03	2	4	5	7	9	11	12	14	16
8.9	79.21	79.39	79.57	79.74	79.92	80.10	80.28	80.46	80.64	80.82	2	4	5	7	9	11	13	14	16

SQUARES (continued)

	0	1	2	3	4	5	6	7	8	9	1	2	3	4	5	6	7	8	9
9.0	81.00	81.18	81.36	81.54	81.72	81.90	82.08	82.26	82.45	82.63	2	4	5	7	9	11	13	14	16
9.1	82.81	82.99	83.17	83.36	83.54	83.72	83.91	84.09	84.27	84.46	2	4	5	7	9	11	13	15	16
9.2	84.64	84.82	85.01	85.19	85.38	85.56	85.75	85.93	86.12	86.30	2	4	6	7	9	11	13	15	17
9.3	86.49	86.68	86.86	87.05	87.24	87.42	87.61	87.80	87.98	88.17	2	4	6	7	9	11	13	15	17
9.4	88.36	88.55	88.74	88.92	89.11	89.30	89.49	89.68	89.87	90.06	2	4	6	8	9	11	13	15	17
9.5	90.25	90.44	90.63	90.82	91.01	91.20	91.39	91.58	91.78	91.97	2	4	6	8	10	11	13	15	17
9.6	92.16	92.35	92.54	92.74	92.93	93.12	93.32	93.51	93.70	93.90	2	4	6	8	10	12	14	15	17
9.7	94.09	94.28	94.48	94.67	94.87	95.06	95.26	95.45	95.65	95.84	2	4	6	8	10	12	14	16	18
9.8	96.04	96.24	96.43	96.63	96.83	97.02	97.22	97.42	97.61	97.81	2	4	6	8	10	12	14	16	18
9.9	98.01	98.21	98.41	98.60	98.80	99.00	99.20	99.40	99.60	99.80	2	4	6	8	10	12	14	16	18

NATURAL SINES

	0′	6′	12′	18′	24′	30′	36′	42′	48′	54′	1′	2′	3′	4′	5′
0°	0.0000	0017	0035	0052	0070	0087	0105	0122	0140	0157	3	6	9	12	15
1	0.0175	0192	0209	0227	0244	0262	0279	0297	0314	0332	3	6	9	12	15
2	0.0349	0366	0384	0401	0419	0436	0454	0471	0488	0506	3	6	9	12	15
3	0.0523	0541	0558	0576	0593	0610	0628	0645	0663	0680	3	6	9	12	15
4	0.0698	0715	0732	0750	0767	0785	0802	0819	0837	0854	3	6	9	12	14
5	0.0872	0889	0906	0924	0941	0958	0976	0993	1011	1028	3	6	9	12	14
6	0.1045	1063	1080	1097	1115	1132	1149	1167	1184	1201	3	6	9	12	14
7	0.1219	1236	1253	1271	1288	1305	1323	1340	1357	1374	3	6	9	12	14
8	0.1392	1409	1426	1444	1461	1478	1495	1513	1530	1547	3	6	9	11	14
9	0.1564	1582	1599	1616	1633	1650	1668	1685	1702	1719	3	6	9	11	14
10	0.1736	1754	1771	1788	1805	1822	1840	1857	1874	1891	3	6	9	11	14
11	0.1908	1925	1942	1959	1977	1994	2011	2028	2045	2062	3	6	9	11	14
12	0.2070	2096	2113	2130	2147	2164	2181	2198	2215	2233	3	6	9	11	14
13	0.2250	2267	2284	2300	2317	2334	2351	2368	2385	2402	3	6	8	11	14
14	0.2419	2436	2453	2470	2487	2504	2521	2538	2554	2571	3	6	8	11	14
15	0.2588	2605	2622	2639	2656	2672	2689	2706	2723	2740	3	6	8	11	14
16	0.2756	2773	2790	2807	2823	2840	2857	2874	2890	2907	3	6	8	11	14
17	0.2924	2940	2957	2974	2990	3007	3024	3040	3057	3074	3	6	8	11	14
18	0.3090	3107	3123	3140	3156	3173	3190	3206	3223	3239	3	6	8	11	14
19	0.3256	3272	3289	3305	3322	3338	3355	3371	3387	3404	3	5	8	11	14
20	0.3420	3437	3453	3469	3486	3502	3518	3535	3551	3567	3	5	8	11	14
21	0.3584	3600	3616	3633	3649	3665	3681	3697	3714	3730	3	5	8	11	13
22	0.3746	3762	3778	3795	3811	3827	3843	3859	3875	3891	3	5	8	11	13
23	0.3907	3923	3939	3955	3971	3987	4003	4019	4035	4051	3	5	8	11	13
24	0.4067	4083	4099	4115	4131	4147	4163	4179	4195	4210	3	5	8	11	13

NATURAL SINES (continued)

	0′	6′	12′	18′	24′	30′	36′	42′	48′	54′	1′	2′	3′	4′	5′
25°	0.4226	4242	4258	4274	4289	4305	4321	4337	4352	4368	3	5	8	11	13
26	0.4384	4399	4415	4431	4446	4462	4478	4493	4509	4524	3	5	8	10	13
27	0.4540	4555	4571	4586	4602	4617	4633	4648	4664	4679	3	5	8	10	13
28	0.4695	4710	4726	4741	4756	4772	4787	4802	4818	4833	3	5	8	10	13
29	0.4848	4863	4879	4894	4909	4924	4939	4955	4970	4985	3	5	8	10	13
30	0.5000	5015	5030	5045	5060	5075	5090	5105	5120	5135	3	5	8	10	12
31	0.5150	5165	5180	5195	5210	5225	5240	5255	5270	5284	2	5	7	10	12
32	0.5299	5314	5329	5344	5358	5373	5388	5402	5417	5432	2	5	7	10	12
33	0.5446	5461	5476	5490	5505	5519	5534	5548	5563	5577	2	5	7	10	12
34	0.5592	5606	5621	5635	5650	5664	5678	5693	5707	5721	2	5	7	10	12
35	0.5736	5750	5764	5779	5793	5807	5821	5835	5850	5864	2	5	7	9	12
36	0.5878	5892	5906	5920	5934	5948	5962	5976	5990	6004	2	5	7	9	12
37	0.6018	6032	6046	6060	6074	6088	6101	6115	6129	6143	2	5	7	9	12
38	0.6157	6170	6184	6198	6211	6225	6239	6252	6266	6280	2	5	7	9	11
39	0.6293	6307	6320	6334	6347	6361	6374	6388	6401	6414	2	4	7	9	11
40	0.6428	6441	6455	6468	6481	6494	6508	6521	6534	6547	2	4	7	9	11
41	0.6561	6574	6587	6600	6613	6626	6639	6652	6665	6678	2	4	7	9	11
42	0.6691	6704	6717	6730	6743	6756	6769	6782	6794	6807	2	4	6	9	11
43	0.6820	6833	6845	6858	6871	6884	6896	6909	6921	6934	2	4	6	8	11
44	0.6947	6959	6972	6984	6997	7009	7022	7034	7046	7059	2	4	6	8	10
45°	0.7071	7083	7096	7108	7120	7133	7145	7157	7169	7181	2	4	6	8	10
46	0.7193	7206	7218	7230	7242	7254	7266	7278	7290	7302	2	4	6	8	10
47	0.7314	7325	7337	7349	7361	7373	7385	7396	7408	7420	2	4	6	8	10
48	0.7431	7443	7455	7466	7478	7490	7501	7513	7524	7536	2	4	6	8	10
49	0.7547	7559	7570	7581	7593	7604	7615	7627	7638	7649	2	4	6	8	9
50	0.7660	7672	7683	7694	7705	7716	7727	7738	7749	7760	2	4	6	7	9
51	0.7771	7782	7793	7804	7815	7826	7837	7848	7859	7869	2	4	5	7	9
52	0.7880	7891	7902	7912	7923	7934	7944	7955	7965	7976	2	4	5	7	9
53	0.7986	7997	8007	8018	8028	8039	8049	8059	8070	8080	2	3	5	7	9
54	0.8090	8100	8111	8121	8131	8141	8151	8161	8171	8181	2	3	5	7	8
55	0.8192	8202	8211	8221	8231	8241	8251	8261	8271	8281	2	3	5	7	8
56	0.8290	8300	8310	8320	8329	8339	8348	8358	8368	8377	2	3	5	6	8
57	0.8387	8396	8406	8415	8425	8434	8443	8453	8462	8471	2	3	5	6	8
58	0.8480	8490	8499	8508	8517	8526	8536	8545	8554	8563	2	3	5	6	8
59	0.8572	8581	8590	8599	8607	8616	8625	8634	8643	8652	1	3	4	6	7
60	0.8660	8669	8678	8686	8695	8704	8712	8721	8729	8738	1	3	4	6	7
61	0.8746	8755	8763	8771	8780	8788	8796	8805	8813	8821	1	3	4	6	7
62	0.8829	8838	8846	8854	8862	8870	8878	8886	8894	8902	1	3	4	5	7
63	0.8910	8918	8926	8934	8942	8949	8957	8965	8973	8980	1	3	4	5	6
64	0.8988	8996	9003	9011	9018	9026	9033	9041	9048	9056	1	2	4	5	6

NATURAL SINES (continued)

	0′	6′	12′	18′	24′	30′	36′	42′	48′	54′	1′	2′	3′	4′	5′
65°	0.9063	9070	9078	9085	9092	9100	9107	9114	9121	9128	1	2	4	5	6
66	0.9135	9143	9150	9157	9164	9171	9178	9184	9191	9198	1	2	4	5	6
67	0.9205	9212	9219	9225	9232	9239	9245	9252	9259	9265	1	2	3	4	6
68	0.9272	9278	9285	9291	9298	9304	9311	9317	9323	9330	1	2	3	4	5
69	0.9336	9342	9348	9354	9361	9367	9373	9379	9385	9391	1	2	3	4	5
70	0.9397	9403	9409	9415	9421	9426	9432	9438	9444	9449	1	2	3	4	5
71	0.9455	9461	9466	9472	9478	9483	9489	9494	9500	9505	1	2	3	4	5
72	0.9511	9516	9521	9527	9532	9537	9542	9548	9553	9558	1	2	3	3	4
73	0.9563	9568	9573	9578	9583	9588	9593	9598	9603	9608	1	2	3	3	4
74	0.9613	9617	9622	9627	9632	9636	9641	9646	9650	9655	1	2	2	3	4
75	0.9659	9664	9668	9673	9677	9681	9686	9690	9694	9699	1	1	2	3	4
76	0.9703	9707	9711	9715	9720	9724	9728	9732	9736	9740	1	1	2	3	3
77	0.9744	9748	9751	9755	9759	9763	9767	9770	9774	9778	1	1	2	2	3
78	0.9781	9785	9789	9792	9796	9799	9803	9806	9810	9813	1	1	2	2	3
79	0.9816	9820	9823	9826	9829	9833	9836	9839	9842	9845	1	1	2	2	3
80	0.9848	9851	9854	9857	9860	9863	9866	9869	9871	9874	0	1	1	2	2
81	0.9877	9880	9882	9885	9888	9890	9893	9895	9898	9900	0	1	1	2	2
82	0.9903	9905	9907	9910	9912	9914	9917	9919	9921	9923	0	1	1	1	2
83	0.9925	9928	9930	9932	9934	9936	9938	9940	9942	9943	0	1	1	1	2
84	0.9945	9947	9949	9951	9952	9954	9956	9957	9959	9960	0	1	1	1	1
85	0.9962	9963	9965	9966	9968	9969	9971	9972	9973	9974	0	0	1	1	1
86	0.9976	9977	9978	9979	9980	9981	9982	9983	9984	9985	0	0	1	1	1
87	0.9986	9987	9988	9989	9990	9990	9991	9992	9993	9993	0	0	0	1	1
88	0.9994	9995	9995	9996	9996	9997	9997	9997	9998	9998	0	0	0	0	0
89	0.9998	9999	9999	9999	9999	0000	0000	0000	0000	0000	0	0	0	0	0

NATURAL COSINES

											Subtract Differences				
	0′	**6′**	**12′**	**18′**	**24′**	**30′**	**36′**	**42′**	**48′**	**54′**	**1**	**2**	**3**	**4**	**5**
0°	1.0000	0000	0000	0000	0000	0000	9999	9999	9999	9999	0	0	0	0	0
1	0.9998	9998	9998	9997	9997	9997	9996	9996	9995	9995	0	0	0	0	0
2	0.9994	9993	9993	9992	9991	9990	9990	9989	9988	9987	0	0	0	1	1
3	0.9986	9985	9984	9983	9982	9981	9980	9979	9978	9977	0	0	1	1	1
4	0.9976	9974	9973	9972	9971	9969	9968	9966	9965	9963	0	0	1	1	1
5	0.9962	9960	9959	9957	9956	9954	9952	9951	9949	9947	0	1	1	1	1
6	0.9945	9943	9942	9940	9938	9936	9934	9932	9930	9928	0	1	1	1	2
7	0.9925	9923	9921	9919	9917	9914	9912	9910	9907	9905	0	1	1	1	2
8	0.9903	9900	9898	9895	9893	9890	9888	9885	9882	9880	0	1	1	2	2
9	0.9877	9874	9871	9869	9866	9863	9860	9857	9854	9851	0	1	1	2	2

Appendix 2

NATURAL COSINES (continued)

	0′	6′	12′	18′	24′	30′	36′	42′	48′	54′	Subtract Differences				
											1′	2′	3′	4′	5′
10°	0.9848	9845	9842	9839	9836	9833	9829	9826	9823	9820	1	1	2	2	3
11	0.9816	9813	9810	9806	9803	9799	9796	9792	9789	9785	1	1	2	2	3
12	0.9781	9778	9774	9770	9767	9763	9759	9755	9751	9748	1	1	2	2	3
13	0.9744	9740	9736	9732	9728	9724	9720	9715	9711	9707	1	1	2	3	3
14	0.9703	9699	9694	9690	9686	9681	9677	9673	9668	9664	1	1	2	3	4
15	0.9659	9655	9650	9646	9641	9636	9632	9627	9622	9617	1	2	2	3	4
16	0.9613	9608	9603	9598	9593	9588	9583	9578	9573	9568	1	2	3	3	4
17	0.9563	9558	9553	9548	9542	9537	9532	9527	9521	9516	1	2	3	3	4
18	0.9511	9505	9500	9494	9489	9483	9478	9472	9466	9461	1	2	3	4	5
19	0.9455	9449	9444	9438	9432	9426	9421	9415	9409	9403	1	2	3	4	5
20	0.9397	9391	9385	9379	9373	9367	9361	9354	9348	9342	1	2	3	4	5
21	0.9336	9330	9323	9317	9311	9304	9298	9291	9285	9278	1	2	3	4	5
22	0.9272	9265	9259	9252	9245	9239	9232	9225	9219	9212	1	2	3	4	6
23	0.9205	9198	9191	9184	9178	9171	9164	9157	9150	9143	1	2	4	5	6
24	0.9135	9128	9121	9114	9107	9100	9092	9085	9078	9070	1	2	4	5	6
25	0.9063	9056	9048	9041	9033	9026	9018	9011	9003	8996	1	2	4	5	6
26	0.8988	8980	8973	8965	8957	8949	8942	8934	8926	8918	1	3	4	5	6
27	0.8910	8902	8894	8886	8878	8870	8862	8854	8846	8838	1	3	4	5	7
28	0.8829	8821	8813	8805	8796	8788	8780	8771	8763	8755	1	3	4	6	7
29	0.8746	8738	8729	8721	8712	8704	8695	8686	8678	8669	1	3	4	6	7
30	0.8660	8652	8643	8634	8625	8616	8607	8599	8590	8581	1	3	4	6	7
31	0.8572	8563	8554	8545	8536	8526	8517	8508	8499	8490	2	3	5	6	8
32	0.8480	8471	8462	8453	8443	8434	8425	8415	8406	8396	2	3	5	6	8
33	0.8387	8377	8368	8358	8348	8339	8329	8320	8310	8300	2	3	5	6	8
34	0.8290	8281	8271	8261	8251	8241	8231	8221	8211	8202	2	3	5	7	8
35	0.8192	8181	8171	8161	8151	8141	8131	8121	8111	8100	2	3	5	7	8
36	0.8090	8080	8070	8059	8049	8039	8028	8018	8007	7997	2	3	5	7	9
37	0.7986	7976	7965	7955	7944	7934	7923	7912	7902	7891	2	4	5	7	9
38	0.7880	7869	7859	7848	7837	7826	7815	7804	7793	7782	2	4	5	7	9
39	0.7771	7760	7749	7738	7727	7716	7705	7694	7683	7672	2	4	6	7	9
40	0.7660	7649	7638	7627	7615	7604	7593	7581	7570	7559	2	4	6	8	9
41	0.7547	7536	7524	7513	7501	7490	7478	7466	7455	7443	2	4	6	8	10
42	0.7431	7420	7408	7396	7385	7373	7361	7349	7337	7325	2	4	6	8	10
43	0.7314	7302	7290	7278	7266	7254	7242	7230	7218	7206	2	4	6	8	10
44	0.7193	7181	7169	7157	7145	7133	7120	7108	7096	7083	2	4	6	8	10
45°	0.7071	7059	7046	7034	7022	7009	6997	6984	6972	6959	2	4	6	8	10
46	0.6947	6934	6921	6909	6896	6884	6871	6858	6845	6833	2	4	6	8	11
47	0.6820	6807	6794	6782	6769	6756	6743	6730	6717	6704	2	4	6	9	11
48	0.6691	6678	6665	6652	6639	6626	6613	6600	6587	6574	2	4	7	9	11
49	0.6561	6547	6534	6521	6508	6494	9481	6468	6455	6441	2	4	7	9	11

NATURAL COSINES (continued)

											Subtract Differences				
	0′	6′	12′	18′	24′	30′	36′	42′	48′	54′	1′	2′	3′	4′	5′
50°	0.6428	6414	6401	6388	6374	6361	6347	6334	6320	6307	2	4	7	9	11
51	0.6293	6280	6266	6252	6239	6225	6211	6198	6184	6170	2	5	7	9	11
52	0.6157	6143	6129	6115	6101	6088	6074	6060	6046	6032	2	5	7	9	12
53	0.6018	6004	5990	5976	5962	5948	5934	5920	5906	5892	2	5	7	9	12
54	0.5878	5864	5850	5835	5821	5807	5793	5779	5764	5750	2	5	7	9	12
55	0.5736	5721	5707	5693	5678	5664	5650	5635	5621	5606	2	5	7	10	12
56	0.5592	5577	5563	5548	5534	5519	5505	5490	5476	5461	2	5	7	10	12
57	0.5446	5432	5417	5402	5388	5373	5358	5344	5329	5314	2	5	7	10	12
58	0.5299	5284	5270	5255	5240	5225	5210	5195	5180	5165	2	5	7	10	12
59	0.5150	5135	5120	5105	5090	5075	5060	5045	5030	5015	3	5	8	10	12
60	0.5000	4985	4970	4955	4939	4924	4909	4894	4879	4863	3	5	8	10	13
61	0.4848	4833	4818	4802	4787	4772	4756	4741	4726	4710	3	5	8	10	13
62	0.4695	4679	4664	4648	4633	4617	4602	4586	4571	4555	3	5	8	10	13
63	0.4540	4524	4509	4493	4478	4462	4446	4431	4415	4399	3	5	8	10	13
64	0.4384	4368	4352	4337	4321	4305	4289	4274	4258	4242	3	5	8	11	13
65	0.4226	4210	4195	4179	4163	4147	4131	4115	4099	4083	3	5	8	11	13
66	0.4067	4051	4035	4019	4003	3987	3971	3955	3939	3923	3	5	8	11	13
67	0.3907	3891	3875	3859	3843	3827	3811	3795	3778	3762	3	5	8	11	13
68	0.3746	3730	3714	3697	3681	3665	3649	3633	3616	3600	3	5	8	11	13
69	0.3584	3567	3551	3535	3518	3502	3486	3469	3453	3437	3	5	8	11	14
70	0.3420	3404	3387	3371	3355	3338	3322	3305	3289	3272	3	5	8	11	14
71	0.3256	3239	3223	3206	3190	3173	3156	3140	3123	3107	3	6	8	11	14
72	0.3090	3074	3057	3040	3024	3007	2990	2974	2957	2940	3	6	8	11	14
73	0.2924	2907	2890	2874	2857	2840	2823	2807	2790	2773	3	6	8	11	14
74	0.2756	2740	2723	2706	2689	2672	2656	2639	2622	2605	3	6	8	11	14
75	0.2588	2571	2554	2538	2521	2504	2487	2470	2453	2436	3	6	8	11	14
76	0.2419	2402	2385	2368	2351	2334	2317	2300	2284	2267	3	6	8	11	14
77	0.2250	2233	2215	2198	2181	2164	2147	2130	2113	2096	3	6	9	11	14
78	0.2079	2062	2045	2028	2011	1994	1977	1959	1942	1925	3	6	9	11	14
79	0.1908	1891	1874	1857	1840	1822	1805	1788	1771	1754	3	6	9	11	14
80	0.1736	1719	1702	1685	1668	1650	1633	1616	1599	1582	3	6	9	11	14
81	0.1564	1547	1530	1513	1495	1478	1461	1444	1426	1409	3	6	9	11	14
82	0.1392	1374	1357	1340	1323	1305	1288	1271	1253	1236	3	6	9	12	14
83	0.1219	1201	1184	1167	1149	1132	1115	1097	1080	1063	3	6	9	12	14
84	0.1045	1028	1011	0993	0976	0958	0941	0924	0906	0889	3	6	9	12	14
85	0.0872	0854	0837	0819	0802	0785	0767	0750	0732	0715	3	6	9	12	15
86	0.0698	0680	0663	0645	0628	0610	0593	0576	0558	0541	3	6	9	12	15
87	0.0523	0506	0488	0471	0454	0436	0419	0401	0384	0366	3	6	9	12	15
88	0.0349	0332	0314	0297	0279	0262	0244	0227	0209	0192	3	6	9	12	15
89	0.0175	0157	0140	0122	0105	0087	0070	0052	0035	0017	3	6	9	12	15

NATURAL TANGENTS

	0′	6′	12′	18′	24′	30′	36′	42′	48′	54′	1′	2′	3′	4′	5′
0°	0.0000	0017	0035	0052	0070	0087	0105	0122	0140	0157	3	6	9	12	15
1	0.0175	0192	0209	0227	0244	0262	0279	0297	0314	0332	3	6	9	12	15
2	0.0349	0367	0384	0402	0419	0437	0454	0472	0489	0507	3	6	9	12	15
3	0.0524	0542	0559	0577	0594	0612	0629	0647	0664	0682	3	6	9	12	15
4	0.0699	0717	0734	0752	0769	0787	0805	0822	0840	0857	3	6	9	12	15
5	0.0875	0892	0910	0928	0945	0963	0981	0998	1016	1033	3	6	9	12	15
6	0.1051	1069	1086	1104	1122	1139	1157	1175	1192	1210	3	6	9	12	15
7	0.1228	1246	1263	1281	1299	1317	1334	1352	1370	1388	3	6	9	12	15
8	0.1405	1423	1441	1459	1477	1495	1512	1530	1548	1566	3	6	9	12	15
9	0.1584	1602	1620	1638	1655	1673	1691	1709	1727	1745	3	6	9	12	15
10	0.1763	1781	1799	1817	1835	1853	1871	1890	1908	1926	3	6	9	12	15
11	0.1944	1962	1980	1998	2016	2035	2053	2071	2089	2107	3	6	9	12	15
12	0.2126	2144	2162	2180	2199	2217	2235	2254	2272	2290	3	6	9	12	15
13	0.2309	2327	2345	2364	2382	2401	2419	2438	2456	2475	3	6	9	12	15
14	0.2493	2512	2530	2549	2568	2586	2605	2623	2642	2661	3	6	9	12	15
15	0.2679	2698	2717	2736	2754	2773	2792	2811	2830	2849	3	6	9	13	16
16	0.2867	2886	2905	2924	2943	2962	2981	3000	3019	3038	3	6	10	13	16
17	0.3057	3076	3096	3115	3134	3153	3172	3191	3211	3230	3	6	10	13	16
18	0.3249	3269	3288	3307	3327	3346	3365	3385	3404	3424	3	6	10	13	16
19	0.3443	3463	3482	3502	3522	3541	3561	3581	3600	3620	3	7	10	13	16
20	0.3640	3659	3679	3699	3719	3739	3759	3779	3799	3819	3	7	10	13	17
21	0.3839	3859	3879	3899	3919	3939	3959	3979	4000	4020	3	7	10	13	17
22	0.4040	4061	4081	4101	4122	4142	4163	4183	4204	4224	3	7	10	14	17
23	0.4245	4265	4286	4307	4327	4348	4369	4390	4411	4431	3	7	10	14	17
24	0.4452	4473	4494	4515	4536	4557	4578	4599	4621	4642	4	7	11	14	18
25	0.4663	4684	4706	4727	4748	4770	4791	4813	4834	4856	4	7	11	14	18
26	0.4877	4899	4921	4942	4964	4986	5008	5029	5051	5073	4	7	11	15	18
27	0.5095	5117	5139	5161	5184	5206	5228	5250	5272	5295	4	7	11	15	18
28	0.5317	5340	5362	5384	5407	5430	5452	5475	5498	5520	4	8	11	15	19
29	0.5543	5566	5589	5612	5635	5658	5681	5704	5727	5750	4	8	12	15	19
30	0.5774	5797	5820	5844	5867	5890	5914	5938	5961	5985	4	8	12	16	20
31	0.6009	6032	6056	6080	6104	6128	6152	6176	6200	6224	4	8	12	16	20
32	0.6249	6273	6297	6322	6346	6371	6395	6420	6445	6469	4	8	12	16	20
33	0.6494	6519	6544	6569	6594	6619	6644	6669	6694	6720	4	8	13	17	21
34	0.6745	6771	6796	6822	6847	6873	6899	6924	6950	6976	4	9	13	17	21
35	0.7002	7028	7054	7080	7107	7133	7159	7186	7212	7239	4	9	13	18	22
36	0.7265	7292	7319	7346	7373	7400	7427	7454	7481	7508	5	9	14	18	23
37	0.7536	7563	7590	7618	7646	7673	7701	7729	7757	7785	5	9	14	18	23
38	0.7813	7841	7869	7898	7926	7954	7983	8012	8040	8069	5	9	14	19	24
39	0.8098	8127	8156	8185	8214	8243	8273	8302	8332	8361	5	10	15	20	24

NATURAL TANGENTS (continued)

	0′	6′	12′	18′	24′	30′	36′	42′	48′	54′	1′	2′	3′	4′	5′
40°	0.8391	8421	8451	8481	8511	8541	8571	8601	8632	8662	5	10	15	20	25
41	0.8693	8724	8754	8785	8816	8847	8878	8910	8941	8972	5	10	16	21	26
42	0.9004	9036	9067	9099	9131	9163	9195	9228	9260	9293	5	11	16	21	27
43	0.9325	9358	9391	9424	9457	9490	9523	9556	9590	9623	6	11	17	22	28
44	0.9657	9691	9725	9759	9793	9827	9861	9896	9930	9965	6	11	17	23	29
45	1.0000	0035	0070	0105	0141	0176	0212	0247	0283	0319	6	12	18	24	30
46	1.0355	0392	0428	0464	0501	0538	0575	0612	0649	0686	6	12	18	25	31
47	1.0724	0761	0799	0837	0875	0913	0951	0990	1028	1067	6	13	19	25	32
48	1.1106	1145	1184	1224	1263	1303	1343	1383	1423	1463	7	13	20	27	33
49	1.1504	1544	1585	1626	1667	1708	1750	1792	1833	1875	7	14	21	28	34
50	1.1918	1960	2002	2045	2088	2131	2174	2218	2261	2305	7	14	22	29	36
51	1.2349	2393	2437	2482	2527	2572	2617	2662	2708	2753	8	15	23	30	37
52	1.2799	2846	2892	2938	2985	3032	3079	3127	3175	3222	8	16	24	31	39
53	1.3270	3319	3367	3416	3465	3514	3564	3613	3663	3713	8	16	25	33	41
54	1.3764	3814	3865	3916	3968	4019	4071	4124	4176	4229	9	17	26	34	43
55	1.4281	4335	4388	4442	4496	4550	4605	4659	4715	4770	9	18	27	36	45
56	1.4826	4882	4938	4994	5051	5108	5166	5224	5282	5340	10	19	29	38	48
57	1.5399	5458	5517	5577	5637	5697	5757	5818	5880	5941	10	20	30	40	50
58	1.6003	6066	6128	6191	6255	6319	6383	6447	6512	6577	11	21	32	43	53
59	1.6643	6709	6775	6842	6909	6977	7045	7113	7182	7251	11	23	34	45	57
60	1.732	1.739	1.746	1.753	1.760	1.767	1.775	1.782	1.789	1.797	1	2	4	5	6
61	1.804	1.811	1.819	1.827	1.834	1.842	1.849	1.857	1.865	1.873	1	3	4	5	6
62	1.881	1.889	1.897	1.905	1.913	1.921	1.929	1.937	1.946	1.954	1	3	4	5	7
63	1.963	1.971	1.980	1.988	1.997	2.006	2.014	2.023	2.032	2.041	1	3	4	6	7
64	2.050	2.059	2.069	2.078	2.087	2.097	2.106	2.116	2.125	2.135	2	3	5	6	8
65	2.145	2.154	2.164	2.174	2.184	2.194	2.204	2.215	2.225	2.236	2	3	5	7	8
66	2.246	2.257	2.267	2.278	2.289	2.300	2.311	2.322	2.333	2.344	2	4	6	7	9
67	2.356	2.367	2.379	2.391	2.402	2.414	2.426	2.438	2.450	2.463	2	4	6	8	10
68	2.475	2.488	2.500	2.513	2.526	2.539	2.552	2.565	2.578	2.592	2	4	7	9	11
69	2.605	2.619	2.633	2.646	2.660	2.675	2.689	2.703	2.718	2.733	2	5	7	9	12
70	2.747	2.762	2.778	2.793	2.808	2.824	2.840	2.856	2.872	2.888	3	5	8	10	13
71	2.904	2.921	2.937	2.954	2.971	2.989	3.006	3.024	3.042	3.060	3	6	9	12	14
72	3.078	3.096	3.115	3.133	3.152	3.172	3.191	3.211	3.230	3.251	3	6	10	13	16
73	3.271	3.291	3.312	3.333	3.354	3.376	3.398	3.420	3.442	3.465	4	7	11	14	18
74	3.487	3.511	3.534	3.558	3.582	3.606	3.630	3.655	3.681	3.706	4	8	12	16	20
75	3.732	3.758	3.785	3.812	3.839	3.867	3.895	3.923	3.952	3.981					
76	4.011	4.041	4.071	4.102	4.134	4.165	4.198	4.230	4.264	4.297					
77	4.331	4.366	4.402	4.437	4.474	4.511	4.548	4.586	4.625	4.665	Mean Differences no longer sufficiently accurate				
78	4.705	4.745	4.787	4.829	4.872	4.915	4.959	5.005	5.050	5.097					
79	5.145	5.193	5.242	5.292	5.343	5.396	5.449	5.503	5.558	5.614					

NATURAL TANGENTS (continued)

	0′	6′	12′	18′	24′	30′	36′	42′	48′	54′	1′	2′	3′	4′	5′
80°	5.671	5.730	5.789	5.850	5.912	5.976	6.041	6.107	6.174	6.243					
81	6.314	6.386	6.460	6.535	6.612	6.691	6.772	6.855	6.940	7.026					
82	7.115	7.207	7.300	7.396	7.495	7.596	7.700	7.806	7.916	8.028					
83	8.144	8.264	8.386	8.513	8.643	8.777	8.915	9.058	9.205	9.357					
84	9.514	9.677	9.845	10.02	10.20	10.39	10.58	10.78	10.99	11.20	Mean Differences no longer sufficiently accurate				
85	11.43	11.66	11.91	12.16	12.43	12.71	13.00	13.30	13.62	13.95					
86	14.30	14.67	15.06	15.46	15.89	16.35	16.83	17.34	17.89	18.46					
87	19.08	19.74	20.45	21.20	22.02	22.90	23.86	24.90	26.03	27.27					
88	28.64	30.14	31.82	33.69	35.80	38.19	40.92	44.07	47.74	52.08					
89	57.29	63.66	71.62	81.85	95.49	114.6	143.2	191.0	286.5	573.0					

For angles > 76° the reciprocal of the tangent of the complement may be taken.

Answers

UNIT 1: Managing your money

Core

Lesson 1 – About percentages (page 5)

1 a D 40% b C 62% c B 69%
d A 99% e C 40%
2 a K15 b K45 c K90
d K6 e K300
3 a i 0.7 b i 0.4
ii $\frac{7}{10}$ ii $\frac{2}{5}$
c i 0.38 d i 0.85
ii $\frac{19}{50}$ ii $\frac{17}{50}$
e i 1.25 f i 0.66
ii $1\frac{1}{4}$ ii $\frac{33}{50}$
g i 1.32 h i 0.375
ii $\frac{18}{25}$ ii $\frac{5}{8}$
i i 0.0625 j i $0.05\dot{6}$
ii $\frac{5}{80}$ ii $\frac{17}{300}$
4 a 80 b 37.5 c 125 d $58.\dot{3}$
e 270 f 350 g 87.5 h 95
i 162.5 j 82
5 a 83.33 b 27.27 c 66.67 d 11.11
e 26.32 f 42.86 g 433.33 h 216.67
i 15.38 j 114.29

Lesson 2 – Finding a percentage (pages 6–7)

1 a K600 b K4.12 c K4.40
d K161 e K44.85 f K562.50
g K996 h K3 740 000 i K750
j K350 k K452 l K25.83
2 a K2.02 b K28.86
3 K1440
4 a K119 950 b K101 957.50
5 a K600 b K3040 c K20 050
6 a K10 855.60 b K37 994.60
7 a K7 b K25 c K18
d K32 e K60
8 a K1456 b K5896.80
9 K259.20
10 K462 500

Lesson 3 – Smart shopping (pages 8–9)

1 a K466.65 b K169.15 c K679.15
d K509.15 e K424.15
2 a K13.35 b K17.75 c K22.50
d K27.30 e K51.50 f K33.85
g K8.95 h K52.80 i K219.90
3 K41.65 4 Store B
5 a Store C b K15
6 a 50 b K40
c no d printing, K5 cheaper
7 a K10.85 b 20%
8 a K4.80 b K82.08 c K13.92
9 a K14.40
b The 10% figure in Question 8 is a smaller amount after 5% has been subtracted.

Lesson 4 – Percentage increase and decrease (pages 11–12)

1 a i 1.1 ii 1.15 iii 1.2
iv 1.5 v 1.025
b i 0.88 ii 0.94 iii 0.75
iv 0.30 v 0.91.5
2 a i increase ii 25%
b i decrease ii 5%
c i increase ii 8%
d i increase ii 100%
e i decrease ii 12.5%
3 a 25 b 20 c 11
d 22 e 10
4 a K71.10 b K141.55 c K131.12
d K189.05 e K137.08 f K210.32
g K34.30 h K196.15 i K2024
j K299.25
5 a i K60 ii 11%
b i K50 ii 7%
c i K10 ii 5%
d i K6 ii 15%
e i K90 ii 7%
6 a i K105 ii K117.60
b i K437.50 ii K490
c i K26 775 ii K29 988
d i K14 000 ii K15 680
e i K1 820 000 ii K2 038 400
7 a i no ii yes
b 8.6%

Lesson 5 – Mark-up, profit and loss (pages 14–15)

1 a K7.20
b K19.20
2 a i K45 ii K195
b i K180 ii K780
c i K255 ii K1105
d i K87 ii K377
e i K172.50 ii K748.50
f i K126 ii K312
3 a 45t b 17.6%
c i 15t profit ii 2t profit iii 15t loss
d 37.2 e 37
4 a K17 b 42%
5 a K10.50 b K1.50 loss c 12.5%
6 12% profit
7 a K1.50 b 15% c K1.99
d no, 20% off price would be K7.96
8 K105
9 a K16.80 b K128.80
c K109.48 d K2.52 loss

10 a K7000 b K44 000
c K57 200

Lesson 6 – Working backwards (page 18)

1 K300 2 K350
3 K35 000
4 a K30.56 b K21.86
5 K85 6 K250 000
7 a K794.12 b K562.50
8 a 8.428t b K4.64
9 K36 10 K300

Lesson 7 – Budgets (pages 19–21)

1 a K4.50 b PMV
c i 50% ii 25% iii 7%
d 14 weeks e i K26.25
ii Yes, he could cut back on savings to pay the board.
2 a K6 b food
c $\frac{1}{31}$, 3.23% d K29.76
3 a K182.60 b K60.87
c K191.73
4 a K6890 b K413.40
5 a K3.50 b K3.07
6 a

Income		Expenses	
Wages	572	Rent	171.60
		Food	228.80
		Bills	114.40
		School fees	11.44
		Savings	28.60
		Other	17.16

b K5.01 more
7 a K59 418 b K2285
c

Income per fortnight K		Expenses per fortnight	%	K
Salary and allowances	……	House payment	45	1028.25
		Food	35	799.75
		Electricity	2	45.70
		Sewage/Garbage	0.5	11.42
		Entertainment	2.5	57.12
		Clothes and household goods	4	91.40
		School fees	2	45.70
		Newspapers and magazines	1	22.85
		Savings	3.5	79.98
		Telephone	1.5	34.28
		Medical expenses	3	68.55

d Entertainment, perhaps newspapers and magazines could be gone without. Savings could be made on clothes and household goods and food, by buying cheaper items or fewer of them. Savings might be possible on electricity and telephone by using them less often.

Lesson 8 – Depreciation and appreciation (page 24)

1 a K2000 b K1600
c K1280 d K1024
2 a K437.40 b K286.98
c K209.22
3 K4 408 992
4 a K972.41 b K1241.04
5 K1705.86
6 a K2200 b K13 200
c K14 434.20 on sliding scale, which is K1234.20 more than the flat rate value.
7 During year 5 (it is worth K4072.50 at the end of year 4 and K3869 at the end of year 5).
8 a 18 333.33 b 6.66%

Lesson 9 – Working with large sums of money (pages 25–6)

1 a sales 4.07×10^8, depreciation 3.56×10^7, interest 1.17×10^6, capital expenditure 7.77×10^7
b 3.8 million c K300 000
d 10.84% e depreciation of assets
f sales
2 a K26 678 000 b 56%
3 K276 434.24
4 a 84.5% b 22.43%
c 596 700
5 a 2 billion (2 000 000 000)
b i 10%
ii 4237.5 million, 847.5 million, 565 million
c 16534.5 millions of pounds
d K92593 million

Lesson 10 – Revision and assessment (pages 27–9)

1 B 2 E 3 D 4 C 5 C
6 a 40% b 25%
c 31.25% d 12.5%
e 18.18%
7 a 5% decrease b 10% increase
c 17% decrease d 22% increase
e 200% increase
8 a K401.35 b K720 800
c K7 187 500
9 a K403 b K8219.15
c K1.49 billion
10 a 13% b 35%
c 22% d 26%
e 8%
11 a K6.96 b K83.60
12 Store B, by K82.48
13 a K60 b i 58.3% ii 20%
c K200 d 6%

14 K1067.15 15 93t
16 a K450 b 39.3
17 a K21.50 b i 52.4% ii 34.4%
18 a 0.75 b K52.46
c K110
19 a K9100 b K6740.74
20 K290
21 a K231 b i 44.4% ii 12.4%
c K6.80
22 K397.80
23 K120 788
24 a K1024 million b 12.2%

Saving money

Lesson 2 – Simple interest calculations (page 32)

1 a K240 b K11 500
c K656.60 d K170
e K505.48
2 a K1062.50 b K436.64
3 K912
4 a K195 b K3900
c K10 400
5 a K574 b K24 108
6 a K57.75 b K2.75

Lesson 3 – More simple interest calculations (pages 33–4)

1 5% 2 5 years
3 K900 4 K2500
5 $4\frac{1}{2}$% 6 8.9%
7 16.7 years 8 5%
9 K270.10 10 4.25 years

Lesson 4 – Savings accounts (pages 35–6)

1 K248.83
2 a 0.5% b 0.375%
c 0.25%
3 K21.44 4 2.4%
5 K166.50 6 K4944
7 a

Date	Debit	Credit	Balance
1 July			K3296.00
4 Sept		K120.00	K3416.00
18 Dec	K784.00		K2632.00
22 Mar	K235.70		K2396.30
19 Apr		K400.00	K2796.30
17 Jun		K1415.30	K4211.60

b K102.09
8 a 28 b K69.04
c K20 069.04 d K76.70

Lesson 5 – Term deposits (pages 37–8)

1 a i K136 ii K14.38
iii K197.36 iv K8750
b K5189.88
2 a K2115 b K11 160
c K945 d K3524.86
e K5850
3 a Answers will vary. b K111
c i K53.26 ii K6053.26 iii K53.73
iv 180 day option; it earns K4 more interest.

Lesson 6 – Compound interest (pages 39–40)

1 a 4% b 0.667%
c 2% d 0.154%
e 0.022%
2 a K141 778.13 b K2805.10
c K8487.20 d K13 375.46
e K12 682.42 f K217 361.25
3 a K1125.22 b K75.22
4 a K90
b Amount of savings at the end of six months = K600
Interest on savings = K19.50
Balance in account = K619.50
Amount of additional savings at the end of year 1 = K600
Total in account = K1219.50
Interest on savings = K39.63
c March in the second year
d Answers will vary.

Lesson 7 – Compound interest formula (pages 41–2)

1 a K5955 b K13788
c K10 204 d K16 959
e K1920
2 K955 3 K8713
4 a 3% b 8
c K1267 d K267
5 a K2690 b K49 170
c K8492
6 a K47 907 b K1218
*7 a K10 500 b K10 509
c K10 506 *8 K742
*9 K9070.30 *10 9%

Lesson 8 – Inflation, appreciation and depreciation (pages 44–5)

1 E 2 B
3 a K2952 b K1548
4 K24 825
5 a K3.36 b K3.53
c K3.89 d K5.21
6 K166
7 a K179 108 b K29 108
c K25 000 d K4108 profit
8 D 9 K12.34
10 9

Lesson 9 – Comparing investments (page 46)

1 K200 2 K7.93

3 a Compound interest = K586.86,
Simple interest = K675

b Simple interest by K88.

4 K585, Compound interest is now better, by K1.86.

5 a K9511.80 b K2011.80

c 9%

Lesson 10 – Revision and assessment (pages 46–7)

1 A 2 E

3 B 4 K150

5 K2160 6 K2560

7 K512.50 8 K520.35

9 a

Date	Debit	Credit	Balance
01 May			312.67
08 May		103.07	415.74
23 May		67.90	483.64
26 May	59.23		424.41
31 May	29.52		394.89

b 78t

10 a K420 b K1.85

c K26 000 11 K53 491.05

12 K653 13 K8

14 Savings account by K9.39.

Borrowing money

Lesson 2 – Loans and simple interest (pages 49–50)

1 a K265.60 b K1170

c K14 105

2 K23 482.50 3 K2617.21

4 K113

5 a K13 360 b K33 360

c K347.50

6 K1006.58

7 a Nationwide bank b K215

8 13.44% 9 6 months

Lesson 3 – Hire purchase (pages 51–2)

1 a K2000 b 26

c K2773 d K574

e 28.7%

2 a K1500 b K24

c K1999 d K450

e 15%

3 a K160

b In 1 year he has paid K180. Interest in 1 year is K19.20 so he will owe K179.20. Yes, he can pay it off.

4 a K360 b K248.89

5 a i 26 ii 13 iii ???

b i K12 389 ii K12 870 iii K15 522

c i 47.78% ii 28.7% iii 36.8%

d Answers will vary. e Answers will vary.

6 14.25%

7 K14.55

8 a 9000 b 79800

c 19.3% p.a.

Lesson 4 – Credit cards (pages 54–5)

1 a K1.24 b K9.00

c K41.96

2 a i K10 ii K10 iii K17.64

b K200.00

3 K5.29

4 a A = 192.54, B = 401.53 and C = 1133.98

b K15.48

Lesson 5 – Compound interest and reducing balance loans (pages 57–8)

1 a i K1295.02 ii K295.02

b i K24 333.05 ii K4 333.05

c i K10 615.20 ii K615.20

d i K2 771 717.40 ii K771 717.40

2

End of year	Balance of loan (K)	Interest	Repayment	Amount owing at end of year
1	5000.00	0.09 × 5000 = 450	1200.00	4250
2	4250	0.09 × 4250 = 382.50	1200.00	3432.50
3	3432.50	0.09 × 3432.50 = 308.92	1200.00	2541.42
4	2541.42	.09 × 2541.42 = 228.72	1200.00	1570.14

a K1570.14 b K1370.14

3

Amount owing at end of quarter
7480
7459.52
7438.54
7417.06

4 a 2.1% b K12 172.48

c K1172.48

5

End of year	Balance of loan (K)	Interest	Repayment	Amount owing at end of year
1	10 000.00	700	2000.00	8700
2	8700	609	2000.00	7309
3	7309	511.63	2000.00	5820.63
4	5820.63	407.44	2000.00	4228.07
5	4228.07	295.96	2000.00	2524.03

b 7 years c K749.75

d K2749.75

6 a 0.625% b K196 528.39

Earning money

Lesson 1 – How people earn (pages 59–60)

1 a K39.5 b K94.8
c K221.20 d K169.85
e K136.27 f K177.75
2 K222.30 3 K304.80
4 K1305 5 K918.60
6 a K3833 b K885
7 K5300
8 a K33 526 b K2794
9 a K40 925 b K386 500
c K504 616
10 a K2.62
b Her husband earns K2.83 an hour which is more than Nerilee earns.
11 a 20% b K807.69
12 253.5 kilos

Lesson 2 – More about earnings (pages 61–2)

1 K26.25 2 K249.60
3 a K1600 b K96
4 a 20.8% b i K5590 ii K6760
5 a K2420 b 3%
6 a K27 000 b K2250
7 a K461 b K210
c K1078 d K548.96
e i K576 ii K14 976
8 a i K1541.67 ii K355.77 iii K8.89
b i K2133.33 ii K492.31 iii K12.31
c i K4616.67 ii K1065.38 iii K26.63
d i K11 408.33 ii K2632.69 iii K65.82
9 C 10 K5640.60

Lessons 3 & 4 – Tax and other deductions (pages 67–8)

1 a K583 b K269
2 a i K1686.74 ii K3226.74
b i K19 965.60 ii K21 505.6
c i 0 ii K1440.78
d i K85 510 ii K87 050
e i K6488 ii K8028
f i K150 334 ii K151874
g i K3821 ii K5361
h i K33 294 ii K34 834
3 a K2090 b K14410
c K1200.83
4 a K402.50 b K22.80
5 a K181.75 b K10.90
c K170.85 d K4442.10
e 0
6 K15.26
7 K66.25
8 a K455.35 b K27.32
c K483.03 d K380.69
e K38.25
9 a K39.66 b K621.34
c K57.10 d K564.24
10 a Employee A's tax is K4.61; Employee B's tax is K56.77
b Employee A pays 1.49%; Employee B pays 9.16% tax
c No, their tax is not in the same proportion as their salaries.
11 K554.45
12 a K332.50 b K19.95
c K312.55 d K7.94
e K429.61

Lesson 5 – Revision and assessment (Borrowing and earning) (pages 69–70)

1 D 2 C
3 A 4 12.5%
5 a K12 265.85 b K15 925.20
c K10 360.02
6 a 6.24 ÷ 1200
b A = K310.24, B = K59322.24
c

Month	Amount (beginning of month)	Interest	Repayment	Balance (end of month)
3	59322.24	308.48	650	58980.72

7 a K2000 b K2700
c K250 d 12.5% p.a.
8 a K1740 b 39
c i K5940 ii K5270
d Store B, 17.83%
9 K3.81 10 K7540.98
11

End of month	Balance of loan (K)	Interest	Repayment	Amount owing at end of year
6	10 000	620	800	9820
12	9820	608.84	800	9628.84
18	9628.84	596.98	800	9425.82
24	9425.82	584.40	800	9210.22

12 K58.75
13 Glen earns K5.77 more per week.
14 K2248.67
15 a

Name	Pay
Apelis	344.50
Joab	551.25
Justina	177.75
Patti	163.24
Bob	409.75
Victor	58.28

b K1704.77
16 Raymond is paying 6.02% while Neema is paying 4.40%.
17 K11 896.30
18 a K24 710 b K4461.67

UNIT 2: Functions and graphs

Lesson 1 – Revision of directed numbers (pages 83–4)

1 a -7.2, -1, 0, 5 b -101, -99, -98, -50
c -0.26, -0.25, -0.05, 0.5, 0.56

2 a $-0.67 < 0.6$ b $0 > -0.82$
c $-0.983 < -0.97$

3 a 5, 8, 11 b -199, -298, -397
c 4, -2, 1

4 a 12 b -11 c 16
d -5 e -11 f 1
g -0.1 h -1.1 i 6

5 a 9 b -3 c -0.18
d 16 e 4 f -90
g -25 h 4 i 10.4

6

×	+4	-11	80
-6	-24	66	-480
+0.4	1.6	-4.4	32
-1.2	-4.8	13.2	-96

7

×	+7	-5	-0.5
-9	-63	45	4.5
0.06	0.42	-0.3	-0.03
-1.2	-8.4	6	0.6

8 a i -8 ii 4 iii -200 iv 0.3
b i 48 ii -24 iii 1200 iv -1.8

9

÷	+72	-48	+0.36
-6	-12	8	-0.06
+40	1.8	-1.2	0.009
-1.2	-60	40	-0.3

10

÷	+10	-7.2	-0.24
-2	-5	3.6	0.12
8	1.25	-0.9	-0.03
-1.2	-8.33	+6	0.2

Lesson 2 – Mixed operations (pages 85–7)

1 a -1 b 2 c 7 d 0
e 18 f -11 g 6 h 27
i -8 j 1 k 1 l -1

2 81.8 kg

3 a K7 b K6

4

Date	Balance (K)
01/05/09	504.82
08/05/09	616.82
11/05/09	231.32
19/05/09	-168.68
22/05/09	-56.68
25/05/09	143.32
28/05/09	-6.68
31/05/09	93.32

5 a 21
b i 12 ii -3
c $15 - 7 + 2 - 12$
d Level 2 below ground level.

6 a 50 b Stop 3
c Yes d 68

Lesson 3 – Index form (pages 88–9)

1 a b^4 b 2^2x^2y
c a^3b^3 d $5m^2n^4$

2 a base 2, index 8
b base *ab*, index 5

3 a $2^3 \times 3$ b $2 \times 3 \times 5^2$
c $2 \times 3 \times 17$ d $3 \times 11 \times 13$

4

n	2^n	3^n	10^n
1	$2^1 = 2$	3	10
2	4	$3^2 = 9$	100
3	8	27	1000
4	16	81	$10^4 = 10\,000$
5	32	243	100 000
6	64	729	1 000 000

5 a 8 b 36
c 10 800 d 30

6 a $5a^3$ b $14x^5$
c $-32a^5b^9$ d x^6
e $\frac{m^4}{n^3p}$ f $\frac{27}{2}$ $\frac{k^3}{lm^2}$
g $\frac{p}{2}$ h $9p^2$
i $x^4y^7z^{10}$

7 a 2^43^2 b $2^{15}3^9$
c $2^4 \times 3$

Lesson 4 – Working with index laws (pages 90–1)

1 a $\frac{1}{x^7} = x^{-7}$ b $y^{-4} = \frac{1}{y^4}$
c $\frac{6}{z^3} = 6z^{-3}$ d $\frac{5^2}{5^3} = 5^{-1}$
e $\frac{w^3}{w^5} = w^{-2}$ f $\frac{1}{x^3y^4} = x^{-3}y^{-4}$

2 a $\frac{1}{7}$ b $\frac{1}{1000}$
c 1 d 64
e $\frac{1}{16}$ f $\frac{4}{3}$
g $\frac{1}{16}$ h $\frac{4}{9}$
i $\frac{1}{25}$ j $\frac{3}{2}$
k $\frac{6}{5}$ l $\frac{1}{81}$

3 a 2^1 b 2^5
c 2^{-5} d 2^{-8}
e 2^{-2} f $2^{\frac{1}{2}}$
g $2^{\frac{3}{2}}$

4 a 10^3 b 10^2
c 10^1 d 10^0

e 10^{-4} f 10^{-3}
g 10^{-2} h 10^{-1}

5 a $25^{\frac{3}{2}} = (5^2)^{\frac{3}{2}} = 5^3$
b $8^{\frac{2}{3}} = (2^3)^{\frac{2}{3}} = 2^2$
c $64^{\frac{3}{2}} = (2^6)^{\frac{3}{2}} = 2^9$
d $27^{\frac{4}{3}} = (3^3)^{\frac{4}{3}} = 3^4$

6 a 2 b 3
c $\frac{1}{3}$ d 2
e $\frac{1}{2}$ f 3

Lesson 5 – Scientific notation (page 93)

1 32.65, 56.23×10^2, 10×10^{-2}, 0.2348×10^{-1}
2 a 0.2785 b 8001
c 1 754 000 000 000
d 0.000 000 988 3
3 a 9.54×10^{-3} b 1.567×10^8
c 3.401×10^{-6} d 5.5×10^{-1}
e 6.7×10^3
4 0.00653, 2.468×10^{-1}, 4.8609, 5600, 7.856×10^4
5 1.254×10^{-9}, 0.000 000 011 11, 0.000 499 8, $\frac{1}{2000}$, 0.009 64, 1 278 000, 6.780×10^6
6 a 7.2×10^{-12} b 7×10^2
c 2.064×10^{14} d 8.64×10^{-1}
e 3×10^{-10}
7 a 2.7×10^1 b 2×10^5
c 1×10^1 d 1×10^{-5}
8 a 4.165×10^{-6} b 3.0277×10^{-16}
9 2.0456×10^9

Algebra

Lesson 6 – Review of algebraic expressions (pages 95–6)

1 a i 2 ii -5 iii 3
b i 3 ii 6 iii -24
c i 4 ii 15 iii -15
2 a i 4 ii 18 iii $-2bc$, $5bc$
b i 3 ii – iii –
c i 5 ii $-\frac{5}{7}$ iii –
d i 6 ii 3 iii $2hj^2$, j^2h
3 a $12d$ b $13d - 13$
c $8 - 5cd$ d $6b^2cd + 6bc^2d - 3$
e $12 - 3d$ f $10d^2 - 2d$
g d h $4b^2cd - 7bcd + 18$
i $3d^2 + d$ j $16bcd - 5bc^2d$
k $-3d^2 - d$ l $20cd + 3$
m $-18cd$ n $6b^2c^3d^2 + 2b^2c^2d^2$
4 a $3a + 15$ b $7m - 14$
c $7x + 7y$ d $48 - 12c$
e $10a + 5$ f $50 + 40m$
g $45 - 6y$ h $27a - 18b$
5 a $a^2 + 2a$ b $m^2 - 9m$
c $x^3 + 6x^2$ d $4c - c^2$
e $6a^2 + 15a$ f $5m^2 + 4mn$
g $7x^2 - 14xy$ h $27a - 18a^2$
6 a $-3a - 18$ b $-a - b$
c $-4c + c^2$ d $-6a + 8b$
e $-30y + 4y^2$
7 a $5a + 23$ b $10m - 8$
c $3x + 3y$ d $27 - 18c$
e $-2a + 9$ f $-4m^2 - 6$
g $-6y^2 + 57y + 4$ h $-41 - 22c$

Lesson 7 – Factorisation (page 97)

1 a $3a$ b $2m$
c bc d $4q$
e $-4mn$ f $-p$
g $7a^2b$ h 5
i $2b$
2 a 6 b $2a$
c $7b$ d $4a$
e x f -3
g ab h $3mn$
i $2p$
3 a $2a + 1$ b $a + 5$
c $2m - 1$ d $p - 2$
e -2 f ab
g $3uv$ h $3x - 2$
4 a $2(a + 3)$ b $3(3x - 2)$
c $3(4 - m)$ d $6(4d + 3)$
e $9(5 - 7m)$ f $5(4x - 5)$
g $a(2a + 1)$ h $m(m - 3)$
i $y(x - z)$ j $x(y - 3)$
k $p(3q - r)$ l $x(2y - 1 + z)$
5 a $2a(b + 2a)$ b $3x(3x - 1)$
c $3n^2(4m - 1)$ d $2p(4 - q)$
e $7a(b + 4)$ f $2x(3x^2 - 1)$
g $m(1 - m)$ h $ab(b - 1)$
i $a(2 - b + c)$ j $a(a^2 + a + 1)$
k $3xy(5 - 3xy)$ l $p^2(p - 1)$
6 a $-3(a + b)$ b $-2m(m + 1)$
c $-5(a - 2)$ d $-2(5m + 2n)$
e $-2b(3b - 1)$ f $-p(1 + q)$
g $-x(y - z + 3)$ h $-4(x^2 - 4)$
i $-3c(5c - 4)$
7 a $a(2a + b)$ b $-5a(1 + 3b)$
c $2x(1 - 4x - 2x^2)$ d $mn(m - 2 + 5n)$
e $-3a(1 + 2b + 3a)$ f $9m^2(7 - 5m)$
g $p(q - p + 1)$ h $2a(4a^2 + 2a + 1)$
i $x(x - 1)$ j $-x^2(2x + 5)$
k $4b^2(5a^2b + 3a - 2)$ l $-3(2y^2 + xy + 1)$

Lesson 8 – Solving equations (page 99)

1 a $d = 8$ b $m = -\frac{9}{4}$
c $x = -6$ d $f = 44$
e $m = 4$ f $m = \frac{1}{7}$

g $n = {}^-3$ h $y = 23$
i $m = \frac{10}{3}$ j $n = 10$
k $a = \frac{7}{3}$ l $d = {}^-1$
m $x = 48$ n $x = 24$
o $d = 8$ p $x = \frac{3}{2}$
q $t = -\frac{2}{7}$ r $d = \frac{35}{11}$
s $f = {}^-2$ t $x = 23$

Lesson 9 – Solving equations with unknown on both sides (page 101)

1 a $m = \frac{15}{13}$ b $x = -\frac{1}{6}$
c $m = {}^-6$ d $a = 7$
e $c = 7$ f $y = \frac{13}{5}$
g $y = 4$ h $x = \frac{6}{7}$
i $z = \frac{11}{15}$ j $x = 4$
k $x = 5$ l $f = \frac{5}{3}$
m $m = {}^-2$ n $t = \frac{11}{4}$
o $x = 1$ p $n = \frac{25}{2}$
q $m = \frac{7}{3}$ r $x = -\frac{8}{11}$
2 a $g = 2$ b $y = -\frac{1}{5}$
c $x = 8$ d $n = {}^-1$
e $x = 4$ f $p = {}^-1$
g $y = 1$ h $x = 3$

Lesson 10 – Substituting in formulae (pages 102–3)

1 a 9 b 2
c 7 d 41
e 2 f 0
2 a 132 cm b 216 cm
c 104°F d 180 N
e K20 f 341 cm^2
3 a 11.25 b 1.8
c 0.45
4 9.94 seconds 5 550 cm^2
6 a 24 amps b 5 amps
7 5.04 mL 8 K1822.50

Lesson 11 – Transposing formulae (pages 104–5)

1 a $r = 70$ cm b $l = 10$ cm
c $C = 37.8°$
2 a i $a = \frac{F}{m}$ ii 12.5 m/s^2
b i $h = \frac{A - 2\pi r^2}{2\pi r}$ ii 29.77 cm
c i $T = \frac{100I}{PR}$ ii 1.25
d i $P = \frac{A}{(1 + \frac{r}{100})^n}$ ii K13 605.44
e i $r = \sqrt{\frac{3V}{\pi h}}$ ii 9.23 cm
3 a $w = \frac{t - 30}{45}$ b 1.11 kg
4 a $d = \sqrt{\frac{k}{l}}$ b 0.49 m
5 a 180 mm b $t = \frac{4}{5}(200 - l)$
6 a No, weight is not represented in the formula.
b 1.89 seconds c 0.675 metres

Lesson 12 – Practical applications of graphs (pages 107–10)

1 a about 155 beats/min
b about 40 minutes c 80 beats/min
d 10 min e about 41 min
2 a i 1.7 s ii 2.6 s
b i 0.9 s ii 3.2 s
c Because it would take 0 seconds to fall 0 metres.
d No because it must go through (0, 0).
3 a 190 km/h b 240 secs
c 4 d 90 km/h
e 40 secs f about 40%
4 a 3050
b The number of new cases each year has been increasing over these years.
c 1750 d 3500
5 a 28 km b 7 hours 20 minutes
c 3.20 p.m. d 4.5 hours
e It took longer (1.5 hours) to get to his friend's place than it did to return home (1 h 15 min).
f No
6 a i K30 ii K20 iii K20 iv K10 v K45
b 20 km c K155
7 a More is paid off the loan each year as the years increase.
b K140 000 c 12 years
d 13 years e year 20
f year 1

Lesson 13 – Graphing on a Cartesian plane (page 113)

1 $A(5, 3)$, $B(6, {}^-1)$, $C({}^-3, 2)$, $D(0, 4)$, $E(2, 0)$, $F(2, {}^-4)$, $G({}^-3, {}^-4)$

2
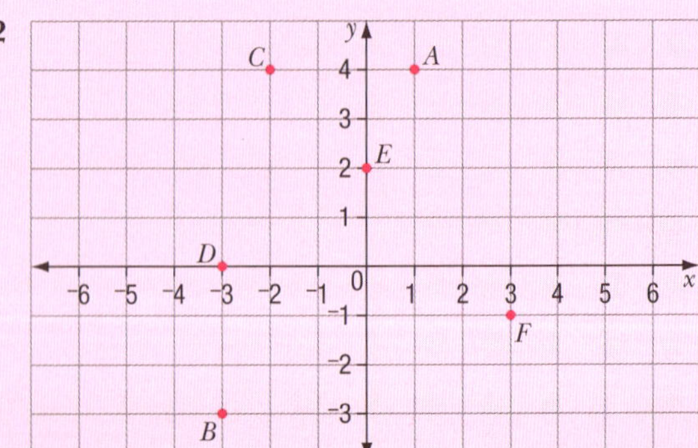

3 i infinite number ii one

4 a and b

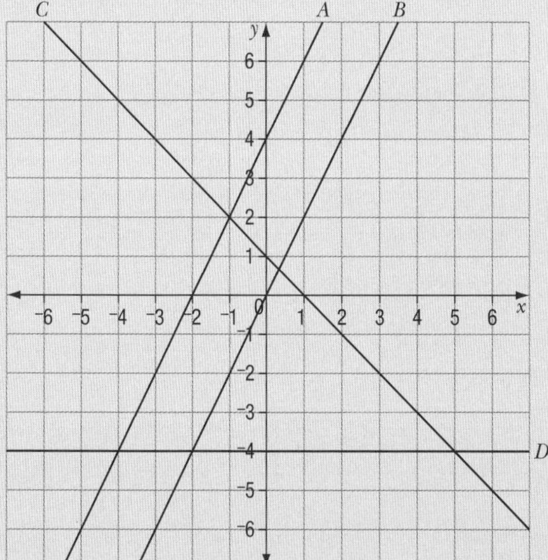

c ($^-1$, 2); second

d i ($^-2$, 0) ii (0, 4)

e i (1, 0) ii (0, 1)

5 a and b

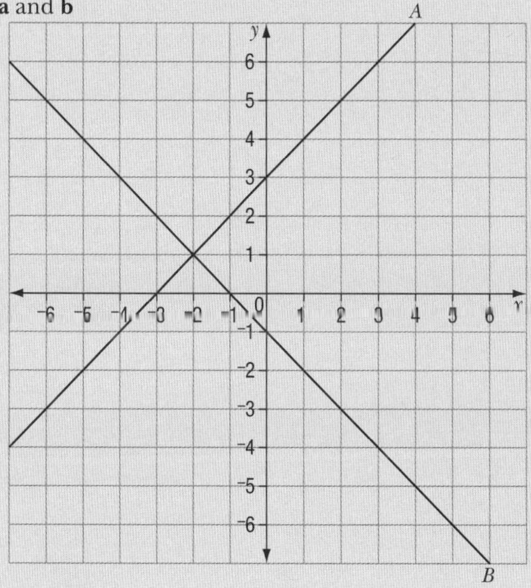

c ($^-2$, 1)

d ($^-3$, 0) and (0, 3)

e ($^-1$, 0) and (0, $^-1$)

Lesson 14 – Gradient of a straight-line graph (pages 116–17)

1 a $^-1$ b $\frac{3}{2}$

c $-\frac{5}{3}$ d 2

e 0 f indeterminate

2 a *B* and *D*

b i *B* ii *F* iii *E* iv *A* v *C* vi *D*

3 a 3 b 5

c 3 d 2

e $^-1$ f 0

g 0 h indeterminate

4 Some possibilities:

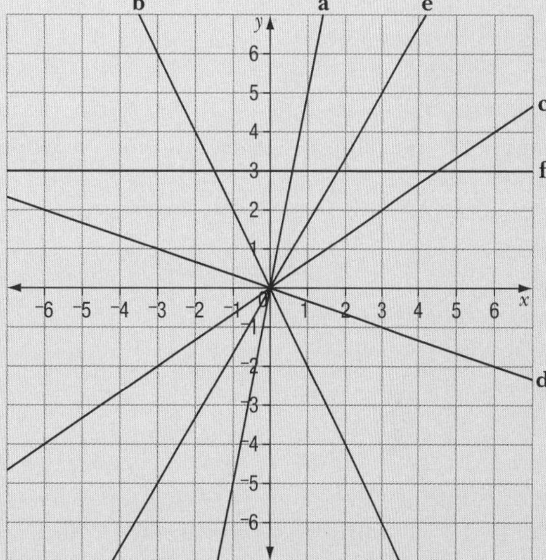

Lesson 15 – Finding the gradient from two points (pages 117–18)

1 a 3 b 3

c $^-2$ d $\frac{7}{2}$

e $^-5$ f $-\frac{1}{3}$

g $-\frac{3}{2}$ h $-\frac{3}{8}$

i 17

2 e i $^-4$ ii 5 iii $\frac{5}{4}$

f gradient, *y*-intercept

Lesson 16 – Identifying the gradient and *y*-intercept (page 120)

1 a i 2 ii $^-3$

b i $^-3$ ii 4

c i $\frac{5}{6}$ ii 4

d i $\frac{1}{3}$ ii $^-2$

e i 12 ii $^-3$

f i $\frac{5}{2}$ ii $^-2$

g i $^-3$ ii 7

h i 2 ii $-\frac{2}{3}$

i i $\frac{4}{3}$ ii 0

2 a $y = 4x + 6$ b $y = {}^-2x - 3$

c $y = \frac{2}{3}x - 2$ d $y = \frac{1}{3}x$

e $y = 4$

3 **a** i -1 ii 5
b i -2 ii 3
c i 3 ii 5
d i $\frac{2}{3}$ ii -2
e i $-\frac{1}{2}$ ii 4
f i $\frac{5}{4}$ ii 2
g i $-\frac{3}{8}$ ii 3
h i 0 ii $\frac{3}{5}$

Lesson 17 – Finding an equation from a graph (pages 122–3)

1 **a** $y=\frac{1}{4}x+3$ **b** $y=-2x$
c $y=\frac{2}{5}x+3$ **d** $y=-\frac{5}{3}x-5$
e $y=-\frac{1}{3}x-2$ **f** $y=x-1$
g $y=\frac{-3}{4}x+3$ **h** $y=3$
i $y=\frac{4}{5}x-4$

2 **i** *F* **ii** *A* **iii** *E* **iv** *D*
v *C* **vi** *B* **vii** *G*

Lesson 18 – Finding the equation from two points (page 124)

1 **a** $y=2x-1$ **b** $y=x-3$
c $y=-2x$ **d** $y=\frac{2}{3}x-5$
e $y=\frac{1}{2}x+\frac{5}{2}$ **f** $y=-\frac{3}{2}x-2$
g $y=-3$ **h** $y=10x+600$

Lesson 19 – Graphing a line from an equation (pages 126–7)

1

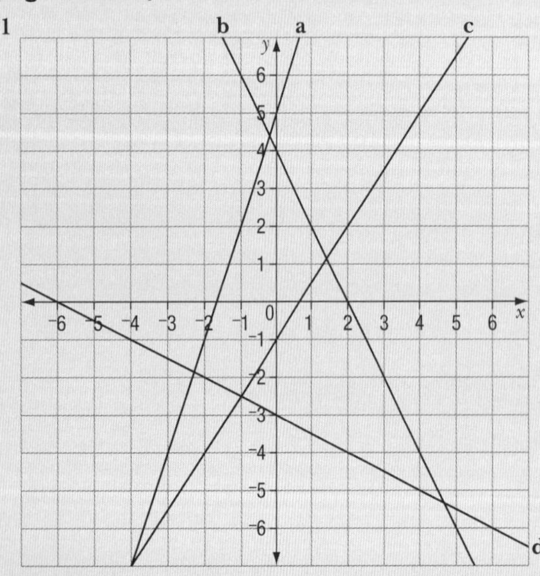

2

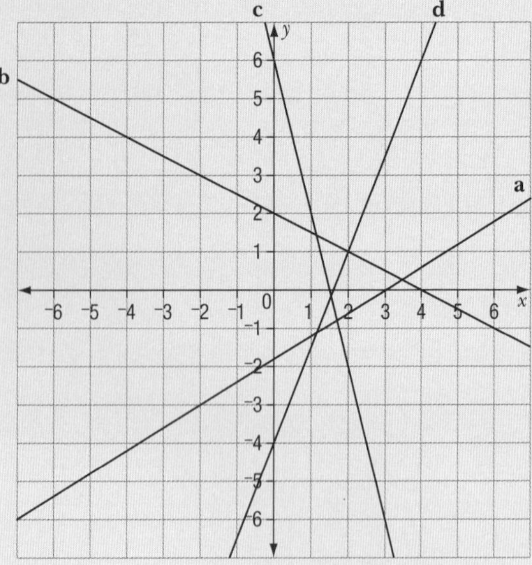

3 i and ii

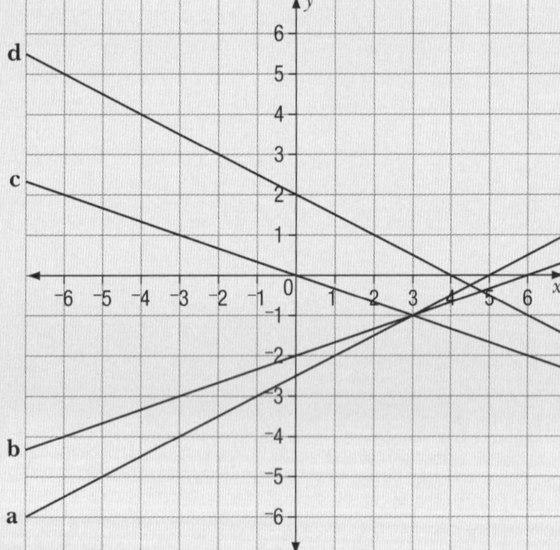

iii **a** (6, 0) and (0, -2)
b (-5, 0) and (0, -2.5)
c (0, 0)
d (4, 0) and (0, 2)

iv **a** $\frac{1}{2}$ **b** $\frac{1}{3}$
c $-\frac{1}{3}$ **d** $-\frac{1}{2}$

4 a, c, e

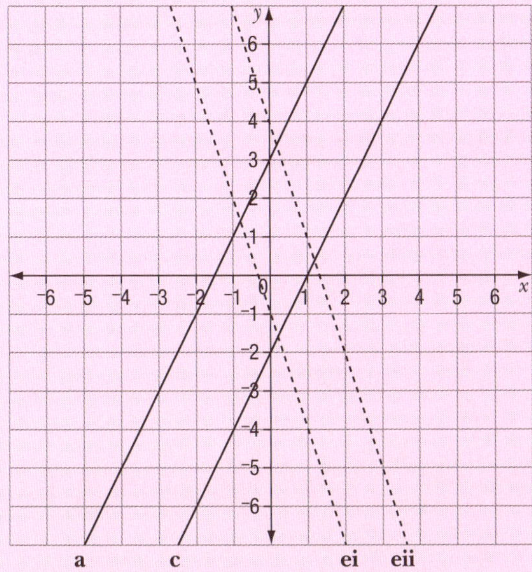

b 2 **d** $y = 2x - 2$

f Straight lines that are parallel have the same gradient.

5 If straight lines have the same gradient then they are parallel.

6 a i and iii **b** i and iii

c ii and iii **d** i and iii

e ii and iii

Lesson 20 – Modelling everyday problems (pages 129–31)

1 a 15 km

b 16 min

c After 6 minutes the car has travelled 9 kilometres.

d i 1.5 km/min **ii** 90 km/h

e *distance travelled*; because the *distance travelled* depends on the *time* spent travelling.

f $d = 1.5t$

2 a cooking time **b** 100 minutes

c 1 kilogram

d The smallest chicken that could be cooked using this guide is 0.8 kg and the largest would be 2.5 kg.

e 50; For each increase of 1 kg in the weight of the chicken the cooking time increases by 50 minutes. OR
For each increase of 100 g in the weight of the chicken the cooking time increases by 5 minutes.

f $C = 50w + 30$

3 a The candle is 30 cm high before it is lit.

b After 50 minutes the candle has burnt down to a height of 15 cm.

c 67 minutes **d** 80%

e gradient = ⁻0.3. For each minute of burning time the candle loses 0.3 cm in height.

f $H = {}^-0.3t + 30$ **g** $0 \leqslant t \leqslant 30$

4 a Salary **b** K520

c If Anton sells 22 phones in a week then his salary will be K640.

d If Anton sell no phones in a week then his salary will be K200.

e 20; Anton's weekly salary increases by K20 for each phone sold.

f $S = 20n + 200$

5 a Amount of electricity used

b If 1000 kWh of electricity is used then the cost will be K360.

c 0.32. For each kWh of electricity used the cost will increase by K0.32.

d $C = 0.32E + 40$

6 a After 2 years the value of the machine is K1440.

b After 5 years the value of the machine is zero.

c ⁻480; The value of the machine decreases by K480 each year.

d $V = {}^-480n + 2400$ **e** $0 \leqslant n \leqslant 5$

7 a K5000 **b** K7000

c After 1 year the value of Maureen's investment is K5400.

d $V = 400t + 5000$

Lesson 21 – Graphing linear relations that model problems (pages 133–4)

1 a $C = 1.5x + 2.2$

b

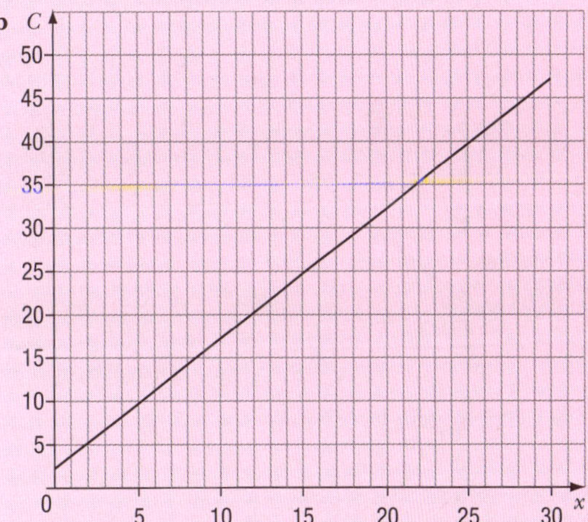

c 1.5; For each kilometre travelled the cost goes up K1.50.

d K32.20 **e** approx 9 km

2 a *speed* and *time*; *time*

b At a constant rate

c

speed (km/hr)

time (secs)

d $\frac{8}{3}$; For each second the speed increases by $\frac{8}{3}$ km/h (the gradient represents the acceleration).
0; The speed is not changing from 40 km/h for the time from 15 seconds to 35 seconds.
$^-4$; For each second from 35 to 45 seconds the speed slows by 4 km/h.

3 a K340
b K580
c $C = 80x + 100$
d

C

x

e 1.25 hours (1 h 15 min)

4 a

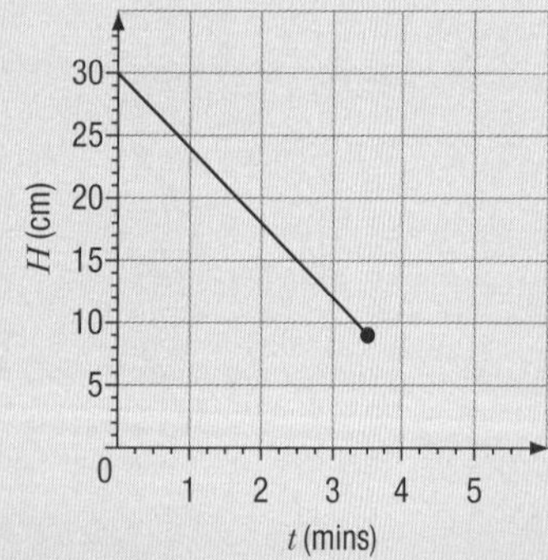

b 2 hours 30 minutes
c $^-6$; The candle burns down 6 cm each hour.
d 5 hours

5 a weekly pay and number of computers sold; weekly pay
b (3, 428) and (8, 608)
c

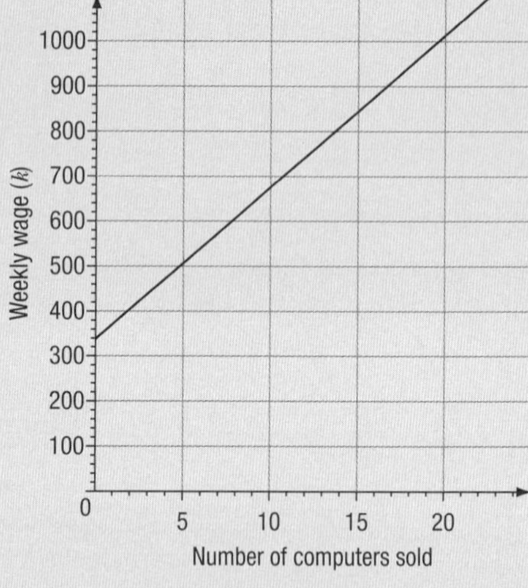

d 36; Anton receives K36 for each computer sold.
e K860
f No; he earns less than K1000 for selling 18 computers and more than K1000 for 19 computers.

6 a cost
b (262, 162.44), (284, 176.08)
c We assume that the cost of the electricity is the same (uniform) for any amount used.

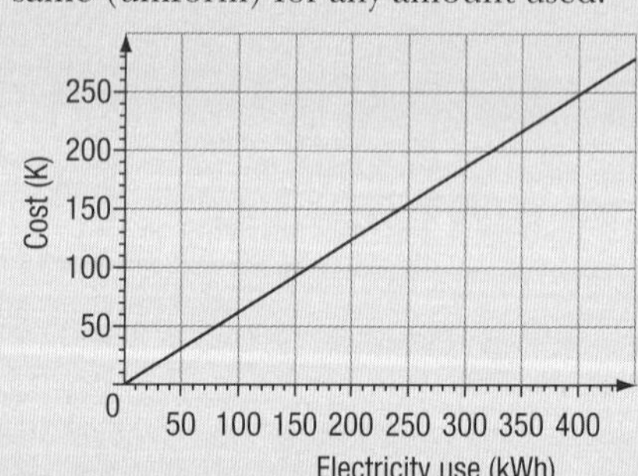

d cost = 0.62 × electricity use
e 0.62; Each kilowatt-hour of electricity costs K0.62.
f K186.00 g 325 kWh

7 a 15 b 70

c

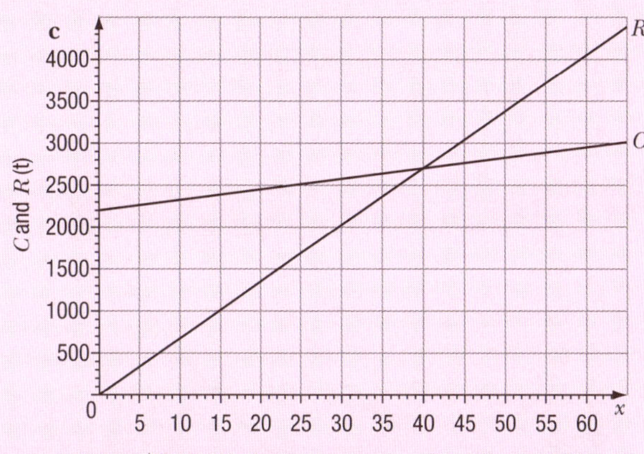

d (40, 2700); 40 doughnuts e $P = 55x - 2200$

f K22.00

Revision and Assessment

Multiple-choice questions

Directed numbers, indices and basic algebra (page 135)

1 B 2 A 3 B 4 D 5 C
6 D 7 C 8 A 9 C 10 B

Graphs (pages 136–7)

1 B 2 A 3 C 4 A 5 D
6 A 7 C 8 D 9 D 10 C

Short-answer questions

Directed numbers, indices and basic algebra (pages 137–8)

1 a 6 b 10
c $^{-}22$ d 1

2 a 1 b $\frac{1}{16}$
c 3

3 a 5.68×10^{-6} b 3.995×10^{11}

4 1.08×10^4

5 a $6x^3$ b $27a^6b^3$
c $32m^8$

6 a $2^{\frac{3}{2}}$ b $\frac{3}{4}$
c 9

7 a $17 - a$ b $41m - 70$

8 a $3a(2a + 3)$ b $^{-}2p(5pq + 1)$

9 a $x = 5$ b $y = 3$
c $x = 33$ d $y = 5.5$
e $x = 4$ f $y = \frac{18}{7}$

10 a 33 b $a = \frac{2(s - ut)}{t^2}$

Graphs (pages 138–40)

1 a $(^{-}2, 0)$ b $(0, 6)$
c $(2, 4)$ d $(^{-}1, 1)$, others possible
e $(7, ^{-}1)$, others possible

2 a $-\frac{3}{2}$ b 2
c 0

3

4 a $-\frac{1}{3}$ b indeterminate

5 a i 7 ii $^{-}4$
b i 2 ii $^{-}13$
c i $-\frac{5}{8}$ ii 2

6 a i $(2, 0)$, $(^{-}2, ^{-}6)$, others
b i $(2, 0)$, $(0, 2.5)$, others

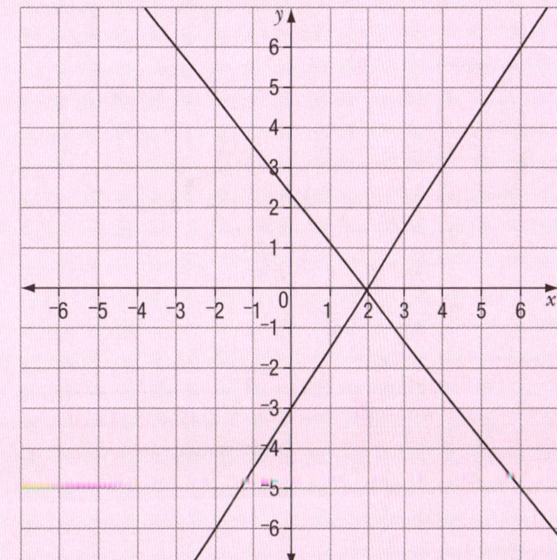

7 $y = 1.5x$

8 *A* $y = -\frac{3}{4}x + 3$ *B* $y = {}^{-}3x$
C $y = \frac{2}{3}x - 3$ *D* $y = {}^{-}4$

9 a $y = \frac{1}{2}x + 2$ b $y = -\frac{4}{7}x + 4$
c $y = 5$

10 a 2.7 m b 0.2 m
c 9 hours d 5 p.m.

Option A: Algebra and graphs

Lesson 1 – Finding the line-of-best-fit for a graph (pages 142–5)

1 a Volume b Yes
c No
d i 400 mL ii about 1540 mL

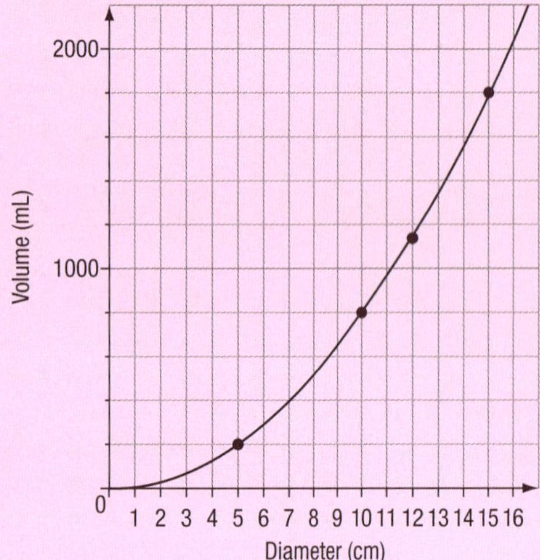

2 a Depth **b** Yes

c No **d i** 740 cm^2 **ii** 1150 cm^2

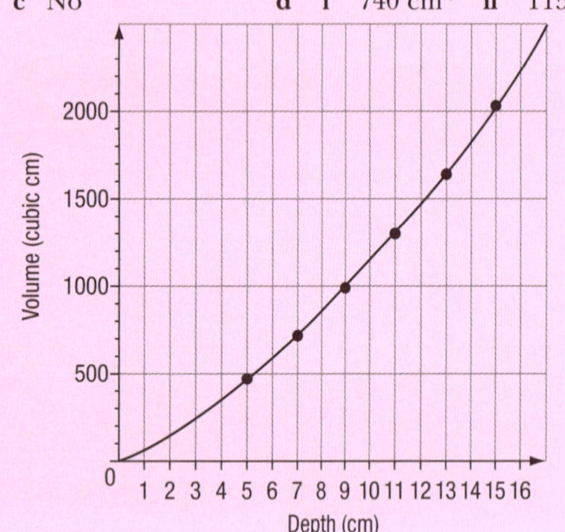

3 a Length and Period (time)

b Period

c and f

d Yes **e** No

f i 0.35 s **ii** 1.07 s

g 12 cm

4 a Speed and Braking distance

b Braking distance

c and f

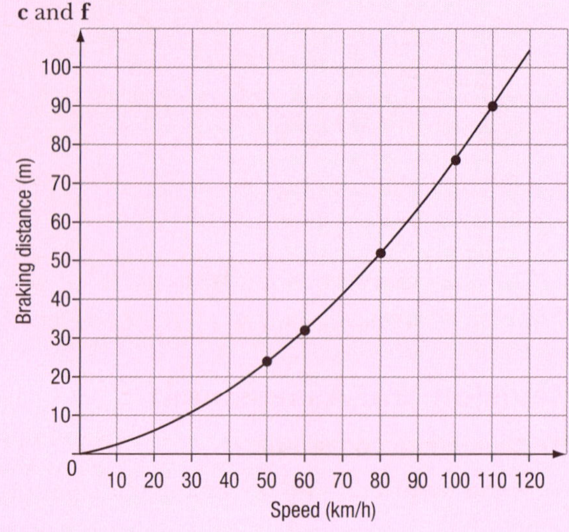

d Yes **e** No

f i 64 m **ii** 10 m

g 87 km/h

5 a Year (time) and Amount owing

b Amount owing

c and f

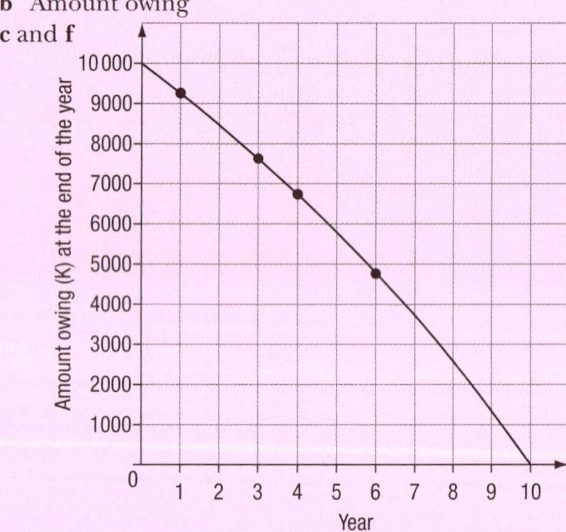

d (0, 10000) and (10, 0)

e No

f i K8400 **ii** K2500

g 5.7 years

6 a Weight and Extension

b Extension

c and **f**

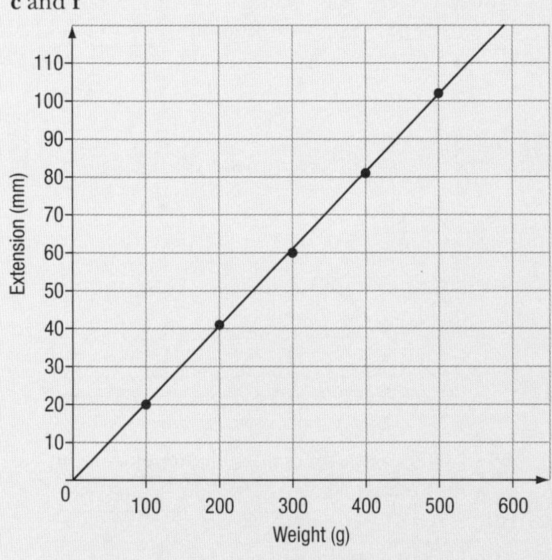

d Yes   **e** Yes

f 50 mm   **g** 240 g

h There would be a (large) weight that would extend the spring to its maximum length, and 5 kg would probably do this. A very large weight could distort the spring and it would not return to its original length.

7 a Time (month) and Weight

b Time (month)

c and **f**

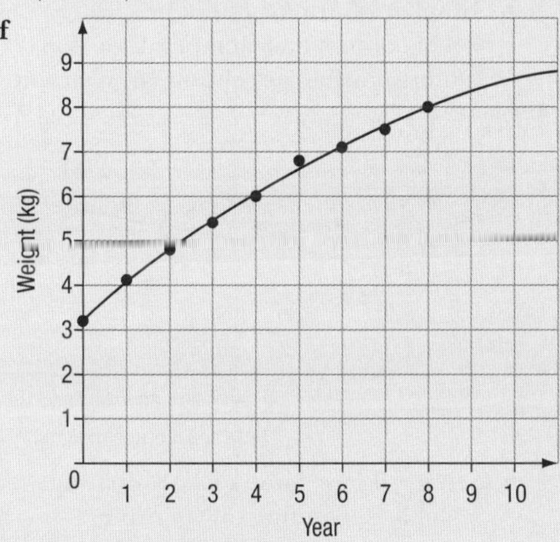

d Rose weighed 3200 g at birth.

e No   **f** 8.5 kg

g In the 5th month.

h No, the rate of growth is greatest in the first year and changes for a child over 12 months.

Lesson 2 – Constructing and interpreting scatter graphs (pages 149–51)

1 a i negative   **ii** strong

b i positive   **ii** moderate

c i negative   **ii** moderate

d i positive   **ii** strong

2 a i positive   **ii** linear   **iii** moderate

b i negative   **ii** linear   **iii** strong

c i no association   **ii** –   **iii** –

d i positive   **ii** not linear   **iii** strong

e i negative   **ii** not linear   **iii** weak

f i positive   **ii** linear

iii moderate (two outliers)

3 a increases

b decreases

c randomly scattered

4 a

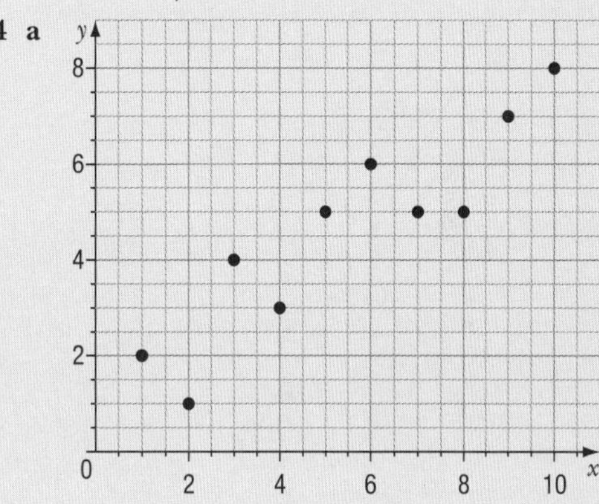

b i linear   **ii** positive   **iii** moderate

5 a Number of drinks sold.

b

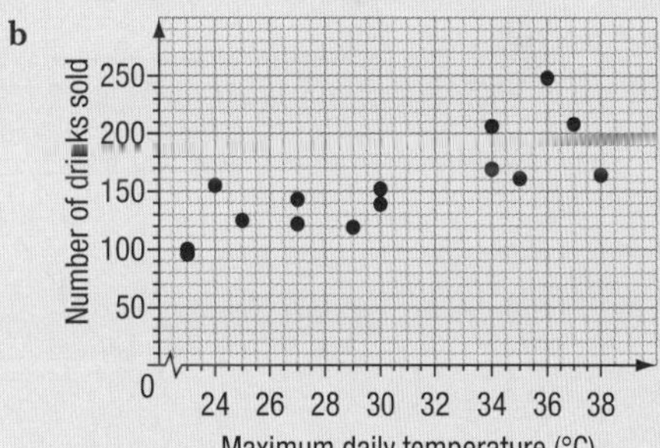

c Positive, moderate, linear, no outliers. As the maximum daily temperature increases, the number of drinks sold also increases.

d No. As the temperature increases thirst also increases. As thirst increases the tendency to buy more drinks also increases.

6 a Time spent preparing

b

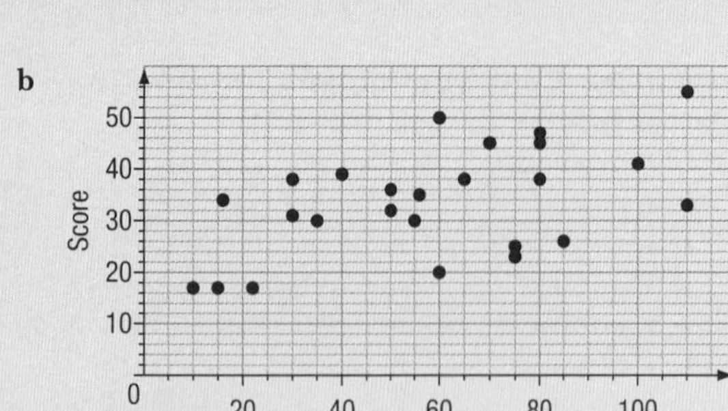

c Positive, moderate to weak, linear, no outliers. As the time spent preparing increases the score also increases.

Lesson 3 – Fitting a line 'by eye' to a scatter graph and finding its equation (pages 153–4)

1 a i and ii

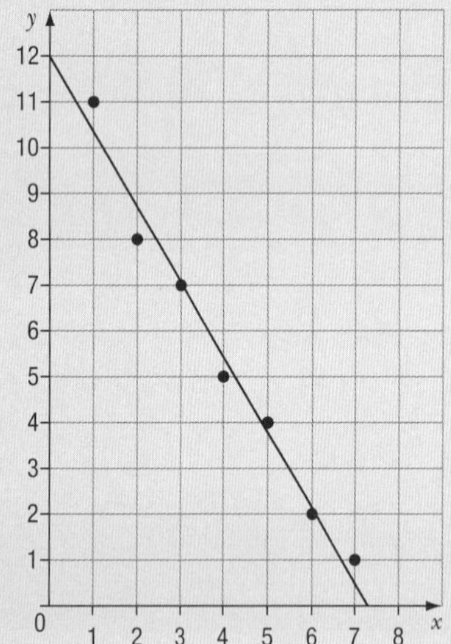

iii $y = {}^{-}16x + 12$; others possible depending on your line.

b i and ii

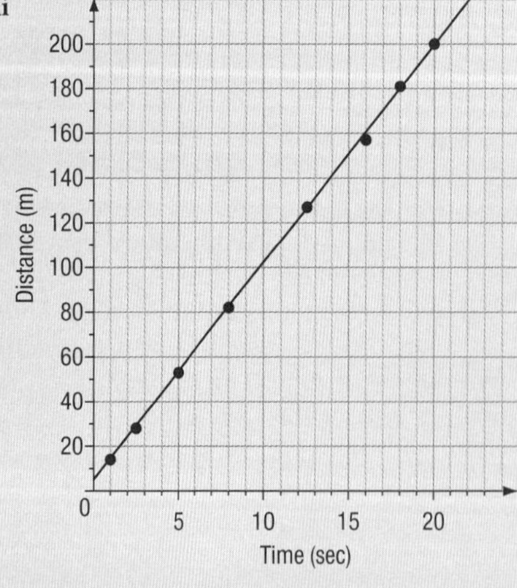

iii Distance = 9.8 × Time + 3.6; others possible depending on your line.

2 a

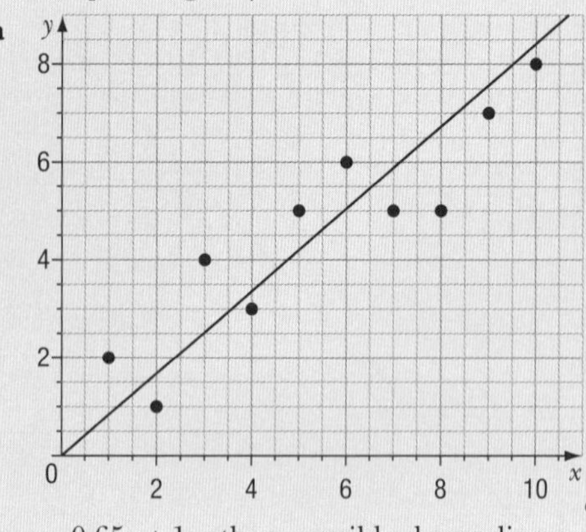

b $y = 0.65x + 1$; others possible depending on your line.

3 a

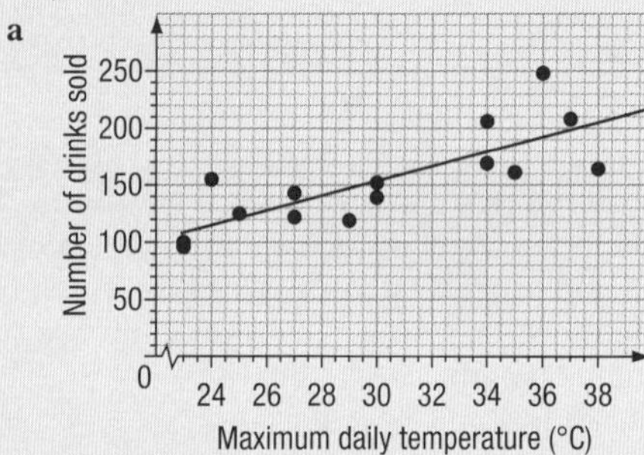

b Number of drinks sold =
7 × Maximum daily temperature – 58.
Others possible depending on your line.

4 a

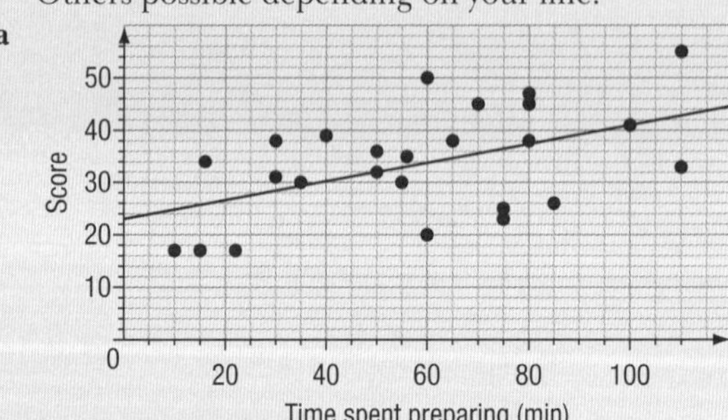

b Score = 0.19 × Time spent preparing + 23. Others possible depending on your line.

Lesson 4 – Interpreting the gradient and intercept (pages 155–7)

1 Line *C*

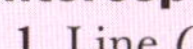

2 a i

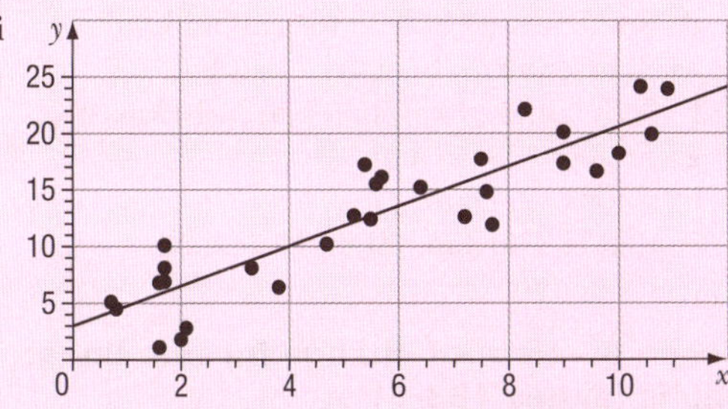

ii $y = 1.8x + 2.7$. Others possible.

iii For each increase of one unit in the *x* value, the *y* value will increase by 1.8 units. When the *x* value is zero the *y* value is 2.7.

b i

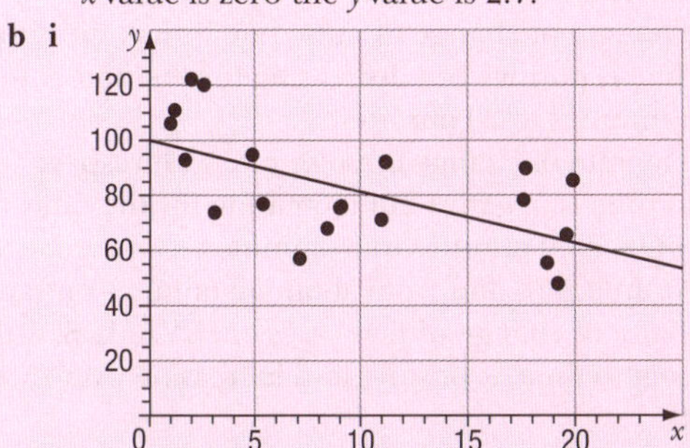

ii $y = {}^{-}1.8x + 100$

iii For each increase of one unit in the *x* value, there would be a decrease of 1.8 units in the *y* value. When the *x* value is zero the y value is 100.

c i

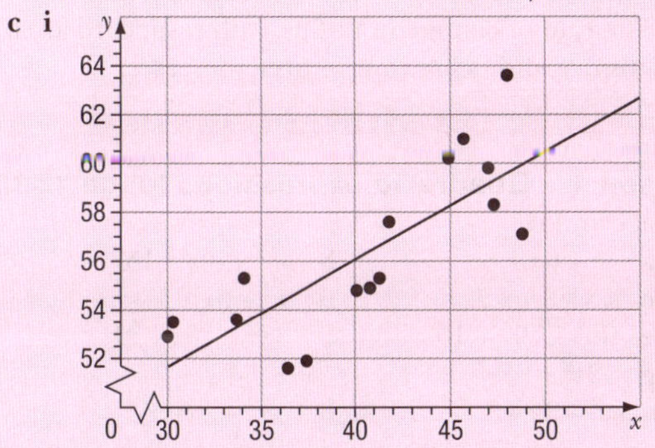

ii $y = 0.44x + 39$

iii For each increase of one unit in the *x* value, there would be an increase of 0.44 units in the *y* value. When the *x* value is zero, the *y* value is 39.

3 a

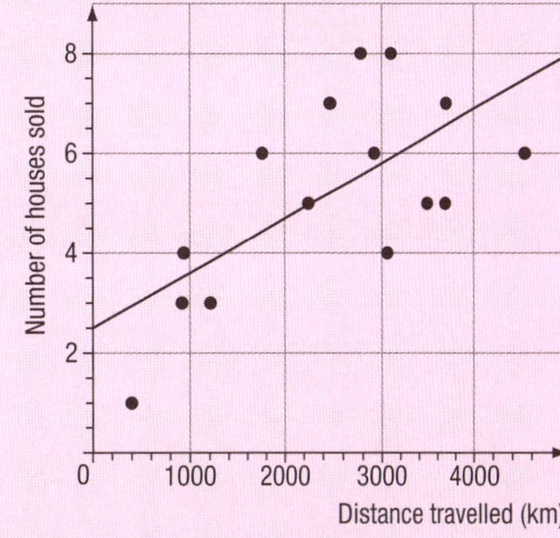

b Number of houses sold = 0.001 × distance travelled + 2.6. Others possible.

c For each kilometre travelled the number of houses sold increases by 0.001. OR For each 1000 kilometres travelled the number of houses sold increases by 1. When the distance travelled is zero, the number of houses sold is 2.6.

d 4.6, interpolation

4 a A

b C

5 Using the equation:
Score = 0.19 × Time spent preparing + 23 (others possible), for each minute of time spent preparing the score will increase by 0.19. If you spend no time preparing then your score will be 23.

Lesson 5 – Constant rates of change (pages 159–61)

1 a 4 km/h b 32 mins c 5 mins
d 13 mins e 1.8 km f 27 mins
g 4 km/h

2 a A b K2.2 per km c K1.8 per km
d 0 up to 10 km e K2 less with A

3

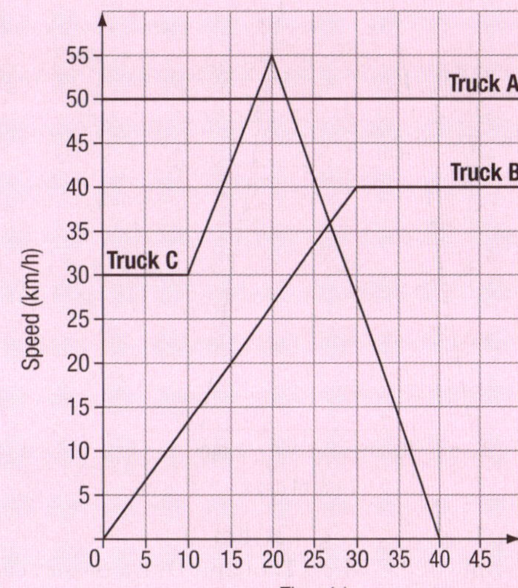

a i C   ii C

b i $\frac{4}{3}$ ii $-\frac{11}{4}$

4 **a** 3.5 km

b

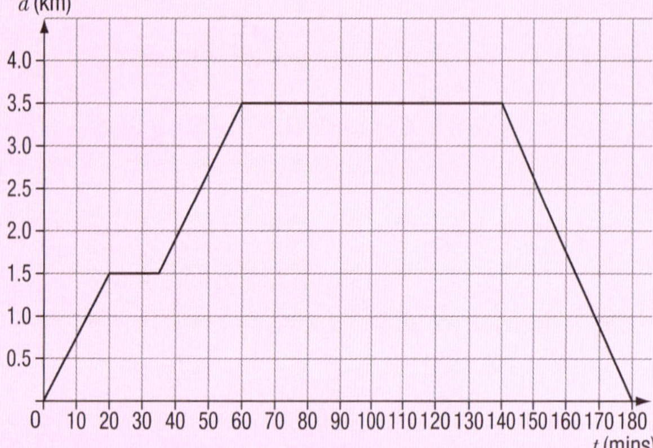

c 4.5   **d** 5.25   **e** 1.30 p.m.

5 **a**

b

Week	1	2	3	4
Rate of growth (cm/week)	8	6.8	5.3	4.6

c The rate of growth is decreasing as the weeks increase.

6 **a**

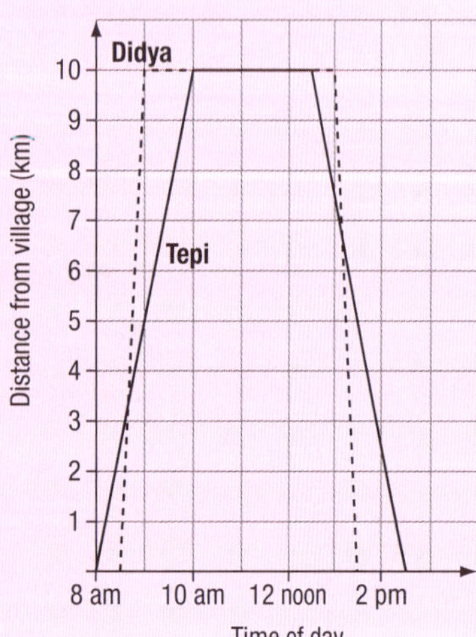

b 5 km/h   **c** 20 km/h

d The time of day when Tepi and Didya passed each other on the road.

e About 8.40 a.m. and 1.15 p.m.

7 **a** 3 km   **b** 15 km

c Railila left at 8 a.m. and Eleni left at 8.48 a.m.

d Railila rested at 9.30 for half an hour.

e Before: 6 km/h; after: 2 km/h

f 3.75 km/h, 0 km/h, 5 km/h, 10 km/h

g At about 11.45 a.m.

h About 12.5 km

i Railila 8.40 a.m.; Eleni 10.48 a.m.

j One hour

Lesson 6 – Rate of change for non-linear graphs (pages 163–5)

1 **a** Negative but increasing

b The rate of change of volume with respect to pressure is negative but increasing as the pressure increases.

2 **a** The rate of change of pulse rate with respect to time is positive but decreasing for the time interval 5 minutes to 20 minutes.

b The rate of change of pulse rate with respect to time is negative but increasing for the time interval 30 minutes to 60 minutes.

c At time zero and from about 52 minutes onwards.

3 The rate of change of new cases of HIV/AIDS, with respect to time, is positive and increasing over these years.

4 The rate of change of the weight of the piglet, with respect to time, is positive and increasing as time increases.

5 The rate of change of debt, with respect to time, is negative and decreasing as time increases.

6 The rate of change of radius, with respect to area, is positive and decreasing as area increases.

Option B: Quadratic equations

Lesson 1 – Quadratic expressions (page 168)

1 **a, c** and **d**

2 **a** $x^2 + 6x + 8$ **b** $m^2 + 8m + 15$

c $n^2 + 8n + 7$ **d** $x^2 + 6x$

e $6x^2 + 19x + 10$ **f** $3x^2 + 16x + 21$

g $12y^2 + 31y + 20$

3 **a** $x^2 - 2x - 8$ **b** $4x^2 - 5x - 6$

c $y^2 + 3y - 28$ **d** $10y^2 - y - 3$

e $3x^2 - 24x$ **f** $^-y^2 - y + 12$

g $8m^2 + 10m - 3$

4 **a** $x^2 - 10x + 16$ **b** $2x^2 - 11x + 12$

c $^-x^2 + 8x - 15$ **d** $^-2y^2 + 3y$

e $15n^2 - 38n + 7$ **f** $^-7x^2 + 41x - 30$

g $x^2 - 11x + 30$

5 **a** $3x^2 - 7x - 20$ **b** $4x^2 - 1$

c $25x^2 - 45x + 18$ **d** $2x^2 - 6x + 20$

e $y^2 + 8y + 16$ **f** $4x^2 + 6x$

g $2y^2 - 12y$ **h** $21y^2 - 42y + 8$

Lesson 2 – Expansion of perfect squares (page 169)

1 **a** $x^2 + 6x + 9$ **b** $4x^2 + 4x + 1$
c $p^2 + 2pq + q^2$ **d** $16m^2 + 56m + 49$
e $36x^2 + 12x + 1$ **f** $9y^2 + 66y + 121$
g $64n^2 + 48n + 9$ **h** $9p^2 + 24pq + 16q^2$
i $9y^2 + 30y + 25$ **j** $25x^2 + 2x + 0.04$

2 **a** $x^2 - 8x + 16$ **b** $4x^2 - 12x + 9$
c $m^2 - 2mn + n^2$ **d** $36y^2 - 12y + 1$
e $25x^2 - 120x + 144$ **f** $4p^2 - 20pq + 25p^2$
g $x^2 - 10x + 25$ **h** $4y^2 - 4y + 1$
i $9x^2 - 36x + 36$ **j** $x^2 - 1.6x + 0.64$

3 **a** $0.64x^2 + 0.48x + 0.09$
b $a^2x^2 - 2ax + 1$
c $36x^2 - 1200x + 10\,000$
d $x^4 - 4x^2 + 4$
e $576y^2 + 1200y + 625$
f $x^2 + 5x + \frac{25}{4}$
g $\frac{1}{4}x^2 - 5x + 25$
h $0.25x^2 - 0.8x + 0.64$
i $1.96y^2 + 14y + 25$
j $a^2x^2 + 2abx + b^2$

4 **a** 1681 **b** 1521
c 2704 **d** 1 002 001
e 996 004 **f** 7225
g 15 129 **h** 902 500
i 36.6025 **j** 23.04

Lesson 3 – Expansions with a difference of two squares (page 170)

1 **a** $x^2 - 9$ **b** $m^2 - n^2$
c $4x^2 - 9$ **d** $49 - x^2$
e $25x^2 - 9$ **f** $1 - y^2$
g $9 - 16x^2$ **h** $16 - 25y^2$
i $49m^2 - 9y^2$ **j** $3x^2 - 75$

2 **a** $x^2 - 5x - 24$ **b** $2m^2 - mn - n^2$
c $2x^2 - 7x - 39$ **d** $y^2 - 18y + 81$
e $25x^2 + 120x + 144$ **f** $-15x^2 - x + 2$
g $0.49x^2 - 0.64$ **h** $25y^2 + 60y + 36$
i $-12x^2 + 5x + 2$ **j** $36y^2 + 60y + 25$
k $4x^2 + 9x + 3.5$ **l** $a^2b^2 - c^2$

Lesson 4 – Factorising quadratic expressions (pages 173–4)

1 **a, c, e and g**

2 **a** $(a + 2b)(a - 2b)$ **b** $(x - 6)(x + 6)$
c $(3 - 2a)(3 + 2a)$ **d** $(5x - 2y)(5x + 2y)$

3 **a** $(p + q)(p - q)$ **b** $(x + 2)(x - 2)$
c $(3m + 1)(3m - 1)$ **d** $(y - 6)(y + 6)$
e $(4 - 3y)(4 + 3y)$ **f** $(2x + 1)(2x - 1)$
g $(5x - 3y)(5x + 3y)$ **h** $(8p - 5q)(8p + 5q)$
i $2(5x - 1)(5x + 1)$ **j** $2(5 + 3k)(5 - 3k)$
k $25(2x + 1)(2x - 1)$ **l** $(10x - 11)(10x + 11)$
m $3(4 + 5a)(4 - 5a)$ **n** $x^2(x + 1)(x - 1)$
o $x(2 + x)(2 - x)$ **p** $5p^2(1 + q)(1 - q)$

4 **a, d, g**

5 **a** $(x + 2)^2$ **b** $(y - 3)^2$
c $(x - 6)^2$ **d** $(x + 4)^2$
e $(x - 7)^2$ **f** $(m - 5)^2$
g $4(x + 1)^2$ **h** $3(p - 2)^2$
i $2(x + 3)^2$ **j** $3(x - 4)^2$
k $2(x - 6)^2$ **l** $a(x + 1)^2$

Lesson 5 – Factorisation of quadratic trinomials of form $ax^2 + bx + c$ (page 177)

1 **a** 2, 8 **b** 2, 16
c $-2, -14$ **d** $-6, -8$
e 1, 64 **f** 2, 32
g $-8, -9$ **h** $-6, -9$

2 **a** $(x + 2)(x + 3)$ **b** $(m + 1)(m + 4)$
c $(p + 3)(p + 8)$ **d** $(q + 2)(q + 12)$
e $(x + 6)(x + 6)$ **f** $(x + 1)(x + 10)$
g $(x + 3)(x + 7)$ **h** $(a + 3)(a + 12)$
i $(b + 4)(b + 12)$ **j** $(x + 7)(x + 9)$
k $(t + 40)(t + 60)$ **l** $(s + 9)(s + 9)$

3 **a** $(x - 1)(x - 2)$ **b** $(m - 3)(m - 4)$
c $(p - 4)(p - 9)$ **d** $(q - 2)(q - 6)$
e $(x - 5)(x - 9)$ **f** $(x - 1)(x - 12)$
g $(t - 8)^2$ **h** $(a - 3)(a - 12)$
i $(b - 11)(b - 12)$ **j** $(x - 3)(x - 21)$
k $(x - 7)^2$ **l** $(x - 3)(x - 7)$

Lesson 6 – Factorisation of quadratic trinomials of form $x^2 + bx - c$ or $x^2 - bx - c$ (page 180)

1 **a** $(x - 4)(x + 2)$ **b** $(m - 3)(m + 5)$
c $(p - 8)(p + 4)$ **d** $(q + 4)(q - 5)$
e $(x - 3)(x + 7)$ **f** $(x + 6)(x - 7)$
g $(x - 9)(x + 11)$ **h** $(a + 4)(a - 9)$
i $(b - 5)(b + 7)$ **j** $(x - 3)(x + 8)$
k $(m + 6)(m - 11)$ **l** $(p - 2)(p + 11)$
m $(a - 8)(a + 4)$ **n** $(b - 3)(b + 15)$
o $(x - 7)(x + 8)$ **p** $(t - 6)(t + 10)$

2 **a** $(x - 8)(x - 1)$ **b** $(m + 3)(m - 5)$
c $(p - 2)(p + 16)$ **d** $(q + 6)(q + 6)$
e $(x + 3)(x - 7)$ **f** $(x - 3)(x + 14)$
g $(x - 3)(x + 13)$ **h** $(a + 3)(a - 12)$
i $(b + 1)(b + 35)$ **j** $(x - 4)(x - 1)$
k $(m - 7)(m + 12)$ **l** $(p - 5)(p + 12)$
m $(a + 4)(a - 30)$ **n** $(b + 15)(b + 3)$
o $(x - 9)(x + 10)$ **p** $(t + 12)(t - 20)$
q $(s + 10)(s - 40)$ **r** $(x - 12)^2$
s $(b + 7)(b - 8)$ **t** $(x - 10)(x - 12)$
u $(m - 7)(m + 9)$ **v** $(p + 3)(p - 24)$
w $(a + 2)(a + 35)$ **x** $(b - 8)(b + 9)$
y $(x + 3)(x - 16)$ **z** $(t - 3)(t - 14)$

Lesson 7 – Factorisation of quadratic trinomials of form $ax^2 + bx + c$ ($a \neq 1$) (page 184)

1 a $(3x+2)(x+3)$ b $(5c+4)(c+3)$
c $(7m+1)(m+4)$ d $(2x+5)(x+2)$
e $(11p+4)(p+1)$

2 a $(2x-1)(x-3)$ b $(5c-7)(c-2)$
c $(7m-6)(m-2)$ d $(13x-12)(x-1)$
e $(3p-4)(p-8)$

3 a $(3x+2)(x-4)$ b $(3c-5)(c+2)$
c $(2m-7)(m+3)$ d $(5x-2)(x+10)$
e $(17p-14)(p+1)$ f $(2x-3)(x+3)$
g $(7x+8)(x-4)$ h $(13b+32)(b-2)$

4 a $(3x+4)(2x+5)$ b $(4c+3)(c+4)$
c $(5m+6)(2m+3)$ d $(4x+3)(3x+8)$
e $(4p+3)(3p+5)$

5 a $(5x-1)(3x+2)$ b $(7c-3)(2c-3)$
c $(4m-5)(3m+4)$ d $(6x+11)(2x-3)$
e $(6p+5)(2p+3)$ f $(4x-3)(2x-9)$
g $(9q-32)(q+2)$ h $(4t-5)(4t+9)$

Lesson 8 – Solving quadratic equations that can be factorised (page 186)

1 a $x = {}^-4, {}^-2$ b $x = 0, 6$
c $y = 1, 4$ d $x = {}^-1, 2$
e $x = {}^-3, -\frac{1}{2}$ f $m = -\frac{4}{5}, \frac{5}{2}$

2 a $(x+5)(x+3)$; $x = {}^-5, {}^-3$
b $(x+5)(x-6)$; $x = {}^-5, 6$
c $(x-5)(x+9)$; $x = {}^-9, 5$
d $3(x+3)(x-3)$; $x = 3, {}^-3$

3 a $x = 2, 16$ b $x = 1, 15$
c $x = 1, 12$ d $x = {}^-4, 3$
e $x = {}^-9, 2$ f $x = {}^-3, 8$

4 a $x = -\frac{1}{2}, {}^-3$ b $x = \frac{1}{3}, 4$
c $x = \frac{5}{3}, {}^-1$ d $x = \frac{3}{2}, 3$
e $x = \frac{1}{6}, -\frac{3}{4}$ f $x = \frac{3}{4}, -\frac{5}{3}$

Lesson 9 – Solving quadratic equations that cannot be factorised (pages 188–9)

1 a $\frac{{}^-2 \pm \sqrt{84}}{10}$; 0.717, ⁻1.117
b $\frac{{}^-5 \pm \sqrt{17}}{2}$; ⁻0.438, ⁻4.562
c $\frac{{}^-6 \pm \sqrt{60}}{{}^-4}$; ⁻0.436, 3.436
d $\frac{3 \pm \sqrt{105}}{8}$; 1.656, ⁻0.906
e $\frac{6 \pm \sqrt{120}}{{}^-14}$; 1.211, ⁻0.354
f $\frac{5}{3}, {}^-1$

2 a $(x+5)^2$, $x = {}^-5$ b $(x-3)^2$, $x = 3$

3 a One b It is zero.

4 a No solution b No solution
c No solution

5 a None b It is negative.

6 a two b one
c negative

Lesson 10 – Graphing quadratic functions (page 191)

1 a ii

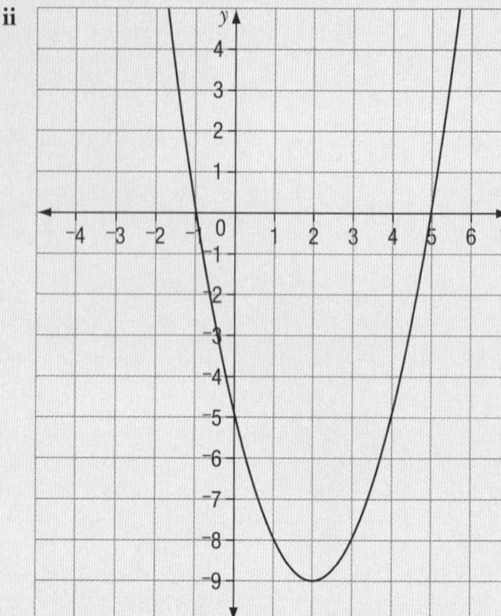

iii minimum at (2, ⁻9) iv $x = 2$
v (⁻1, 0), (5, 0) vi (0, ⁻5)

b ii

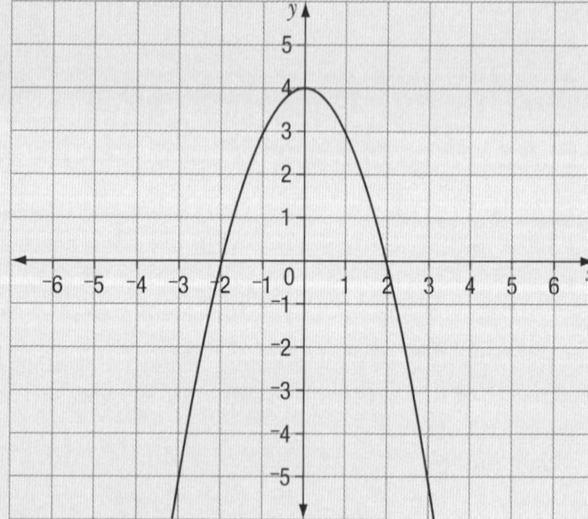

iii maximum at (0, 4) iv $x = 0$
v (⁻2, 0), (2, 0) vi (0, 4)

c ii

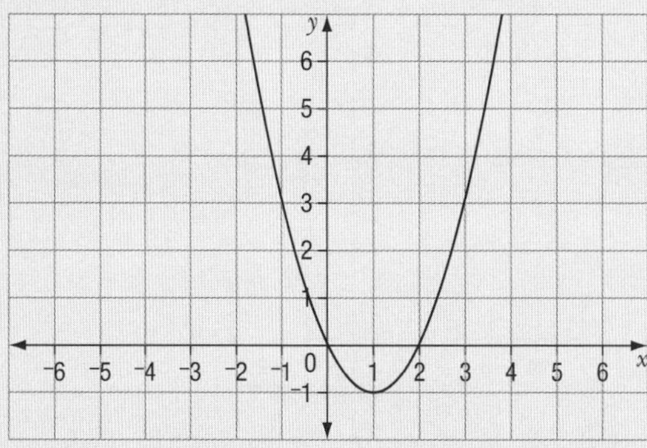

iii minimum at (1, ⁻1)
iv $x = 1$
v (0, 0), (2, 0)
vi (0, 0)

d ii

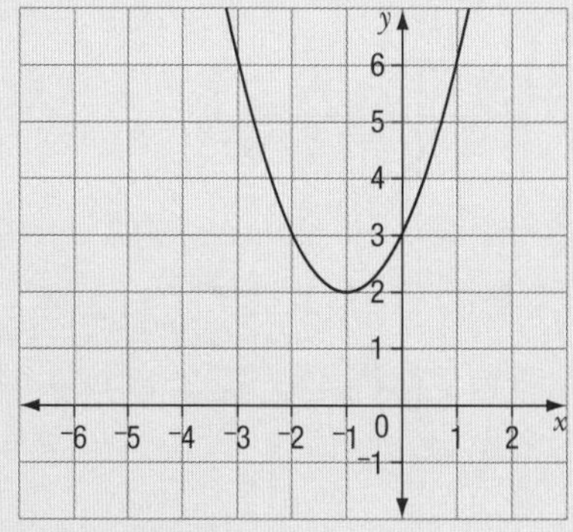

iii minimum at (⁻1, 2)
iv $x = ⁻1$
v None
vi (0, 3)

e ii

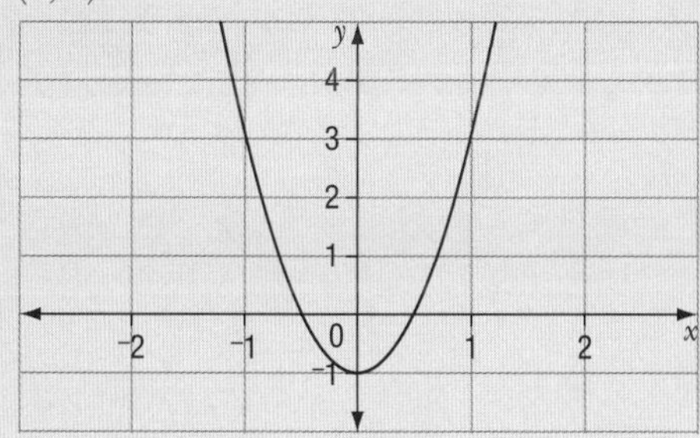

iii (0, ⁻1)
iv $x = 0$
v $(-\frac{1}{2}, 0)$, $(\frac{1}{2}, 0)$
vi (0, ⁻1)

f ii

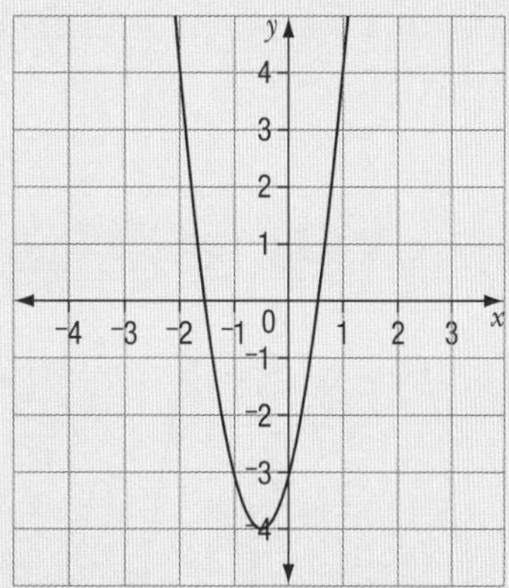

iii minimum at (⁻0.5, ⁻4)
iv $x = ⁻0.5$
v (⁻1.5, 0), (0.5, 0)
vi (0, ⁻3)

Lesson 11 – Using intercepts and turning point to sketch a quadratic (page 192)

1 a

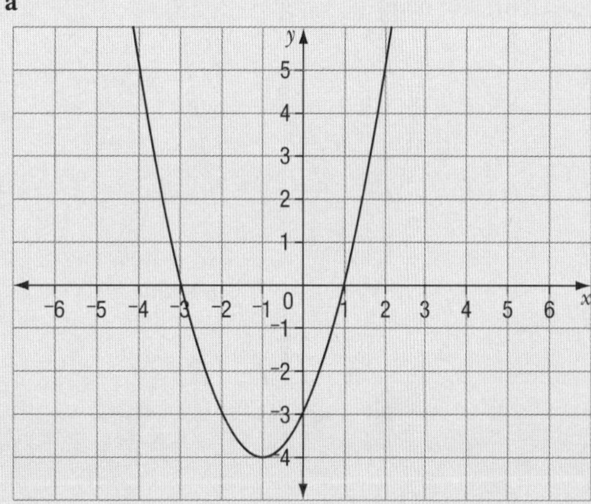

b

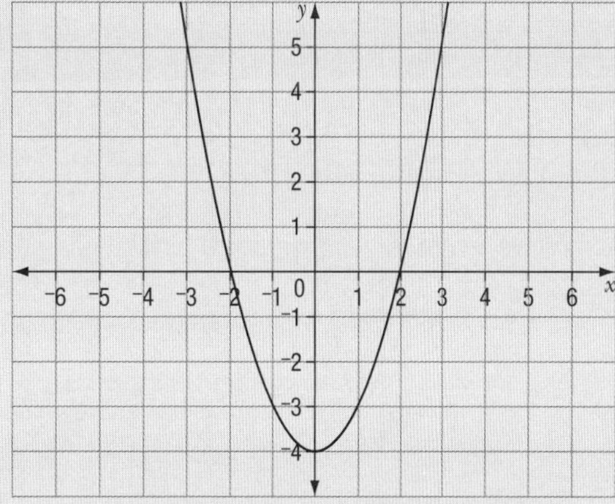

c

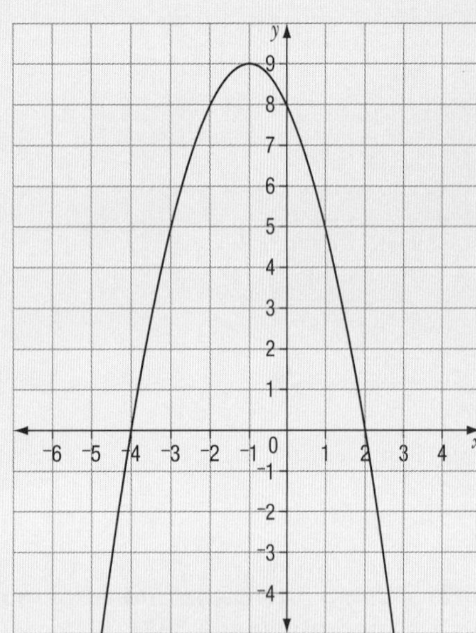

d Turning point (1.5, −2.25)

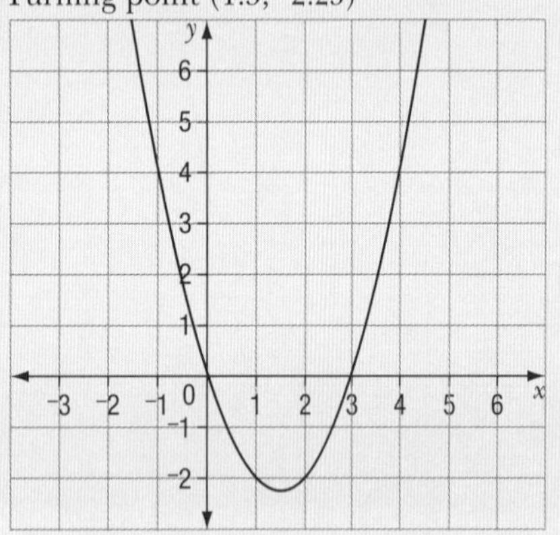

e Turning point (0.5, −4)

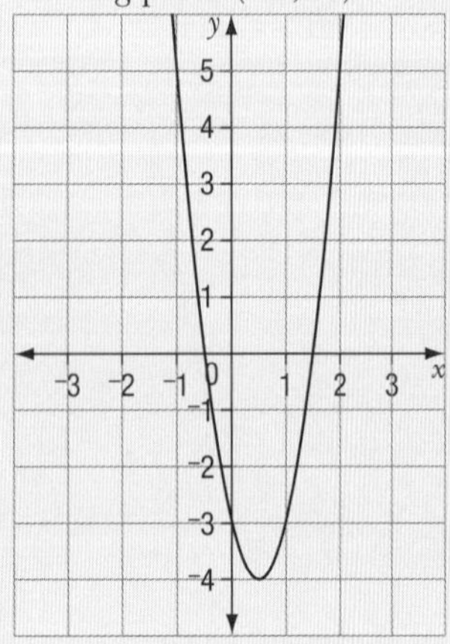

2 a Turning point (−0.75, −3.125)

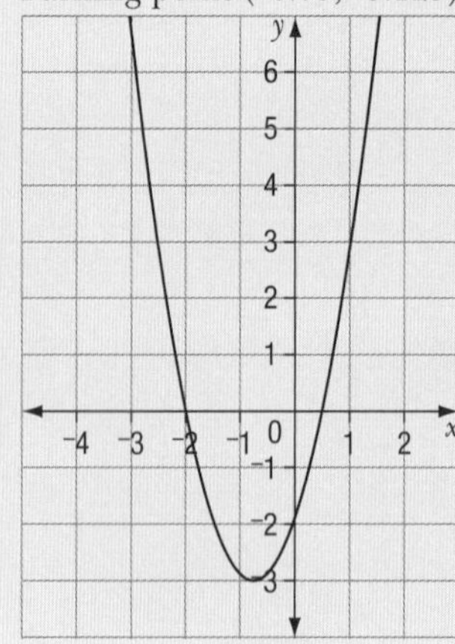

b Turning point $(-\frac{1}{3}, -\frac{25}{3})$

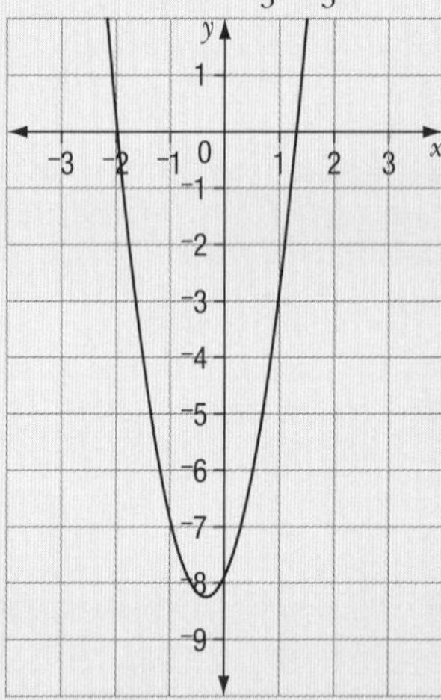

c Turning point $(-\frac{1}{12}, -\frac{361}{24})$

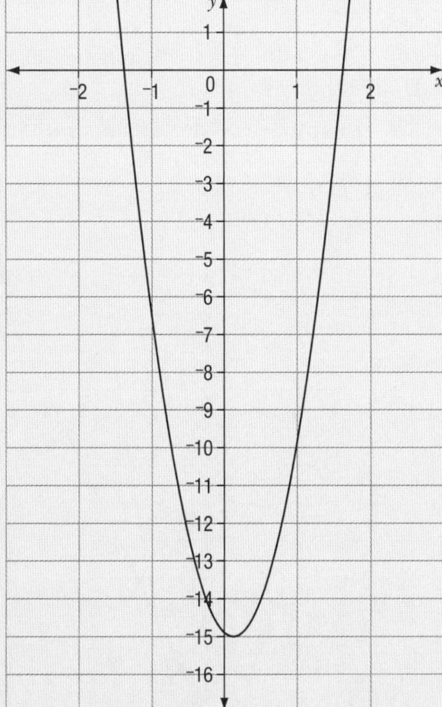

3 a Turning point (1, 4)

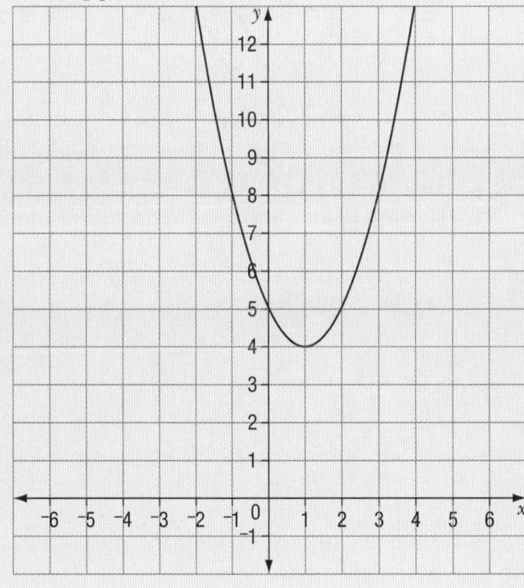

b Turning point (⁻1, 2)

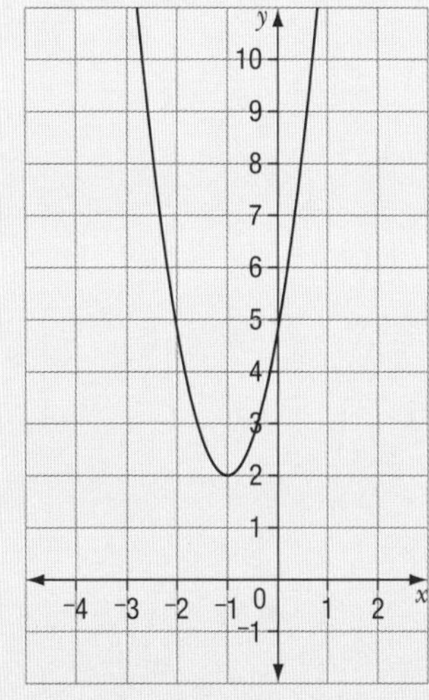

c Turning point (2, ⁻1)

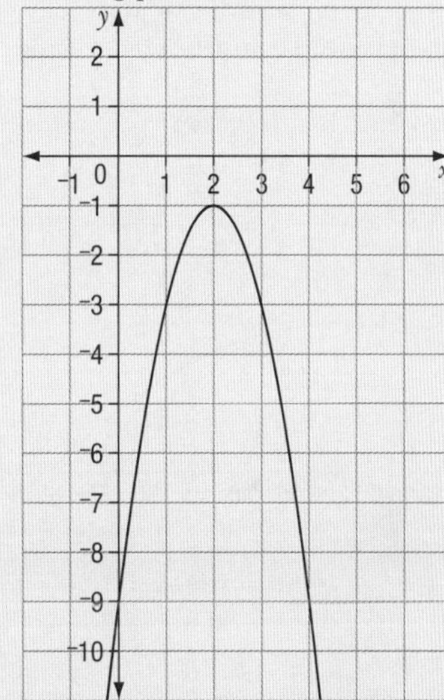

Lesson 12 – Solving simultaneous equations to find a quadratic (pages 195–6)

1 $y = {}^{-}2x^2 + 5$ **2** $y = {}^{-}8x^2 + 15$

3 $y = 2(x - 4)(x - 2)$ **4** $y = \frac{2}{3}x^2 - \frac{5}{3}x - 3$

5 $y = 3x^2 + 2x - 4$ **6** $y = {}^{-}2x^2 + 5x + 1$

7 $y = -\frac{1}{2}x^2 + x - 1$ **8** $y = 3(x - 2)^2$

9 $y = -\frac{8}{9}(x + 3)^2$ **10** $y = 2x^2 - 32$

Lesson 13 – Practical problems involving quadratic functions (pages 198–200)

1 a $24 - x$ **b** $P = x(24 - x)$

c

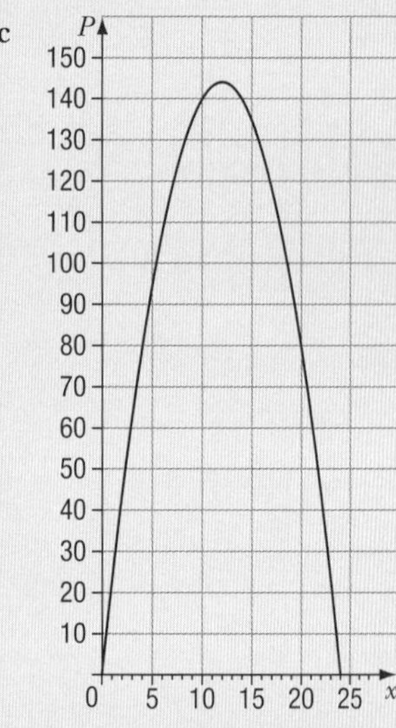

d Maximum P is 144 when $x = 12$.

2 a $x + 1$ b Product = $x(x + 1)$

c The two numbers are 16 and 17 OR $^-16$ and $^-17$.

3 a 66 b 25

4 a 125 cm² b 159π cm²

5 a $h = 0$ m b 60 m

c

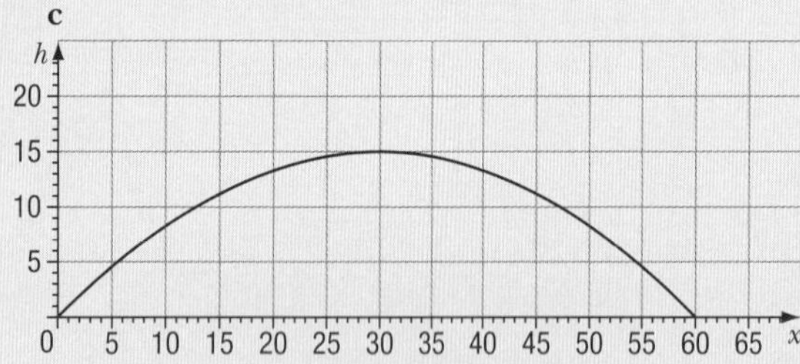

d 15 m

6 a $y = 80 - 2x$ b $A = x(80 - 2x)$

c 800 m² d $x = 20$ m, $y = 40$ m

7 a Student to show.

b $A = x(48 - 2x)$

c

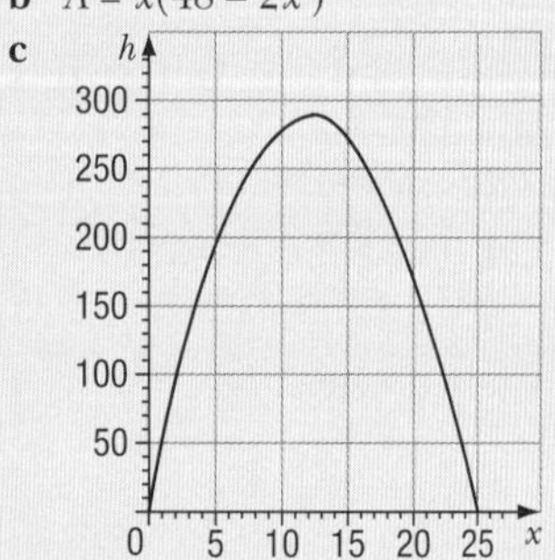

d (12, 288); maximum area 288 m²

e 12 × 8

8 a $x - 2$, $x + 4$

b $x^2 + 2x - 48 = 0$; $x = 6$, $x = {}^-8$

c 6 years

9 a (0, 0), (0.2, 1), (1.2, 1), (1.4, 0)

b $q = 1.4$ c $-\frac{25}{6}$

d $x = 0.7$, maximum height = $\frac{49}{24}$ m

10 a K19.50 b K195.00

c $x(20 - 0.05x)$ e $x = 10$ and $x = 350$

f (180, 1445), K1445, 180

Lesson 14 – Practice for the assessment task (pages 200–1)

1–6 Students to investigate.

7 a (1, $^-6$) and ($^-2$, $^-12$)

b

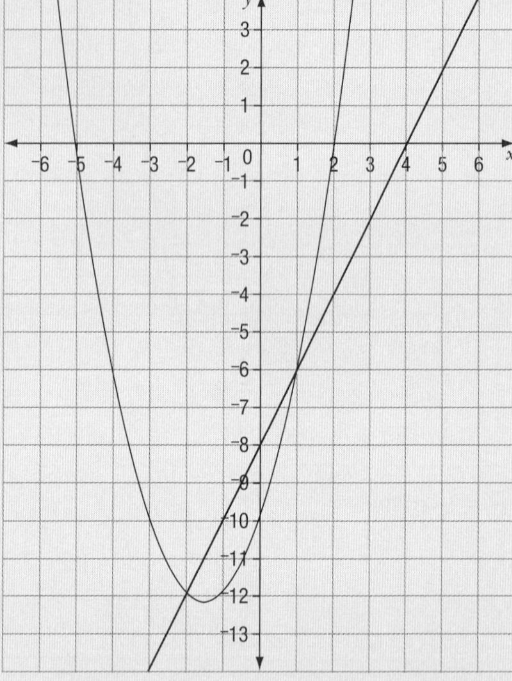

8 a $a = 2$, $b = {}^-1$, $c = {}^-10$

b Turning point $(\frac{1}{4}, {}^-10\frac{1}{8})$

c

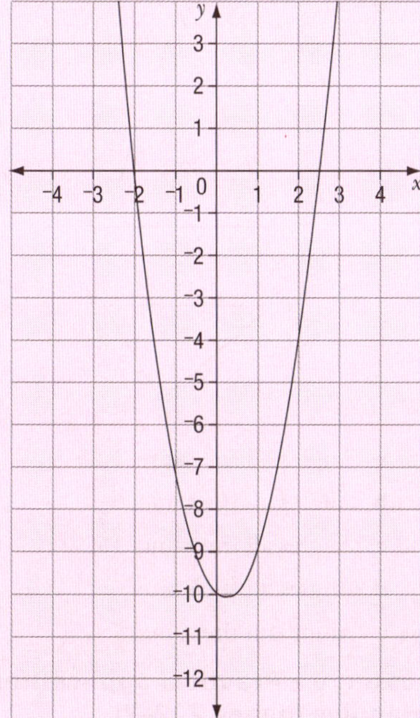

9 a

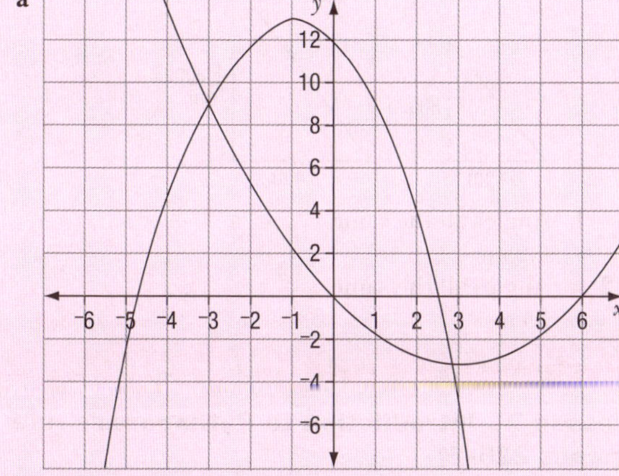

b i and ii (−3, 9) and (3, −3)

10 a 6 m   b 2 s

c

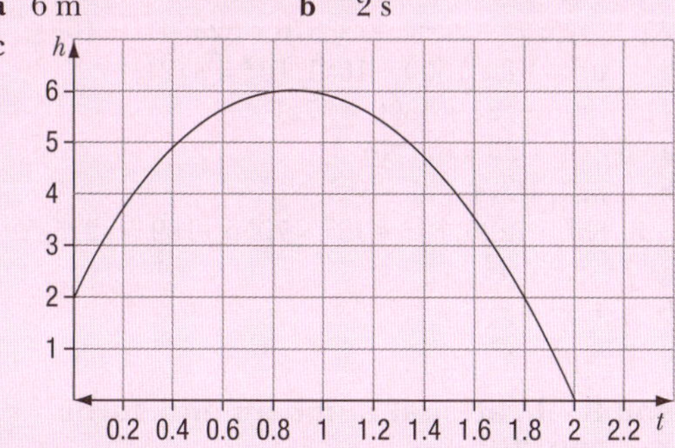

d 6.05 m

UNIT 3: Trigonometric applications

Triangle properties and terminology

Lesson 1 – Triangle terminology and properties (pages 205–7)

1 a equilateral   b isosceles
  c scalene

2 a acute-angled   b right-angled
  c obtuse-angled

3 a Isosceles; two sides the same length with the opposite angles the same size.
  b Right-angled; the longest side opposite the right angle and one angle of size 90°.
  c Equilateral; all angles 60° and all sides the same length.

4 a

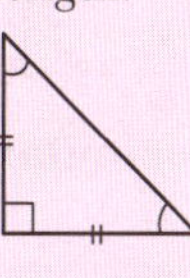

b

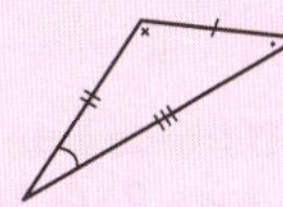

c

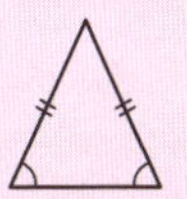

d

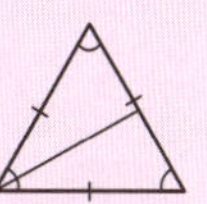

5 a scalene, obtuse-angled
  b scalene, obtuse-angled
  c scalene, right-angled
  d scalene, acute-angled
  e isosceles, acute-angled
  f scalene, right-angled
  g scalene, acute-angled
  h scalene, acute-angled

6 a i 51   ii 11   iii AC
    iv BC and AC v scalene, obtuse-angled
  b i 72.9   ii 62   iii 72.9
    iv QRP v perpendicular bisector
    vi 90   vii 33.4
    viii isosceles, acute-angled
  c i 33   ii hypotenuse
    iii perpendicular   iv $\angle LNM$
    v right-angled scalene

Lesson 2 – Working with properties of triangles (pages 208–9)

1 a, b, c, d

2 a 40   b 35
  c 17   d 79, 101
  e 36, 54, 54   f 15
  g 139, 131

3 a 84, 48   b 60, 60
  c 45, 45   d 20, 90
  e 22, 22   f 69, 111, 34.5

4 a 90, 70, 20 b 100, 40
c 60, 70, 80 d 58, 122, 68
e 35, 35, 55
5 124°

Lesson 3 – Similar triangles (pages 211–13)

1 **c & j, d & f, g & h, e & k**
2 a A & C; corresponding sides in the same ratio
b A & C; angles the same.
c B & C; angles the same.
d A & C; two sides in the same ratio and the included angle the same.
e B & C; two sides in the same ratio and the included angle the same.
3 Answers will vary.
4 a squares; and isosceles, right-angled triangles. All squares are similar; all triangles have the same angles.
b 6
c i 2 : 1 ii 2 : 1 iii 4 : 1 iv $\sqrt{2}$: 1

Lesson 4 – Using scale factors for similar triangles (pages 214–16)

1 a $\frac{1}{2}$
b $\frac{1}{3}$
c 3
d $\frac{4}{3}$
2 a 8.5 b 12.8
c 9 d 5.6

Lesson 5 – Applications of similar triangles (pages 219–20)

1 a 1.25 b 16
c 7.5 d 15
e 2.05 f 4
2 a 1.8

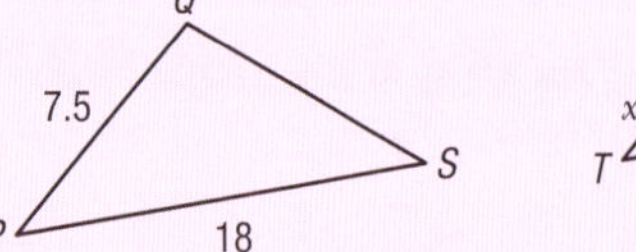

b 2.5

c 6.25

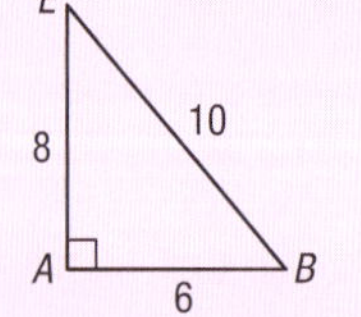

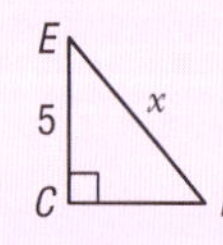

d 3

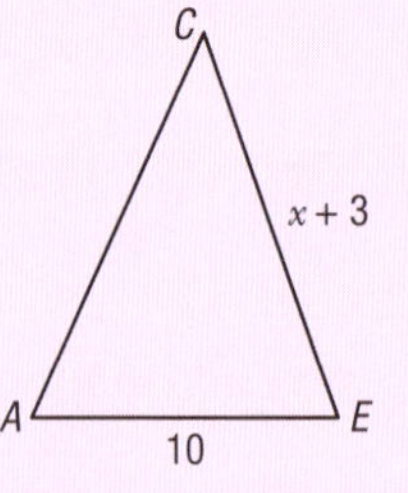

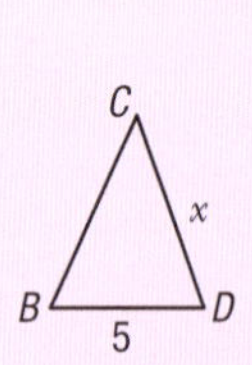

e 3

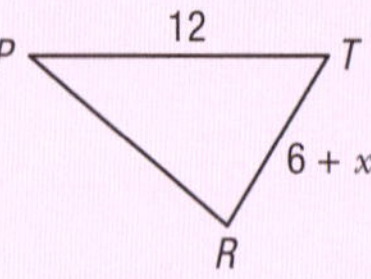

3 a Angles all the same; 1.25
b Angles all the same; $3\frac{1}{3}$
c Angles all the same; 15
d Angles all the same; $\frac{48}{7}$
e Angles all the same; 7

Lesson 6 – Practical applications of similar triangles (pages 222–3)

1 a

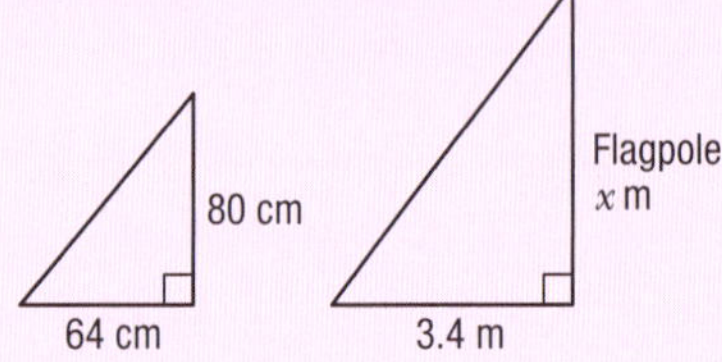

b Angles all the same.
c 4.25 m
2 a Angles all the same.
b 33 m

Pythagoras's rule

Lesson 7 – Introduction to Pythagoras's rule (pages 225–6)

1 a & b Student to draw.
c 13, 13, 169; $5^2 + 12^2$, 25 + 144, 169
2 $25^2 = 625$; $7^2 + 24^2 = 49 + 576 = 625$; $41^2 = 1681$, $9^2 + 40^2 = 81 + 1600 = 1681$; $61^2 = 3721$, $11^2 + 60^2 = 121 + 3600 = 3721$
3 a 64; 36 + 25 = 61; No
b i 324, 169 + 144 = 313, No
ii $85^2 = 7225$, $84^2 + 13^2 = 7056 + 169 = 7225$, Yes
4 a $f^2 = e^2 + d^2$
b $y^2 = x^2 + z^2$
c $p^2 = r^2 + q^2$

Lesson 8 – Irrational numbers and surds (pages 229–30)

1 $\sqrt{101}$, $\sqrt{17}$, $\sqrt{8}$, $\sqrt{83.4}$
2 a 8, 9.95, 12, 6.32, 9, 20, 7.07, 1.3, 30
b 64, 144, 81, 400, 1.69, 900

3 a 15.588 b 11
c 0.806 d 31.623
e 8.185

4 a 11 b 9
c 7 d 10
e 6

5 a i $\sqrt{10}$ ii 3.162
b i $\sqrt{79}$ ii 8.888
c i $\sqrt{3054}$ ii 55.263
d i $\sqrt{65}$ ii 8.062
e i $\sqrt{73}$ ii 8.544

6 a 2, 4; $1.2^2 + 1.6^2$; 4; is; 1.2 : 1.6 : 2; 12 : 16 : 20; 3 : 4 : 5
b 12.3, 151.29; $2.7^2 + 12^2$; 151.29; is; 2.7 : 12 : 12.3; 9:40 : 41
c 234, 54 756, $90^2 + 216^2$; 54 756; is; 90 : 216 : 234; 5 : 12 : 13

Lesson 9 – Calculating the hypotenuse using Pythagoras's rule (pages 232–3)

1 a $\sqrt{65}$, 8.062 b $\sqrt{103.33}$, 10.165
c $\sqrt{61}$, 7.810 d $\sqrt{128}$, 11.314
e $\sqrt{45.25}$, 6.727 f $\sqrt{279.625}$, 16.722

2 a $\sqrt{18}$, 4.243 b $\sqrt{346}$, 18.601
c $\sqrt{62.72}$, 7.920 d $\sqrt{625}$, 25

3 a $\sqrt{389}$, 19.723 b $\sqrt{144.89}$, 12.037
c $\sqrt{45}$, 6.708

4 a 40 cm b 98.985 cm
c 57.941 cm

5 292.023 cm

6 $a = 15$, $b = 25$, $c = \sqrt{1250}$

Lesson 10 – Using Pythagoras's rule to calculate length of shorter side (page 236)

1 a 16.25 b 0.94
c 7.08 d 14.01
e 6.35 f 123.37

2 a 12.69 b 24
c 13.23 d 28.91
e 16.58 f 15.49
g 41.43

Lesson 11 – Applications of Pythagoras's rule (pages 238–40)

1 a 9.44 b 18.76
c 34.47 d 144.92
e 8.29 f 120.13

2 3536 mm

3 a $2x$ b $x = 13.42$, 13.42, 26.84

4 30.02 cm 5 54.92 m

6 2.41 m 7 2.06 km

8 a 19.66 km
b 14.75 min

9 14.28 m

10 a i 7.113 ii 9.81
b i 3 ii 15.72
c i 16.279 ii 20.27
d i 8.307 ii 17.15
e i 6.708 ii 19.65

Lesson 12 – Pythagoras's rule in three dimensions (pages 241–4)

1 Student to draw.

2 43.27 m

3 16.16 cm

4 a 7.81 m b 8.77 m
c *EC*, *BH*, *FD*

5 a Student to draw. b 10 m
c 5 m d 9.43 m

6 a Student to draw. b 42.72 cm
c 73.65 cm

7 a 6.40 m b 7.07 m
c 5.83 m

8 a 24 m b 48 m
c 33.94 m

9 17.32 cm

Activity

a Shortest distance 5.408 m
b 2.333 m from *X*

Lesson 13 – Distance between two points on the Cartesian plane (page 245)

1 a $\sqrt{2}$, 1.41 b 6
c 5 d $\sqrt{20}$, 4.47

2 a scalene b scalene
c isosceles

3 $\overline{AB} = \sqrt{8}$, $\overline{BC} = \sqrt{18}$, $\overline{AC} = \sqrt{26}$

4 $AC = CB = BD = DA = \sqrt{17}$; diagonals $AB = 8$, $CD = 2$

5 0, 4

6 a (6, 2), (1, 7), others possible.
b circle
c $\sqrt{(y-2)^2 + (x-1)^2} = 5$; $(y-2)^2 + (x-1)^2 = 25$

Introduction to the trigonometric ratios

Lesson 14 – The sine ratio (pages 247–8)

1 a sin 45° = 0.7071 b sin 80° = 0.9848
c sin 8° = 0.1392 d sin 23° = 0.3907
e sin 72° = 0.9511 f sin 67° = 0.9205
g sin 90° = 1 h sin 0° = 0

2 All the values are between 0 and 1 inclusive. Smallest value is 0 and largest value is 1.

3 a hypotenuse b adjacent
c adjacent d opposite
e hypotenuse f opposite

Lesson 15 – Using the sine ratio to find length of sides (page 250)

1 a 0.6 m b 1.1 m
c 11.8 cm d 15.9 m
e 7.0 m f 5.4 cm
2 a 29.3 cm b 100.4 m
c 40.0 m d 70.0 m
e 20.0 cm f 30.0 cm
3 8.73 m
4 7.55 m
5 3.70 km

Lesson 16 – Finding the angle from opposite side and hypotenuse (page 252)

1 a 42°36′ b 79°6′
c 18°45′ d 55°14′
2 a 15°30′ b 41°32′
c 46°33′ d 90°0′
e 73°14′ f 7°5′
g 0°42′
3 a i 0.7 ii 44°26′
b i 0.6232 ii 38°33′
c i 0.764 ii 49°49′
d i 0.512 ii 30°48′
e i 0.6373 ii 39°36′
f i 0.5883 ii 36°2′
4 40°32′
5 34°6′

Lesson 17 – The cosine ratio (pages 254–5)

1 a 0.8090 b 0.9135
c 0.1977 d 1.0000
e 0.9909 f 0
g 0.5000 h 0.7071
i 0.0756 j 0.9563
2 All the values are between 0 and 1 inclusive. Smallest value is 0 and largest value is 1.
3 a i $\cos 29° = \frac{x}{9}$ ii 7.9 cm
b i $\cos 62° = \frac{x}{10}$ ii 4.7 cm
c i $\cos 50° = \frac{x}{156}$ ii 100.3 mm
d i $\cos 56° = \frac{x}{23}$ ii 12.9 cm
e i $\cos 65° = \frac{x}{40}$ ii 16.9 m
f i $\cos 34° = \frac{x}{24}$ ii 19.9 mm
4 a i $\cos 59° = \frac{103}{x}$ ii 200.0 mm
b i $\cos 60° = \frac{30.2}{x}$ ii 60.4 m
c i $\cos 32° = \frac{424}{x}$ ii 500.0 mm
d i $\cos 40° = \frac{38.3}{x}$ ii 50.0 cm
e i $\cos 77° = \frac{18}{x}$ ii 80.0 cm
f i $\cos 24° = \frac{1827}{x}$ ii 1999.9 mm
5 17.2 cm
6 5.0 m

Lesson 18 – Finding the angle from adjacent side and hypotenuse (pages 256–7)

1 a 6°1′ b 25°41′
c 0° d 67°24′
e 68°35′ f 90°
g 83°27′ h 60°
2 a i 0.6667 ii 48°11′
b i 0.6 ii 53°8′
c i 0.4286 ii 64°37′
d i 0.6250 ii 51°19′
e i 0.3900 ii 67°3′
f i 0.4400 ii 63°54′
3 45°34′
4 52°
5 23°
6 i 16.07 m ii 34°19′

Lesson 19 – The tangent ratio (pages 259–60)

1 a 0.4245 b 11.43
c 1.0000 d 0.1125
e 1.4281 f 0.5774
g indeterminate h 3.431
2 The values can be any real number. They can be greater than 1.
3 a i $\tan 32° = \frac{x}{9}$ ii 5.62 cm
b i $\tan 64° = \frac{x}{10}$ ii 20.5 cm
c i $\tan 65° = \frac{x}{12}$ ii 25.7 cm
d i $\tan 18° = \frac{x}{23}$ ii 7.5 cm
e i $\tan 25° = \frac{x}{40}$ ii 18.7 m
f i $\tan 74° = \frac{x}{50}$ ii 174.4 mm
4 a i $\tan 60° = \frac{86.6}{x}$ ii 50 mm
b i $\tan 31° = \frac{15}{x}$ ii 25.0 m
c i $\tan 48° = \frac{90}{x}$ ii 81.0 mm
d i $\tan 39° = \frac{32.4}{x}$ ii 40.0 cm

e i $\tan 77° = \frac{39}{x}$ ii 9.0 cm

f i $\tan 22° = \frac{808}{x}$ ii 2000.0 mm

5 50.8 cm
6 21.7 cm
7 11.0 m

Lesson 20 – Finding the angle from opposite side and adjacent side (pages 261–2)

1 a 18°6′ b 34°24′
c 53°54′ d 64°34′
e 19°51′ f 0
g 39°52′ h 84°42′
i 69°11′

2 a i 0.4444 ii 23°58′
b i 1.25 ii 51°20′
c i 1.3333 ii 53°8′
d i 0.7143 ii 35°32′
e i 3.448 ii 73°50′
f i 0.4400 ii 23°45′

3 a 56°19′ b 33°41′

4 20°36′

Solving right triangles using trigonometry

Lesson 21 – Using all three trig ratios (pages 264–6)

1 a 9.7 cm b 68°54′
c 7.2 mm d 30°58′
e 91.3 cm f 50.1 mm

2 17.3 cm
3 $\theta = 22°37'$; $\alpha = 38°40'$
4 30.8 cm
5 2671 mm
6 9.85 m
7 90.8 m
8 a 58° b 13.2 cm
c 29.5 cm
9 a 7.8 km b 518 m
10 a i 56.31° (56°19′) ii 61.69°
iii 3.195 m iv 1.803 m
v 0.855 m vi 1.821 m
b 17.348 m

Lesson 22 – Applications of Pythagoras's rule in three dimensions (pages 267–8)

1 a 69 m b 24°
2 a 6.35 m b 6.78 m
c 20.55° (20°33′)
3 a $PR = 4.243$ m, $AC = 5.659$ m
b 0.707 m c 84°24′
d 7.23 m

4 a 8.485 m b 6.185 m
c 46°41′ d 56°19′
5 a 39°37′ b 35°56′

Revision and Assessment

Similar triangles and Pythagoras's rule

Multiple-choice questions (pages 269–70)

1 D	2 C	3 A	4 C	5 C
6 A	7 C	8 B	9 C	10 D
11 B				

Short answer questions (pages 271–2)

1 0.5
2 a 12 m b 22 m
3 a

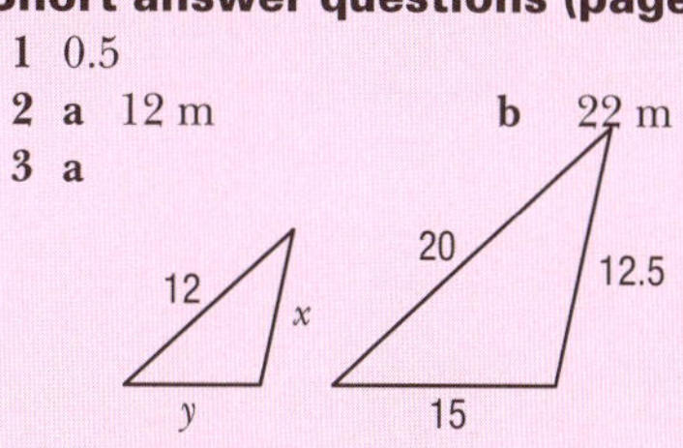

b i 7.5 m ii 9 m
4 a $\sqrt{194} \approx 13.93$ cm b $\sqrt{175} \approx 13.23$ cm
5 7.7 cm
6 47.83 m
7 8.94 m, 17.33 m
8 12.99 cm
9 19.21 m
10 a $AB = BC = \sqrt{17}$; $AC = \sqrt{34}$
b $AB^2 + BC^2 = 17 + 17 = 34$
$AC^2 = 34$
11 21.66 m

Trigonometric ratios

Multiple-choice questions (pages 273–4)

1 B
2 D
3 C
4 C
5 B
6 B
7 D
8 D
9 D
10 C
11 A
12 C

Short answer questions (pages 274–6)

1 a

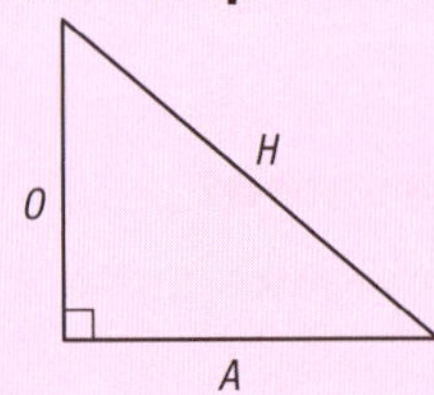

b–d Student to answer.
e sine f tangent
g cosine

2 a 0.829 b 0.978
c 2.144

3 a 60° b 31°53′
c 80°24′

4 2.2906° (2°17′)

5 56°15′

6 112°37′

7 a 23°35′ b 52°26′

8 a 37° b 14.38 m
c 29.91 m

9 9.32 km

10 56 m

Option A: Surveying

Lesson 1 – Angles of elevation and depression (pages 278–80)

1 6.50 m
2 65.96 m
3 90.84 m
4 163.9 m
5 12867 m

Lesson 2 – The sine rule (pages 284–5)

1 a 38.81 m b 51°12′

2 a 10.59 cm b 8.49 cm
c 24.46 m d 44.45 cm

3 a A = 18°2′, B = 29°58′
b A = 47°11′, C = 74°49′
c A = 41°5′, B = 110°55′

4 B = 45°23′, C = 24°37′, c = 14.63 m

5 a 11.30 b 25.01
c 19.39

6 a 27°37′ b 62°2′
c 32°46′

Lesson 3 – The cosine rule (page 289)

1 a 20.44 m b 60.88 m
c 4.10 m

2 42.35 cm

3 a A = 34°3′, B = 101°32′, C = 44°25′
b Z = 41°25′, Y = 55°46′, X = 82°49′

4 36°58′

5 41.32 cm

6 θ = 40°32′, S 49°28′ W

7 29°41′

Lesson 4 – Using sine and cosine in more complex situations (pages 291–3)

1 39.60 cm
2 66°; 4.55 cm
3 14.28 m
4 a 108° b 16.18 cm
c 36°
5 a 23.71 m b 97°11′
6 x = 33.07 m, y = 17.70 m
7 x = 17.44 cm
8 a AD = 84.52 m b AC = 188.57 m
9 47.30 m

Lesson 5 – Area of a triangle (pages 295–6)

1 a 100.37 mm² b 61.72 m²
c 16.96 cm² d 422.86 cm²
e 74.15 m² f 62.35 mm²

2 a 1333.00 mm² b 39.45 m²
c 131.91 m² d 129.63 m²
e 227.48 m² f 370.41 m²

3 a 196.60 cm² b 88.94 m²
c 177.11 m²

Lesson 6 – Maps and scale drawings (pages 297–8)

1 a 25 km
b

100 km

c i 55 mm ii 19 mm
d i 137.5 km ii 47.5 km

2 a Markings 1 cm long.

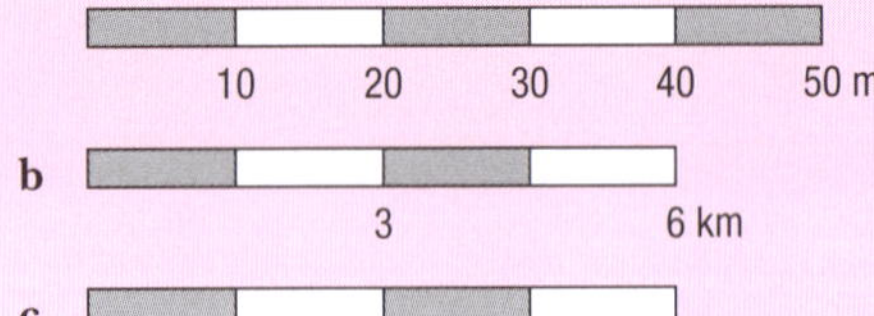

b

3
6 km

c

1 km

3 a 1 : 250 000 b 1 : 5 000 000
c 1 : 4 000 000

4 7.2 cm
5 12.2 cm
6 1 : 3 500 000
7 a Student to draw. b Student to measure.
c 46.25 cm
8 a 337 m b 6920 m²

Lesson 7 – The clinometer (page 301)

Students to compare and discuss results.

Lesson 8 – Using a plane table for surveying (page 303)

Students to compare and discuss results.

Lesson 9 – Triangulation (page 305)

1 a 221.82 m b 207.69 m
c 11517.41 m²

2 **a** 26.01 m **b** 22.75 m

3 **a** 117.76 m **b** 140.46 m

c 256.5 m **d** 237.61 m

e 23650.28 m^2

4 **a** 234.45 m **b** 264.88 m

c 12629.32 m^2 **d** 11131.13 m^2

e 242.49 m

Lesson 10 – Traverse surveys (pages 308–9)

1 Students to complete.

2 **a** 5240 m^2

b **i** 75.29 m

ii 70.10 m

iii 72.45 m

3 **a & b** Students to complete.

c 214.96 m

d 2700 m2

Lesson 11 – Surveying: practice for the project (pages 309–10)

1

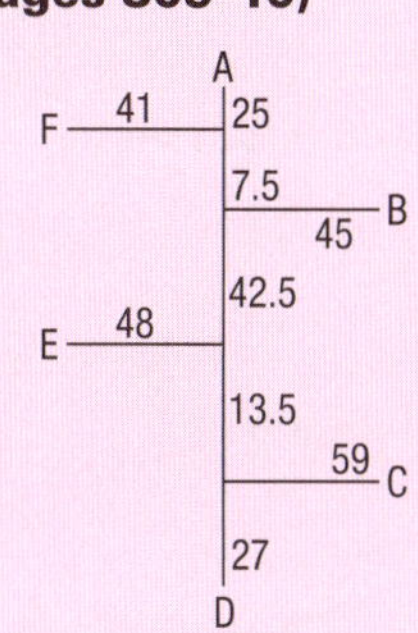

b Area = 7242 m^2 Perimeter = 340 m

2 **a**

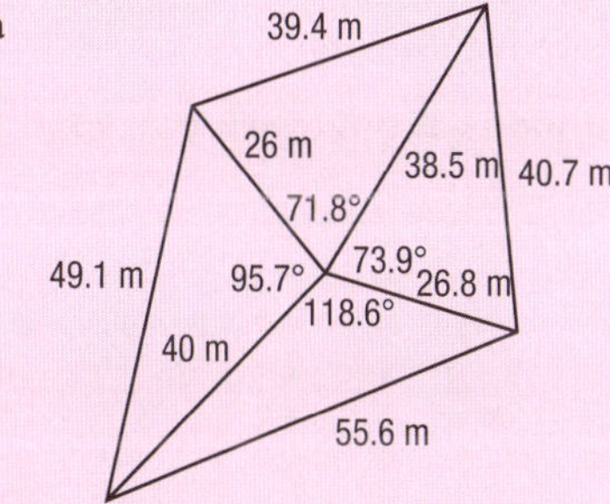

b 184.8 m **c** 187.2 m

d 1959.2 m^2

3 **a** $AL = 66.16$ m, $BL = 84.80$ m

b 59.96 m **c** 45.12 m

d 28°1′

4 **a** $\angle AQB = 36°$, $\angle APB = 15°$, $\angle ARB = 10°$

b **i** 73.01 m

ii 49.06 m

iii 145°

iv 116.65 m

c 72.60 m **d** 118.86 m

e 4064.45 m^2

Option B: Navigation

Answers to Lessons 1–6 are in Option A.

Lesson 7 – Bearings (pages 313–14)

1 **a**

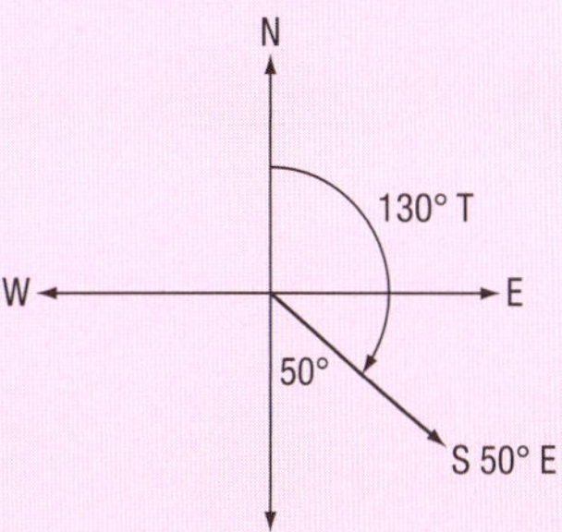

b

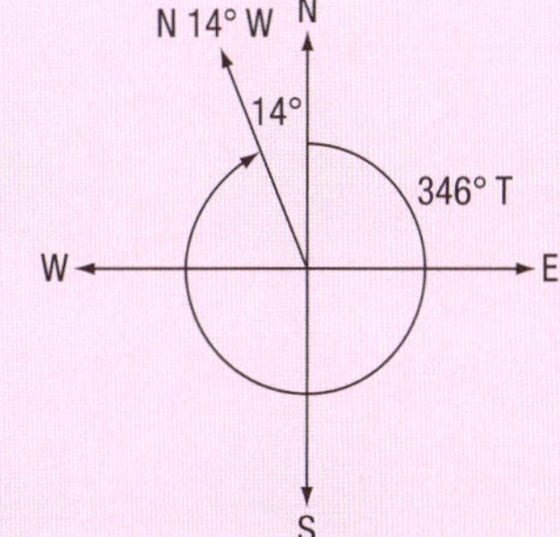

c

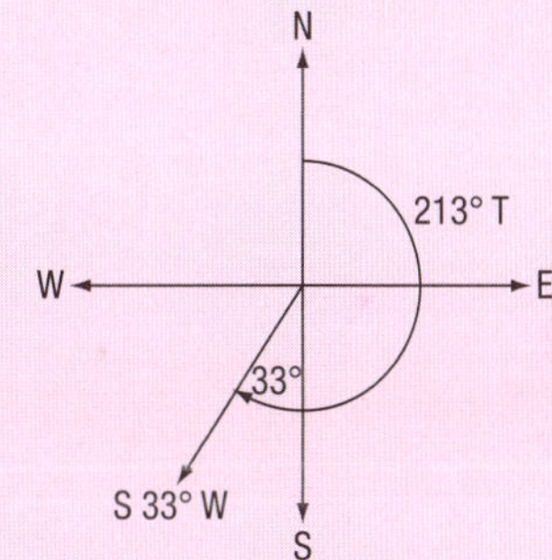

2 **a** S 83° E

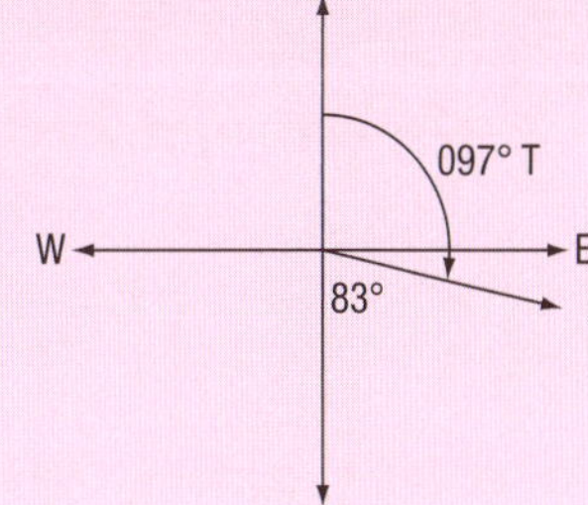

b S 16° W

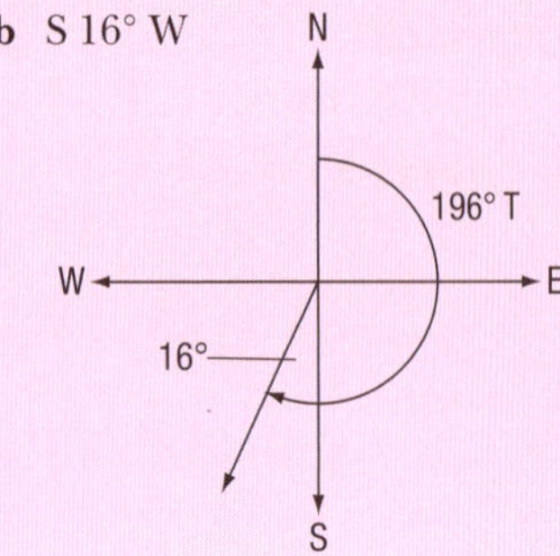

c N 58° W

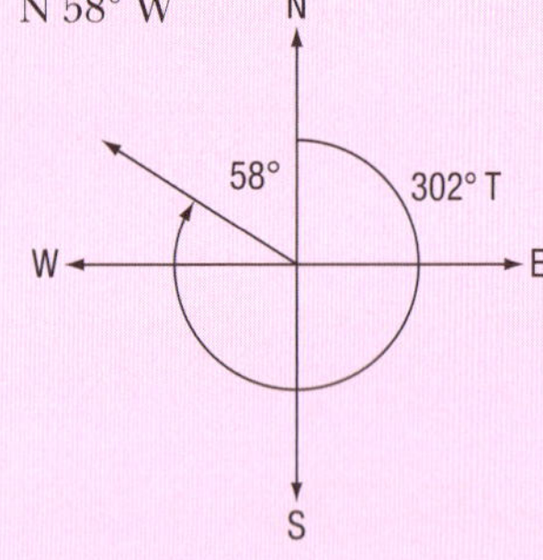

3 a 078° **b** 340°
c 258° **d** 294°

4 a 000° **b** 060°
c 120° **d** 180°
e 240° **f** 300°
g 090° **h** 120°

5 a 123° **b** 17.68 km
c 34.70° (34°42′) **d** 277.30° (277°18′)

6 a

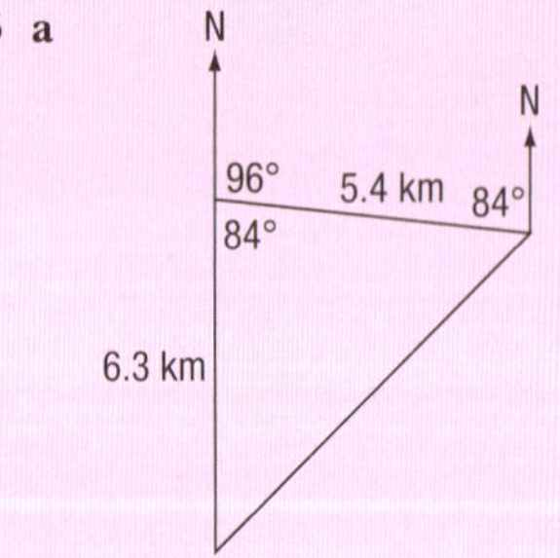

b 7.86 km **c** 223.12°T (223°7′ T)

7 4.94 km

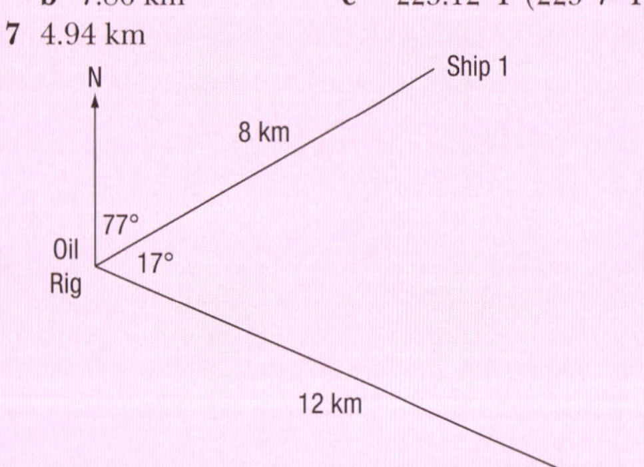

8 133°

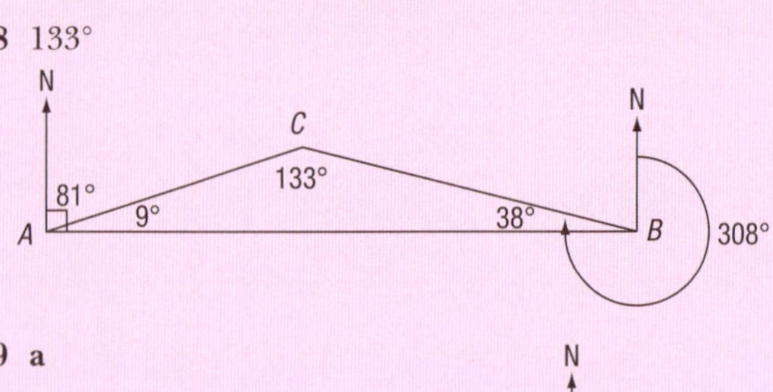

9 a

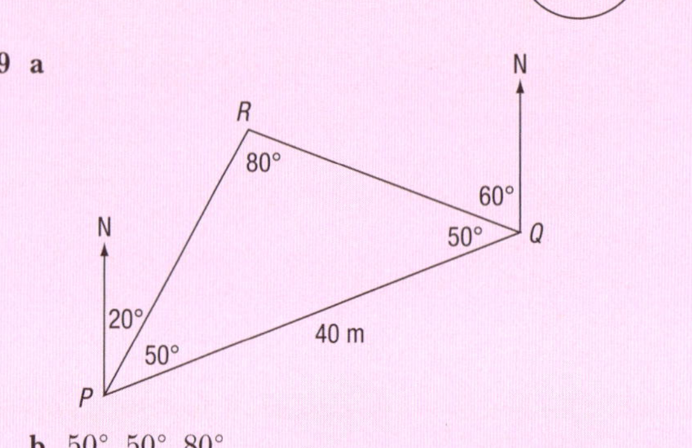

b 50°, 50°, 80°
c 31.11 m

Lesson 8 – Position on the Earth's surface (pages 317–18)

1 a (5°S, 148°E) **b** (6°S, 144°E)
c (6°S, 156°E) **d** (4°S, 144°E)
e (4°S, 152°E)

2 a Kandep (others) **b** Mendi (others)
c Buin

3 a Kerema **b** Esa'ala

4 a Rabaraba **b** Marawaka
c Tapini

5 a Volograd, Stuttgart (others)
b Warsaw, Amsterdam (others)

Lesson 9 – Distances on navigation maps (page 322)

1 a 3.704 km **b** 46.3 km
c 407.44 km **d** 0.7408 km

2 a 18.52 km/h **b** 22.22 km/h
c 9.26 km/h **d** 12.038 km/h

3 a 12.6 n miles **b** 23.335 km

4 a 955 n miles
b 1768.7 km

5 a i 8 knots **ii** 14.82 km/h
b i 15 km **ii** 27.78 km/h
c i 1.6 km **ii** 2.96 km/h

6 Using the markings on the vertical scale: 55′ ...35°...5′ must mean 34°55′ ...35°...35°5′ ... as you go down. This means that the angle is increasing as you go down so that this map is of a location south of the equator.

7 a (35°2.8′S, 151°59.8′E)
b (35°7′S, 151°57.6′E)
c (35°0.7′S, 151°50.2′E)

8 a i 8.3 ii 15.37
b i 4.3 ii 8.52
c i 8.2 ii 15.19

9 Points *X*, *Y* and *Z* to be located on Chart 1.

Lesson 10 – Bearings on nautical maps (page 325)

1 a 302° b 025° c 295° d 310°

2 a 111° b 056° c 205° d 268°

3 and 4

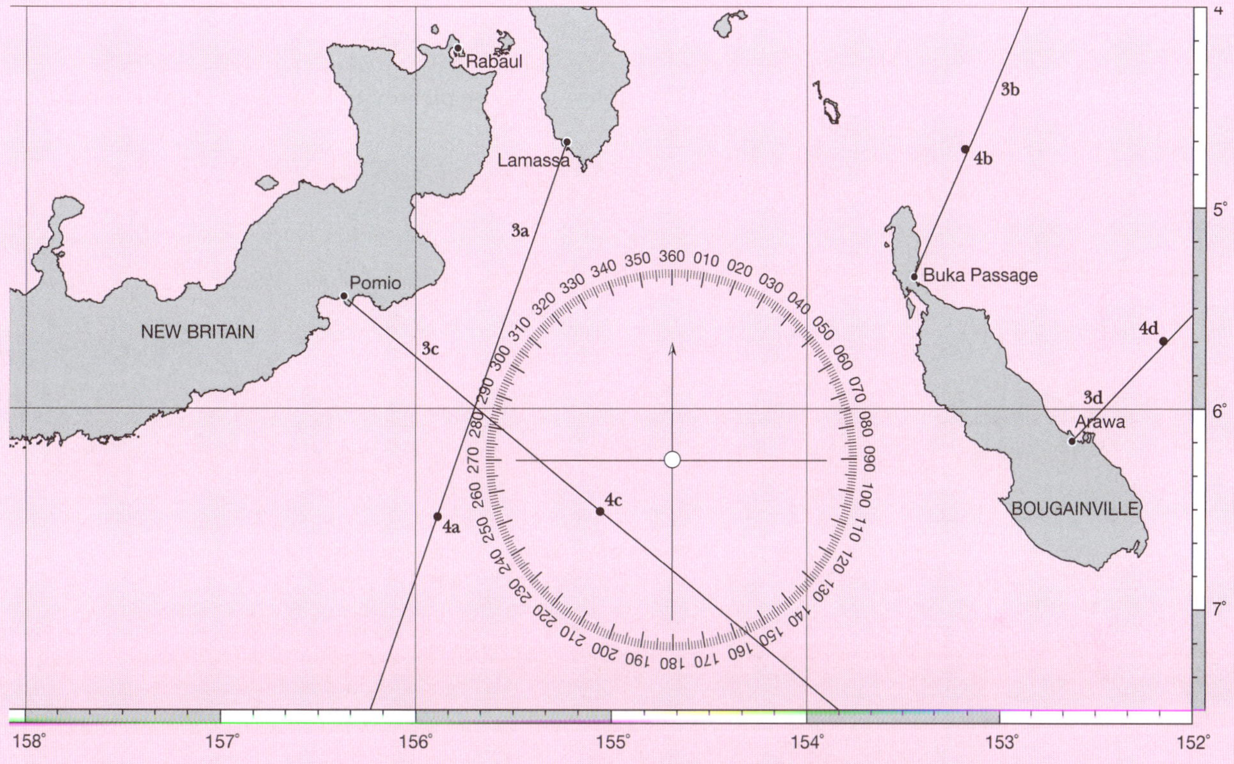

5 (5°47′S, 155°30′E)

6 6.40 p.m.

Lesson 11 – Plotting straight-line courses (page 327)

1 a 342° b 492 km c 59 min

2 a 137° b 546 km c 1 h 49 min

3 a 180° b 300 km c 1 h 15 min

4 a 067° b 188 km c 38 min

Lesson 12 – Navigation at sea (page 331)

Students to provide.

Lesson 13 – Navigation: practice for the project (pages 331–2)

1 a Students to draw. b 75°
c 7.475 n mile d 40.25° (40°15′)
e 028.25° f 19.475 n miles, 36.07 km
g 1 h 18 min

2 a i 122°
ii about 420 km
iii 1 h 24 min
b Students to complete.

Acknowledgments

The authors and the publisher wish to thank the following copyright holders for reproduction of their material.

Getty Images/Bates Littlehales, p. 7; iStockphoto/Bim, p. 25; iStockphoto/DIGITALproshots, p. 277; iStockphoto/mevans, p. 52; iStockphoto/RonTech2000, p. 246; Photolibrary/Alamy/Sebastian, p. 62; Photolibrary/Alamy/Wild Places Photography/Chris Howes, p. 86; Sawczak Irene, pp. 2, 9, 14, 19, 59; Shutterstock/bbbb, p. 80 (top right); Shutterstock/Rekindle Photo and Video, p. 26; Shutterstock/Vereshchagin Dmitry, p. 105.

Every effort has been made to trace the original source of copyright material contained in this book. The publisher will be pleased to hear from copyright holders to rectify any errors or omissions.

Notes